ANNUAL REVIEW OF EARTH AND PLANETARY SCIENCES

ANNUAL REVIEW OF EARTH AND PLANETARY SCIENCES

FRED A. DONATH, *Editor*
University of Illinois—Urbana

FRANCIS G. STEHLI, *Associate Editor*
Case-Western Reserve University

GEORGE W. WETHERILL, *Associate Editor*
University of California—Los Angeles

VOLUME 2

1974

ANNUAL REVIEWS INC. 4139 EL CAMINO WAY PALO ALTO, CALIFORNIA 94306

ANNUAL REVIEWS INC.
Palo Alto, California, USA

International Standard Book Number 0-8243-2002-6
Library of Congress Catalog Card Number 72-82137

Assistant Editor	Jean McComish
Indexers	Mary Glass
	Susan Tinker
Compositor	Typesetting Services Ltd, Glasgow, Scotland

PRINTED AND BOUND IN THE UNITED STATES OF AMERICA

PREFACE

In this volume we have again attempted to provide the reader with critical reviews covering research in a number of important areas in the earth and planetary sciences. These reviews are not intended to be either research papers or in-depth surveys of narrow, specialized areas; nor are they expected collectively to represent, in a single volume, extensive coverage of any one discipline within these sciences. As we noted in the Preface to Volume 1 (1973), the coming of age of the earth and planetary sciences has brought a great diversity in the perception of and the approaches to the solution of many challenging problems. It has also brought to individual scientists the realization of the virtual impossibility of keeping current in a vast and burgeoning literature. We recognized the need for a vehicle for broadly based critical reviews, and we expressed the hope that the breadth of coverage of this new review series would help each of us to appreciate the vastness and diversity of effort now directed toward achieving understanding in the earth and planetary sciences.

Volume 2 represents a reaffirmation of our thinking. We direct this broad collection of critical reviews especially to the attention of scientists who might typically expect to find in this kind of volume an integrated collection of papers treating a narrow area of specialization. In this regard, we commend to the reader Professor Rubey's remarks in the introductory chapter concerning the positive consequences of interaction among the separate disciplines and the need for a greater breadth of learning and an increased awareness, appreciation, and understanding of current and recent work *throughout* the earth and planetary sciences.

We thank the authors of Volume 2 for having contributed an excellent set of reviews, and also acknowledge with thanks the guest participation of Charles Helsley and John Hower in planning this volume. We again express our desire to be responsive to the thoughts of the entire scientific community we serve. We will always be glad to receive suggestions for topics that seem to be particularly appropriate for review, and to respond to constructive criticism.

THE EDITORIAL COMMITTEE

CONTENTS

Fifty Years of the Earth Sciences, *William W. Rubey* 1

Iceland in Relation to the Mid-Atlantic Ridge, *Gudmundur Pálmason and Kristján Sæmundsson* 25

Evolution of Arc Systems in the Western Pacific, *Daniel E. Karig* 51

Growth Lines in Invertebrate Skeletons, *George R. Clark II* 77

The Physical Chemistry of Seawater, *Frank J. Millero* 101

Geophysical Data and the Interior of the Moon, *M. Nafi Toksoz* 151

Low Grade Regional Metamorphism: Mineral Equilibrium Relations, *E-an Zen and Alan B. Thompson* 179

Regional Geophysics of the Basin and Range Province, *George A. Thompson and Dennis B. Burke* 213

Clays as Catalysts for Natural Processes, *J. J. Fripiat and M. I. Cruz-Cumplido* 239

Marine Diagenesis of Shallow Water Calcium Carbonate Sediments, *R. G. C. Bathurst* 257

Earthquake Mechanisms and Modeling, *James H. Dieterich* 275

Solar System Sources of Meteorites and Large Meteoroids, *George W. Wetherill* 303

The Atmosphere of Mars, *Charles A. Barth* 333

Current Views of the Development of Slaty Cleavage, *Dennis S. Wood* 369

Phanerozoic Batholiths in Western North America, *Ronald W. Kistler* 403

Satellites and Magnetospheres of the Outer Planets, *W. I. Axford and D. A. Mendis* 419

Some Related Articles Appearing in Other Annual Reviews 475

Reprint Information 476

Cumulative Index of Contributing Authors 477

Cumulative Index of Chapter Titles 478

William W. Rubey

FIFTY YEARS OF THE EARTH SCIENCES—A RENAISSANCE

✖10014

William W. Rubey
Department of Geology and Institute of Geophysics and Planetary Physics, University of California, Los Angeles, California 90024

I have been invited to prepare an introductory chapter which may be of a more or less philosophical nature, approaching present-day earth and planetary science in the light of my own earlier days and attempting to analyze some of the current trends and attitudes of students and investigators. I have been assured in this very flattering invitation that, among other tempting possibilities, this approach would permit the author to say just about anything he likes regarding the quality and quantity of present-day research, whether or not the preparation of students is adequate for the research to be pursued, and, in general, to comment on one's experience in working with the passing array of scientific investigators.

I have resisted these tempting possibilities and decided to try to tell about some of the things that seem to me most interesting and important in the very active fifty years since I was a graduate student. I have since realized that my choice of a topic was overly ambitious and that I have bitten off more than I can chew gracefully. Nevertheless, I have tried and the chewing has been good exercise and a lot of fun, some of which I hope the reader will share with me. In this article I shall not endeavor, as the title implies, to review the entire field of interests in the earth sciences, but rather to discuss briefly the newest and to me most exciting developments. This means, regrettably, that the subjects selected for mention must reflect largely my own personal preferences and prejudices.

My earlier days in geology and earth science have, of course, greatly influenced what I have tried to do since then. I was a lucky beneficiary of one of those uncommon combinations of ideas, facilities, and people that doubtless have greatly influenced many other lucky scientists in various disciplines. As a graduate student I was drawn to the geology department at Yale by the presence there of Schuchert, Knopf, Ford, Bateman, Lull, Gregory, Longwell, and Dunbar (and by the offer of an instructorship, without which I could not have made it). When I got to New Haven in the fall of 1922 I found a small group of graduate students there who had been attracted by pretty much the same combination of people and facilities that had brought me. It was these other graduate students there at Yale who were to teach me most of what I learned in geology and earth science in the next few years and for that matter in the rest of my days. James Gilluly, Thomas B. Nolan, J. Frank Schairer, and George Gaylord Simpson were there, to mention only those who influenced me most deeply. To this day I can recall, in almost word-for-word detail, discussions, scientific and otherwise, that I had with some of these fellow students. Meaning no disrespect whatever, I cannot recall in comparable detail any of the discussions, scientific or otherwise, I had at that time with members of the highly respected faculty there at Yale. I suspect that my experiences regarding the relative effectiveness of learning from fellow students and from faculty may not be unprecedented. As I have grown older and become a faculty member myself, I have noticed that graduate students tend to take more stock in the impressions and opinions of their student peers than they do in those of their professors.

Others, besides fellow graduate students, from whom I learned a lot were close friends and colleagues—Bradley, Hewett, Matthes, McKelvey, Oriel, Reeside, Spencer, and Williams—while I was a member of that remarkable organization, the U.S. Geological Survey.

In those early days we graduate students believed firmly that a major renaissance of the earth sciences was dawning. Our ideas about this coming renaissance were rosy but vague. We expected it to be a change from the dominantly descriptive phase of geology that had prevailed in the past to an increasing emphasis on measurement, numbers, and quantitative relationships; but we certainly did not anticipate computers or the part they would come to play in the earth and all other sciences. We assumed, of course, that new subdisciplines would take shape as new techniques and new concepts evolved, but we had no very clear idea what these new subdisciplines would be.

As it has turned out, we were at least partly right in what we anticipated for the future. The expected renaissance has, in fact, materialized, and is still materializing. This past half century has been a remarkable one in the development of the earth sciences; and the rate at which new developments are coming to the fore continues to accelerate decade after decade.

Although we were partly right in what we anticipated fifty years ago, we were only partly right. As we have lived through this renaissance which we had predicted and which we welcomed, we found ourselves hard put to keep abreast of what was going on around us. We had anticipated changes, to be sure, but not the continually increasing blizzard of publications that accompanied these changes and on which, in fact, they depended. I doubt that any of us who in the early 1920s talked hopefully about the coming renaissance in the earth sciences has managed to stay abreast of the growing literature in more than a very few of the fields that have been undergoing these remarkable developments.

The challenge and excitement of new techniques, concepts, and problems have brought with them, however, some less welcome corollaries. The continuing flood of new literature has made it nearly impossible for anyone today to become anything other than a specialist in some particular subdiscipline of earth science.

This was not so before the renaissance overtook us. Then it was possible for a young earth scientist to be trained in several subdisciplines and to maintain himself respectably thereafter as something of a generalist. I count myself lucky that I grew up when I did, for today I am a member of a truly endangered, but I hope not a vanishing, species—the general geologist—more specifically the general field geologist, a jack-of-all-trades as an associate of later date once called me in a manner not intentionally complimentary. The young people of today are much better trained in the fundamentals of physics, chemistry, and mathematics than were most of the students of my generation. Nevertheless, if they are to find a job, keep it, and attain professional recognition they are virtually compelled to concentrate their attention and research in some relatively narrow field of specialization.

A geophysicist friend tells me he has noted that scientists are of two kinds—those who keep up with the literature and those who do research. The situation is becoming truly intolerable. What we need is a Utopia where only one out of every three or four manuscripts that gets written is ever published and where promotion and honors committees give double credit for manuscripts embalmed in a permanent mortuary or sanctified in a crematorium. Such a Utopia, however, seems unlikely in the near future.

However, encouraging signs are beginning to appear that the engulfment of scientists in the sheer volume of their publications and the consequent trend toward ever narrower specialization may possibly be drawing to a close. In nearly all scientific fields, the growing number of journals devoted to review articles is evidence that scientists have found a tool that may help them keep their heads above water in the torrent of new literature, and thus have some time left for their own research.

I am aware that some people believe that information storage and retrieval by electronic methods will eventually solve the publication problem and that computers and data banks are destined to replace publications and libraries. However, on the basis of some experience with a major, well-financed project dealing with chemical and biological data, I am convinced that scientific data can be stored readily enough by such means, but I remain skeptical about the effectiveness and reliability of the retrieval part of the process. Furthermore, the human brain, imperfect though it is, has that precious ability possessed by no computer of forgetting that which is trivial. For these reasons, I am unconvinced that computers will really solve our publication problem in the foreseeable future.

Of much potential consequence is the enhanced interest in the origin of the atmosphere and ocean, the earth and solar system, and of life itself which has been spreading widely through many fields of science for the past few decades, and which has been further stimulated by the discoveries that have resulted from the recent manned and unmanned ventures into space. Problems such as these all involve many scientific disciplines and call for an increased breadth of learning for each individual scientist.

One of the trends in the earth sciences that has been most conspicuous during the past half century has been the rise and growth to maturity of the fields of geophysics, geochemistry, and now lunar and planetary science, fields independent of and outside the confines of geology in the classical or traditional sense. At first the establishment and growth of these separate disciplines appeared to be a breaching and fragmentation of the essential unity of the earth sciences. But the long-term consequences have been precisely the opposite of fragmentation. As scientists from different backgrounds of training and experience came to work together on closely related problems they learned to recognize and respect the contributions that workers with other techniques and different points of view can bring to common problems of the constitution and history of the earth. Marine geophysicists in particular have brought together many diverse disciplines and broken down traditional barriers between disciplines. Even more than the increasing interest in the origin of life and of the solar system mentioned above, the mutual stimulation that has resulted from geologists, geophysicists, and geochemists working together is operating to offset, at least in part, the stifling tendency toward ever narrowing specialization.

I shall attempt to mention briefly, informally, and without the documentation of specific references some of what seem to be the more outstanding developments and newer fields of emphasis in the earth sciences during the past half century. Ideally such summaries should embrace the entire range of disciplines and subdisciplines involved, but it is inevitable that the topics I have chosen reflect disproportionately my own personal experiences and interests. To undertake summaries of developments in such a wide range of fields it is convenient to organize and subdivide the selected subjects into the conventionally recognized disciplines and subdisciplines. This subdivision is a man-made classification, however, and nature recognizes no such artificial boundaries. In fact, the subject matter of each discipline in the earth sciences grades isomorphously into that of one or more of the others. Yet it would be confusing and serve no useful purpose to try to avoid a conventional order of treatment.

MINERALOGY AND PETROLOGY

Starting with mineralogy, this field has progressed during the past fifty years from the stage of a largely descriptive to a highly exact science. Probably the outstanding developments that have affected the field have been: 1. the advances made by Larsen, Wright, Merwin, and their colleagues in the techniques of optical study by means of the petrographic microscope and 2. the development of X-ray diffraction methods under the leadership of von Laue, W. H. and W. L. Bragg, Debye, Wyckoff, and others to determine the internal structure—the space lattice and unit cell dimensions—of crystals.

One of the more spectacular revolutions has been in the mineralogy of clays. Because of their small particle size and variable composition and optical properties, clays were for many years widely thought to be a variable and intractable mixture of amorphous colloids. In the mid-1920s Swedish and German investigators published the first X-ray diffraction analyses of clay materials. By the early 1930s this new technique was being widely used and it was soon learned that the great majority of clays are actually composed of definite crystalline minerals. Pauling's pioneer study of the crystal structure of micas in 1930 laid the ground work for the study and classification of the structures of the layered clay minerals. At about the same time French workers demonstrated that differential thermal analyses were a powerful tool in the study of the composition of clays. Ten years later it was realized that the hundred-thousandfold magnifications made possible by the electron microscope shed essential new light on the poorly known morphological nature of clay minerals. As a result of these instrumental advances, clay mineralogy under the leadership of Ross, Hendricks, Kerr, Grim, and their colleagues has become a rigorous discipline and the refinement of these determinative techniques and the geologic occurrence of the different mineral species is the subject of a large and varied literature. The application of these new tools to more strictly petrologic problems such as the origin, diagenesis, and stability relations of the different clay minerals and the low-grade metamorphism in tectonically disturbed regions is only now getting under way.

In petrology, major new developments have profoundly modified the course of research on problems of igneous, metamorphic, and sedimentary rocks. Perhaps the most far-reaching development, in its consequences, has been the application of the principles of physical chemistry to the study of igneous, and later, metamorphic rocks. The Geophysical Laboratory of the Carnegie Institution of Washington was founded in 1905 under the directorship of A. L. Day. The basic principle guiding the program at this laboratory has been that all experimental work is done on materials of known purity and under accurately known conditions. At the outset, research was restricted to simple anhydrous systems. Some petrologists expressed doubts that geologically significant results could be attained under artificially imposed requirements so unlike those of the complex conditions under which real rocks were formed, and cautioned that phase equilibrium studies are of questionable value in geology because under natural conditions equilibrium is rarely attained. However, as fundamental data on the simpler systems accumulated, the stability fields of more complex three-component systems came under study and progress toward understanding some of the processes of formation of igneous rocks began to appear.

In successive stages in 1915, 1922, and 1928, N. L. Bowen announced his epoch-making theories of the reaction principle and magmatic differentiation by fractional crystallization—theories based upon the field occurrence of igneous rocks and the vast fund of experimental evidence he and his associates at the Geophysical Laboratory had been accumulating. Subsequently, experimental work at the Laboratory has been successfully extended to include systems that contain water and to include the stability fields of minerals that compose the metamorphic rocks.

Bowen's differentiation theory has had tremendous influence in this country and abroad, although questions still remain about how universally it may be applied. However, it is probably fair to state that the science of petrology has been made over by the results of experimental work at the Geophysical Laboratory and by Bowen's theory. Perhaps an even greater contribution of the Laboratory is one that extends beyond petrology into many other fields of science. This is the demonstration that problems of seemingly hopeless complexity may be divided into simpler, more tractable parts and brought into the laboratory for analysis and solution.

Several other major developments have influenced igneous and metamorphic petrology during the past fifty years. In 1930 Bruno Sander presented the results of his revolutionary studies on petrofabrics. He and W. Schmidt and their co-workers had developed a technique for microscopic analysis of preferred orientation of crystal grains in metamorphic rocks. These concepts and methods were soon applied to igneous rocks by Hans Cloos and his students. Since then petrofabric analysis, or structural petrology as it is sometimes called, has become the standard method of study of the internal fabric of deformed metamorphic and igneous rocks. Petrofabrics is a field that might with equal justification be treated as petrology or as structural geology.

In the 1930s and 1940s, many petrologists in English-speaking countries became greatly interested in the concept of granitization which had been advocated by eminent French petrologists for more than a century and by many Scandinavian petrologists for nearly as long. This concept that many large masses of granite or granite-like rock are the result not of injection of molten magma but of metasomatic replacement became a subject of lively dispute. After much new field investigation, it has now been demonstrated reasonably well that some granites clearly have been crystallized from magmas and that others are the result of metasomatic transformation; but significant differences of opinion still persist about the relative abundance of the two types.

In the period 1911–1915 the principle was established by Goldschmidt, Eskola, and others that the minerals of a metamorphic rock are determined by the bulk composition of the rock and a given set of pressure-temperature conditions. In 1920, Eskola announced the important concept of metamorphic facies—an individual facies includes rocks of diverse chemical composition and hence also mineral composition that have attained chemical equilibrium under a particular set of physical conditions. The original concept has since been modified somewhat and extended by Eskola, Turner, and others and is now widely accepted as the basis for classification of metamorphic rocks. With the new data on stability fields of a number of individual minerals, it is becoming possible to characterize quite accurately the P-T conditions under which different metamorphic facies originated.

It is difficult to draw a sharp boundary between the fields of sedimentary petrology and sedimentation. I have tended to equate sedimentary petrology with study of the rocks themselves and with use of the petrographic microscope, and sedimentation with laboratory experiments and logical analysis of the processes by which these rocks were formed. But the basis for this distinction is not readily defensible and it is probably rather idle word juggling to pursue the niceties of differences between the two fields any further. A major development that has affected these two fields, sometimes lumped together as sedimentology, in the past half century has been the tremendous increase in number of workers drawn to the study of sedimentary rocks; the resulting development of new methods and subjects of study, such as the sole marks on the base of sandstone beds; and the consequent expansion of publications in the entire field. This increase has been caused partly by the practical value of knowledge about sedimentary rocks in the search for oil, and partly by the publication in 1926 of the book, a *Treatise on Sedimentation,* prepared under the auspices of a committee of the National Research Council, by W. H. Twenhofel and collaborators, which attracted many students into what had been a relatively neglected field.

GEOCHEMISTRY

The concept of geochemistry as a distinct entity involving both geology and chemistry goes back more than a century, but the concept made scant headway until around 1910–1920. Then, within a decade, Clarke in the United States, Goldschmidt and associates in Norway, and Vernadsky and Fersman in Russia began a series of masterful contributions that were to set the scope and tone of the subject of geochemistry which subsequently took form. The program of experimental petrology then getting under way at the Geophysical Laboratory in Washington has here been treated as a major development in petrology, but in all fairness it can also be considered one of the major influences in the development of the new science of geochemistry.

The surge in geochemical research that became evident in the 1920s can be attributed in large measure to the availability of new methods of optical and X-ray spectrography which made possible the quantitative determination of major and trace elements in very low concentrations. The new electron microprobe is now stimulating renewed analytic work on the occurrence of the rarer elements.

Although the field of geochemistry, narrowly construed, is restricted to the earth, the methods and results are being applied to the study of other parts of the solar system. It has even been suggested that geochemistry is a branch of the general chemistry of the universe which might be called cosmochemistry.

The abundance of the elements and of nuclides in the materials of the earth and in meteorites has been recognized from the outset as one of the main tasks of geochemistry. The information on cosmic abundances of elements that has been assembled as the result of geochemical research has contributed notably to the evolution of modern physics and astrophysics.

An interesting and important aspect of geochemistry has been the attention given by Conway, Hutchinson, Urey, myself, Poldervaart, Holland, Ronov, Sillén, Garrels, and others to the composition of the earth's early atmosphere and ocean and to some of the details of the geochemical balance that has been maintained between the masses of rocks eroded and sediments deposited. These studies have been an integral part of the interdisciplinary interest in the origin of the earth, the solar system, and life. To test some of the rival hypotheses concerning the nature and composition of the early atmosphere and ocean, careful petrographic and chemical studies are needed of the very oldest known fine-grained sedimentary rocks.

Urey and associates in 1951 developed a powerful technique for determining the temperature of seawater in the past by using the ratio of O^{16} to O^{18} in carbonate rocks. Emiliani, applying this technique to pelagic foraminifera, in 1954 detected the temperature maxima of Pleistocene interglacial stages and a progressive fall in temperature of the deep waters of the Pacific from Oligocene to the Pleistocene.

Other important developments of the past few decades include: 1. geochemical and geobotanical prospecting for mineral deposits using highly sensitive analytic techniques and observing the distribution of plant species that are known to concentrate certain elements in their tissues; and 2. studies of the beneficial and toxic effects of certain trace elements on plants and animals of agricultural interest. A great amount of data of this nature has been accumulated in the past forty years. The Society for Environmental Geochemistry and Health has recently been formed for study of the many and usually obscure relationships between trace element concentrations and human health. Helen Cannon has been a foremost leader in both geobotanical prospecting and environmental geochemistry.

A field that has proven to be of great significance is one that might, with almost equal justification, be classed either as geochemistry or as nuclear geophysics. This is the dating of the age of rocks and other materials by methods that depend upon the radioactive decay of uranium, thorium, potassium, and rubidium. In 1907, Boltwood used the lead-uranium ratio to determine the age of a pitchblende, but it was not until the late 1920s that the isotopes of ordinary and radiogenic lead were determined by means of the mass spectrograph. Thereafter the determination of rock ages increased markedly. However, application of the technique was difficult when it depended upon finding U or Th minerals in rocks that one wished to date. But with the extension of these techniques by Aldrich, Wetherill, and colleagues in the 1950s to K-Ar and Rb-Sr decay, age determinations of common minerals such as zircon, mica, feldspar, and hornblende became possible and consequently the number of dated rocks has increased greatly. The result has been an increasing realization of the great length and complexity of the Precambrian record and of the evidence of repeated stages of metamorphism in the geologic history of many regions. The oldest rock now known is an amphibolite gneiss approximately 3700 million years old, found in 1971 by McGregor, Black, and associates in southwestern Greenland.

Another important method of radioactive dating was invented in 1947 by W. F. Libby. It is based upon the decay of radiocarbon (C^{14}) in plant and animal tissue and shell material that has been separated for some time from contact with the atmosphere. Carbon-fourteen ages are not reliable for samples of ages more than about 70,000 years, but C^{14} dating has been of tremendous value in establishing the chronology of climatic changes during the later part of the Pleistocene and, subsequently, of many areas in anthropology and archeology.

GEOPHYSICS

The segment of earth sciences that has grown most and over the widest areas in the past half century has been the group of more or less related subjects known collectively as geophysics. This expansion has been so vigorous that only a selection of representative developments can be mentioned here.

Many subjects formally included within geophysics are not considered here—magnetic fields and charged particles and radiation in the upper atmosphere and space, solar-terrestrial relations, ionospheric physics, auroral and airglow phenomena, atmospheric electricity, tidal theory, meteorology, and much of geodesy and physical oceanography. But the subjects here considered under Geophysics—the nature and structure of the crust and interior of the earth, gravity, seismology, geophysical studies of the oceans and glaciers, hydrology—include a lot of territory ranging from the center of the earth into the lower atmosphere.

Many of these solid earth and earth-atmosphere-hydrosphere interface fields of study were definitely established before 1920. The Pratt and Airy models of isostasy had been formulated 65 years before and the Hayford-Bowie method of surveying the degree of isostatic balance in the United States had been completed in the 1910s. The principal ideas concerning the theory of seismic waves, the origin of earthquakes, and the internal structure of the earth had been outlined by 1910. Methods of geophysical prospecting for minerals and oil by magnetic, gravimetric, electric, and seismic methods had been developed. However, except for prospecting for iron ore by magnetic methods, they had not yet been used with much success.

Then, in the mid-1920s a remarkable expansion in the scope and intensity of geophysical work began and it is still continuing. In 1924 the first edition of Jeffreys' *The Earth* appeared. This monumental work provided a theoretical treatment and viewpoint for many separate areas of geophysical research and served as a stimulus to new observation and theoretical research. In 1925 five salt domes were discovered in the Gulf Coast by use of the torsion balance. Geophysical prospecting promptly became respectable and these discoveries were quickly followed by the widespread use of gravimeters and magnetometers. and by especially vigorous expansion of refraction and then reflection seismic methods. Electrical logging of wells came into active use in the early 1930s.

In response to a recommendation by the Interdivisional Borderlands Committee of the National Research Council, a critical tabulation of a wide range of quantities and physical constants needed in geological and geophysical calculations, the *Handbook of Physical Constants,* was compiled by Birch, Schairer, and Spicer and published in 1942. As may be judged by the many references to it in the subsequent literature, this compilation proved to be of great usefulness. A revised edition, edited by Clark, was published in 1966.

In the years following 1920, seismology made great strides toward depicting, in ever greater detail, the structure of the earth's interior. Turner demonstrated the existence of deep-focus earthquakes in 1922. Maps of the distribution of deep, medium-depth, and shallow shocks in Japan and in the East Indies were prepared by Wadati and by Berlage, and in 1937 of the west coast of South America and elsewhere by Gutenberg and Richter. These showed that the earthquake foci fall into zones that, starting from the oceanic trenches, incline steeply downward under the landward, or island arc, side. Benioff later studied this systematic distribution of quakes in greater detail and the term Benioff zone has now become part of the standard nomenclature.

The adaptation of seismic refraction apparatus for use at sea by Ewing, Worzel, and Press revealed that the Mohorovicic discontinuity is approximately 12 km below sea level, thus indicating a crustal thickness of only about 7 km beneath the ocean floor in contrast to 35 to 40 km beneath the continents.

Analysis of surface waves by Press, Ewing, Anderson, Knopoff, and colleagues has demonstrated, among other things, what had been suggested earlier by Gutenberg, that a low velocity zone exists at a depth of 70 to 200 km beneath the oceans and under nearly all of the continental areas except the older shields. This discovery has revived interest in Daly's early concept of a relatively weak asthenosphere underlying a stronger lithosphere.

Not only is the existence of this low velocity zone well established, but seismic observations in the past few decades have disclosed increasing evidence of abrupt increases in seismic velocities with depth. This unexpected fine structure in the velocity-depth relations in the rocks of the mantle extends from the Mohorovicic discontinuity down to a depth of about 900 km. Evidence is growing that these changes in velocity are the result of transformations from low-pressure to high-pressure phases of the materials that constitute the mantle.

It was suggested by J. F. Lovering, G. C. Kennedy, and others that the Moho itself is caused not by a difference in chemical composition but by a phase transformation, but this view is not widely held today. Interest now centers on the changes in seismic velocities that lie at greater depths.

Bernal in 1936 and Jeffreys in 1937 suggested that an inhomogeneity thought to be located near 400-km depth might be caused by a phase transformation of olivine to a denser spinel structure. In a pioneering paper in 1952 Birch suggested that the elastic properties of the upper mantle are consistent with a composition of familiar minerals such as olivine, pyroxene, and garnet but that the properties of the mantle below 900 km are very different and resemble those of close-packed minerals such as spinel, rutile, periclase, and corundum. He proposed that the transition zone of elastic properties between the Moho and 900 km is most readily accounted for as a series of pressure-induced changes in crystal structure.

At that time no direct means of testing the hypothesis were available. However, indirect methods depending upon comparative crystal chemistry and studies of germanate analogs of silicates indicated that a number of such phase transformations should be expected. Shock waves produced by high explosives or by meteorite impact generate instantaneous pressures up to several thousand kilobars, and study of rocks that have been subjected to shock metamorphism demonstrated the presence of a number of high-pressure phases. Laboratory apparatus capable of generating higher pressures than had previously been possible were under development and in 1958 Ringwood by direct experiment discovered the olivine to spinel transformation. Since then, apparatus capable of generating pressures of approximately 200 kbar, simultaneously with high temperatures, has been developed and a number of the other transformations predicted by Birch have been observed. Experimental work at ever higher pressures is continuing, but even now it may be said that Birch's hypothesis of 1952 has been verified in all essential respects.

New instruments such as tiltmeters and strain gauges have broadened the scope of seismic research and new techniques of analyzing seismic data have been developed. The disposal of waste fluids down a deep well near Denver evidently set off a swarm of small earthquakes in the early 1960s. Study of this series of quakes led to the interpretation that the intensity of fluid pressure affects significantly the release of stored energy in rocks and has raised the hope that some day it may become possible to control the release of energy of shallow earthquakes by controlling the fluid pressure in the rocks of an area.

Arrays of instruments for monitoring the occurrence of small earthquakes and of creep have led to recent advances in Russia, Japan, and the United States which indicate that small premonitory phenomena, such as changes in the ratio between pressure and shear wave velocities from small shocks, changes in radon emission and in the electrical conductivity of rocks, and tilting movements, may serve as grounds for predicting major earthquakes weeks to months before their occurrence.

Data on the remanent magnetism of rocks of different ages and the positions of magnetic poles deduced therefrom have been collected by many observers starting in the 1930s. These data had become sufficiently numerous by the early 1950s to lead a number of workers (Blackett, Bullard, Runcorn, Irving, Nairn, and associates) to conclude that the magnetic poles have wandered extensively relative to the positions of the present continents and that these polar positions do not coincide for rocks of the same age from different continents. Although several possible explanations of these observed facts have been proposed, it is probable that most geophysicists today take these observations to indicate that the continents have separated from one another and drifted apart along different paths. Many details of the paths of polar wandering are still obscure, however, and serious problems of interpretation remain.

At the same time that data were accumulating on the positions of paleomagnetic poles, it was found that about half the rocks studied have directions of magnetization oriented directly opposite or 180° away from the others. Research in the field and laboratory has shown that a small number of these reversed magnetic fields are caused by the presence in a rock of two or more magnetic minerals that did not crystallize simultaneously. But from the work of Cox, Doell, Nagata, Einnarsson, McDougall, and others, it is now known that the great majority of changes from normal to reversed magnetization and vice versa that can be dated closely took place at the same time in the United States, Iceland, England, Scotland, France, Russia, Japan, Australia, and New Zealand, and thus must have been worldwide reversals of the earth's magnetic field. The history of these magnetic reversals is now well established for the past five million years, and from incomplete evidence and indirect methods it may be extended back to as much as 200 million years ago. This new magnetic polarity time scale is a stratigraphic tool which may prove to be of great usefulness.

In 1923 Vening Meinesz invented a pendulum for the measurement of gravity from within a submarine. In the next ten years he conducted gravity surveys in the East and West Indies. In both regions long narrow curving zones of large negative gravity anomalies were discovered that follow closely along the island arcs and bordering oceanic trenches. Similar zones of negative gravity anomalies have since been found associated with other island arcs elsewhere. Vening Meinesz and his co-workers proposed a downward buckling of the crust to explain the presence of these zones. Ewing and Worzel suggested as an alternative, based on seismic and gravity studies in the Puerto Rico Trench, that thick accumulations of light sediments in the trenches may readily account for the zones of negative anomalies.

It would be difficult to exaggerate the far-reaching consequences that have followed from the application of a number of different geophysical techniques to study of the ocean basins. The development of echo sounding during and after World War I made possible the preparation of bathymetric charts of an areal coverage and quality of detail not possible before. These charts displayed relationships not previously recognized and they opened up a host of new unanswered questions. They stimulated the formulation of many new generalizations about the nature and origin of the ocean floor and the rocks beneath it and they have guided the studies later undertaken by sea-going geophysicists and geologists on numerous oceanographic expeditions.

Veatch and Smith in 1939 published a series of bathymetric charts of the region off the east coast of the United States that showed a number of submarine canyons cutting the edge of the continental shelf and extending to depths of 3000 m or more. As other continental shelves and slopes were examined, similar canyons were found to be common and widely distributed. Their interpretation of these features as the result of subaerial erosion during a Pleistocene lowering of sea level aroused great differences of opinion. In 1936 Daly suggested that the canyons were cut by streams of muddy water on the ocean floor (density currents or turbidity currents) that flowed down the continental slopes during the periods of Pleistocene emergence of the continental shelves. Kuenen in 1937 demonstrated the reality of such currents by laboratory models. Nevertheless, the proposal did not immediately resolve the controversy because of doubts that such currents would have sufficient force to erode rock. However, in 1952 Heezen and Ewing attributed a series of submarine telegraph cable breaks to a fast-flowing turbidity current set off by the Grand Banks earthquake of 1929. Multiple cable breaks elsewhere have since been attributed to the same cause and the existence of powerful turbidity currents capable of eroding the submarine canyons is now widely, although not universally, accepted.

An interesting byproduct of the submarine canyon controversy was the emergence of the concept of turbidites, layers of graded sediment laid down in deep water by turbidity currents. The concept took hold and raged as a scientific fashion for a number of years. Attempts were made to apply it to almost all known sedimentary rocks. Like many such fashions in science, it brought about a healthy reconsideration of many venerable but outworn interpretations—for example, that nearly all sedimentary rocks now exposed on continents were laid down in shallow water—but the momentum of the fashion also carried it beyond the bounds of proof and reasonableness in some areas. In spite of the excess zeal exhibited in some attempted applications, the residue of the fad has been a net gain for science.

Ewing in 1935 pioneered the measurement of sediment thicknesses on the floor of the deep ocean by seismic refraction methods and Piggott in 1936 developed a sediment corer in which a charge of explosive drove a sampling barrel as much as 3 m into the bottom sediments. But it was not until after World War II that the geophysical exploration of the ocean basins really got under way. Methods for collecting core samples of sediment up to 23 m long were developed by Kullenberg in 1947. Ewing and associates built cameras for photographing the sea floor that produced unexpected evidence of strong water currents and living organisms at great depths. Bullard in 1949 designed a thermal probe that made possible the measurement of heat flow from the ocean floor. By 1950 research vessels of the Lamont (now Lamont-Doherty) Geological Observatory, and to a lesser extent also of the Woods Hole Oceanographic Institution, were using these various new geophysical methods principally in the Atlantic Ocean, and vessels of the Scripps Institution of Oceanography were doing the same in the Pacific.

Present day expeditions of these and other oceanographic institutions are now equipped to make continuous sounding records, determine crustal thicknesses by explosion seismology and the attitude of layers within the sediments by the seismic reflection profiler, measure heat flow, photograph the bottom, collect samples, tow a magnetometer, and measure gravity while under way. One of the more promising of the new instruments now coming into use is the deep diving submersible, such as the *Alvin* of the Woods Hole Oceanographic Institution, which permits examination of the details of submarine topography and the sampling of rocks to depths of 2 km or more.

Great numbers of heat flow measurements from the ocean floor by Bullard, Maxwell, Revelle, and co-workers have revealed that the quantity of heat flow from oceanic areas is very closely the same as that from continental areas. The significance of this unexpected equality is still under vigorous debate.

The ambitious Mohole project which was planned to drill through the oceanic crust and sample the underlying mantle fell victim to a series of political mishaps and engineering and scientific misjudgments. However, the current JOIDES program (now called the Deep Sea Drilling Project) which succeeded it has drilled from the *Glomar Challenger* about 450 holes well distributed in all major ocean basins except the Arctic to depths as much as 1000 m below the sea floor, with extremely rewarding scientific results. Now that the problem of re-entry of the drilling tools into the hole after changing bits appears to have been solved, it has become possible to drill through basalt layers (both flows and sills) that may be encountered, and perhaps to determine the nature and position of bed-rock below the oceanic sediments.

By no means has all the modern exploration of the ocean basins been in the deep sea. Extensive bathymetric sounding and bottom sampling projects have been carried out off the east and west coasts of the United States, the Sea of Okhotsk, the China Sea, the waters off New Zealand, the west coast of Africa, the Mediterranean, and the Red Sea. The 70% of the earth's surface covered by the oceans is finally yielding its secrets to this worldwide campaign of exploration; and the earth sciences are undergoing significant modifications of viewpoint and concepts as a result.

The veritable revolution that had been taking place in geophysics was brought to focus in the designation of the International Geophysical Year (IGY), 1957–1958, during which much previously uncoordinated record gathering and research in many countries was intensified and concentrated upon specific problems. The IGY was responsible for a rich harvest of results and it gave impetus to many fields of geophysical research, as is demonstrated by the programs that it inspired, such as the Indian Ocean Expedition and the Upper Mantle Project and its successor, the Geodynamics Project.

The first scientific traverse in the interior of Antarctica was made by the Norwegian-British-Swedish Antarctic Expedition during the 1951–1952 field season. Seismic soundings were taken, rates of snow accumulation determined, and temperatures in the snow measured. This first traverse served as a model for the similar expeditions that proliferated during the IGY. As a result of this concentration of activity in Antarctica many glaciological and other data were gathered on ice thicknesses, snow accumulation rates, snow-ice stratigraphy, and polar climates for the past several thousand years. Ice thicknesses of nearly 3 km have been measured, indicating that earlier estimates had been much too low. Seismic and gravity data have shown that the continent is in essential isostatic balance and that crustal thicknesses average about 30 km in western Antarctica and 40 km in eastern Antarctica.

Observational and theoretical advances in glaciology have not been restricted to Antarctica alone. Core drilling has yielded a continuous stratigraphic record of several thousand years of relatively unchanging atmospheric composition and climate in northern Greenland. The ice islands in the Arctic Ocean have come under intensive examination. Detailed studies of the deformation and flow of ice in glaciers in Switzerland, Norway, Russia, and Alaska by Perutz, Nye, Kamb, and associates have furnished the basis for new concepts of glacier geophysics involving the application of rigorous rheological principles to the phenomena of ice movement in glaciers.

STRUCTURAL GEOLOGY

Structural geology or tectonics—to my ear the two terms are used almost but not quite interchangeably—is concerned with the causes, processes, and effects of rock deformation at all scales from the microscopic and hand specimen to mountain ranges and continents. Its study involves field observation, laboratory experimentation, and theoretical analysis, and includes such a variety of topics that only those points that have seemed to me most important can be touched on here.

The gross aspects of rock folding and fracturing have been familiar since the earliest days of the earth sciences. But the accumulation of data and growth of new concepts and sophisticated methods of analysis during the past half century, and especially during the past decade, have been such that structural geology is virtually a new discipline today, one which seems to herald a higher stage of knowledge about the earth. The great increase in geologic mapping in nearly all parts of the world, which has been carried out for economic and scientific purposes, has resulted in an enormous mass of facts about the forms of rock deformation and the tectonic features of the earth.

Bucher's immensely stimulating *Deformation of the Earth's Crust,* published in 1933, endeavored to assemble on a worldwide scale the geological facts that bear on crustal deformation and to derive therefrom a series of inductive generalizations or laws. It has greatly influenced the course of thinking on tectonic matters. The Tectonic Map of the United States, prepared under the auspices of a committee of the National Research Council, appeared in 1952 and was a useful tool for further tectonic studies. It brought together masses of scattered data which had been accumulating for decades.

In 1923 Schuchert published a monumental work on North American geosynclines, which was extended and revised by Kay in 1951. These elongate zones of very thick sedimentary rocks have been found on every continent and they have had an intimate but not yet well understood relation to the formation of mountain systems.

Overthrust faulting with large displacements had been recognized in Germany, the Swiss Alps, the Scottish Highlands, the Scandinavian Peninsula, the Appalachians of the eastern United States and Quebec, the Canadian Rockies of Alberta, and the northern Rockies of the United States before 1920. The discovery of other large overthrust faults in many mountain ranges elsewhere continued to be reported and such reports became virtually a fad in the 1920s and 1930s. Many of these reported overthrusts were confirmed by later work but some were found to be based on overzealous enthusiasm. Nevertheless, the net results of the fashion were scientifically beneficial because of increased understanding of the extent and nature of these almost incredible natural phenomena.

Rodgers focused attention on the interpretation of Appalachian structure as thin-skinned, that is, involving only the sedimentary cover but not the basement rocks beneath. This concept of basal shearing planes or planes of detachment separating the sedimentary cover from its substratum continues to be fruitful in unraveling the deformational history of many mountain ranges.

An effort was made by Hubbert and the writer in 1959 to help account for the mechanism of overthrust faulting by attributing to the internal fluid pressure in rocks an important role that had previously been neglected.

The behavior of rock materials under stress has been studied extensively since 1934 by Griggs and his students and associates, making use of and extending the high pressure techniques developed by Bridgman. The experimental deformation of rocks has become a well-recognized field of structural geology that is actively supported at many leading academic and governmental institutions. A paper by Griggs, "A theory of mountain making," published in 1939, remains influential today in tectonic thinking about how convection currents in the earth's mantle could conceivably account for the building of mountain ranges.

Two great advances in analysis of the observed facts of rock deformation came with the publication in 1931 of Nadai's book, *Plasticity*, and an article by Hubbert published in 1937 on the theory of scale models applied to structural problems.

It had been long and widely held in Europe and in America that diastrophic and orogenic events are of worldwide, or at least continent-wide, extent and that they occur in brief periods of time separated by long periods of quiescence. This concept was emphatically and effectively challenged by Gilluly in 1949. However, with the great increase in recent years in the numbers of accurately dated rocks the concept of orogenic periodicity has been revived among those workers who equate orogenies with magmatic events. It is fair to state that the question is not yet finally resolved.

Others long before had noted the similar shape of the east and west sides of the Atlantic Ocean, but a specifically formulated theory of continental drift was proposed by Wegener in 1915 and independently by Taylor in 1910 and 1923. This theory was suggested by the striking similarity in shape of the coast lines of North and South America on one side and Europe and Africa on the other, and by similarities of geology on opposite sides of the Atlantic. The theory was supported by Argand, Staub, Holmes, and a few others in Europe and vigorously by DuToit in Africa, but, except for Daly and Grabau, it received very little acceptance by geologists and geophysicists in America.

Interest in the theory was renewed somewhat when detailed bathymetric charts, which became available in the late 1930s, showed that the two sides of the Mid-Atlantic Ridge extending down the middle of the Atlantic fit just as closely to the western and eastern continental shore lines as do the continental shore lines themselves. It was no longer easy to attribute such a series of neat fits—North and South America to the Mid-Atlantic Ridge, Mid-Atlantic Ridge to Europe and Africa, and North and South America to Europe and Africa—to mere coincidence.

The continental drift theory underwent a sharp revival of interest from many earth scientists when the results of paleomagnetic studies became available, which seemed to indicate that the different continents had separated from one another and moved apart along independent courses.

The most convincing evidence, however, that continents have drifted apart came unexpectedly as new evidence from the geophysical and geological exploration of the ocean basins became known. Magnetic surveys of the west coast of North America by Mason, Raff, and Vacquier in 1958 to 1965 revealed striking and puzzling bands or stripes of alternately enhanced and reduced magnetic field. The presence of these same striking patterns of linear magnetic stripes in the North and South Atlantic, the South Pacific, and the northwestern Indian Ocean was demonstrated by Heirtzler and colleagues a few years later. The stripes are traceable for thousands of kilometers but are offset laterally by transverse fractures, some places by as much as 1000 km. The stripes extend parallel to the mid-ocean ridges and are roughly symmetrically spaced on each side of them.

Hess in 1960 and 1962 and Dietz in 1961 proposed that the oceanic crust and the sea floor are simply the near-surface expression of a mass of convecting mantle material that wells up beneath mid-oceanic ridges. The sea floor, according to this proposal, thus spreads laterally in both directions from the mid-ocean ridges, and the continents on each side of the growing ocean move farther and farther apart. The high heat flows on the ridges were thought to be consistent with the submarine vulcanism associated with the formation of new crust. Dietz suggested the term sea-floor spreading for the concept and this has taken hold.

Vine and Matthews in 1963 and Morley and Larochelle independently in the following year suggested that the rocks of the new ocean floor are magnetized in the direction of the earth's magnetic field. The pattern of alternately enhanced and reduced magnetic stripes (normal and reversed directions of magnetic polarity) is then formed as the newly magnetized rocks move outward from the ridge to make place for the still newer rocks that are being magnetized, first in one direction then the other, as the earth's field reverses.

The concept of sea-floor spreading outward from the axes of the ridges demanded as corollary a new interpretation of the transverse fractures that offset the mid-ocean ridges. Wilson in 1965 suggested that the lines of offset be called transform faults which are characterized by their direction of relative horizontal movement being opposite to that of the traditional or familiar transverse faults. In an analysis of the first motions of earthquakes along mid-ocean ridges, Sykes in 1967 found that the relative motion on these transverse fractures is, in fact, that which had been predicted by Wilson.

When the concept of sea-floor spreading was first proposed it was hardly more than an interesting suggestion. Within the next three years, however, new and unexpected evidence from oceanic sediments and continental lavas virtually confirmed it. Opdyke and associates at the Lamont Geological Observatory in 1966 found that the sediments in three cores from Antarctic waters show a sequence of normal and reversed magnetic polarities which duplicates exactly the magnetic stratigraphy of the past $4\frac{1}{2}$ million years that had been worked out by Cox, Doell, and Dalrymple in 1964, using K-Ar dating. Vine in 1966 demonstrated to the satisfaction of many earth scientists that the history of reversals of the earth's magnetic field for at least the past few million years is laid out like a map on the sea floor and that it, too, agrees in relative spacing with the history of reversals established from lavas on land. These three lines of evidence on magnetic reversals—that from lava flows, from oceanic sediments, and from the patterns of anomalies on the sea floor—are in agreement and thus seem to confirm Vine's proposed explanation of the magnetic stripes.

At the outset the concept of sea-floor spreading and the evidence on which it was based centered largely on mid-ocean ridges, the source from which the new oceanic crust came. But as the concept acquired a firmer status, major interest shifted to another aspect of the problem: what eventually happens to the oceanic crust that is continually being generated at the ridges? Oliver and Isacks in 1967 discovered anomalous seismic zones that appear to correspond to segments of lithosphere that have been underthrust into the mantle beneath island arcs. Further investigations have strongly indicated that oceanic trenches mark the sites where oceanic crust dives down along what are called subduction zones—the Benioff zones of earlier years.

Morgan in 1967 undertook an analysis of the geometry of spreading sources, transform faults, spreading rates, and subduction zones. He proposed that the vast areas of oceanic crust between ridges and trenches behave as rigid plates that rotate slowly about distant poles. McKenzie carried out three-dimensional analyses of the motion of these plates and Le Pichon in 1968 made the first successful global analysis based on the simplifying assumption that the entire surface of the earth may be considered to consist of only six rigid plates. Thus was born the all-embracing new concept of global or plate tectonics. Within a year articles began appearing in all earth science journals of English-speaking countries on one aspect or another of the new tectonics. Attempts, led by such able geologists as Menard, Dickinson, Ernst, Dewey, Bird, and Atwater, and followed by dozens of others, are being made to apply the concept to nearly all known tectonic features of whatever age. It is an exciting time of innovative thinking and critical reevaluation of traditional interpretations. Truly, we are now in the midst of a major revolution in the earth sciences. As a consequence of all this, someone has noted that the volume of printed matter devoted to the subject must be comparable to that of the new rocks being generated each year at the mid-ocean ridges.

This revolution in the earth sciences is an interesting phenomenon in the sociology of science. The theory of continental drift, which had been rejected by nearly all geologists and geophysicists for forty or more years, has suddenly become what already in 1968 was being called the bandwagon. As has happened before in revolutions of various kinds, scientific or otherwise, a significant number of new converts defend the concept with evident emotion against criticisms of any kind, or even against minor reservations regarding some aspect of it. Undoubtedly, much of permanent value has come and is continuing to come from the efforts being made to unify and explain by means of plate tectonics what were previously only unrelated empirical facts. However, the momentum of enthusiasm is also responsible for a number of contributions of lesser value which are disarmingly naive in matters such as the physical properties of rocks or the distribution and composition of rock types in a given region. Perhaps the only serious risk being run during the height of this scientific revolution is that, with the current emphasis on exciting new ideas and possible new applications of the plate tectonics concept, a generation of young geologists and geophysicists is coming to maturity without having had adequate experience in some of the more prosaic but essential disciplines that characterize science—for example, an insistence upon the testability of conjectures.

In spite of many successes thus far, a host of important and critical questions still remain. Some of these questions are now under active investigation and may be answered soon. The answers to others probably must wait upon the gathering of new data and perhaps the formulation of new hypotheses. The search for answers to such questions as those listed below assures us that the immediate future will be a period of excitement and discovery in the earth sciences.

Will the patterns of magnetic anomalies and the ages of oceanic sediment from other areas continue to fall into the same simple relationships as those thus far found? And, if not, how crucial will this be to the validity of the concept in its present form?

In what way is the concept of plate tectonics to be brought into harmony with the geological evidence of great vertical movements up and down of parts of the continents and with the presence of great thicknesses of sediments in geosynclines?

Can the concept be used to explain the deformation of all intensely folded and overthrust mountain ranges whether, like the Andes, they are near continental margins, or as far inland as the Rockies?

How far back into the past has sea-floor spreading and subduction been operating—into the Paleozoic, the later Precambrian, or all of geologic time?

Has the process of plate tectonics been one in which the continents have been in continual relative movement, sometimes separating, sometimes colliding and overriding, and sometimes simply rubbing and grinding past one another?

And what is the mechanism by which plate tectonics operates—as a conveyor belt carried on the top of a convection cell, by down-slope gravity sliding from hot spots on mid-ocean ridges, or by some other process?

GEOMORPHOLOGY

Geomorphology, or physiography as it was called fifty years ago, is the study of the occurrence and origin of landforms, those resulting from erosion by streams, waves, wind, and ice, and those built up from deposits of erosional debris. Inasmuch as it is concerned with depositional as well as erosional landforms, geomorphology may be said to grade indistinguishably into sedimentation. The record of alpine and continental glaciation is written not only in glacial deposits but perhaps even more clearly in characteristic glacial landforms. For this reason, geomorphology is a principal tool for unraveling the complexities of Pleistocene history, a tool which has been made much sharper by the advent of C^{14} dating.

The development of physiography in America had for many years depended largely on the work and concepts of W. M. Davis. His system of describing landforms genetically in terms of a central concept of the erosion cycle was widely and highly esteemed. This concept had been developed to explain the geomorphologic features observed in the humid climate of the Appalachians. In response to growing criticism of the restrictions imposed by its original formulation, Davis in 1930 modified and extended the concept of the erosion cycle to apply it to landforms in areas of arid and semi-arid climates.

However, this modification did not suffice to hold back the rising tide of dissatisfaction with the Davisian system of interpreting landforms. In 1927 there appeared a book, *Morphologische Analyse,* by W. Penck which seriously challenged some of the basic assumptions of the Davisian concepts. Penck emphasized the important role of denudation during the gradual uplift of a landmass whereas the classic Davis cycle of erosion began after an undissected landmass had been uplifted.

Davis had held that, because of weathering and soil creep, hill slopes flatten during progress of the erosion cycle until there remains only a land surface of low relief which he called a peneplain. Penck, on the other hand, asserted that slopes retain their original steepness as they retreat laterally, leaving behind them a beveled rock surface of low relief much resembling the pediments commonly found at the base of steep mountain fronts in arid and semi-arid regions.

Subsequent studies by King in southern Africa and Hack in the Appalachians have tended to support the Penck idea of lateral retreat of slopes rather than the Davis idea of gradual flattening of land surfaces. Many surfaces once considered dissected remnants of widespread peneplains may, in fact, have been formed as pediments cut on bed rock by the action of rills, channels, and rain wash. But sharp differences of opinion still persist and the final answers are not yet all in. Future work may show that true Davisian peneplains have formed under certain as yet unspecified conditions and extensive pedimented surfaces under certain other conditions.

Until recently the treatment of geomorphological problems has been largely qualitative. The most significant and promising development in geomorphological research in recent years has been the efforts that have been made to apply quantitative methods to field observations, laboratory experiments, and theoretical analyses. These efforts, led by Leopold and his associates, including this writer, have been most extensive in the area of stream behavior—alluvial geology as it is sometimes called.

In the area of the geomorphic effects of the wind, an outstanding but rather isolated example of the new trend toward quantitative treatment of field and laboratory data is the work of Bagnold in the 1930s on the effect of wind on the transport of sand. In the area of ice work the studies of Nye and others on the mechanics of glacier flow have been mentioned above, under the heading of Geophysics. Nothing strictly comparable to the quantitative studies regarding streams, wind, and ice is under way on the geomorphic effects of beach processes. However, important work on wave erosion and shoreline processes and deposits is being done by sedimentologists.

In summary it may be said that geomorphology is moving toward becoming a completely revitalized science. It is now at a stage which might be compared with that of petrology in the first two decades of this century when the Geophysical Laboratory was getting started.

PALEONTOLOGY AND STRATIGRAPHY

As an undergraduate I started out in earth science hoping to become a paleontologist. I later fell by the wayside but have always had many close friends who were paleontologists. Nevertheless, in spite of such ties of amity with the field, I feel myself even farther at sea in the subject matter of paleontology than in that of the other fields mentioned in preceding paragraphs.

Fossils new to science and better representatives of poorly known forms are constantly being discovered and it is the primary duty of the paleontologist to place these new facts on record. For this reason, purely descriptive work continues to be prominent in paleontological research. However, the paleontologist is also under obligation to undertake monographic treatment of the ecologic assemblage of faunas or floras in particular geological formations and to work out systematic and phylogenetic sequences of the families and orders of fossils that are his speciality.

Interesting new developments of the past fifty years in paleontology which deserve mention include the rapid expansion of micropaleontology since 1920, due primarily to its usefulness as a tool in making detailed correlations of rocks drilled through in the search for oil. For this purpose many different types of fossils have been found useful, among which should be mentioned especially the foraminifera, ostracods, conodonts, and more recently, diatoms, radiolaria, and coccoliths.

Palynology, the study of spores and pollen, was inaugurated as a special discipline about 1920 by Scandinavian students of Pleistocene and Recent peat bogs. The finding of alternations of spruce-pine and mixed hardwood assemblages became an effective tool for deciphering the history of glacial advances and retreats and of other climatic changes. More recently palynological research has proven successful in stratigraphic dating and correlation of rocks of Tertiary, Mesozoic, and Paleozoic age.

A significant new development has been an increasing application of statistical methods to the concepts of species and genus and the concept of biological communities. This statistical trend was led by Simpson in vertebrate paleontology and it has since been extended into invertebrate paleontology and paleobotany.

Over the past fifty years paleontologists have become increasingly interested in considering animal and plant communities and their ecological significance. Paleoecology had its beginnings in an increasing emphasis on quantitative methods in marine biology which may be said to have started with Edward Forbes in England in the mid-1850s. The work was continued by investigators in France, Denmark, and Germany and in 1940 Vaughan initiated a movement in the United States that led to the establishment of a National Research Council committee on the ecology of marine organisms. In 1957 the massive two volume *Treatise on Marine Ecology and Paleoecology*, prepared by this committee under the chairmanship of H. S. Ladd, was published.

A startling zoological-paleontological event was the discovery in 1939 in deep water off the eastern coast of Africa of a living coelacanth, a primitive fish not known in the fossil record later than the Cretaceous. The living genus *Latimeria* has changed little from its Mesozoic ancestors. Other living fossils found since 1920 were *Metasequoia*, the Chinese redwood, discovered in 1941, and *Neopilina*, a mollusk dredged from the deep floor of the Pacific off Central America in 1952 during a Danish oceanographic expedition. Closely related forms of this mollusk had been thought to be extinct since the middle of the Paleozoic. Additional closely related forms have subsequently been discovered in the Peru-Chile Trench and in the Indian Ocean.

In 1954 Tyler and Barghoorn discovered fossil remains of simple organisms comparable to algae and fungi in chert in the Gunflint iron formation (age approximately 1800 million years) along the shore of Lake Superior. Searches by Barghoorn, Schopf, Cloud, Engel, and Macgregor since then for other Precambrian fossils have yielded algae, algal-like, and bacteria-like forms in rocks of 1000 to 3000 million years age in the United States, Canada, Australia, South Africa, and Rhodesia. Barghoorn and Schopf in 1966 described fossil bacteria from chert in the Fig Tree series in South Africa (age approximately 3100 million years), the oldest fossils now known.

A new and interesting development in the geochemistry of organic matter might logically have been treated under the heading of Geochemistry, except that it is closely related to the discoveries of Precambrian fossils mentioned above and that, after all, it is a new aspect of paleontology. In the mid-1950s Abelson initiated a program of study of the organic constituents of fossils. Hoering, an associate of Abelson's, and Meinschein, Oro, and Eglinton at other institutions in the United

States and England have participated actively in these studies. Much of interest has emerged from these studies concerning the types of chemical processes that were used by the most primitive organisms. The studies have indicated that the organic matter preserved in old sediments contains many of the same constituents as like material today, that photosynthesis was conducted 2000 million years ago by reactions similar to those being used today, and that small amounts of the most stable amino acids—the essential building blocks from which living proteins are built—have been found in very ancient sediments. About 95% of the organic matter in sedimentary rocks is in the form of kerogen, a residual material of complex and variable composition. A better knowledge of the processes involved in the formation and transformation of kerogen will be helpful in understanding the origin of petroleum and the evolution of life. The search to find unmetamorphosed early Precambrian rocks that contain significant amounts of organic matter seems certain to continue.

Stratigraphy is the study of stratified or layered rocks. It primarily concerns sedimentary rocks but its principles apply also to lava flows and volcanic tuffs and to sedimentary and volcanic rocks that have been metamorphosed. Stratigraphy merges almost indistinguishably into sedimentation and sedimentary petrology which, as mentioned above, are closely related to one another. New developments in any one of these three fields contribute to progress in the others.

Stratigraphy has three principal objectives. First is the description of the local sequence of rocks in each of many areas. Knowledge of the kinds of rocks exposed in many individual places is the very substance of geology and a firm grasp of the details of the local rock sequence is an essential tool for the preparation of a geologic map. The second objective is to correlate or to determine, largely by means of fossil evidence, the time relations of the local rock sequences in widely separated areas. The third objective is to interpret from these correlated local rock sequences the geologic history of these local areas and therefore of the earth.

Because it must deal with an almost unlimited number of details in many local areas, stratigraphy is in danger of drowning in the sea of facts that it accumulates. To help keep these nearly endless facts in some order, geologists have developed systems for the naming and classification of rock units. Somewhat different systems of stratigraphic nomenclature have been developed in Europe and North America—that in Europe being based primarily on time equivalence, as judged by fossil content, and that in North America being based on the lithologic character of different rock units. To clarify the principles of stratigraphic nomenclature as used in the United States, a widely representative committee of geologists of the Federal and State geological surveys, the Geological Society of America, and the American Association of Petroleum Geologists formulated an elaborate code of rules and recommendations which was published in 1933. In the rather extended deliberations of this committee it was the responsibility of J. B. Reeside Jr. and the writer to compile, annotate, and attempt to reconcile the sometimes conflicting views of other committee members. An American Commission on Stratigraphic Nomenclature, established in 1946, prepared and, in 1961, published a revision of the 1933 stratigraphic code.

Stratigraphy is of great practical value and for this reason it has expanded rapidly along with the great growth of petroleum exploration. Not only has stratigraphic information been extremely useful in the search for oil but the drilling of many wells has contributed enormously to stratigraphic knowledge. Subsurface geology, consisting of the study of cores, cuttings, and drillers' and electric well logs, has revealed many stratigraphic features that were unknown on the basis of surface geology alone.

The burdensome quantity of detailed stratigraphic information that has accumulated in recent decades would be almost useless had there not been invented a number of different kinds of maps by which these data can be synthesized and interpreted.

Paleogeographic maps which show the inferred distribution of land and sea at different times in the past were being compiled in Germany, France, and the United States in the first decade of this century. In 1910 Schuchert published a collection of 54 such maps and these have gone through successive revisions by Schuchert, Dunbar, and others since then.

Emphasis has since shifted from the relatively simple paleogeographic maps to a wide variety of other maps that permit meaningful syntheses of the enormous quantities of stratigraphic detail that have been gathered. Among these stratigraphic maps, as they are sometimes called collectively, are isopach maps and convergence maps of the thickness of a particular series of strata or of the interval between two stratigraphic marker beds. The rocks within a stratigraphic interval may differ appreciably from place to place. Maps showing the ratios of rocks of different composition are called lithofacies maps. An example is a map of the ratio of dolomite to limestone in a stratigraphic unit. Other examples are a map of the ratio of clastic to nonclastic rocks and a map of the proportion of coarse to fine detrital components in a rock unit under investigation. A biofacies map shows the areal distribution of different fossil assemblages. Paleogeologic maps show the areal geology of an ancient surface as it was exposed before burial under a later formation. It is evident that a series of such stratigraphic maps affords a multidimensional view and an understanding of a stratigraphic unit that would not be possible by earlier noncartographic methods of study. Leaders in the development of this wide variety of maps, the objectives of which have been primarily economic, have been Levorsen, Sloss, Krumbein, and their associates.

OTHER SUBJECTS

In this listing of major developments in the earth sciences in the past fifty years, many subjects of very great importance have been omitted because of space limitations.

Nearly all the earth sciences (vertebrate paleontology is the only exception that comes to mind) have depended greatly for their support and growth upon their economic applications. Yet, except for some discussion of geophysical exploration and petroleum geology, scarcely any mention has been made here of such important fields as mining geology, mineral resources (including the search for strategic minerals and for nuclear and geothermal energy sources), ground-water hydrology, engineering and environmental geology (including waste disposal underground and problems of perma-frost), military geology, and remote sensing from spacecraft by use of multispectral photography, which some think may prove to be a valuable adjunct to reconnaissance geologic exploration for mineral resources.

Nor have exciting new aspects of lunar and planetary science that have been very much in the fore in the past decade been mentioned, even though for several years this writer was a participant of sorts in the late, lamented lunar program. For these and other important omissions I can only offer my sincerest apologies.

THE LOOK AHEAD

From among the list of many developments in the earth sciences that have been considered here, the truly big things that have happened in the past fifty years in my opinion have been: 1. the impact of the program of experimental petrology initiated at the Geophysical Laboratory, 2. the far-reaching consequences of the geophysical exploration of the ocean floor, and 3. the enthusiastic (some might say unguarded) acceptance of the plate tectonics concept that has swept the earth sciences. Other observers, admittedly, may appraise the relative significance of developments during these years quite differently.

So much for what has happened in the past fifty years. What about the look ahead? Predictions about the future in any field of human endeavor are notoriously dangerous and subject to bitter error. Nevertheless, on the basis of the directions we have been traveling, what kinds of projections into the future can reasonably be made?

It seems to me that, except for the problem arising from the publication explosion which we hope may be on the way toward some amelioration, the immediate future promises to be a period of excitement and new advances in the earth sciences. To cite some obvious examples: it seems reasonable to expect important new developments in the petrology and stability relations of clay minerals; in the JOIDES program of deep drilling at sea; in the many unanswered questions concerning plate tectonics; in quantitative geomorphology; in the geochemistry of organic matter in fossils; and, of course, in an unknown number of even more exciting fields not now foreseen. The earth sciences appear now to be in a satisfactory state of turbulent activity and healthy confusion.

ICELAND IN RELATION TO THE MID-ATLANTIC RIDGE

10015

Gudmundur Pálmason and Kristján Sæmundsson
Orkustofnun (National Energy Authority), Reykjavík, Iceland

INTRODUCTION

The location of Iceland astride the axis of the Mid-Atlantic Ridge gives to it, in many ways, a unique role in the study of processes taking place at the mid-ocean ridge crests. If ocean-floor spreading is taking place in the way envisaged by plate tectonics then Iceland must be splitting apart. The geological evidence for such a process is suggestive but opinions are still divided on whether it is conclusive or not as far as Iceland is concerned. In any case the process would probably be more complicated there than in the adjoining submarine segments to the north and southwest of Iceland. This is clearly indicated by the prominent topographic high of Iceland as well as by the belt of earthquake epicenters (Figure 1) which follows the crests of the Reykjanes and Iceland-Jan Mayen Ridges but takes a detour on crossing Iceland.

Iceland has traditionally been described as a part of the Brito-Arctic, or Thulean basalt province, of which the basalt areas in Greenland, the Faeroes, and Scotland are also a part. With the knowledge gained in the past decade or two about the basaltic nature of the oceanic crust, it appears more natural to consider Iceland as a part of the oceanic basalt province, a part which by virtue of the nature of the underlying crust and mantle is elevated to its subaerial position. The two prominent neighboring topographic features, the seismically and volcanically active Mid-Atlantic Ridge crest and the aseismic Greenland-Iceland-Faeroes Ridge, must enter into any hypothesis to explain the elevation of Iceland relative to the deeper basins to the south and northeast.

The hypothesis of ocean-floor spreading and plate tectonics has in recent years given a great impetus to geonomic studies of Iceland and the surrounding ocean areas. Bathymetric, magnetic, gravity, and seismic reflection surveys have been carried out on the Reykjanes, Iceland-Jan Mayen, and Iceland-Faeroes Ridges (3, 19, 35, 45, 49, 56, 57, 68, 105, 118, 120) and such surveys are in progress on the insular shelf surrounding Iceland. Seismic refraction studies have been made on the Reykjanes and Iceland-Faeroes Ridges (19, 34, 105) and on parts of the insular

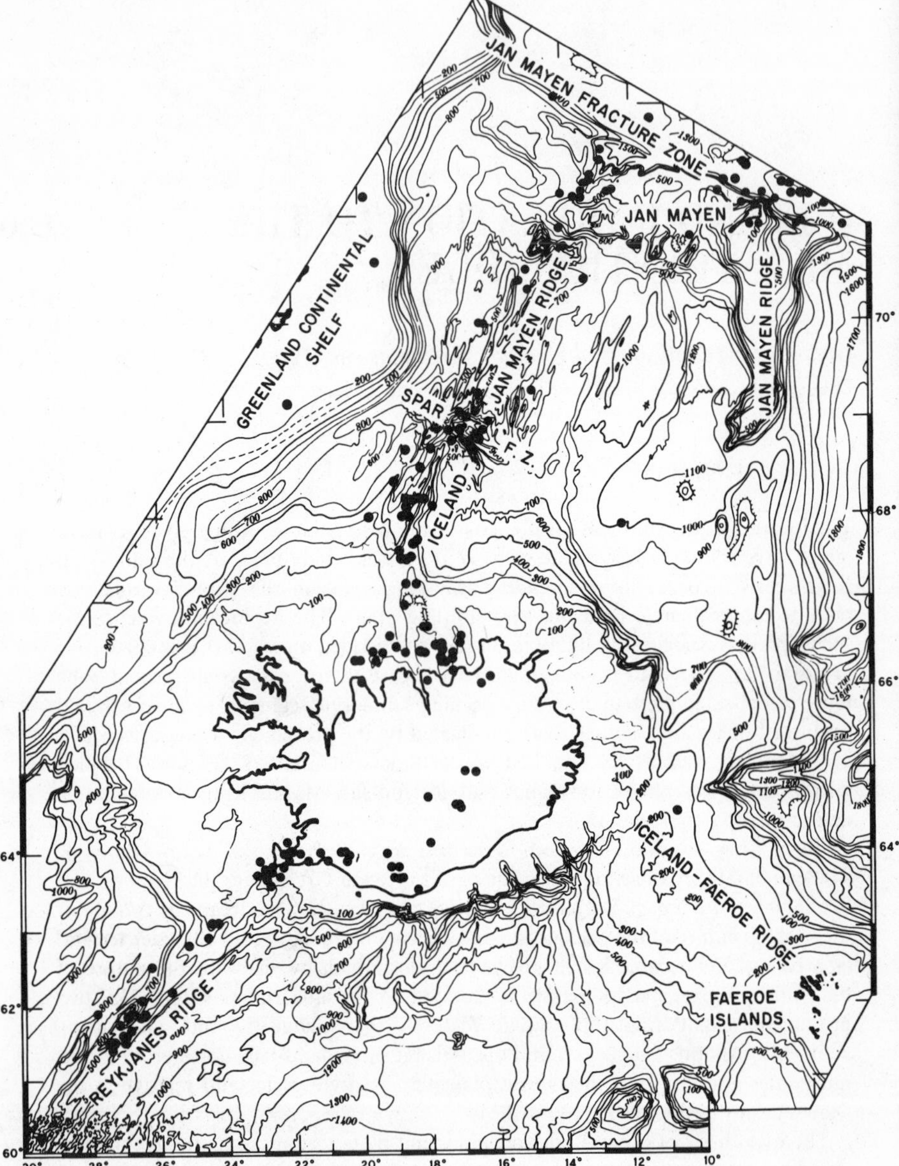

Figure 1 The earthquake epicenter belt through Iceland. Earthquakes of magnitude greater than about 4 in the period January 1955 to March 1972 are shown. Based on Sykes (102), U.S.C.G.S., and N.O.A.A. determinations of epicenters. Bathymetry from Johnson et al (57). Depths are in nominal fathoms (1/400 sec travel time).

shelf (77). Aeromagnetic surveys have also been carried out over large parts of the ocean between Greenland and Europe (6). Heat flow measurements (53, 105) and rock dredgings (23, 62) have been made on the Reykjanes Ridge and the chemistry of the rock samples studied (87). Various lines of research have been followed on land. Some of the earlier work has been summarized by Thorarinsson (108) and Sigurgeirsson (94) and in Björnsson (12). Geological studies comprise mapping in active zones of rifting and volcanism as well as mapping of older rocks in various parts of the country (13, 27, 32, 44, 58, 60, 81, 84, 85, 91, 92, 122, 123, 127). Central volcanic complexes are being studied in both environments—in active zones particularly in relation to geothermal areas, which are energy sources of potential economic importance (2, 14, 21, 39, 46, 47, 81, 84, 91, 109, 124). Petrological and geochemical studies of basalts and acid rocks are in progress (54, 55, 74, 96, 97, 110). Regional heat flow has been mapped (76, 78). Crustal structure and thickness has been mapped by refraction seismology (7, 75, 77). Aeromagnetic surveys are in progress (89, 95). A new systematic gravity survey has been completed (unpublished observations). Crustal and upper mantle electrical conductivity has been mapped (50, 50a, 50b). Studies of magnetic properties of basalts are in progress (63, 64). Absolute age determinations are accumulating (22, 33, 40, 48, 67, 70, 73, 93, 106). Four different groups are attempting to measure crustal movements in the active zones (24, 42, 115, 116) directly. Studies of microearthquakes have been carried out on an increasing scale in recent years (61, 129, 131). The above studies are carried on by both Icelandic and other scientists, sometimes as cooperative projects.

The purpose of this paper is to review the present state of knowledge pertaining to geological processes at the crest of the Mid-Atlantic Ridge in Iceland. The presentation will be problem-oriented rather than by disciplines. The surface geology will be discussed first, then evidence on the nature of the crust and thereafter of the mantle. Finally the problem of crustal drift in Iceland will be discussed.

SURFACE GEOLOGY

Iceland is composed almost entirely of subaerial Cenozoic basalts, with some 10% of acid and intermediate rocks. The active zone of rifting and volcanism (Figure 2), also loosely called the Neovolcanic zone or the median zone, crosses the country in a complex pattern from the southwest, where it connects with the Reykjanes Ridge axis, to the northeast, where it connects along an oblique offshore zone with the Iceland-Jan Mayen Ridge (also called the Kolbeinsey Ridge). The median zone is flanked by strips of Quaternary flood basalts followed by Tertiary flood basalts which often have gentle dips towards the active zone. The presently available K/Ar age determinations (Figure 3) indicate a trend of increasing age with distance from the active zone of rifting and volcanism, the oldest rocks dated at 16 m.y. being found in northwestern Iceland and in eastern Iceland.

The zone of rifting and volcanism displays a great variety of volcanic forms, cut by numerous faults and open fissures running mainly NE-SW in southern Iceland with a more northerly trend in northern Iceland. The predominant type of volcano

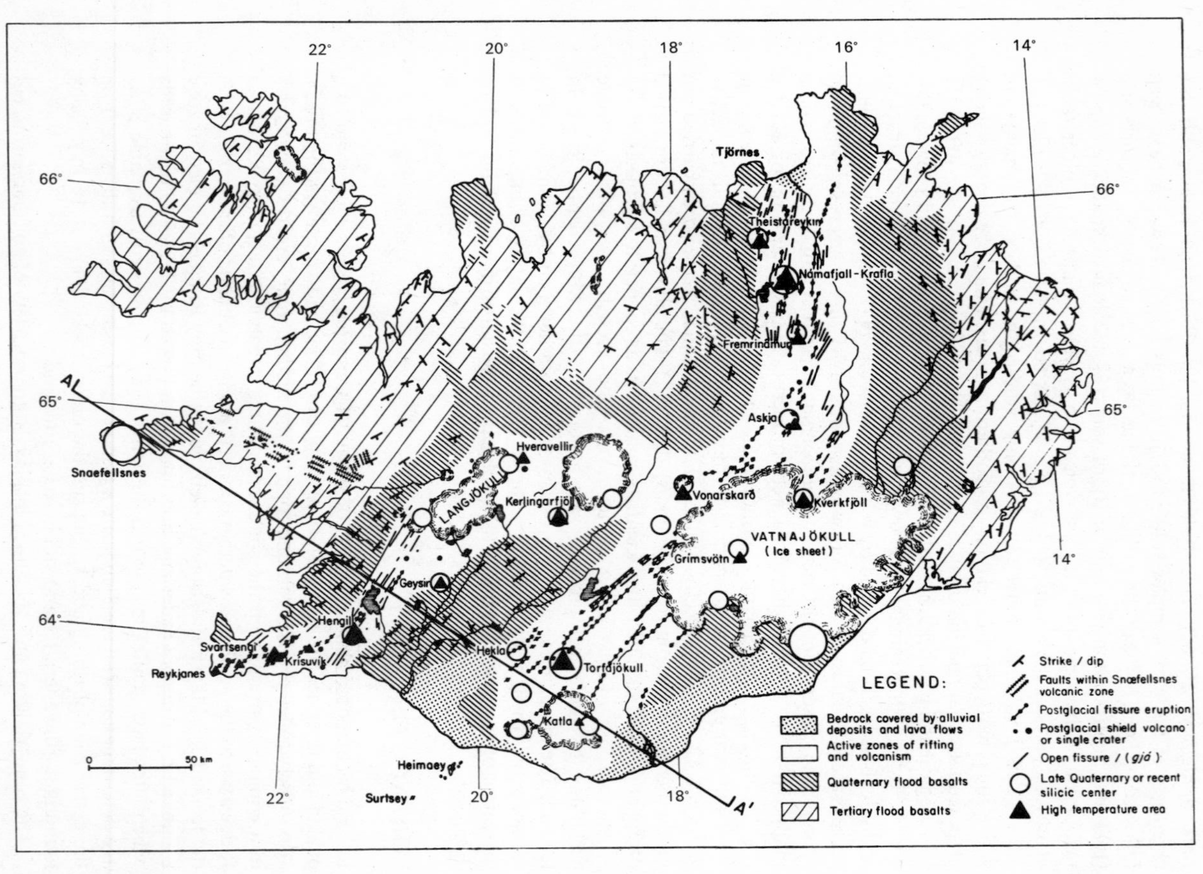

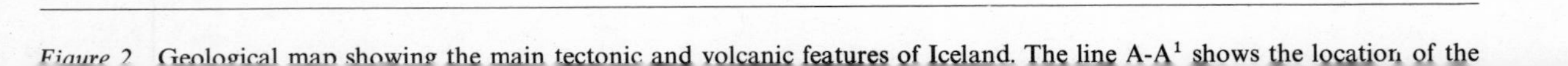
Figure 2 Geological map showing the main tectonic and volcanic features of Iceland. The line A-A[1] shows the location of the

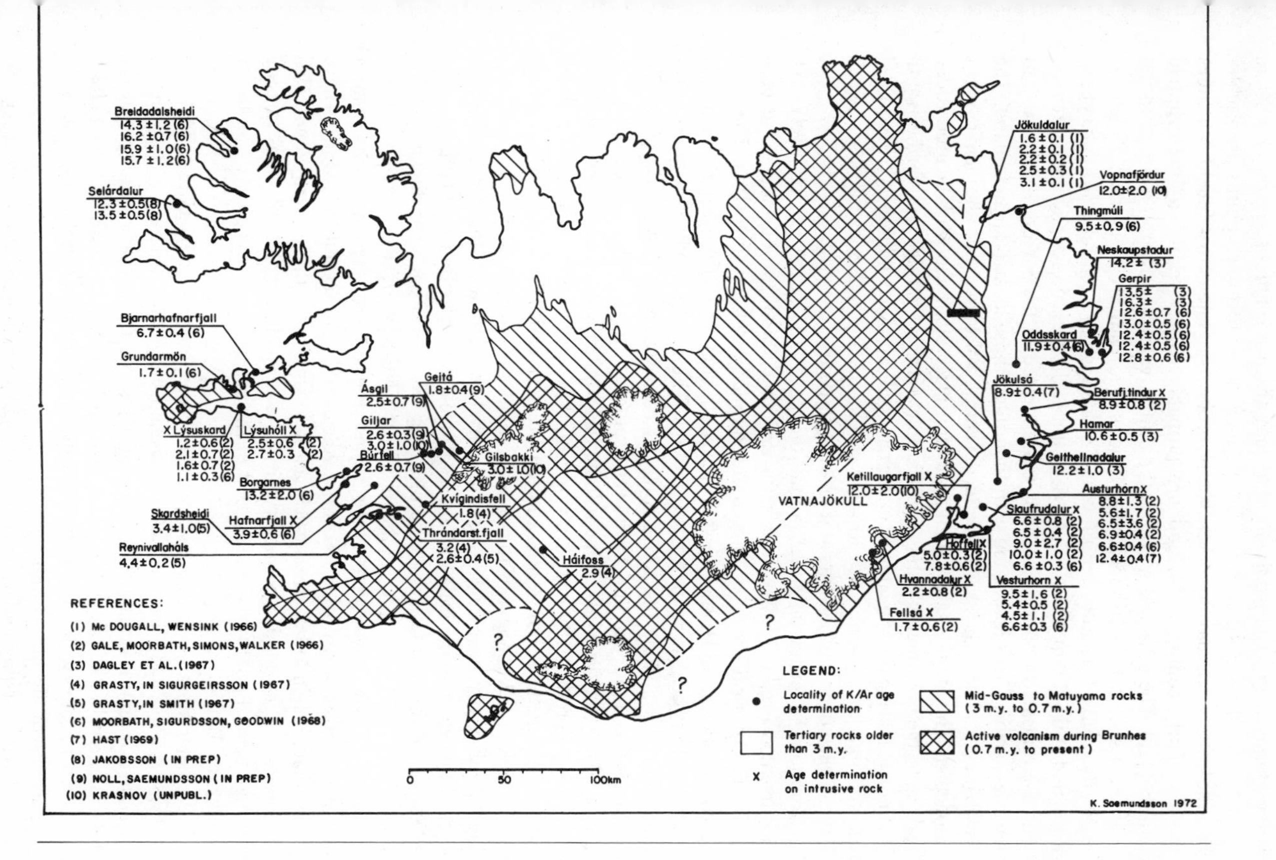

Figure 3 A compilation of presently available K/Ar age determinations from Iceland. (From references 22, 40, 48, 67, 70, 73, 93, 98, and unpublished work by S. P. Jakobsson and by A. Krasnov.)

is the monogenetic eruptive fissure that may reach a length of several kilometers or even a few tens of kilometers. The products of the fissure eruptions are usually basaltic lavas which may flow over distances of tens of kilometers. Shield volcanoes are also fairly common (58) and are probably closely related to the eruptive fissures (128). The other main type is the central volcano, which in contrast to the eruptive fissure is a site of repeated eruptions in a relatively small area, sometimes distinctly grouped around a central vent. The life span of a central volcano has been estimated at 0.5–1.0 m.y. (73, 79). Most of the acidic rocks in Iceland are associated with the central volcanoes. However, those located in the SW and NE appear to have less acidic rocks associated with them than those located farther inland. The Reykjanes peninsula lacks well-defined central volcanoes and the acidic rocks altogether. This might reflect changes taking place along the Mid-Atlantic Ridge leading to the presumably purely basaltic volcanism of the ridge crest farther south and north. During the glacial periods the changed external conditions have modified the volcanic forms, giving rise to table mountains and hyaloclastite ridges which have formed over central vents and fissures, respectively.

Eruptions in historical times (about the past 1100 years) are relatively well documented (e.g. 107, 109). It is estimated that about 30 volcanoes have been active in historical times, and about 200 during postglacial time (108). The most recent eruption is on Heimaey just off the south coast (55, 110). At the time of this writing it has been going on for two months. The rate of production of eruptives and its variation along the active zone has been studied by Jakobsson (54), who estimates that about 480 km^3 have been erupted during postglacial time or on the average 0.048 km^3/yr. Most of this volume was erupted within the eastern zone with a maximum productivity just south of central Iceland where it may be 4–5 times higher than at the southwestern and northeastern ends of the zone. The average productivity per kilometer of the zone is about 1.4×10^{-4} km^3/yr. The relatively high productivity near central Iceland correlates well with the relatively great crustal thickness on the Iceland-Faeroes Ridge deduced by Bott et al (19).

Studies of the northeastern zone of rifting and volcanism have shown that the central volcanoes are located in the middle of NNE-SSW trending swarms of faults and eruptive fissures which have their highest intensity near the center but diminish in both directions away from the central volcano (83, 85). Several such swarms have been identified in other areas such as Reykjanes and the eastern zone. This distribution is in good agreement with the swarm distribution of dikes passing through central volcanoes as observed in the eastern Iceland Tertiary basalts by Walker (123).

Lenticular structure in the lava pile has been pointed out by Gibson (43) and Gibson & Piper (44). The controlling mechanism of this structure is seen in the grouping of central volcanoes and dike swarms into units which would produce a lenticular shield-like pile of lavas in a given period of time. The structural pattern of the active zones with their central-volcano, fissure-swarm couples emphasizes the significance of this observation.

The zone of rifting and volcanism is a locus of high heat flow as evidenced by the 15–20 high-temperature geothermal areas distributed more or less uniformly along the zone (Figure 2).[1] Many, but not all of the high-temperature areas are associated with central volcanoes, the Reykjanes peninsula being a notable exception. The heat source is probably the general long-term heating of the crust by dikes and other intrusions associated with the volcanism. The conductive heat flow within the zone usually cannot be deduced directly from geothermal gradients in shallow holes because the subsurface temperature field is disturbed by water movement in the relatively permeable zone of rifting. Meteoric water may flow for distances on the order of 50 km (4), driven mainly by a hydrostatic pressure difference due to elevation variations.

Of particular significance with respect to mid-ocean ridge processes is the Reykjanes high-temperature thermal area, located in the axial zone at the tip of the Reykjanes peninsula (13, 112). Exploration by drilling to 1750 m, where a temperature of about 290°C has been measured, has shown that seawater percolates to a depth of a few kilometers. This process will modify the temperatures in the upper part of the crust, and it may be responsible for the relatively large scatter in heat flow values at the mid-ocean ridges, as was suggested by Pálmason (76).

Rough estimates of the total heat output of the high-temperature areas in Iceland (15) indicates that per unit length of the volcanic zone it amounts on the average to 15 MW/km (1 MW $= 10^6$ watt). The average heat output of extrusive volcanism is about 20 MW/km.

The most detailed studies of the deeply incised older flood basalts have been made in eastern Iceland by Walker and his collaborators, and a similar survey by the second author of this paper is in progress in southwest Iceland. The basalt lavas have a regional dip towards the volcanic zone increasing from near zero at the top of the pile to 5–10° at sea level (Figure 4). This was interpreted by Walker (122, 123) as being due to a gradual sagging down of the pile by the accumulation of volcanics in an active zone. The basalt lavas are cut by numerous dikes distributed unevenly in swarms and decreasing in number upwards in the pile. From a study of secondary mineral zoning and an upwards extrapolation of the dike intensity to a zero value, Walker (123) deduced the original top of the lava pile to have been at an elevation of about 1500 m in the area studied.

Einarsson (27) has studied the structure of the Tertiary lava pile in other parts of Iceland. In many respects his results are similar to those of Walker for eastern Iceland. The dips are usually smallest in the uppermost part of the pile, increasing downwards, sometimes jumpwise across discordances. The directions of the dips do not, however, show as clear a relationship to the presently active volcanic zones as do the dips in most of southwestern and eastern Iceland (cf Figure 2). Differing from Walker, Einarsson (31) concludes that the relatively steep dips in the lower part of the Tertiary basalt pile were caused by a major tectonic phase after the main bulk of the pile had been formed. There are certain difficulties associated with this interpretation as will be discussed later.

It appears likely that the volcanism has been continuous from the Tertiary

[1] A high-temperature geothermal area is here defined as a hydrothermal circulation system with a subsurface temperature over 200°C at less than 1000 m depth.

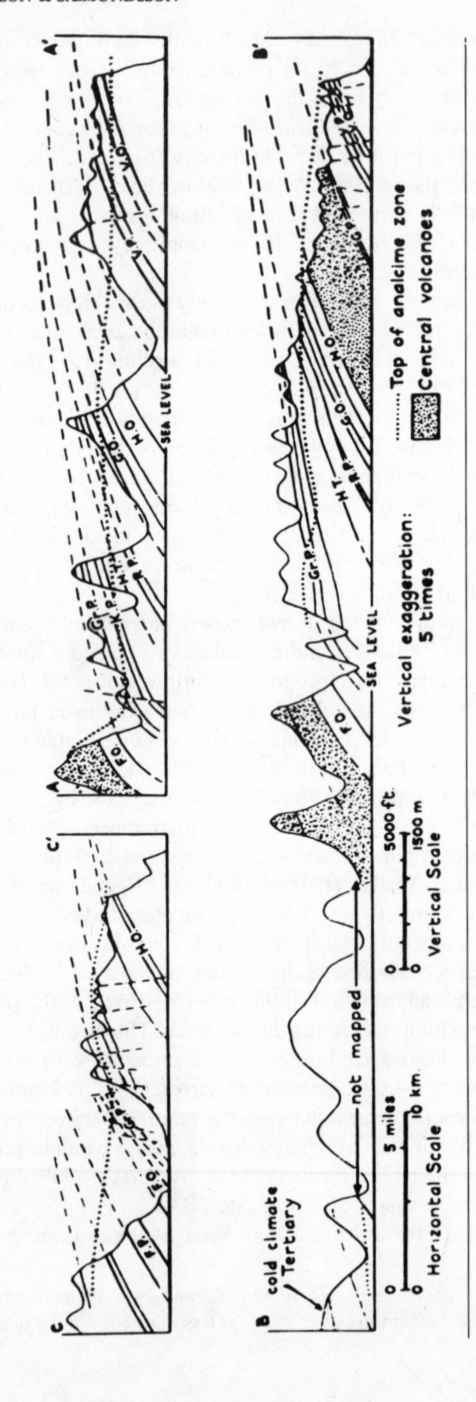

Figure 4 Sections from west (left) to east through the Tertiary basalt pile in eastern Iceland according to Walker (16).

through Pleistocene and postglacial times (108), although the pattern of the active zones may have changed. A prominent flexure of the Tertiary flood basalts in eastern Iceland (125), traceable all along the western border of the Tertiary outcrops, has been identified with a major discontinuity of the stratigraphic succession in this area (85). Stratigraphic correlations, age determinations, and evidence from detailed studies in northeastern Iceland across this area suggest that the eastern volcanic zone is a relatively young feature formed perhaps 4 m.y. ago. Low heat flow values in eastern Iceland also support this conclusion. They indicate, when interpreted in terms of a cooling lithospheric plate moving away from a zone of crustal accretion, that the eastern Iceland lava pile is older than corresponds to its distance from the present eastern zone (78). The general pattern of dips of the Tertiary flood basalts (Figure 2) shows a certain synclinal symmetry about two or perhaps three zones. This points to a more complicated history of crustal accretion than indicated on the Reykjanes Ridge. A similar shifting of the spreading axis in the area between Iceland and Jan Mayen is indicated by bathymetric and magnetic data in that area (57, 68).

The Snaefellsnes volcanic zone in western Iceland is somewhat of an anomaly with regard to the main zones of rifting and volcanism. The Recent and Pleistocene volcanics form an east-west lenticular pile resting unconformably on tilted Tertiary basalts (92). Volcanic activity has been confined mainly to three lines arranged en echelon with a WNW-ESE direction. Tensional open fissures, which are common in the main SW-NE zones of rifting, are scarce or nonexistent, but numerous dip-slip faults occur. The Snaefellsnes zone is aseismic and heat flow is low as evidenced by drillhole measurements (78) and lack of major geothermal areas. The alkalic and transitional basalt volcanism (54, 91) in contrast to the tholeiitic basalts of the main zones also points to a different state of the upper mantle.

Sigurdsson (92) has suggested that the Snaefellsnes zone is a transcurrent fault-zone generated by a differential spreading rate in north and south Iceland. So far, however, the evidence given by magnetic anomaly patterns southwest and north of Iceland does not indicate greater changes in spreading rate with latitude than expected on the basis of the assumed pole position for the movement of the North-American-European plates (57). It is likely, nevertheless, that the Snaefellsnes zone is in some way related to the prominent change in strike of tectonic features taking place at about 65° N latitude. A major change in crustal thickness across this zone is indicated by the available seismic refraction data (77).

Several active fracture zones have been suggested in Iceland (86, 103, 129, 130), mainly on the basis of earthquake epicenter distribution and changes in the tectonic pattern and strike of the volcanic zones. Of these the Tjörnes fracture zone near the north coast appears to be the best founded. Its existence is supported by earthquake distribution, submarine topography, strike-slip faults on land, and offset of the volcanic zone (85).

CRUSTAL STRUCTURE

Evidence for the deeper structure of the crust comes mainly from geophysical data which must be interpreted with due regard to surface geology. The geology of

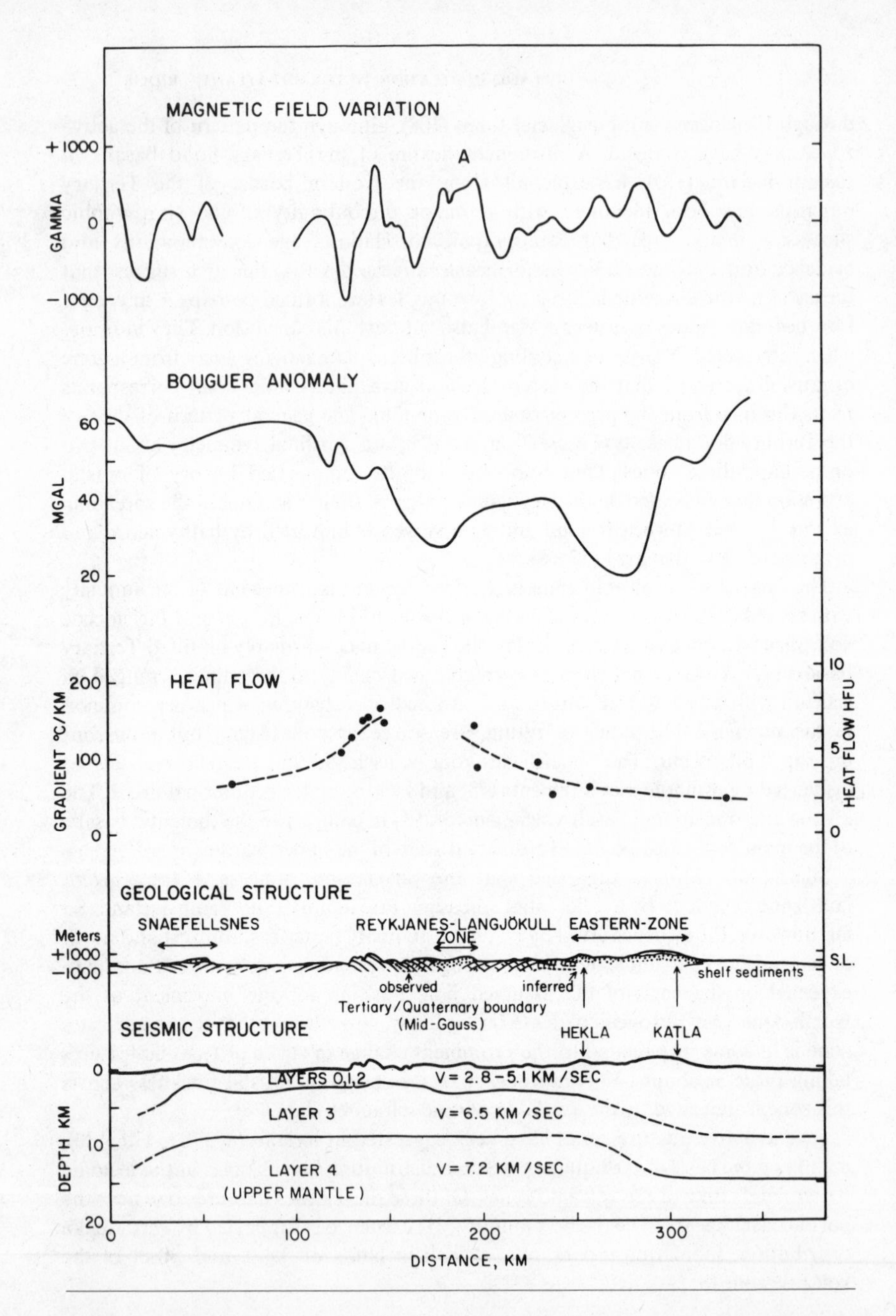

Figure 5 Profiles along line A-A[1] in Figure 2 showing seismic structure (77), geological cross section, heat flow (78), Bouguer anomaly [from (26) and later unpublished work], and magnetic field (Th. Sigurgeirsson and L. Kristjánsson). The central axial magnetic anomaly is denoted by A.

Iceland suggests that the crust is basaltic, although the relative abundance of acidic rocks and the elevation relative to the surrounding ocean floor has led to suggestions that the visible basalt pile might be underlain by a continental sialic fragment (9, 10, 52, 117).

Relatively detailed seismic refraction measurements have been carried out in the last 15 years to study the seismic velocity structure of the crust and its thickness (7, 77). A characteristic layering has been found, resembling the oceanic crust in velocity values, but thicker (cf Figure 5). The lowest seismic velocities, 2.0–3.3 km/sec, are found in the active zone of rifting and volcanism for near-surface rocks apparently consisting of a mixture of recent lava flows, hyaloclastic tuffs, and breccias. This formation, termed layer 0, reaches a maximum thickness of about 1000 m. Drillholes to a maximum depth of 1750 m in the Reykjanes thermal area show that this formation, seismically determined to be about 900 m thick, consists mainly of hyaloclastic tuffs and breccias and tuffaceous sediments, with basalt lavas as a minor component (13, 112). These rocks are believed to have been erupted under subaerial conditions or at shallow water depth.

The Tertiary and Quaternary surface basalts on both sides of the active zone have distinctively higher velocity values, averaging about 4.1 km/sec (layer 1). This velocity group is also found beneath the low-velocity surface layer in the active zone. Surface geology indicates that it consists mainly of basalt lavas which may, however, contain an appreciable amount of intercalated sediments and clastic volcanics. Drillholes on the Reykjanes peninsula indicate that it may contain up to 50% of tuffaceous rocks. The thickness of this layer is usually 0.5–2.0 km with an average value about 1.0 km.

At greater depth the velocity increases to about 5.2 km/sec on the average (layer 2). This group is exposed only in a small area in southeastern Iceland where it consists of basaltic lavas mixed with basic and acid intrusions. From its relatively shallow depth in many of the old flood basalt areas it appears likely that it is composed mainly of flood basalts. Its thickness is usually in the range 1–3 km with an average value close to 2.1 km.

The three seismic layers discussed above can apparently be correlated with known surface or near-surface rock formations. The underlying layer 3 with a P-wave velocity of about 6.5 km/sec is found beneath the whole of Iceland but nowhere at the surface. The depth to its upper boundary has been mapped in some detail (Figure 6) and found to be quite variable, usually in the range 1–5 km but in one area reaching 10 km. No simple relationship to the active volcanic zones is evident, but in several cases a shallow depth to layer 3 coincides with major central volcanoes.

Layer 3 in Iceland is probably to be equated to the oceanic layer, although the average P-wave velocity in the oceanic layer is commonly given as 6.7–6.8 km/sec. The lower velocity in the Icelandic crust cannot be explained wholly by higher temperatures. On the Reykjanes Ridge, Talwani et al (105) showed the existence of layer 3 with a velocity close to 6.5 km/sec in agreement with the Icelandic results. The thickness of layer 3 in Iceland is usually in the range 4–5 km, but from a limited amount of data a larger thickness is indicated in northern Iceland (77).

There seems little doubt that layer 3 in Iceland is essentially basaltic in nature.

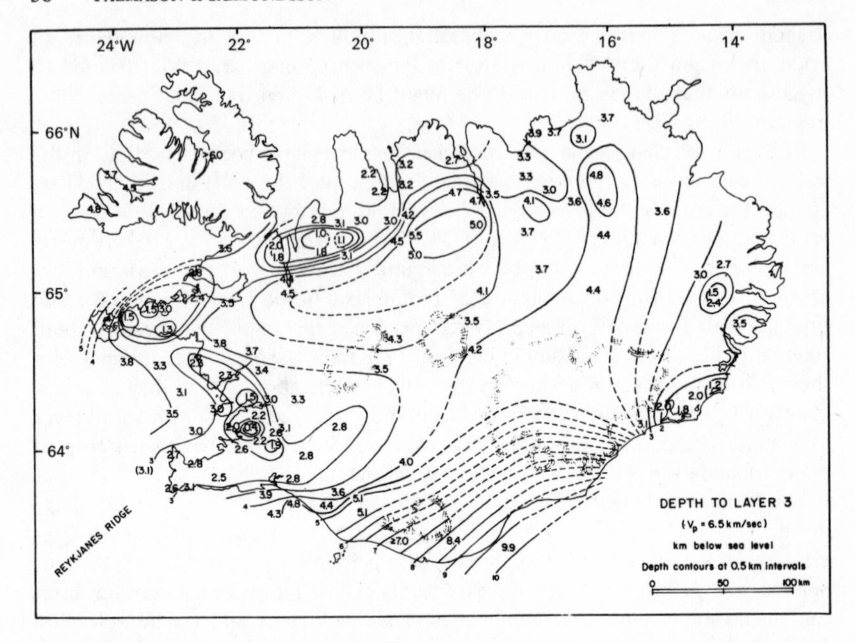

Figure 6 Depth to layer 3 in Iceland, based on about 80 refraction profiles (77).

Earlier ideas about a sialic substratum receive no support from recent geophysical and geochemical evidence, rather the contrary. Poisson's ratio of layer 3 (75, 77), Sr isotope ratios of basic and acid rocks (69, 74, 90), and Pb isotope ratios (132) all argue against a sialic substratum beneath Iceland. The acid component of central volcanism appears best explained by fractionation from a basaltic parent, either by fractional crystallization (21, 91) or by fractional melting of the lower crust (46, 97). All these results support the decision of Bullard et al (20) to omit Iceland when fitting together the continents across the northern Atlantic.

The cause of the higher velocity in layer 3 in Iceland relative to the overlying rocks is not well understood. From a comparison of the depth to the upper boundary of layer 3 with crustal temperatures as inferred from borehole data, Pálmason (77) suggested that a temperature-dependent process, perhaps metamorphism of the basaltic rocks, might be responsible for the increase in velocity in layer 3. A similar view had earlier been expressed by Einarsson (28) who pointed out that the seismic boundaries, in many places nearly horizontal, cut across the stratigraphic horizons given by the surface lavas dipping 5–10°.

Another possibility is that layer 3 in Iceland is largely composed of intrusives associated with the volcanism (16, 43, 44). There is little doubt that the volume fraction of intrusives, mainly in the form of dikes, increases downwards in the crust. Model calculations of crustal growth by dike injection and surface lavas show that the lower crust in Iceland should consist almost entirely of intrusives (78) (cf Figure 8). The steady-state model, however, requires the intrusive fraction to be elevated in the zone of volcanism relative to the adjacent lithospheric plates. No such elevation in the upper boundary of layer 3 in the active zone in Iceland is indicated by the available seismic data. This throws some doubt on the hypothesis that the intrusives as such are causing the increase in seismic velocity.

Perhaps the correct geological interpretation of layer 3 is to be sought in a combined effect of intrusives and some temperature-dependent effect, such as metamorphism or filling of pore space in the surface lava formations as they subside under the load of new lavas erupted in the active zone. It may be significant that with many central volcanoes, both Tertiary and younger, a positive gravity anomaly and a relatively shallow depth to layer 3 are associated. The density difference between layer 3 and the overlying flood basalts has been estimated to be about 0.2 g/cm^3 (77). It is well known that the intrusives fraction beneath central volcanoes is relatively high and the strong metamorphism associated with them (123, 125) also indicates rising of the isothermal surfaces in the crust during their period of activity.

It should be remarked here that the hypothesis of Hess (51), that layer 3 (the oceanic layer) consists of serpentinized peridotite, is contradicted by the available evidence on crustal temperatures in Iceland. According to Hess the serpentinization cannot take place at temperatures above about 500°C. The lower part of layer 3 is in many parts of Iceland inferred to be at a temperature of 500–1000°C on the basis of extrapolated borehole temperatures (77, 78).

Magnetic surveys have played a key role in the development of ideas of sea-floor spreading and plate tectonics. Iceland might offer a clue to what geological structures may possibly be causing the anomaly patterns commonly found. The well-known Reykjanes Ridge magnetic pattern (49, 105) continues towards the SW corner of Iceland, and the strong positive central anomaly can be followed continuously along the volcanic rift zone on the Reykjanes peninsula towards Langjökull (cf Figure 5). This continuity has been shown clearly by the aeromagnetic surveys of Sigurgeirsson (95). Strong linear anomalies are also observed over the eastern volcanic zone (Sigurgeirsson, unpublished observations). It should be noted also that on both sides of the active zone northeast of the Reykjanes peninsula, Matuyama and Gauss epoch rocks (79) correlate fairly well with the corresponding magnetic anomalies on Sigurgeirsson's map (95). Correlations of linear anomalies with magnetic stratigraphy attempted by Piper (80) for northeastern Iceland on the basis of Serson et al (89) are dubious in the light of recent studies in that area (85).

Magnetic surveys over the Iceland-Jan Mayen Ridge (68, 119, 120) show that a magnetic pattern with a strong central anomaly exists there also. The continuation of the axial anomaly into Iceland (68, 89) is not as clear as in southwest Iceland. South of about 67°N two or three strong positive linear anomalies appear to be present (68).

Talwani et al (105) concluded on the basis of correlations of magnetic field variations with topography on the Reykjanes Ridge that the magnetization responsible for the anomalies resides mainly in the top 400 m of the basalt layer.

To explain the strength of the central anomaly it is then necessary to assume a stronger magnetization in the axial zone. Some evidence in support of this is available from dredged rock samples (23). Observations in Iceland do not appear to support the assumption that such a thin, highly magnetized layer is responsible for the magnetic anomalies. Kristjánsson (64) studied the magnetic properties of drill chips from up to 2000 m deep boreholes in southwest Iceland. The magnetic properties reside primarily in magnetite, and a lack of a downward trend in magnetite content indicates that a 2 km thick layer at least may be responsible for the magnetic anomalies in Iceland, including the Reykjanes peninsula. The two results, however, may not be incompatible, because a different upper crustal structure may be expected in the two areas, due in part to the larger production rate of extrusives in the Iceland area.

Strong isolated magnetic anomalies are often found associated with major central volcanoes in Iceland. The best known of these is the Stardalur magnetic anomaly in southwest Iceland (38, 95, 101). A study of chips from a 200 m deep borehole revealed the presence of highly magnetic tholeiitic lava flows. A combination of a high magnetite content and a high paleofield strength is considered to be responsible for a 10–20 times higher magnetization than that found on the average in Icelandic basalts of a similar age.

THE UPPER MANTLE

In revealing the state and structure of the upper mantle beneath Iceland, seismic, gravity, heat flow, and magnetotelluric data are significant, as are geographical variations in the chemical composition of extrusive rocks. This evidence will be reviewed below.

The anomalously low seismic velocities in the upper mantle beneath much of the North Atlantic between Greenland and Europe were already evident in the early refraction measurements of Ewing & Ewing (34). Between 56°N and 72°N six profiles gave velocities in the range 6.9–7.7 km/sec and only one profile, located in the Norwegian basin, yielded a value over 8.0 km/sec. The thickness of the overlying crust was in the range 3–5 km. None of these profiles was close to Iceland or the Iceland-Faeroes Ridge. Relatively detailed refraction measurements in Iceland and on the insular shelf off the south and west coasts (7, 77) have shown that the P-wave velocity in the uppermost mantle is close to 7.2 km/sec, and the crustal thickness, i.e. the depth to the 7.2 km/sec velocity, varies from 8–9 km in southwest Iceland to 14–15 km in southeast Iceland (Figure 5) and possibly also in northern Iceland. A single reversed profile about 200 km southeast of Iceland, south of the Iceland-Faeroes Ridge, gave an upper mantle velocity of 7.1 km/sec and a crustal thickness of only 4.6 km (19). On the Reykjanes Ridge upper mantle wave velocites of 7.3–7.4 km/sec are found below a 3–5 km thick crust (105).

Francis (36) used the travel times of body waves from earthquakes to the north and southwest of Iceland to deduce the velocity distribution beneath the ridge axis through Iceland. He deduced a linearly increasing velocity from just over 7.0 km/sec in the uppermost mantle to about 8.0 km/sec at about 250 km depth. From velocity variations along different paths he also estimated the width of the deep anomalous zone to be 300 km. This is significantly narrower than Tryggvason's (113) estimate of 1000 km, which was based on a study of only four earthquakes.

The residuals of arrival times of distant earthquakes to receiving stations in Iceland, as compared with standard travel-time tables, have been used by Tryggvason (114) and Long & Mitchell (65) to study the upper mantle structure. A time delay of at least 1 sec is found to be associated with the upper mantle. This can be explained by a low upper mantle velocity of about 7.4 km/sec extending to a depth of 200–250 km. The study by Long & Mitchell (65), based on four receiving stations, indicates that there are no major variations in the time delay over Iceland itself.

The ratio of P to S velocities may be an indicator of the state of the upper mantle. No detailed studies of this ratio have been made, but the work of Francis (36) indicates a ratio of about 1.89, a value considered typical for the low-velocity zone beneath the lithosphere. A normal value for the oceanic upper mantle is about 1.76. The crust beneath Iceland has a P to S velocity ratio of 1.78–1.80 with no significant variation between the crustal seismic layers (75, 77). The higher value which is indicated in the upper mantle beneath Iceland suggests a state of partial fusion.

The regional gravity field in Iceland and the surrounding ocean gives significant information regarding the upper mantle. Both satellite data and terrestrial gravimetry show that the free-air anomalies are positive over much of the ocean between Greenland and Europe and over Greenland as well (59, 104, 105). The free-air anomalies average about 50–60 mgal on the ridge segments south and north of Iceland (35, 68, 105), and 40–50 mgal on the Iceland-Faeroes Ridge (19, 35). From the available data the free-air anomalies appear to decrease somewhat over the basins south and northeast of Iceland. Over Iceland they are similar or somewhat higher than on the adjoining ridges, indicating that Iceland may be at the center of the regional positive gravity anomaly located in the northern part of the North Atlantic.

The Bouguer anomalies are essentially reflected in the topography since isostatic equilibrium prevails for large-scale topographic features. The lowest Bouguer values are found in central Iceland, about −30 mgal (26). Towards the coast of Iceland the values increase to +40 to 50 mgal. At the crest of the Reykjanes Ridge the Bouguer anomaly increases southwards reaching 180 mgal at 54°N (35, 105). North of Iceland the Bouguer anomaly is about 120–140 mgal between 68° and 70°N (68). On the Iceland-Faeroes Ridge the Bouguer anomaly rises gently to a value of about 110 mgal towards the Faeroe Islands (19). In the deeper basins south and northeast of Iceland the Bouguer values are still higher, in the range 150–200 mgal.

The available seismic refraction measurements indicate that only a minor part of the Bouguer variation can be attributed to variations in crustal thickness. The major part is then most likely caused by inhomogeneities in the density of the upper mantle. The outward increase from central Iceland appears to be relatively smooth, judging from the Iceland-Faeroes profile (19) which does not indicate a deep-seated structure associated with the present coastline of Iceland or the edge of the insular shelf.

To explain the Bouguer minimum Bott (17, 18) suggested that it could be caused by partial fusion in the upper mantle resulting from a decrease in confining pressure in a rising mantle limb. A 10% partial fusion would lead to a density reduction of about 0.03 g/cm^3. If the density contrast extends over a depth range of 200 km, the corresponding Bouguer anomaly could reach about -250 mgal, of the same order as the Bouguer minimum in Iceland relative to the surrounding ocean areas.

An estimate of the viscosity of the upper mantle may be obtained by studying the rate of glacio-isostatic rebound of Iceland at the end of the ice age. From a comparative study of Iceland and Scandinavia Einarsson (29) concluded that the viscosity of the upper mantle beneath Iceland was an order of magnitude lower than that beneath Scandinavia. A still lower subcrustal viscosity is indicated by Tryggvason (116a) for central southern Iceland.

The regional heat flow in Iceland gives direct indications as to the state of the upper mantle. Although the surface heat flow is often disturbed by the movement of water giving rise to the geothermal areas, it nevertheless appears possible to discern a certain pattern of conductive heat flow with high values (up to 7 HFU) near the volcanic zones (cf Figure 5) and lower values (down to 1.7 HFU) at greater distance from them.[2] On the basis of the available regional heat flow data and model calculations of crustal heating by dike intrusions in the volcanic rift zone, Pálmason (77, 78) concluded that the solidus of basalts was reached at a depth of 10 km or less beneath the volcanic zone in southwest Iceland. This implies that the uppermost mantle as defined by the 6.5/7.2 km/sec boundary, located at a depth of 8–9 km, is at a temperature close to the solidus of basalts. It will be noted that the thermal gradient inferred for the active zones in Iceland is similar to the gradient deduced by Gass & Smewing (41) for the Troodos Massif on the basis of secondary mineral zoning. A similar or slightly lower gradient was inferred by Ade-Hall et al (1) for the eastern Iceland lava pile, also on the basis of secondary mineral zoning.

Magnetotelluric measurements (50) in the rift zone in southwest Iceland indicate resistivities of 10–20 Ωm at 10–15 km depth. A comparison of these values with laboratory data for basalt and peridotite leads to the conclusion that crustal temperatures at that depth are in the range of 800–1100°C.

The geophysical data discussed above lead to a remarkably coherent picture of the state and structure of the upper mantle beneath Iceland and the surrounding ocean. The high heat flow, low seismic velocity, and low density, together with a high P to S velocity ratio all show that the upper mantle beneath Iceland is in a state of partial fusion. The gravity anomaly and the teleseismic P-wave delays indicate that this state of partial fusion may extend to a depth of the order of 200 km. The anomalous mantle appears to become thinner away from Iceland.

[2] 1 HFU = 1 μcal/cm^2 sec $\approx$ 42 mW/m^2.

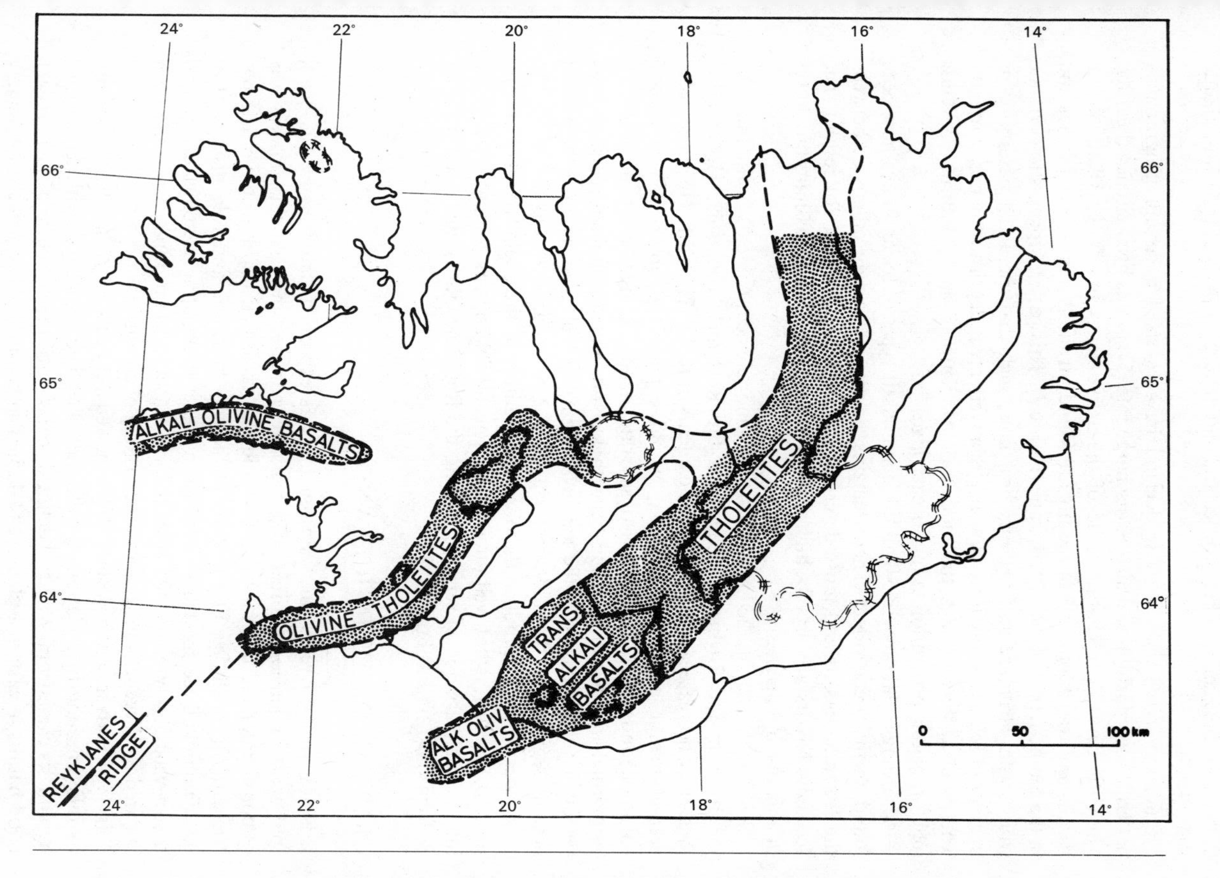

Figure 7 Petrological division of the volcanic zones in Iceland, based on postglacial lavas (54).

The Reykjanes Ridge appears to be underlain by an anomalous mantle intermediate in thickness between that beneath Iceland and the deeper parts of the Mid-Atlantic Ridge (105).

Studies of postglacial lavas in Iceland have revealed a certain geographical pattern of composition and also of discharge rate (54) as has been discussed earlier. Tholeiitic basalts are characteristic of the active zones of rifting. Alkali olivine basalts and transitional alkali basalts are found on the flanks at Snaefellsnes and in the southern part of the eastern zone (Figure 7). This pattern appears to correlate with crustal structure and regional heat flow, such that the alkali basalts occur in areas of relatively low heat flow and relatively great depth to the upper mantle. This is in agreement with the models of Aumento (5) and McBirney & Gass (66) for magma generation at mid-ocean ridges.

According to Jakobsson (54) all the main basalt types of Iceland have now been found on the ocean ridges. There remains, however, a certain difference between the Icelandic basalts and those described from the Mid-Atlantic Ridge (96). The Icelandic basalts are usually higher in iron, titanium, and potassium, but lower in alumina and magnesia.

Schilling (87) has studied rare earth and minor element contents of the tholeiitic basalts from the Reykjanes Ridge and the Reykjanes peninsula. He found a systematic decrease southwards in the contents of K, La, Ti, and P. He investigated several mantle source models that would explain the observed geochemical variation and concluded that two separate mantle sources were necessary: a primordial mantle plume rising beneath Iceland, and the globally existing low-velocity layer assumed to be the source of magma for crustal accretion at the normal oceanic segments of the mid-ocean ridges. The two sources would mix along the Reykjanes Ridge to produce the observed gradient in the chemistry. The mixing process, however, is not well understood. This model appears compatible with the longitudinal variation of seismicity along the Reykjanes Ridge discussed by Vogt & Johnson (121) and Francis (37).

CRUSTAL DRIFT IN ICELAND

It is now over 40 years since the opinion was expressed that Iceland was being split apart by tensional forces of a regional character and that this process was responsible for the fissural volcanism and the associated linear tectonic features. Nielsen (72) concluded after a study of the zone of rifting and volcanism west of the Vatnajökull ice sheet that subsidence was a dominating tectonic feature and that it was closely associated with the volcanism. Nielsen appears to have been inspired by Wegener's ideas about the drifting of continents.

In 1964 Bodvarsson & Walker (16) proposed a process of crustal drift in Iceland by dike injection in a stationary volcanic zone. The geological evidence for this hypothesis came from Walker's detailed mapping of the structure of the Tertiary flood basalts in eastern Iceland (122–125). The geophysical evidence came mainly from early heat flow observations and theoretical studies attempting to explain them (15). Gravity and seismic refraction data were also used, but their interpretation was criticized by Einarsson (28) and has since been modified in the way discussed earlier. Einarsson also pointed out that the regional dips of the flood basalts in northern and western Iceland do not bear as clear a relationship to the presently active volcanic zones as the eastern Iceland dips do, and one should therefore not extrapolate the eastern Iceland structure to the rest of Iceland. As has been discussed earlier, however, recent geological and geophysical evidence indicates that the active volcanic zones have shifted during the geological history of Iceland between 2 or 3 zones (78, 82, 85, 130).

Einarsson has been the chief contender against crustal drift in Iceland (e.g. 28, 31). Many of his arguments are no doubt valid in that one should not extrapolate to Iceland the relatively simple models of crustal spreading usually assumed for the submarine parts of the ridges. There are more constraints imposed by geological observations on the spreading process in Iceland than elsewhere and these have to be taken into account. But it also appears that many of his arguments are not necessarily incompatible with drift if one takes into account the possibility of shifting of the active zones.

An important point in Einarsson's arguments is the pattern of regional dips of the Tertiary flood basalts (cf Figure 2). The apparent symmetry of the dips about three zones is interpreted by him as the result of folding (cf also 25) and breaking up of the pile of more or less horizontal lavas, followed by peneplanation. However, it is difficult to accept this interpretation. The dips are sometimes 5–10° over areas of a few tens of kilometers measured perpendicularly to the strike of the lavas, which means that a thickness of several kilometers has been eroded away in the stratigraphically lowest parts in a few million years. This difficulty has been pointed out by Walker (123) and Einarsson (31, 32). Furthermore, one would expect alteration to increase towards the older tilted strata (82), and the dike intensity to increase also. Neither of these is confirmed by observations.

It appears that Walker's interpretation that the regional dips (16, 123) are due to sagging resulting from a stacking of lavas in an active zone is in better agreement with available observations, if one accepts the shifting of volcanic activity between two or more zones.

Hast (48) has made direct stress measurements in shallow boreholes in rocks outside and adjacent to the active volcanic zones in Iceland. He concluded that horizontal compressive stresses prevailed everywhere. This does not appear to be compatible with ocean-floor spreading processes. The stress field to be expected in the crust adjacent to and within a zone of rifting and volcanism is, however, little known and further studies are required before a definite conclusion can be drawn from the available stress measurements.

The numerous faults and open fissures in the Icelandic zone of rifting and volcanism are considered by most workers as evidence of tensional movement (e.g. 71), although Einarsson (30, 31) considered them to be largely surface expressions of deeper faults with a strike-slip movement. Estimates based on the total integrated width of open fissures in postglacial lava fields, together with the estimated age of the fissured lavas, give a double drift rate from a few millimeters to 2 cm per year in certain parts of the zone of rifting and volcanism (11, 16, 108, 126). The results of direct measurements of horizontal movements across some conspicuous zones of rifting (24, 42) indicate small extensional movements but further measurements over a longer time interval are needed to obtain significant results.

Vertical displacements in the active zones relative to their margins may prove to be as strong an evidence for crustal drift as horizontal movements. According to the models of Bodvarsson & Walker (16) and Pálmason (78), elevation changes as a result of sagging of the lava pile in the active zone may be expected to reach 1–2 mm/yr on the average. Measurements by Tryggvason (115, 116) on selected profiles in southwest Iceland indicate that subsidence of this order is taking place. Schleusener & Torge (88) have made repeated precise gravity measurements along a 100 km long profile which includes crossing of the volcanic zone in northeast Iceland. Between 1965–1970, they found the measured gravity values within the active zone to increase 0.005–0.01 mgal/yr with respect to the Tertiary basalt area to the west. It appears possible to interpret this as a subsidence of the surface amounting to a few cm/yr, depending on how the underlying masses are rearranged.

In addition to direct measurements of the subsidence there is supporting geological evidence also. Drillholes extending 1000–2000 m below sea level at the very tip of the Reykjanes peninsula penetrate rocks which are typical of land or shallow water eruptions (13, 112). A similar result is deduced from a 1565 m deep hole in Heimaey at the southern end of the eastern zone (111). Last but not least,

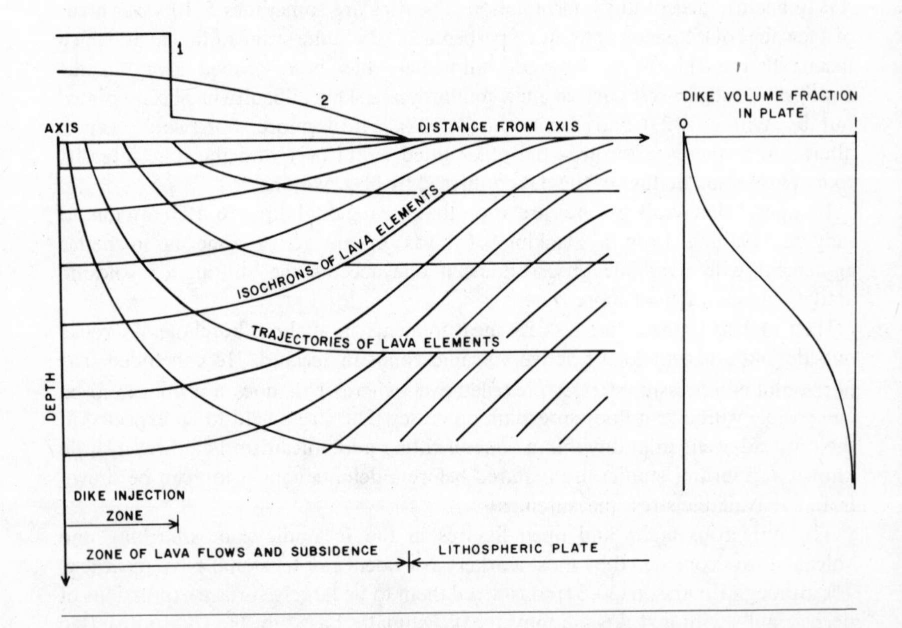

Figure 8 Schematic model of crustal accretion in Iceland by dike intrusions and surface lavas in a single volcanic zone: 1. distribution of dike injection activity, 2. distribution of lava flows. Modified from Pálmason (78).

the regional dips of often 5°–10° at sea level in subaerial basalt lava piles adjacent to the volcanic zones show that the subaerial lavas plunge to a depth of several kilometers below sea level beneath the active zones.

Pálmason (78) has made calculations on a model of crustal accretion, similar to that of Bodvarsson & Walker (16), by dike injection and surface lava flows in a single volcanic zone. Using model parameters based on estimates from Iceland, it is possible to reproduce fairly well the regional structure of the eastern Iceland lava pile as described by Walker, the regional dips as well as the average dike volume relationships without, however, taking the swarm distribution or the central volcanoes into account. Figure 8 shows the model schematically. Dike injection takes place mainly in a column of a certain width, with minor eruptive activity outside this zone. The surface lavas flow over a wider zone, in which subsidence takes place, equivalent to the build-up by lavas, thus keeping the surface at a more or less constant level on the average. The rate of increase of dike volume with depth in the uppermost part of the adjacent plate of the model depends on the relative distributions of dike injection activity and surface lava flows. For a narrow dike injection distribution the dike fraction in the uppermost part of the plate will be small.

This is an important property of the model which makes it applicable not only to the eastern Iceland pile, but also to western and northern Iceland, where dike intensity appears to be smaller than in eastern Iceland. The trend of downward increase of the dips in the basalt pile is a generally observed phenomenon in the eroded lava pile, although the absolute dips vary from one area to another.

A study of microearthquakes in the active zone may lead to a better understanding of the dynamic processes taking place. Several such surveys have been made (61, 129, 131) and an extensive survey of the Reykjanes peninsula is being undertaken at present (S. Björnsson, personal communication). The microearthquake activity appears to be largely confined to certain areas within the active zone, and the intensity varies with time, sometimes reaching over 1000 events per day. The focal depth range appears to be mainly confined to the upper part of the seismic layer 3. Focal mechanism solutions from the Reykjanes peninsula show both strike-slip and dip-slip movements (61), with the axis of maximum tension more or less perpendicular to the main direction of the ridge axis through Iceland. Larger earthquakes of magnitude 6 or more are known to occur only in an east-west zone in southern Iceland from Reykjanes to Hekla, and in a broad east-west zone off the north coast (100). The depth of focus is unknown. Focal mechanism solutions (99, 103, 130) are still few but they indicate strike-slip movements.

DISCUSSION AND CONCLUSIONS

The central question to be answered concerning Iceland and its relationship to the Mid-Atlantic Ridge is whether Iceland fits into the framework of ocean-floor spreading or not. The presently available evidence favors a process of drift. This process, however, appears to be more complicated than is usually envisaged for the submarine parts of the Mid-Atlantic Ridge.

There is increasing evidence that the active zones of rifting and volcanism have not been stationary during the period involved in building up the Icelandic basalt pile. The evidence for a shifting of the active zones comes from stratigraphic and structural studies and from heat flow data. This has a number of implications for the interpretation of data from Iceland in terms of ocean-floor spreading.

The magnetic anomaly pattern, which according to ocean-floor spreading theories originates at the accreting plate margins, would be affected and one would not find undisturbed the symmetrical pattern observed, e.g. on the Reykjanes Ridge. Therefore a correlation of the Reykjanes Ridge pattern with the pattern in Iceland neither proves nor disproves ocean-floor spreading in Iceland. The Icelandic anomaly pattern has to be interpreted independently of the neighboring ocean patterns and with due regard to the history of the volcanic zones on land.

The pattern of regional dips of the flood basalts may be viewed as reflecting the position not only of the presently active zones of rifting and volcanism but also of previously active zones. This invalidates the argument against drift in Iceland that the regional dips do not in general conform to the pattern of dip towards the active zone that is exhibited in the eastern Iceland basalt pile.

The crustal structure of Iceland does not appear to contradict ocean-floor spreading. The evidence provided by seismic velocities, P to S velocity ratios, Sr isotope ratios, lead isotope ratios, and petrogenetic considerations, points to a basaltic crust down to the upper mantle. The continuity of layer 3 as observed across the active zones is also in agreement with ocean-floor spreading, whether the layer 3 seismic velocity is attributed to an increased intrusives fraction, or to metamorphism, or a combination of both.

The regional pattern of conductive heat flow as revealed by about 20 boreholes appears to be compatible with crustal accretion in accordance with plate tectonics. The heat flow is high near the Reykjanes-Langjökull zone, which is the landward continuation of the Reykjanes Ridge crest. It decreases away from this zone in southwest Iceland. The lowest heat flow values found in eastern Iceland appear compatible with a cooling lithospheric plate moving away from a zone of plate accretion. The apparent absence of a widespread conductive heat flow anomaly associated with the eastern zone is in agreement with other evidence for its relative youthfulness.

The hypothesis of ocean-floor spreading provides a good working hypothesis for interpreting data of Icelandic geology, geophysics, and geochemistry. Most of the available evidence appears consistent with it. Various arguments which have been presented as contradicting drift in Iceland (e.g. 8) do not appear to be valid in the light of what has been discussed in this paper.

Literature Cited

1. Ade-Hall, J. M., Palmer, H. C., Hubbard, T. P. 1971. The magnetic and opaque petrological response of basalts to regional hydrothermal alteration. *Geophys. J. Roy. Astron. Soc.* 24:137–74
2. Annells, R. N. 1968. *A geological investigation of a Tertiary intrusive centre in the Vididalur-Vatnsdalur area, northern Iceland.* PhD thesis, Univ. St. Andrews
3. Aric, K. 1970. Über die Struktur des

Reykjanes-Rückens nach den Ergebnissen reflexionsseismischer Messungen. *Z. Geophys.* 36:229–32
4. Arnason, B., Sigurgeirsson, T. 1967. Hydrogen isotopes in hydrological studies in Iceland. In *Isotopes in Hydrology,* 35–47. Vienna: IAEA
5. Aumento, F. 1967. Magmatic evolution on the Mid-Atlantic Ridge. *Earth Planet. Sci. Lett.* 2:225–30
6. Avery, O. E., Burton, G. D., Heirtzler, J. R. 1968. An aeromagnetic survey of the Norwegian Sea. *J. Geophys. Res.* 73:4583–600
7. Båth, M. 1960. Crustal structure of Iceland. *J. Geophys. Res.* 65:1793–807
8. Beloussov, V. V. 1970. Against the hypothesis of ocean-floor spreading. *Tectonophysics* 9:489–511
9. van Bemmelen, R. W., Rutten, M. G. 1955. *Tablemountains of Northern Iceland.* Leiden: E. J. Brill. 217 pp.
10. van Bemmelen, R. W. 1972. *Geodynamic Models; An Evaluation and a Synthesis.* Amsterdam: Elsevier. 267 pp.
11. Bernauer, F. 1943. Junge Tektonik auf Island und ihre Ursachen. In *Spalten auf Island,* ed. O. Niemczyk, 14–64. Stuttgart: Verl. K. Wittwer
12. Björnsson, S., Ed. 1967. *Iceland and Mid-Ocean Ridges,* Rit 38. Reykjavík: Soc. Sci. Island. 209 pp.
13. Björnsson, S., Arnórsson, S., Tómasson, J. 1972. Economic evaluation of Reykjanes thermal brine area. *AAPG Bull.* 56:2380–91
14. Blake, D. H. 1970. Geology of Alftafjordur volcano, a Tertiary volcanic centre in South-Eastern Iceland. *Sci. Iceland* 2:43–63
15. Bodvarsson, G. 1954. Terrestrial heat balance in Iceland. *Tímarit Verkfraedingafelags Isl.* 39:69–76
16. Bodvarsson, G., Walker, G. P. L. 1964. Crustal drift in Iceland. *Geophys. J. Roy. Astron. Soc.* 8:285–300
17. Bott, M. H. P. 1965. Formation of oceanic ridges. *Nature* 207:840–43
18. Bott, M. H. P. 1965. The upper mantle beneath Iceland. *Geophys. J. Roy. Astron. Soc.* 9:275–77
19. Bott, M. H. P., Browitt, C. W. A., Stacey, A. P. 1971. The deep structure of the Iceland-Faeroe Ridge. *Mar. Geophys. Res.* 1:328–51
20. Bullard, E. C., Everett, J. E., Smith, A. G. 1965. The fit of the continents around the Atlantic. *Phil. Trans. Roy. Soc. A* 258:41–51
21. Charmichael, I. S. E. 1964. The petrology of Thingmúli, a Tertiary volcano in eastern Iceland. *J. Petrol.* 5:435–60
22. Dagley, P. et al 1967. Geomagnetic polarity zones for Icelandic lavas. *Nature* 216:25–29
23. De Boer, J., Schilling, J. G., Krause, D. C. 1970. Reykjanes Ridge; implication of magnetic properties of dredged rocks. *Earth Planet. Sci. Lett.* 9:55–60
24. Decker, R. W., Einarsson, P., Mohr, P. A. 1971. Rifting in Iceland: New geodetic data. *Science* 173:530–33
25. Einarsson, Th. 1967. The extent of the Tertiary basalt formation and the structure of Iceland. See Ref. 12, pp. 170–79
26. Einarsson, Tr. 1954. *A Survey of Gravity in Iceland,* Rit 30. Reykjavík: Soc. Sci. Island. 22 pp.
27. Einarsson, Tr. 1962. *Upper Tertiary and Pleistocene Rocks in Iceland,* Rit 36. Reykjavík: Soc. Sci. Island. 197 pp.
28. Einarsson, Tr. 1965. Remarks on crustal structure in Iceland. *Geophys. J. Roy. Astron. Soc.* 10:283–88
29. Einarsson, Tr. 1966. Late and postglacial rise in Iceland and sub-crustal viscosity. *Jökull* (Reykjavík) 16:157–66
30. Einarsson, Tr. 1967. The Icelandic fracture system and the inferred causal stress field. See Ref. 12, pp. 128–39
31. Einarsson, Tr. 1968. Submarine ridges as an effect of stress fields. *J. Geophys. Res.* 73:7561–76
32. Einarsson, Tr. 1971. Magnetic polarity groups in the Fljótsdalsheidi area, including Gilsá. *Jökull* (Reykjavík) 21: 53–58
33. Everts, P., Koerfer, L. E., Schwarzbach, M. 1972. Neue K/Ar-Datierungen isländischer Basalte. *N. Jahrb. Geol. Paläontol.* 1972:280–84
34. Ewing, J. I., Ewing, M. 1959. Seismic refraction measurements in the Atlantic Ocean basins, in the Mediterranean Sea, on the Mid-Atlantic Ridge, and in the Norwegian Sea. *Bull. Geol. Soc. Am.* 70:291–318
35. Fleischer, U. 1971. Gravity surveys over the Reykjanes Ridge and between Iceland and the Faeroe Islands. *Mar. Geophys. Res.* 1:314–27
36. Francis, T. J. G. 1969. Upper mantle structure along the axis of the Mid-Atlantic Ridge near Iceland. *Geophys. J. Roy. Astron. Soc.* 17:507–20
37. Francis, T. J. G. 1973. The seismicity of the Reykjanes Ridge. *Earth Planet. Sci. Lett.* 18:119–24
38. Fridleifsson, I. B., Kristjánsson, L. 1972. The Stardalur magnetic anomaly, SW-Iceland. *Jökull* (Reykjavík) 22: 69–78
39. Fridleifsson, I. B. *Petrology and structure of the Esja Quaternary volcanic*

region, SW-Iceland. D. Phil. thesis, Univ. Oxford. (In preparation)
40. Gale, N. H., Moorbath, S., Simons, J., Walker, G. P. L. 1966. K-Ar ages of acid-intrusive rocks from Iceland. *Earth Planet. Sci. Lett.* 1:284–88
41. Gass, I. G., Smewing, J. D. 1973. Intrusion, extrusion and metamorphism at constructive margins: Evidence from the Troodos Massif, Cyprus. *Nature* 242:26–29
42. Gerke, K. 1967. Ein Beitrag zur Bestimmung rezenter Erdkrustbewegungen. In *Aus der geodätischen Lehre und Forschung,* 66–78. Stuttgart: Verl. Wittwer
43. Gibson, I. L. 1966. The crustal structure of eastern Iceland. *Geophys. J. Roy. Astron. Soc.* 12:99–102
44. Gibson, I. L., Piper, J. D. A. 1972. Structure of the Icelandic basalt plateau and the process of drift. *Phil. Trans. Roy. Soc. A* 271:141–50
45. Godby, E. A., Hood, P. J., Bower, M. E. 1968. Aeromagnetic profiles across the Reykjanes Ridge southwest of Iceland. *J. Geophys. Res.* 73:7637–49
46. Grönvold, K. 1972. *Structural and petrochemical studies in the Kerlingarfjöll region, central Iceland.* D. Phil. thesis, Univ. Oxford. 237 pp.
47. Hald, N., Noe-Nygaard, A., Pedersen, A. K. 1971. The Króksfjördur central volcano in north-west Iceland. *Acta Natur. Island.* II(10). 29 pp.
48. Hast, N. 1969. The state of stress in the upper part of the earth's crust. *Tectonophysics* 8:169–211
49. Heirtzler, J. R., LePichon, X., Baron, J. G. 1966. Magnetic anomalies over the Reykjanes Ridge. *Deep-Sea Res.* 13:427–43
50. Hermance, J. F., Grillot, L. R. 1970. Correlation of magnetotelluric, seismic, and temperature data from southwest Iceland. *J. Geophys. Res.* 75:6582–91
50a. Hermance, J. F. 1973. An electrical model for the sub-Icelandic crust. *Geophysics* 38:3–13
50b. Hermance, J. F. Constraints on temperatures beneath Iceland from magnetotelluric data. *Phys. Earth Planet. Interiors.* In press
51. Hess, H. H. 1962. History of ocean basins. In *Petrologic Studies,* 599–620. Geol. Soc. Am.
52. Holmes, A. 1965. *Principles of Physical Geology.* London: Nelson. 1288 pp.
53. Horai, K., Chessman, M., Simmons, G. 1970. Heat flow measurements on the Reykjanes Ridge. *Nature* 225:264–65
54. Jakobsson, S. P. 1972. Chemistry and distribution pattern of Recent basaltic rocks in Iceland. *Lithos* 5:365–86
55. Jakobsson, S. P., Pedersen, A. K., Rönsbo, J. G., Larsen, L. M. 1973. Petrology of mugearite-hawaiite: Early extrusives in the 1973 Heimaey eruption, Iceland. *Lithos* 6:203–14
56. Johnson, G. L., Tanner, B. 1971. Geophysical observations on the Iceland-Faeroe Ridge. *Jökull* (Reykjavík) 21: 45–52
57. Johnson, G. L., Southall, J. R., Young, P. W., Vogt, P. R. 1972. Origin and structure of the Iceland Plateau and Kolbeinsey Ridge. *J. Geophys. Res.* 77: 5688–96
58. Jónsson, J. 1967. The rift zone and the Reykjanes peninsula. See Ref. 12, pp. 142–48
59. Kaula, W. M. 1972. Global gravity and mantle convection. In *The Upper Mantle,* ed. A. R. Ritsema, 341–59. Amsterdam: Elsevier
60. Kjartansson, G. 1960, 1962, 1965, 1968, 1969. *Geological Map of Iceland,* Sheets 1, 2, 3, 5, 6. Reykjavík: Mus. Natur. Hist.
61. Klein, F. W., Einarsson, P., Wyss, M. Microearthquakes on the Mid-Atlantic plate boundary on the Reykjanes peninsula, Iceland. In press
62. Krause, D. C., Schilling, J. G. 1969. Dredged basalt from the Reykjanes Ridge, North Atlantic. *Nature* 224:791–93
63. Kristjánsson, L. 1970. Paleomagnetism and magnetic surveys in Iceland. *Earth Planet. Sci. Lett.* 8:101–8
64. Kristjánsson, L. 1972. On the thickness of the magnetic crustal layer in southwestern Iceland. *Earth Planet. Sci. Lett.* 16:237–44
65. Long, R. E., Mitchell, M. G. 1970. Teleseismic P-wave delay time in Iceland. *Geophys. J. Roy. Astron. Soc.* 20:41–48
66. McBirney, A. R., Gass, I. G. 1967. Relations of oceanic volcanic rocks to mid-oceanic rises and heat flow. *Earth Planet. Sci. Lett.* 2:265–76
67. McDougall, I., Wensink, H. 1966. Paleomagnetism and geochronology of the Pliocene-Pleistocene lavas in Iceland. *Earth Planet. Sci. Lett.* 1:232–36
68. Meyer, O., Voppel, D., Fleischer, U., Closs, H., Gerke, K. 1972. Results of bathymetric, magnetic and gravimetric measurements between Iceland and 70°N. *Deut. Hydrogr. Z.* 25:193–201
69. Moorbath, S., Walker, G. P. L. 1965. Strontium isotope investigation of igneous rocks from Iceland. *Nature* 207: 837–40

70. Moorbath, S., Sigurdsson, H., Goodwin, R. 1968. K-Ar ages of the oldest exposed rocks in Iceland. *Earth Planet. Sci. Lett.* 4:197–205
71. Nakamura, K. 1970. En echelon features of Icelandic ground fissures. *Acta Natur. Island.* II (8). 15 pp.
72. Nielsen, N. 1933. Contributions to the physiography of Iceland, with particular reference to the highlands west of Vatnajökull. *D. Kgl. Danske Vidensk. Selsk. Skrifter,* Naturv. og Matemat. Afd., 9, Raekke IV. 5. 105 pp., 32 plates, 9 maps. Copenhagen
73. Noll, H., Saumundsson, K. K-Ar ages of rocks from Husafell in West-Iceland. In preparation
74. O'Nions, R. K., Grönvold, K. 1973. Petrogenetic relationships of acid and basic rocks in Iceland: Sr-isotopes and rare-earth elements in late and post-glacial volcanics. *Earth Planet. Sci. Lett.* 19:397–409
75. Pálmason, G. 1963. Seismic refraction investigation of the basalt lavas in northern and eastern Iceland. *Jökull* (Reykjavík) 13:40–60
76. Pálmason, G. 1967. On heat flow in Iceland in relation to the Mid-Atlantic Ridge. See Ref. 12, pp. 111–27
77. Pálmason, G. 1971. *Crustal Structure of Iceland from Explosion Seismology,* Rit 40. Reykjavík: Soc. Sci. Island. 187 pp.
78. Pálmason, G. 1973. Kinematics and heat flow in a volcanic rift zone, with application to Iceland. *Geophys. J. Roy. Astron. Soc.* 33:451–81
79. Piper, J. D. A. 1971. Ground magnetic studies of crustal growth in Iceland. *Earth Planet. Sci. Lett.* 12:199–207
80. Piper, J. D. A. 1973. Interpretation of some magnetic anomalies over Iceland. *Tectonophysics* 16:163–87
81. Saemundsson, K. 1967. Vulkanismus und Tektonik des Hengill-Gebietes in Südwest-Island. *Acta Natur. Island.* II(7). 195 pp.
82. Saemundsson, K. 1967. An outline of the structure of SW-Iceland. See Ref. 12, pp. 151–61
83. Saemundsson, K. 1971. *Relation between geological structure of Iceland and some geophysical anomalies.* (Abstr.). 1st Eur. Earth Planet. Phys. Colloq., Reading, England
84. Saemundsson, K. 1972. Notes on the geology of the Torfajökull central volcano. *Náttúrufrædingurinn* (Reykjavík) 42:81–99. (In Icelandic with an English summary)
85. Saemundsson, K. Evolution of the axial rifting zone in northern Iceland and the Tjörnes fracture zone. *Bull. Geol. Soc. Am.* In press
86. Schäfer, K. 1972. Transform Faults in Iceland. *Geol. Rundsch.* 61(3):942–60
87. Schilling, J. G. 1973. Iceland mantle plume existence and influence along the Reykjanes Ridge: I. geochemical evidence. *Nature* 242:565–69
88. Schleusener, A., Torge, W. 1971. Investigations of secular gravity variations in Iceland. *Z. Geophys.* 37:679–701
89. Serson, P. H., Hannaford, W., Haines, G. V. 1968. Magnetic anomalies over Iceland. *Science* 162:355–57
90. Sigurdsson, H. 1967. The Icelandic basalt plateau and the question of sial. See Ref. 12, pp. 32–46
91. Sigurdsson, H. 1970. *The petrology and chemistry of the Setberg volcanic region and of the intermediate and acid rocks of Iceland.* Ph.D. thesis, Durham Univ. 308 pp.
92. Sigurdsson, H. 1970. Structural origin and plate tectonics of the Snaefellsnes volcanic zone, western Iceland. *Earth Planet. Sci. Lett.* 10:129–35
93. Sigurgeirsson, T. 1967. Discussion. See Ref. 12, p. 159
94. Sigurgeirsson, T. 1970. A survey of geophysical research related to crustal and upper mantle structure in Iceland. *J. Geomagn. Geoelec.* 22:213–21
95. Sigurgeirsson, T. 1970. Aeromagnetic survey of SW Iceland. *Sci. Iceland* 2:13–20
96. Sigvaldason, G. E. 1969. Chemistry of basalts from the Icelandic rift zone. *Contr. Miner. Petrol.* 20:357–70
97. Sigvaldason, G. E. 1973. *The petrology of Hekla and origin of silicic rocks in Iceland.* Science Institute, Univ. Iceland (mimeographed)
98. Smith, P. J. 1967. The intensity of the Tertiary geomagnetic field. *Geophys. J. Roy. Astron. Soc.* 12:239–58
99. Stefánsson, R. 1966. Methods of focal mechanism studies with application on two Atlantic earthquakes. *Tectonophysics* 3:210–43
100. Stefánsson, R. 1967. Some problems of seismological studies on the Mid-Atlantic Ridge. See Ref. 12, pp. 80–89
101. Steinthorsson, S. Kristjánsson, L., Sigvaldason, G. E. 1971. *Studies of drill cores from an unusual magnetic high in SW-Iceland.* (Abstr.). 1st Eur. Earth Planet. Phys. Colloq., Reading, England
102. Sykes, L. 1965. The seismicity of the Arctic. *Bull. Seismol. Soc. Am.* 55:501–18
103. Sykes, L. 1967. Mechanism of earth-

quakes and nature of faulting on the mid-oceanic ridges. *J. Geophys. Res.* 72:2131–53

104. Talwani, M., LePichon, X. 1969. Gravity field over the Atlantic Ocean. In *The Earth's Crust and Upper Mantle,* ed. P. J. Hart, Geophys. Monogr. 13:341–51. Am. Geophys. Union
105. Talwani, M., Windisch, C. C., Langseth, M. G. Jr. 1971. Reykjanes Ridge crest: A detailed geophysical study. *J. Geophys. Res.* 76:473–517
106. Tarling, D. H., Gale, N. H. 1968. Isotopic dating and palaeomagnetic polarity in the Faeroe Islands. *Nature* 218:1043–44
107. Thorarinsson, S. 1964. *Surtsey—The New Island in the North Atlantic.* Reykjavik: Almenna bókafélagid. 110 pp.
108. Thorarinsson, S. 1965. The median zone of Iceland. In *The World Rift System,* UMP Symposium, Ottawa, Canada. Geol. Survey Can. Pap. 66–14:187–211
109. Thorarinsson, S. 1967. The eruptions of Hekla in historical times. A tephrochronological study. In the series *The Eruption of Hekla 1947–1948.* Reykjavik: Soc. Sci. Island. 183 pp.
110. Thorarinsson, S., Steinthorsson, S., Einarsson, Th., Kristmannsdottir, H., Oskarsson, N. 1973. The eruption on Heimaey, Iceland. *Nature* 241:372–75
111. Tómasson, J. 1967. On the origin of sedimentary water beneath Vestmann Islands. *Jökull* (Reykjavík) 17:300–11
112. Tómasson, J., Kristmannsdóttir, H. 1972. High temperature alteration minerals and thermal brines, Reykjanes, Iceland. *Contrib. Miner. Petrol.* 36: 123–34
113. Tryggvason, E. 1961. Wave velocity in the upper mantle below the Arctic-Atlantic Ocean and Northwest Europe. *Ann. Geofis.* 14:379–92
114. Tryggvason, E. 1964. Arrival times of P-waves and upper mantle structure. *Bull. Seismol. Soc. Am.* 54:727–36
115. Tryggvason, E. 1968. Measurement of surface deformation in Iceland by precision leveling. *J. Geophys. Res.* 73:7039–50
116. Tryggvason, E. 1970. Surface deformation and fault displacement associated with an earthquake swarm in Iceland. *J. Geophys. Res.* 75:4407–22

116a. Tryggvason, E. 1973. Surface deformation and crustal structure in the Myrdalsjökull area of south Iceland. *J. Geophys. Res.* 78:2488–97

117. Tryggvason, T. 1965. Petrographic studies on the eruption products of Hekla 1947–1948. In the series *The Eruption of Hekla 1947–1948.* Reykjavík: Soc. Sci. Island. 13 pp.
118. Ulrich, J. 1960. Zur Topographie des Reykjanes-Rückens. *Kieler Meeresforsch.* 16:155–63
119. Vogt, P. R., Schneider, E. D., Johnson, G. L. 1969. The crust and mantle beneath the sea. In *The Earth's Crust and Upper Mantle,* ed. P. J. Hart, Geophys. Monogr. 13:556–617. Am. Geophys. Union
120. Vogt, P. R., Ostenso, N. A., Johnson, G. L. 1970. Magnetic and bathymetric data bearing on sea-floor spreading north of Iceland. *J. Geophys. Res.* 75:903–20
121. Vogt, P. R., Johnson, G. L. 1973. A longitudinal seismic reflection profile of the Reykjanes Ridge: Part II—Implications for the mantle hot spot hypothesis. *Earth Planet. Sci. Lett.* 18:49–58
122. Walker, G. P. L. 1959. Geology of the Reydarfjördur area, eastern Iceland. *Quart. J. Geol. Soc. London* 114:367–91
123. Walker, G. P. L. 1960. Zeolite zones and dyke distribution in relation to the structure of the basalts in eastern Iceland. *J. Geol.* 68:515–28
124. Walker, G. P. L. 1963. The Breiddalur central volcano, eastern Iceland. *Quart. J. Geol. Soc. London* 119:29–63
125. Walker, G. P. L. 1964. Geological investigations in eastern Iceland. *Bull. Volcanol.* 27:1–15
126. Walker, G. P. L. 1965. Evidence of crustal drift from Icelandic geology. *Phil. Trans. Roy. Soc.* 258:199–204
127. Walker, G. P. L. 1966. Acid rocks in Iceland. *Bull. Volcanol.* 29:375–406
128. Walker, G. P. L. 1972. Compound and simple lava flows and flood basalts. *Bull. Volcanol.* 35:579–90
129. Ward, P. L., Pálmason, G., Drake, C. 1969. Microearthquake survey and the Mid-Atlantic Ridge in Iceland. *J. Geophys. Res.* 74:665–84
130. Ward, P. L. 1971. New interpretation of the geology of Iceland. *Bull. Geol. Soc. Am.* 82:2991–3012
131. Ward, P. L., Björnsson. S. 1971. Microearthquakes, swarms and the geothermal areas of Iceland. *J. Geophys. Res.* 76:3953–82
132. Welke, H., Moorbath, S., Cummings, G. L., Sigurdsson, H. 1968. Lead isotope studies on igneous rocks from Iceland. *Earth Planet. Sci. Lett.* 4: 221–31

EVOLUTION OF ARC SYSTEMS IN THE WESTERN PACIFIC

Daniel E. Karig
Department of Geological Sciences, Cornell University, Ithaca, New York 14850

INTRODUCTION

Very shortly after their discovery arcuate island chains, with their associated volcanoes, trenches, and seismicity, were recognized as critical tectonic complexes in some way related to the development of orogenic zones (e.g. 102). However, technical inability to adequately investigate these primarily submarine features, and even worse, the often misleading conclusions drawn from the few segments above water, delayed an adequate understanding of island arc systems until after the Second World War. Since then, high precision echo-sounders, low-frequency acoustic profiling, and better bottom sampling apparatus have clarified arc system morphology and shallow structure. These observations, coupled with seismological and other geophysical studies, have now directly and indirectly indicated the role of island arc systems in the orogenic process and in global tectonics.

Differences in opinion exist concerning many, or most, details of arc system character and evolution, but most workers in western Europe, Japan, and the Americas accept the underlying concept of plate tectonics. In this framework, island arc systems result from the underthrusting, or subduction, of one lithospheric plate beneath another. The following discussions assume the validity of plate tectonics and concentrate on processes by which subduction is manifested in the geologic record. Different concepts of island arc evolution, which deny the concept of subduction, are discussed by Meyerhoff, Meyerhoff & Briggs (74), Carey (13), Beloussov (7), and others.

This paper reviews recent data and the resulting ideas which bear on evolution of arc systems in the western Pacific. Results and conclusions from arc systems in other regions have been introduced where pertinent. Geological and crustal geophysical aspects are stressed, and there is no attempt to review either data or theoretical studies from the subcrustal regions beneath the arc systems, or petrologic studies of island arc igneous activity.

STRUCTURAL FRAMEWORK OF AN ARC SYSTEM

Unfortunately, island arc terminologies which have been presented over the years have not reflected the uniformity of morphology and structure which has generally

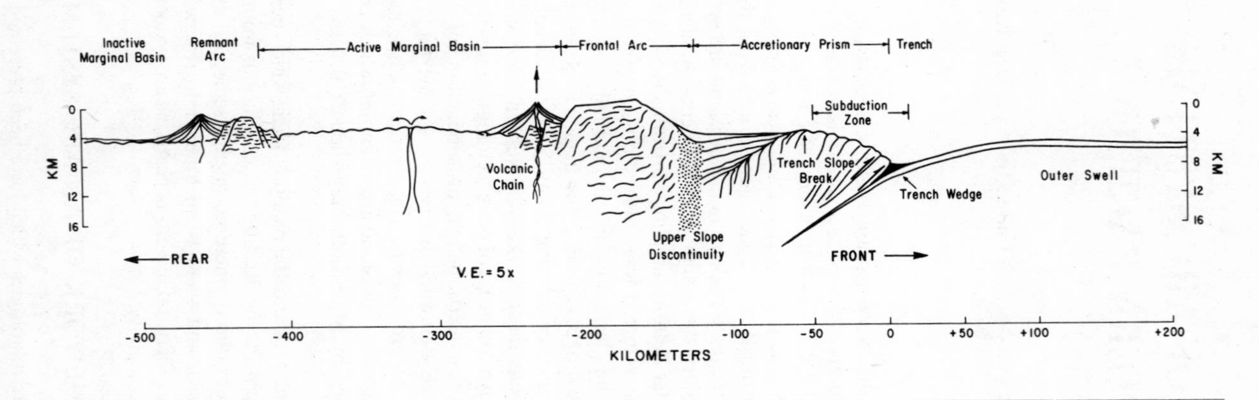

Figure 1 Generalized framework of a western Pacific island arc system after Karig, Le Pichon & Sharman (57) showing the terminology used in this paper.

been observed in arc systems. For instance, the Indonesian terminology as presented by Vening Meinesz (108) called for an outer or tectonic arc between volcanoes and trench and could not be applied successfully to the Tonga system. Although there are islands trenchward of the Tongan volcanoes, these are part of the volcanic arc as defined in Indonesia, and the tectonic arc is deeply submerged and poorly developed.

A new terminology (Figure 1) has been developed from recent geological and geophysical studies in the western Pacific, which employs generalized and descriptive terms although some have genetic connotations. A fully developed island arc system can be divided into three sections across its trend. On the forward side of the system is the inner slope of the trench, which apparently overlies the area of active subduction and accretion. Behind the volcanic chain an extensional terrane may form in which previously developed sections of the arc system are separated from the active system. Between them lies the frontal arc, a relatively passive block of older, thicker crust which undergoes primarily vertical displacements, although important plutonic and metamorphic processes may occur at deeper levels.

This subdivision stresses the processes of arc system growth and destruction. Differences among the arc systems of the western Pacific stem from variations in the subduction process along the leading edge of the system and from the degree to which extension has operated along the trailing edge. Confusing this pattern are secondary effects of oblique subduction, collisions, and arc polarity reversals.

THE LEADING EDGE AND SUBDUCTION PROCESSES

The slope leading down into the trench axis from the frontal arc is one of the most complex areas within an arc system, showing broad variations in morphology, lithologic content, and structural style (Figure 2). It can generally be divided into an upper, undeformed, and sediment-covered section and a lower, steeper slope underlain by deformed sediments or acoustically opaque material. The boundary between the two segments can be expressed as a change in slope, a shelf or terrace edge, or a discrete ridge. This range in aspect has led to a corresponding range in descriptive terms, including tectonic, outer, or nonvolcanic arc, midslope basement high, and trench slope break.

Variations in the inner trench slope, rather than being random, can be reasonably correlated with differences in supply of sediment to the subduction zone from the few major sources. In arc systems where a plate carrying a relatively thin pelagic sediment cover is being subducted and where there are few sources of terrigenous sediment, the inner trench slope is in two sections separated by a simple slope break (19, 47, 56). Good examples of this type are the Tonga, Kermadec, and Mariana trenches. Very limited dredging of the lower slope has recovered predominantly igneous rocks (26, 35, 87). Pelagic constituents have not been reported, but their selective removal during dredging might be expected. The single seismic refraction profile over this variety of trench slope (83) indicated that it is underlain by a crust having seismic velocity units similar to those of oceanic

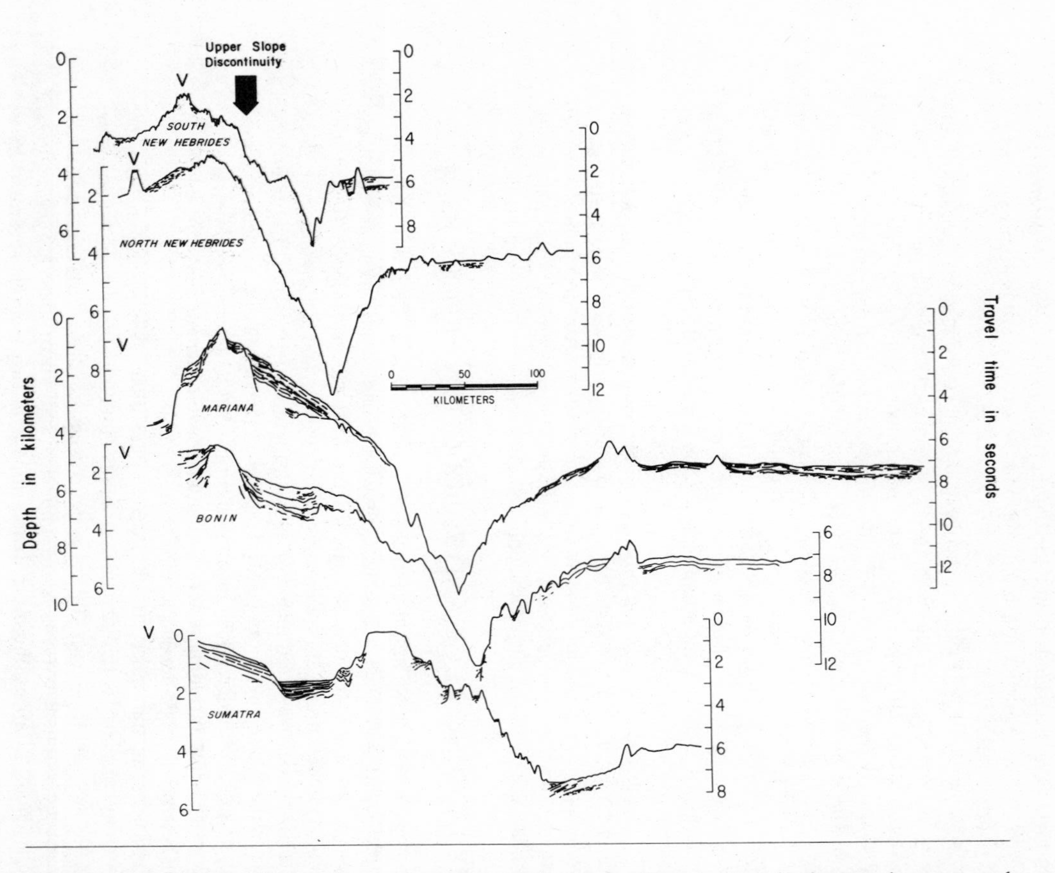

Figure 2 Seismic reflection profiles across inner trench slopes of arc systems having varying ages and sedimentation characteristics (57). These profiles are normalized to the upper slope discontinuity in order to show the wide range in width and shape of the accretionary prism.

crust, but somewhat thickened. On gravity profiles over these trenches, the free air minima nearly coincide with the topographic axis (103, 104, etc). These data suggest that material underlying the inner trench slope is composed of oceanic crustal components, although probably in a structurally disordered state.

With a second variety of inner trench slope, a thick sediment section is fed to the trench, but relatively little is derived from the frontal arc near the point of subduction. The sediments represent either the cover on the downgoing plate or a turbidite wedge in the trench axis which has been transported along the axis from a distant source. The Central Aleutian, Luzon, and Sumatra arc systems represent this style, which displays a strongly developed ridge between the upper and lower slope sections. The distinct trough behind the ridge results from inability of sediment sources on the frontal arc to keep pace with accretionary upbuilding of the trench slope break.

Exposures on this ridge form of the trench slope break (57), together with bottom sampling (110) and seismic refraction profiles (65, 90), indicate that the lower slope, trench slope break, and much of the upper slope trough are underlain by a basement of deformed sediments. The strong landward displacement of free-air minima on gravity profiles across these arc systems also suggests low-density, sedimentary basement (30, 103).

In arc systems where terrigenous sediment influx from the frontal arc is high, the upper slope area is kept filled, forming a shelf or terrace. Often sufficient material passes across the upper slope and into the trench to form a turbidite wedge which effectively supplies the subduction zone with abundant sediments, regardless of the nature of sediments on the downgoing plate. In this case, well represented by the eastern Aleutian area, basement characteristics are similar to those of the second variety (109). In other cases, as in the Japan arc, a smaller turbidite contribution to the trench and thin pelagic sediments on the downgoing plate produces a denser basement which may contain igneous rocks (66, 103).

These categories of inner trench slope are artificial in that all transitions can be found. It should also be noted that the morphologic style and lithologic content of the inner trench slope is independent of its size (57), which closely reflects the duration of subduction and absolute feed rates to the subduction zone.

Structural Style of Accretion

Structures within the accretionary prism which result from subduction have only begun to be investigated. A single Deep Sea Drilling Project (DSDP) hole (110), a few good reflection profiles, reconnaissance geology of the trench slope break on several islands, and a hazardous transfer of observed structures from orogenic zones to their original setting (e.g. 80, 81) constitute the bases for several available models.

The stripping off of sediment cover from the downgoing plate in fold or thrust units and its accretion to the trench wall is suggested by some seismic reflection profiles (6, 14, 99, 100, 109). Where they are next observed, at the trench slope break, the sediments are isoclinally folded with steeply dipping axial surfaces. Most of this deformation is thought to occur on the lowermost trench slope, because sediments usually become acoustically opaque at the base of the slope and because DSDP site 181, part way up the lower slope, recovered highly deformed sediments (110).

The structural process by which material is transferred from oceanic to arc plate probably varies according to the thickness and type of sediment in the trench and also the rate of subduction. In trenches where a turbidite section, deposited in the trench, overlies pelagic or hemipelagic sediments, deposited on the downgoing plate, the turbidites may be systematically removed and accreted while the rest of the section may be underthrust to deeper levels (57). Shearing should be localized in the top of the pelagic section because these sediments have minimum strength and maximum porosity (31). Bedding thrusts, aided by expected high pore pressures would place highly deformed turbidites over the surficial turbidites on the trench floor (Figure 3a), and would explain the observed sharp juxtaposition of very opaque acoustic basement and undeformed sediments at the base of the inner trench wall (30). Slow subduction appears to favor a greater degree of folding in the accretionary process (100; M. S. Marlowe, personal communication).

The mode of deformation in subduction zones involving downgoing plates with thin pelagic covers is more difficult to model. The weak, high porosity cover could again be utilized as a zone of shear, and all but the uppermost sediments might be carried deep below the inner slope of the trench. However, growth of the inner slope area with time in this variety of subduction zone (57) indicates that pieces of oceanic crust and even upper mantle are sheared off the downgoing plate (Figure 3b). This might occur intermittently when large-scale topography, such as one of the normal fault scarps which develop on the flexed outer trench slope (66), or a seamount, impinges on the inner trench slope.

Apparently deformation is most intense at the base of the inner trench slope but continues, at a decreasing rate, to the trench slope break, where the greatest cumulative effects are seen. Sediments of the upper slope which overlie the trench slope break are openly folded and moderately faulted (30, 37, 57, 67).

This deformation and limited geodetic measurements on pertinent islands demonstrate uplift of the trench slope break relative to both sea level and the upper slope area (48, 110). There is little or no evidence to support earlier views of major subsidence in the area of the trench slope break (39, 82).

On the other hand, there is substantial evidence that the upper slope area is subsiding relative to the frontal arc, to the trench slope break, and often, to sea level. Many reflection profiles show faulting or folding on both flanks of the upper slope sediment pile (e.g. 37, 48, 57, 67) and often reveal a downward increase in displacement or in rearward tilt (e.g. 91, 110). The sharp contrast between this zone of subsidence and the emergent frontal arc (e.g. 28, 89) thus requires that a structural discontinuity separate the two units.

The boundary between the two regimes has been designated the upper slope discontinuity (57). In part, the upper slope discontinuity appears to represent the upper continental or insular slope which existed before a pulse of subduction began. With time, an accretionary prism builds the lower slope outward from the remnant upper slope and sediments fill in behind and over a basement of accreted material, but there is also demonstrable faulting and flexure along the upper slope discontinuity (30, 47, 110), especially in the broader accretionary prisms. Other than being steeply dipping and downthrown toward the trench, however, the nature and origin of the faults is not yet known.

Available geodetic and structural evidence can be combined with bathymetric profiles which show changes in the width of the inner trench slope with time to produce a reasonably satisfactory growth model of subduction zones. Initiation or

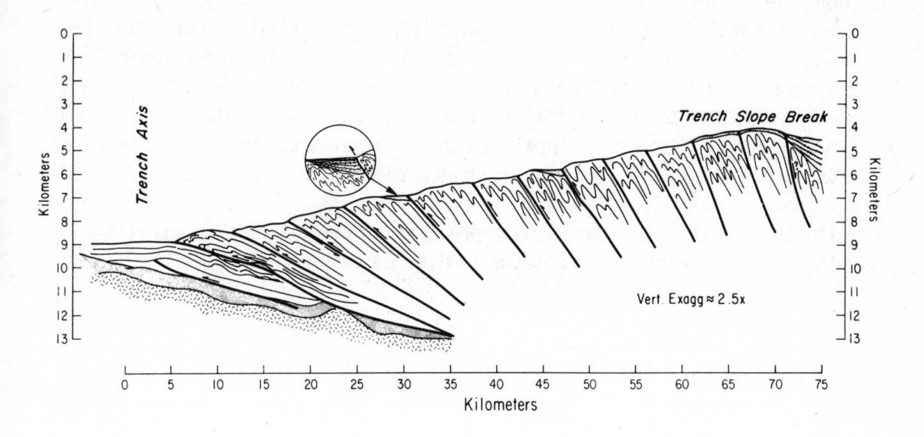

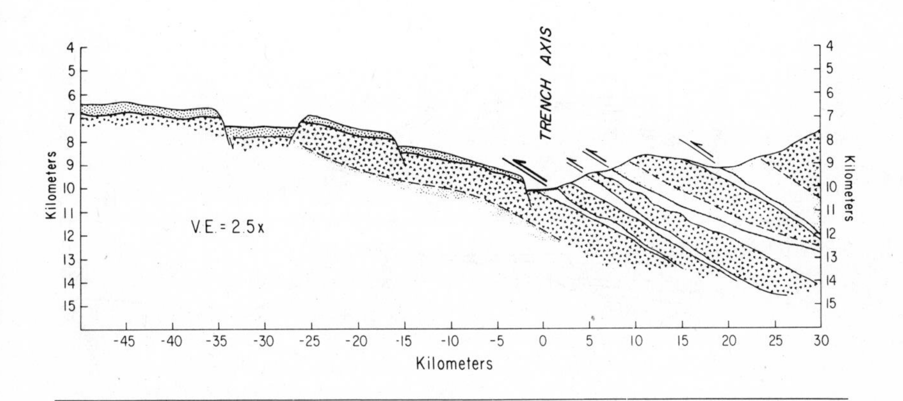

Figure 3 A. Possible subduction mechanism where turbidites are accreted to the trench wall and the pelagic sediments and basement carried to deeper levels. Bedding plane step thrusts sided by high pore pressures in the weak pelagic sediments produce the observed juxtaposition of highly deformed sediments next to and overlying undeformed turbidites. B. Possible subduction mechanism where a very thin sediment cover is involved. Intermittent accretion of basement slabs, intercalated between highly deformed pelagic sediments, is postulated.

rejuvenation of subduction produces a trench near the foot of a pre-existing continental or insular slope. If a significant rise sedimentary apron is present, it is uplifted and deformed into an initial accretionary prism, as in the Shikoku arc system (37, 57).

As material is accreted, either in fold packets or in slabs, previously accreted material is lifted up the lower slope to form the crest of the trench slope break. Then, however, the material subsides and the trench slope break moves forward (Figure 4).

Uplift is attributed to the effects of underplating and internal deformation within the accreted material beneath the lower slope. Subsidence of the upper trench slope is apparently a reflection of the depression of the flatter, upper section of the Benioff zone by the weight of accreted material (57). The cause of the change from subsidence to uplift across the upper slope discontinuity is unknown, but permanent uplift of the frontal arc has been attributed to flexure or irreversible effects of compression (28).

During growth of an accretionary prism, or of several such prisms (11), the volcanic chain remains fixed, or moves slowly to the rear, relative to the crust below

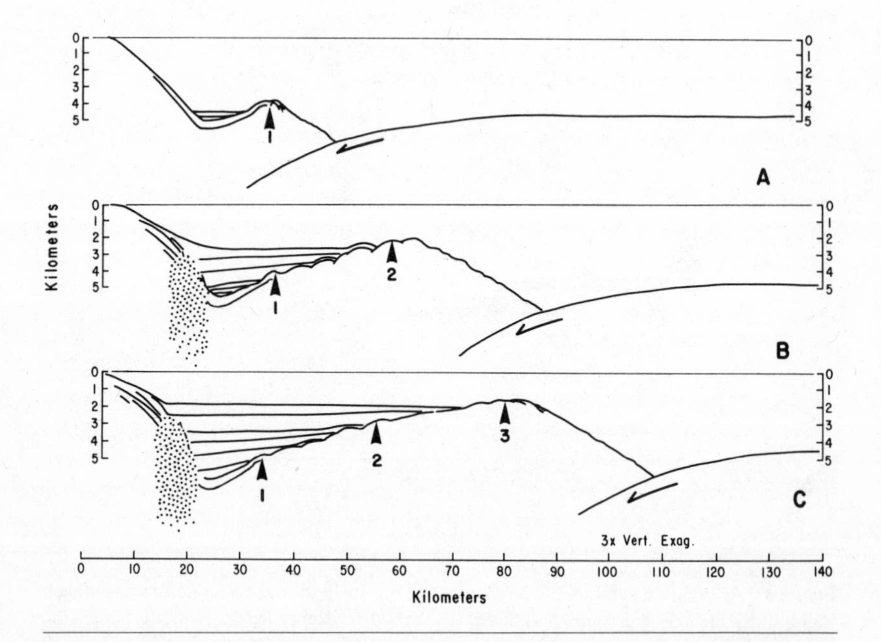

Figure 4 Model of accretion along arc system where a moderate to thick sediment cover occurs (57). Accreted material rises to occupy the trench slope break, then subsides to form basement for upper slope sediments. At the same time, the trench slope break is migrating outward.

(21). This leads to an increasingly broad volcano-trench separation and, because the Benioff zone remains at a fairly stable depth beneath the volcanoes, to a very broad, flattened upper Benioff section (30, 32, 57, 62). The trenchward overstepping of the low P/T metamorphic regime over the high P/T zone, as proposed by Matsuda & Uyeda (72), Oxburgh & Turcotte (85), and others, appears to require a different process; either a longer term, discrete shift of the volcanic chain and Benioff zone, or some form of tectonic disruption.

Tectonic erosion rather than accretion has been suggested to explain the igneous rocks and the morphology on the walls of many western Pacific trenches (30, 82). Growth of the accretionary prism with time in these arc systems and volume calculations indicating the need for basaltic crust within the accretionary prism (57) obviate the need for such erosion.

THE TRAILING EDGE AND CRUSTAL EXTENSION

Behind the volcanic chain in an active island arc system lie one or more marginal basins, whose character and origin are still subjects of debate. Some marginal basins are volcanically and structurally active; these always lie directly behind the volcanic chain of the arc system and are called inter-arc basins (47), or active marginal basins (49). The few investigations in the small number of active marginal basins in the western Pacific have revealed only their general characteristics.

Pillow basalts are erupted, without the formation of central shields, along the axial zones of the active Mariana (33, 48) and Lau (97) basins. In these larger, active marginal basins the axial zone is elevated relative to the basin flanks (50, 51), but shows no clearly defined central rift or axial symmetry (Figure 5). The locus of active volcanism is restricted by the sediment distribution to the central 40 km or so, but there are no data delineating the extent or form of volcanic centers within this band. Fissure eruptions have been suggested (48). Basin morphology consists of linear ridges and troughs which most likely reflect tilted fault blocks, perhaps modified by volcanism. No transform faults have yet been recognized, but this may be due to lack of data.

Although there is a remarkable dearth of sediment cover in active marginal basins, the sedimentation rates are very high. Volcaniclastic aprons, building out from the volcanic chain, accumulate at rates exceeding 100 m/m.y. (12, 25, 48). Montmorillinitic clays and finer volcanic detritus are being hemipelagically deposited further away from the volcanoes at rates decreasing with distance (29, 47, 48). Foraminiferal and calcareous nannofossil oozes are the dominant sediments furthest from the volcanic chain at active marginal basins having no continentally derived sediments. In these basins the overall sediment pattern is one of thicker deposits along the basin flanks, especially on the flank adjacent to the volcanoes, and thinnest deposits in the axial zone. In active marginal basins which lie adjacent to continental margins, as the Okinawa Trough (111), thick sediment sections may be involved in faulting and volcanic activity along the axial zone.

Fault scarp systems generally bound active marginal basins on both flanks,

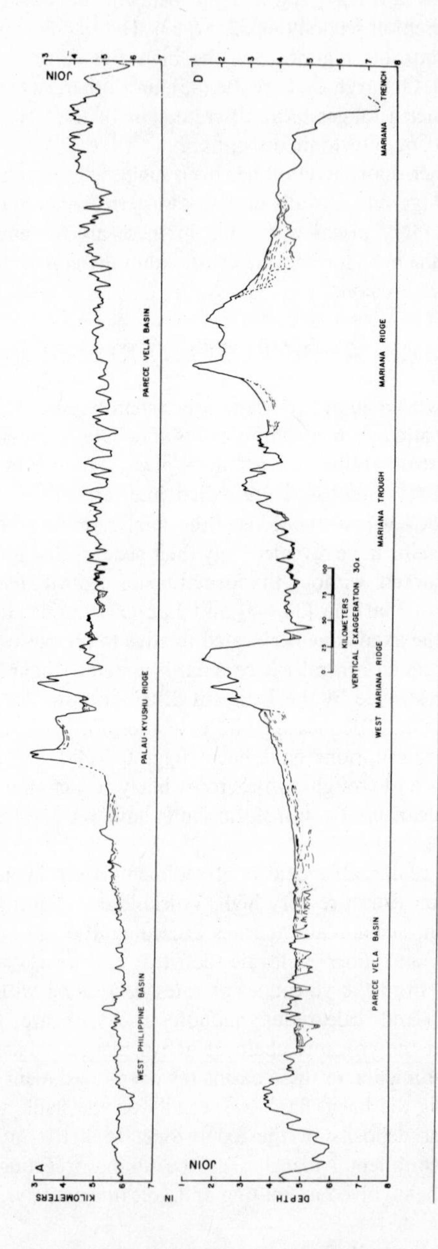

Figure 5 Seismic reflection profile across the Mariana arc system showing the character of the active marginal basin (Mariana Trough) and inactive basins to the west (48).

although the volcanic chain often conceals the scarps on the forward side. The rear flank forms part of a submarine ridge, which is defined as the third arc (107) or, in more general terms, a remnant arc (50). In the simplest examples, remnant arcs appear to represent the rear flanks of frontal arcs, and display primarily volcanic chains and volcaniclastic aprons (Figure 5). In other, more complex situations, remnant arcs may represent entire arc systems which have been abandoned or which have collided with other remnant arcs (50). Some remnant arcs are relatively wide and display broad, flat tops, while others are little more than truncated volcaniclastic aprons. Where more fully developed, they show a basement core, fault bounded on both sides, with the rear faults modified by volcanic activity. There is accumulating evidence that remnant arcs have subsided up to several kilometers from sea level or from much shallower depths (50, 94).

Inactive marginal basins, separated by remnant arcs and lying behind the active basin, if one is present, show basement morphology similar to the active basins (Figures 5 and 6). In most older basins, however, a thick sediment cover obscures the basement. This sediment cover ranges from a few hundred meters of pelagic muds in basins protected from continental sediment sources, like the Philippine Sea, to several kilometers of terrigenous turbidites where a continent forms a basin flank (73, 98).

The depth to basement in marginal basins, after the effects of gravitational loading by sediment are removed, increases with age from approximately 2.5 km in active basins to more than 5.5 km in older basins (49). This increase in depth with age, accompanied by a fall in crustal heat flow values from over 3 HFU to normal (e.g. 96) is qualitatively similar to that observed in the main ocean basins.

Although this similarity suggests that both marginal basins and major oceanic basins are generated by a similar process, there are minor, but perhaps significant, differences in their crustal and lithospheric character. Marginal basins of the Philippine Sea and several other regions are deeper than oceanic basins of the same age and some have higher heat flow at both the equivalent age and depth (96). This may be attributed to a thinner than oceanic lithosphere, as suggested to exist beneath the Philippine Sea (46), which might result from sublithospheric shear and erosion (96). Most of the difference in depth could also be accounted for by thinner than oceanic crustal layers which are suggested by seismic refraction studies in the arc system.

Identification of the basaltic basement as a product of extensive volcanic activity occurring after the basin was originally trapped has been postulated by Scholl et al (93) for the Kamchatka Basin and by Edgar, Ewing & Hennion (23) for some Caribbean basins. The idea that marginal basins originated through the subsidence of continental crust (8, 58, 105) has largely been abandoned. Generation of island arc systems from transform faults has been suggested in some areas (56, 106).

Very few data other than drill holes of depths beyond the present capability of the Deep Sea Drilling Program will unambiguously settle the problem. Closely controlled seismic refraction profiles can place limits on the possible thickness of a basaltic cover and masked sediments which might overlie normal ocean crust. Discrimination between trapping and extensional origins might be made by looking

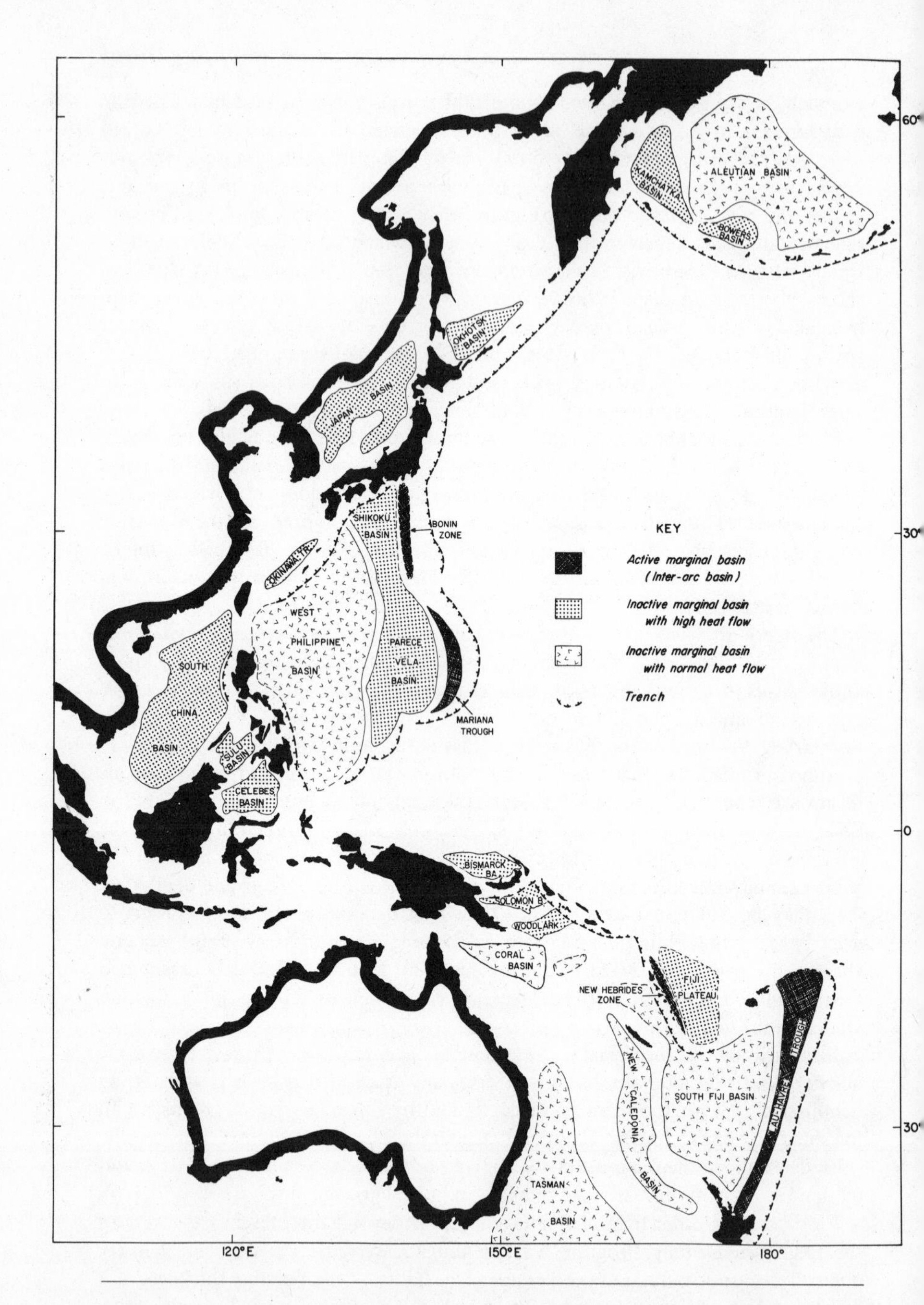

Figure 6 Distribution of marginal basins in the western Pacific, after Karig (49) showing the classification based on increasing age, depth, and crustal heat flow.

for systematic changes in basement age which would result from axial crustal spreading.

Magnetic anomalies in marginal basins do not show well-developed, symmetrical linear patterns which can be correlated with those generated at oceanic rises. Linear anomalies do exist, but in most cases they are not coherent or identifiable. Intensive surveying of the Sea of Japan (41) reveals low amplitude (200 gamma peak to peak), high frequency nonsymmetrical anomalies in comparison to those of the Pacific Ocean east of Japan. Murauchi (82) reports similar low amplitude anomalies, often averaging less than 100 gamma, in the Philippine Sea.

If the magnetic anomalies of marginal basins and oceans are to be compared correctly, however, the remnant fields must have been generated at the same time along spreading ridges with similar orientations and magnetic latitudes. On this basis, the data are ambiguous and scanty. Spreading in both the active Mariana basin (48) and the East Pacific Rise off Central America (36) is producing anomalies with 200–300 gamma amplitudes. In contrast, anomalies of late Oligocene age in the Kamchatka basin (60, 94) have amplitudes only about half that of anomalies generated under similar conditions along the Mid-Atlantic Ridge. Spreading rates, which were not considered in these comparisons, may affect the anomaly amplitude in marginal basins.

While magnetic lineations have often been noted in marginal basins, there are relatively few cases where symmetry or identification of these lineations has been claimed. Luyendyk, MacDonald & Bryan (69) and Ben Avraham, Bowin & Segawa (9) report identifiable anomalies in the Woodlark basin near New Guinea and the West Philippine Basin, respectively. A much more convincing set of anomalies, out to anomaly 3 (5 m.y.) has been reported from the Lau Basin (J. G. Sclater, personal communication), but this organization breaks down further from the spreading axis.

The apparently different magnetic field characteristics in marginal basins and mid-ocean ridges might result from differences in the magnetic properties of the extruded basalts, from degradation of remnant magnetism in marginal basin basalts by injection into thick sediment covers or by hydrothermal alteration (M. Marshall, personal communication), or from a different process of crustal extension.

Basalts from the Mariana basin, however, have as high remnant magnetizations and Konigsberger ratios as do mid-ocean ridge tholeiites (M. Marshall, personal communication). Furthermore, there is very little sediment cover near most of the extensional zones of active marginal basins. Effects of hydrothermal alteration are unknown but, to date, thermally altered basalts are uncommon in marginal basins and have been dredged only from fault zones. A mode of crustal extension in marginal basins different from the passive spreading at mid-ocean ridges emerges as the most reasonable cause for the differences in magnetic field characteristics.

The underlying mechanism responsible for the young crust, high heat flow, and apparent extensional tectonism in marginal basins is presently the focus of considerable attention. Oxburgh & Turcotte (85), Hasebe, Fujii & Uyeda (34), Karig (49), and others favor a rise of heated mantle material from the shear zone

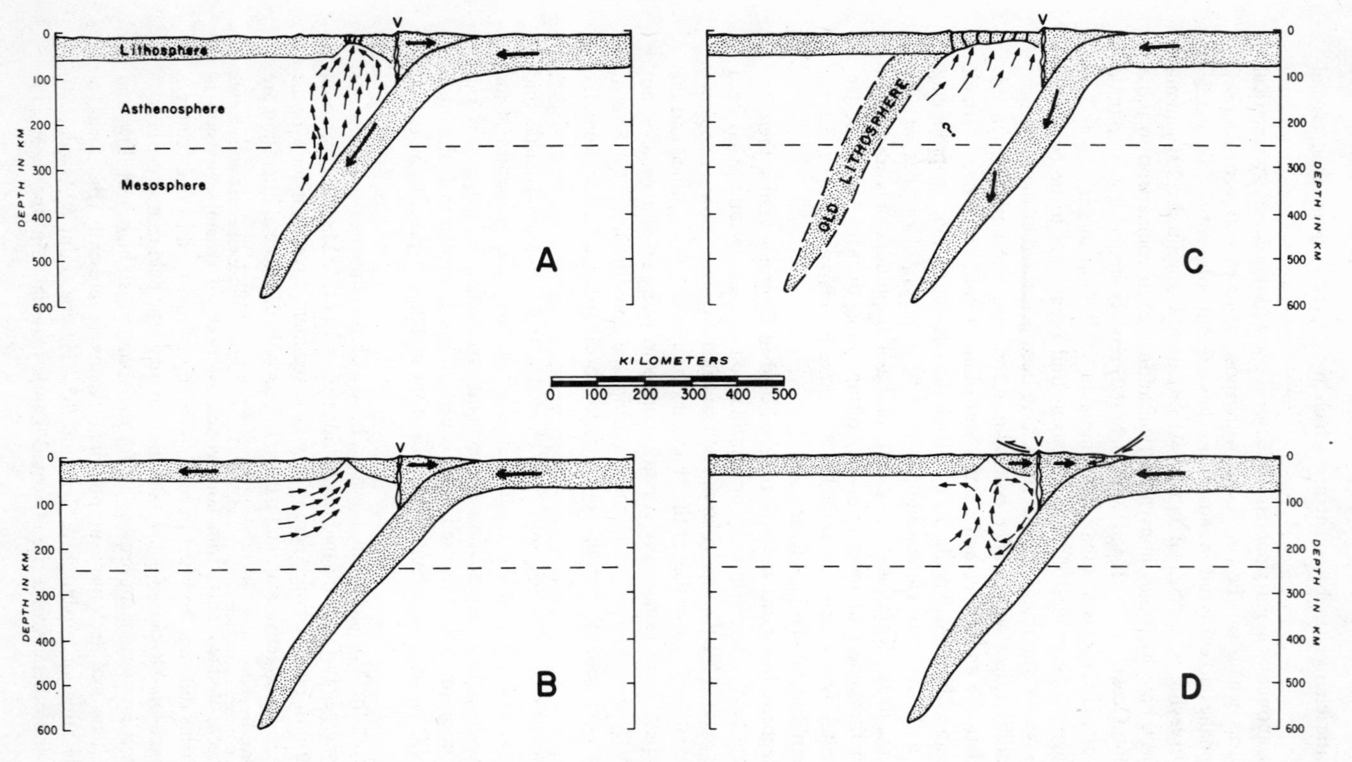

Figure 7 Hypothetical models advanced to explain the young crust and high heat flow in marginal basins. A. Active diapirism resulting from heat and/or water generated along the Benioff zone. B. Passive diapirism in response to regional extensional stress in the lithosphere. C. Stepwise migration of arc system with collapse of the oceanic crust behind the new arc (45). D. Subsidiary convection cells driven by drag along the Benioff zone (38, 101).

along the top of the downgoing plate (Figure 7A). A variant of this model, suggested by studies on the role of water in the mantle (114), would have water, derived from the downgoing plate, cause partial melting and diapiric rise of the immediately overlying mantle. Either stated or implied in both schemes is the active nature of the diapirically rising mass, which would cause extensional features at the surface but produce and transmit compressional stresses at deeper levels of the lithosphere.

A more passive upwelling of asthenospheric material into an extensional zone generated by some sort of plate rearrangement or by cessation of subduction has been suggested by Packham & Falvey (86), Scholz, Barazangi & Shor (95), and Kanimori (45) (Figure 7B, C). Hydrodynamic schemes which utilize the viscous drag along the Benioff zone to drive asthenospheric convection cells (e.g. 101) have also been proposed (Figure 7D).

Hydrodynamic schemes, which recirculate rather than increase the mass of material beneath the island arc region, find little support in geologic data. Although Sleep & Toksöz (101) cite a compressional focal mechanism behind the Tonga arc, other workers have found no similar earthquakes and there is no geologic evidence of sufficient crustal shortening within the arc system to balance the extension in the inter-arc basin. Moreover, no explanation for the seaward migration of the arc system or the increase in volume is offered.

The studies of seismic energy attenuation and the Q structure beneath island arc systems hold great promise for discrimination among possible mechanisms. There already appears to be a close correlation between low Q (high attenuation) zones and active or recently active marginal basins (4, 77, 79; M. Barazangi, personal communication). Preliminary results indicate that the forward edge of the low Q zone, near the volcanic chain, is steep and fairly sharp (4, 77) and that the rear edge is probably also quite steep. These steep boundaries, and the continuity of the low Q zone to the Benioff zone at depths greater than 250 km are difficult to explain in the hydrodynamic models.

The zone of high attenuation can better be explained by diapiric rise, either active or passive. With passive upwelling, the adiabatic rise of asthenosphere up a relatively steep geothermal gradient could cause partial melting, but it is doubtful whether the low Q zone produced would be as steep or extend as deep as observations indicate.

General occurrence of extension within arc systems during rapid subduction (49) rather than following cessation of subduction is further evidence in favor of active diapirism. Only in the Scotia arc system does extension appear to be associated with slow subduction (5). However, there is no low Q zone observed beneath this postulated extensional zone (M. Barazangi, personal communication), which might suggest an origin different from that of western Pacific marginal basins.

Although the extension in the Mariana trough might be related to subduction of the Philippine plate into the second series of trenches, this type of passive extension cannot be generally applied. Overlap of the Tonga and New Hebrides trenches and the geometrically required extension has been suggested as the reason for the Lau and New Hebrides inter-arc basins (4). However, the Lau basin is only the northern end of an extensional zone which continues into New Zealand. In this part of the arc, as in the Japan, Ryukyu, and other arc systems, the frontal arc and trench must be forced forward, increasing the distance between trench and continent.

If rise of mantle material beneath active marginal basins is the cause, rather than the effect, of crustal extension, then low Q zones should be observed where the hot, low-density material has not yet forced its way into the lithosphere to cause extension. Large-scale uplift and silicic volcanism in the Central Andes (84, 88, 92) are hints that mantle diapirism exists below that area. Seismological results are conflicting. James (42) finds no low Q material beneath the volcanic chain and Altiplano, but both the earlier study of Molnar & Oliver (79) and more detailed recent work (M. Barazangi, personal communication) indicate a substantial zone of high attenuation.

Major problems with an active diapir model are that the rate of crustal extension in active basins implies too great a rate of supply of mantle from the Benioff zone (49) and that the zone of extension apparently remains in the center of the active basin, which would be far behind the Benioff zone and volcanic chain in the larger marginal basins.

The most acceptable model calls for collection of heated mantle below the lithosphere behind rapidly subducting arc systems. Whenever the hydrodynamic forces of the diapir overcome regional compressive forces, the material would rise into the lithosphere, and cause crustal extension. Cessation of subduction would be only one example of reduction of regional compression. Locus of upwelling, strongly controlled by existing weakness zones, would begin near the volcanic chain and migrate away as new lithosphere was symmetrically generated in the inter-arc basin.

Active diapirism may be sufficiently different from passive diapirism to produce the variations in tectonic and magnetic characteristics between marginal basins and ocean basins. Moreover, the transition from subduction-related extension to interplate extension in the Gulf of California (55) should serve as a warning that extension begun by active diapirism might be continued by other mechanisms.

Still to be satisfactorily explained is the discontinuous nature of extension in marginal basins, which produces the series of basins and remnant arcs. The direct correlation of extension with subduction pulses (49) is too simplistic, and cessation of extension might better be related to exhaustion of available mantle material, perhaps by reduction of subduction rates. Sleep & Toksöz (101) suggest a cyclic process where extension causes flattening of the Benioff zone which, in turn, causes resistance to extension and cessation of spreading.

Analogous crustal extension within continental margin arc systems or island arc systems with very broad frontal arcs of continental nature seems to be expressed in volcano-tectonic rift zones (49, 55). Geologic syntheses in these zones (2, 49, 70) indicate that extension, in the form of basin-range faulting, is preceded by widespread eruption of ignimbrites. Silicic magmas apparently result from melting within the lower crust (24), which in turn reflects a rise in geothermal gradient. Active diapirism above the Benioff zone is a reasonable source for the excess heat required. In island arc systems, where there is a thinner, more mafic crust, this early volcanism occurs as much smaller eruptions of volatile-rich dacite and rhyolite (54). Following the rupture of the lithosphere in the volcano-tectonic rift zones, bimodal eruptions of both silicic and basaltic volcanic rocks takes place (64).

At least some island arc systems have developed from continental margin systems by evolution of volcano-tectonic rift zones into marginal basins. The Japanese arc is the most obvious example. The spectrum of arc systems around the Pacific suggests a process by which a broad strip of continental crust is intially separated from the continental core. Continued subduction and pulses of extension split this broad arc and create additional basins. The long-term result is to preferentially remove the continental crust from the rear of the frontal arc and leave it behind in the older remnant arcs (51) and to accrete new crust, with oceanic affinities, along the forward edge.

One of the most intriguing of current problems is the nonuniform geographic distribution of marginal basins. Rapid subduction may be a necessary prerequisite for extension, but it is not sufficient, because neither active nor inactive marginal basins exist along the west coast of South and Central America. In some of these areas volcano-tectonic rift zones or low Q zones occur, but since late Mesozoic, marginal basins have been developed most dominantly in the western Pacific region. Prior to that time, there is little evidence for marginal basins in that area and growing evidence of late Paleozoic-early Mesozoic marginal basins along the west coast of North America (e.g. 15).

Over the last 50 years, a number of proposals link the formation of marginal basins with a general westward shift of the crust or lithosphere over the deeper levels. Recent proposals of relatively fixed hot spots (plumes) has led to quantitative analyses of plate motions over the asthenosphere. From one of these, Wilson & Burke (113) deduced that marginal basins develop behind subduction zones flanking plates which are stationary or slow-moving relative to the asthenosphere. An important test of the validity of this and other models is to determine the distribution of marginal basins among the various subduction zones with time and to compare this distribution with plate motions deduced from hot spots.

SECONDARY TECTONIC PROCESSES

The effects of oblique subduction, collisions, and polarity reversals are considered secondary to accretion, extension, and associated processes. They, and perhaps other tectonic effects not yet recognized, are responsible for the development of varied and complex compressive plate boundaries.

Where the trajectory of the downgoing plate is not perpendicular to the trench, oblique subduction occurs. Very often in such cases, the strike-slip component of subduction is not accommodated in the trench, but occurs within the arc system, most commonly within active extensional zones (27, 56) (Figure 8). Longitudinal shearing may combine with extension to produce en-echelon extensional structures (as in the New Hebrides, Tonga, Bonin systems) or less commonly occur as discrete strike-slip faults (as in the Taupo Graben of New Zealand). Strike-slip faulting initiated in this manner may have continued after extension ceased in the Proto-

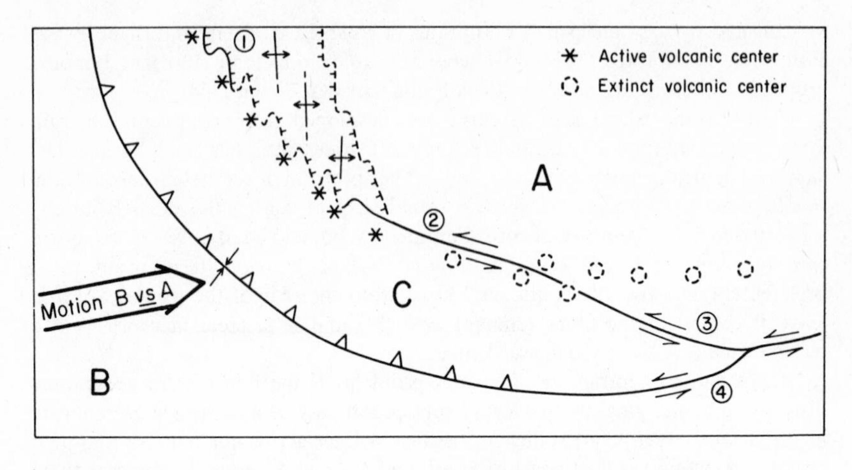

Figure 8 Model showing various methods by which the strike-slip components of oblique subduction can be expressed within the arc system. 1. En echelon extension. 2. Strike-slip faults near the volcanic chain. 3. Strike-slip faulting between the trench and volcanic chain. 4. Strike-slip faulting in the trench. Also shown by light arrows are the distribution of inter-plate motions when a third small plate (C) is created between the major plate (A and B).

Gulf of California (55) and Semanko rift (59, 112). Where no extensional zones exist, strike-slip faulting can occur between the volcanic chain and the trench, as on Median tectonic line of Japan (44), Longitudinal Valley of Taiwan (40), and Atacama Fault (1) or may not be separated from the under-thrust component.

The longitudinal shear component is separated from underthrusting because resistance to rupture along vertical planes within the lithosphere, especially within the extensional zone, is less than frictional resistance along the low-dipping Benioff zone. Thus, between the two major plates, a narrow plate is created which migrates along the subduction zone and causes the direction of underthrusting to be more nearly perpendicular to the plate boundary than the motion vector between the two major plates.

Most island arc systems in the western Pacific have normal polarity (face away from the continents), but a fair number have the opposite, or reversed, polarity. Either the geometry of reversed polarity was an initial condition, generated within an ocean basin (16, 68) or the arc system has since reversed its polarity. Evidence for polarity reversal has been presented for the New Hebrides (56, 76), Solomon (50, 61) and Luzon arc systems (52). To date, there is no substantial evidence that the reversed polarity of any arc system was generated from an initial condition.

In some instances, polarity reversal may have resulted from collision between the arc system and another tectontic unit, perhaps another arc system, a mid-plate rise

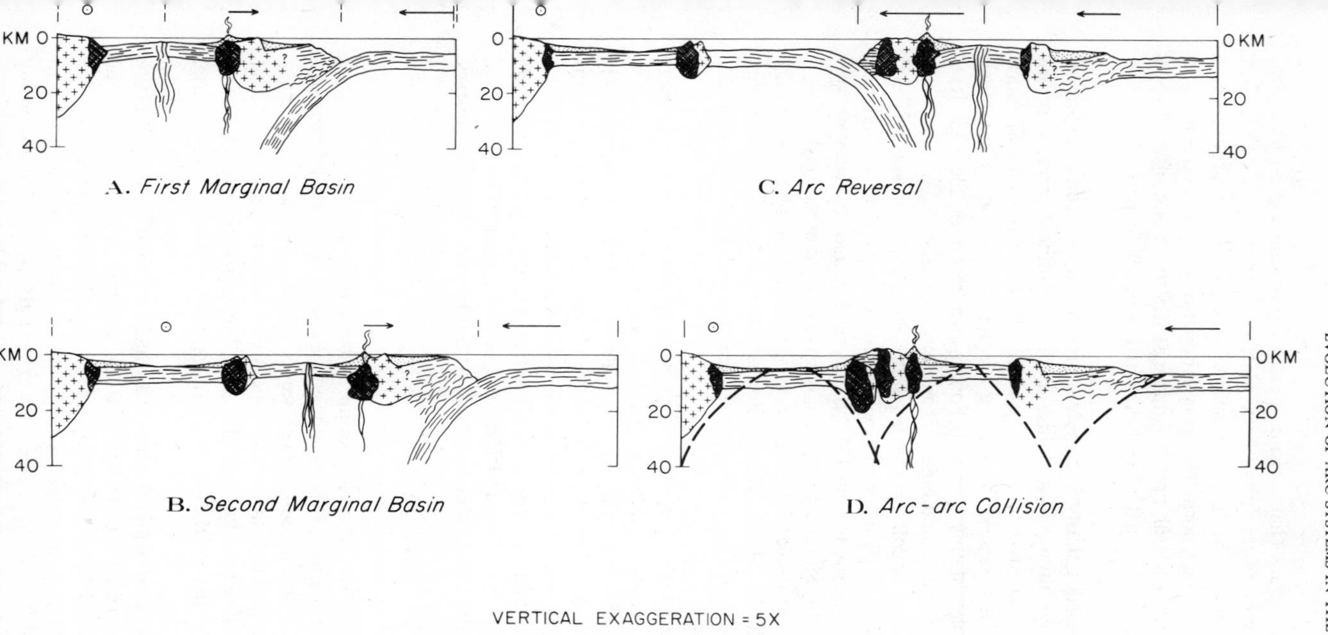

Figure 9 Hypothetical development of a marginal basin sequence, including arc polarity reversal and the emplacement of the ultramafic overthrust sheets. The arrows show qualitative motions of crustal blocks relative to the continental block at the left edge of each figure. Stage A illustrates the opening of the first marginal basin, which, as mentioned in the text, may not be as well organized as shown. Spreading is shown to be symmetric and localized but may be diffused across the basin. In stage B, a remnant arc has been created and a new marginal basin developed. In this and other figures, the consumption rates in the trenches are higher than the closure rates between major blocks by the amount of extension in the active marginal basin. In stage C, arc system polarity has reversed. Extension is shown behind the reversed arc system but need not be present. Collision between the reversed arc system and a remnant arc (stage D) leads to the overthrusting of the oceanic type crust on the leading edge of the frontal arc and requires that continued crustal consumption shift to a new locus, as indicated by the heavy dashed lines (50).

(61), or a very large seamount chain. More often, no such unit can be identified. A relative rearward motion of the lithospheric plate carrying the arc system over the asthenosphere may rotate the Benioff zone into and past the vertical so as to require more energy to continue normal polarity subduction than to reverse polarity (50). An apparent polarity reversal, now beginning at the south end of the Luzon arc system (27, 52), may be related to collision of the northern part of that arc system with Taiwan.

Subduction on reversed polarity arc systems results in the consumption of marginal basin crust, and ultimately in collision with a continent or remnant arc (Figure 9). The Hidaka-Kamuikotan orogen of Hokkaido and Sakhalin (78), much of New Caledonia (10), and several belts along the northern side of New Guinea (17, 43, 50) are examples of this type collision, which result in the high level emplacement of ophiolite sheets and high pressure metamorphic rocks.

These and other varieties of collisions involved in subduction processes (18) lead to further rearrangement of plate boundaries, since continental or arc crust is relatively light and resists subduction (71). If the relative motion between two plates is to remain unchanged, then subduction must shift to another locale.

In better documented examples of rejuvenation or initiation of subduction in the western Pacific (e.g. Shikoku, Manila, New Hebrides, Solomon, and New Britain arc systems), the new trench develops near an interface between oceanic crust and a continental or arc block.

The initiation of a subduction zone or rejuvenation following complete cessation of activity, remains an enigmatic but important process. Inactive subduction zones do not have the mass imbalances which are dynamically sustained in active arc systems, implying that tectonic and structural readjustments have taken place. Upon rejuvenation of subduction the new trench may form slightly forward of the final position of the old oceanic downgoing plate, and account for the large and relatively undeformed sheets of oceanic crust associated with the Franciscan complex of California (3) and with other subduction complexes.

CONCLUSIONS

Patterns of island arc and continental margin systems show superficially broad variations, but an increasing mass of data supports the early ideas which suggested a rather uniform underlying process. There seems to be no reason to create a classification of different types of subduction zone. Variations are important in that they reveal subsidiary or related tectonic processes, but to use them to subdivide arc systems into types masks the underlying continuum in the fundamental processes.

Variations among the relatively few parameters which control the character of arc systems can produce very complicated overall tectonic patterns. In most old orogenic belts these patterns would be further complicated by collisions which close the marginal basins and juxtapose the ridges. The complexities of recent arc patterns in the Melanesian and Indonesian-Philippine regions strongly suggest that unraveling a complex orogenic zone may be possible only in very general terms.

Identification of arc system units within orogens is critical for this interpretation, but depends on criteria that are now only poorly developed. Some indication of arc polarity has been obtained in simple cases from the K_2O-SiO_2 ratios (22), from the contrasting metamorphic regimes (78), and from the spatial relationships between subduction zone and upper slope rocks (63). Marginal basins probably play a significant role in most orogens, and if diagnostic sediment facies and facies patterns could be determined, might be valuable aids in determining ages of subduction pulses and arc polarity at that time. Undoubtedly, a combination of sedimentological and structural data, including the spatial relationships among rock and structural units will be necessary to successfully interpret any orogen. The task must begin with a better understanding of these parameters in contemporary arc systems.

Another major remaining task is to investigate temporal and spatial ordering of all events associated with consuming plate margins. Previous geotectonic concepts assumed orogenic cycles, based on frequently observed sequences of rock units and structures. Many of the proposed causes of these observations have been shown to be erroneous, but the consistent sequence and pattern of geologic units maintain their validity.

Several large-scale geologic syntheses of the subduction process, using both the observations in orogenic zones and the reconnaissance marine data, have been presented (18, 20, 75) as replacements for the orogenic or geosynclinal cycle. However, only intensive, problem-oriented geological and geophysical investigations in active arc systems will produce accurate and sufficiently detailed correlations between the geology observed in orogenic zones and processes active along subducting plate boundaries, and will permit the construction of an adequate orogenic flow sheet.

Literature Cited

1. Allen, C. R. 1965. Transcurrent faults in continental areas. *Phil. Trans. Roy. Soc. London Ser. A* 258:82–89
2. Armstrong, R. L., Ekren, E. B., McKee, E. H., Noble, D. C. 1969. Space-time relations of Cenozoic silicic volcanism in the Great Basin of the western United States. *Am. J. Sci.* 267:478–90
3. Bailey, E. H., Blake, M. C. Jr., Jones, D. L. 1970. On-land Mesozoic oceanic crust in California coast ranges. *US Geol. Surv. Prof. Pap.* 700-C:70–81
4. Barazangi, M., Isacks, B. 1971. Lateral variations of seismic wave attenuation in the upper mantle above the inclined earthquake zone of the Tonga Island Arc: Deep anomaly in the upper mantle. *J. Geophys. Res.* 76:8493–8517
5. Barker, P. F. 1972. A spreading centre in the east Scotia Sea. *Earth Planet. Sci. Lett.* 15:123–32
6. Beck, R. H. 1972. The oceans, the new frontier in exploration. *APEA J.* 12: 5–29
7. Beloussov, V. V. 1970. Against the hypothesis of ocean-floor spreading. *Tectonophysics* 9:489–511
8. Beloussov, V. V., Ruditch, E. M. 1961. Island arcs in the development of the earth's structure. *J. Geol.* 69:647–58
9. Ben Avraham, Z., Bowin, C., Segawa, T. 1972. An extinct spreading centre in the Philippine Sea. *Nature* 240:453–55
10. Brothers, R. N., Blake, M. C. Jr. 1973. Tertiary plate tectonics and high pressure metamorphism in New Caledonia. *Tectonophysics* 17:337–58
11. Burk, C. A. 1973. Uplifted eugeosynclines and continental margins. *Geol. Soc. Am. Mem.* 132:75–86
12. Burns, R. E. et al 1972. Glomar Challenger down under: Deep Sea

Drilling Project leg 21. *Geotimes* 17: 14–16
13. Carey, S. W. 1970. Australia, New Guinea and Melanesia in the current revolution in concepts of the evolution of the earth. *Search* 1:178–89
14. Chase, R. L., Bunce, E. T. 1969. Underthrusting of the eastern margin of the Antilles by the floor of the western North Atlantic Ocean, and origin of the Barbados ridge. *J. Geophys. Res.* 74: 1413–20
15. Churkin, M. 1972. Paleozoic marginal ocean basin-volcanic arc systems in the Cordilleran fold belt. *Conf. Modern Ancient Geosynclinal Sedimentation Univ. Wisc.* 34–35
16. Coleman, R. G. 1971. Plate tectonic emplacement of upper mantle peridotites along continental edges. *J. Geophys. Res.* 76:1212–22
17. Davies, H. L., Smith, I. E. 1971. Geology of eastern Papua. *Geol. Soc. Am. Bull.* 82:3299–3312
18. Dewey, J. F., Bird, J. M. 1970. Mountain belts and the new global tectonics. *J. Geophys. Res.* 75:2625–47
19. Dickinson, W. R. 1971. Clastic sedimentary sequences deposited in shelf, slope, and trough settings between magmatic arcs and associated trenches. *Pac. Geol.* 3:15–30
20. Dickinson, W. R. 1971. Plate tectonic models of geosynclines. *Earth Planet. Sci. Lett.* 10:165–74
21. Dickinson, W. R. 1973. Widths of modern arc-trench gaps proportional to past duration of igneous activity in associated magmatic arcs. *J. Geophys. Res.* 78:3376–89
22. Dickinson, W. R., Hatherton, T. 1967. Andesitic volcanism and seismicity around the Pacific. *Science* 157:801
23. Edgar, N. T., Ewing, J., Hennion, J. 1971. Seismic refraction and reflection in the Caribbean Sea. *Am. Assoc. Petrol. Geol. Bull.* 55:833–70
24. Ewart, A., Green, D. C., Carmichael, I. S. E., Brown, F. H. 1971. Voluminous low temperature rhyolitic magmas in New Zealand. *Contrib. Mineral. Petrol.* 33:128–44
25. Fischer, A. G. et al 1971. *Initial Reports of the Deep Sea Drilling Project.* Washington, DC: GPO 6. 1329 pp.
26. Fisher, R. L., Engel, C. G. 1969. Ultramafic and basaltic rocks dredged from the nearshore flank of the Tonga Trench. *Geol. Soc. Am. Bull.* 80:1373–78
27. Fitch, T. J. 1972. Plate convergence, transcurrent faults and internal deformation adjacent to the southeast Asia and the western Pacific. *J. Geophys. Res.* 77:4432–60
28. Fitch, T. J., Scholz, C. H. 1971. Mechanism of underthrusting in southwest Japan: A model of convergent plate interactions. *J. Geophys. Res.* 76: 7260–92
29. Griffen, J. J., Koide, M., Hohndorf, A., Hawkins, J. W., Goldberg, E. D. 1972. Sediments of the Lau Basin—rapidly accumulating volcanic deposits. *Deep Sea Res.* 19:133–38
30. Grow, J. A. 1973. Crustal and upper mantle structure of the central Aleutian Arc. *Geol. Soc. Am. Bull.* 84:2169–92
31. Hamilton, E. L. 1971. Elastic properties of marine sediments. *J. Geophys. Res.* 76:579–604
32. Hamilton, R. M., Gale, A. W. 1968. Seismicity and structure of North Island, New Zealand. *J. Geophys. Res.* 73: 3859–76
33. Hart, S. R., Glassley, W. E., Karig, D. E. 1972. Basalts and sea floor spreading behind the Mariana island arc. *Earth Planet. Sci. Lett.* 15:12–18
34. Hasebe, K., Fujii, N., Uyeda, S. 1970. Thermal process under island arcs. *Tectonophysics* 10:335–55
35. Hawkins, J. W., Fisher, R. L., Engel, C. G. 1972. Ultramafic and mafic rock suites exposed on the deep flanks of Tonga trench. *Geol. Soc. Am. Abstr. with Programs* 4:167
36. Herron, E. M. 1972. Sea floor spreading and the Cenozoic history of the ea[illegible] central Pacific. *Geol. Soc. Am. Bull.* 1671–92
37. Hilde, T. W. C., Wageman, J. M Hammond, W. T. 1969. The structur of Tosa terrace and Nankai trough off southeastern Japan. *Deep Sea Res.* 16: 67–76
38. Homes, A. 1965. *Principles of Physical Geology.* London: Thomas Nelson. 1288 pp.
39. Hoshino, M. 1969. On the deep sea terrace. *La Mer* 3:222–24
40. Hsu, T. L. 1962. Recent faulting in the longitudinal valley of eastern Taiwan. *Mem. Geol. Soc. China* 1:95–102
41. Isezaki, N., Hata, K., Uyeda, S. 1971. Magnetic survey of the Japan Sea (Part I). *Bull. Earthquake Res. Inst. Univ. Tokyo* 49:77–83
42. James, D. E. 1971. Plate tectonic model for the evolution of the central Andes. *Geol. Soc. Am. Bull.* 82:3325–46
43. Johnson, T., Molnar, P. 1972. Focal mechanisms and plate tectonics of the southwest Pacific. *J. Geophys. Res.* 77: 5000–32

44. Kaneko, S. 1966. Transcurrent displacement along the Median Line, southwest Japan. *N. Z. J. Geol. Geophys.* 9:45–59
45. Kanimori, H. 1971. Great earthquakes at island arcs and the lithosphere. *Tectonophysics* 12:187–98
46. Kanimori, H., Abe, K. 1968. Deep structure of island arcs as revealed by surface waves. *Bull. Earthquake Res. Inst. Tokyo Univ.* 46:1001–25
47. Karig, D. E. 1970. Ridges and basins of the Tonga-Kermadec Island arc system. *J. Geophys. Res.* 75:239–55
48. Karig, D. E. 1971. Structural history of the Mariana island arc system. *Geol. Soc. Am. Bull.* 82:323–44
49. Karig, D. E. 1971. Origin and development of marginal basins in the western Pacific. *J. Geophys. Res.* 76:2542–61
50. Karig, D. E. 1972. Remnant arcs. *Geol. Soc. Am. Bull.* 83:1057–68
51. Karig, D. E. Comparison of island arc-marginal basin complexes in the north-west and south-west Pacific. *The Western Pacific: Island Arcs, Marginal Seas, Geochemistry,* ed. P. J. Coleman. Univ. West Australia Press. In press
52. Karig, D. E. 1973. Plate convergence between the Philippines and the Ryukyu Islands. *Mar. Geol.* 14:153–68
53. Karig, D. E. Tectonic erosion in trenches. *Earth Planet. Sci. Lett.* In press
54. Karig, D. E., Glassley, W. E. 1970. Dacite and related sediment from the West Mariana Ridge, Philippine Sea. *Geol. Soc. Am. Bull.* 81:2143–46
55. Karig, D. E., Jensky, W. 1972. The proto-Gulf of California. *Earth Planet. Sci. Lett.* 17:169–74
56. Karig, D. E., Mammerickx, J. 1972. Tectonic framework of the New Hebrides island arc system. *Mar. Geol.* 12:187–205
57. Karig, D. E., Le Pichon, X., Sharman, G. Accretion and subduction in Pacific trenches. *Geol. Soc. Am. Bull.* In press
58. Kaseno, Y. 1972. On the origin of the Japan Sea basin. *Proc. Int. Geol. Congr. 24th,* Sect. 8:37–42
59. Katili, J. A. 1970. Large transcurrent faults in southeast Asia with special reference to Indonesia. *Geol. Rundsch.* 59:581–600
60. Kienle, J. 1971. Gravity and magnetic measurements over Bowers Ridge and Shirshov Ridge, Bering Sea. *J. Geophys. Res.* 76:7138–80
61. Kroenke, L. W. 1972. Geology of the Ontong Java Plateau. *Hawaii Inst. Geophys. Publ.* HIG: 72–75, 119 pp.
62. Lahr, J. C., Page, R. A. 1972. Hypocentral locations in the Cook Inlet region of Alaska. *Trans. Am. Geophys. Union* 53:1042
63. Landis, C. A., Bishop, D. G. 1972. Plate tectonic and regional stratigraphic-metamorphic relations in the southern part of the New Zealand geosyncline. *Geol. Soc. Am. Bull.* 83:2267–84
64. Lipman, P. W., Prostka, H. J., Christiansen, R. L. 1971. Evolving subduction zones in the western United States, as interpreted from igneous rocks. *Science* 174:821–25
65. Ludwig, W. J. 1970. The Manila Trench and West Luzon Trough-III: Seismic refraction measurement. *Deep Sea Res.* 17:553–71
66. Ludwig, W. J. et al 1966. Sediments and structure of the Japan Trench. *J. Geophys. Res.* 71:2121–37
67. Ludwig, W. J., Hayes, D. E., Ewing, J. I. 1967. The Manila Trench and West Luzon Trough-I: Bathymetry and sediment distribution. *Deep Sea Res.* 14: 533–44
68. Luyendyk, B. P., Bryan, W. B., Jezek, P. A. Shallow structure of the New Hebrides island arc. *Geol. Soc. Am. Bull.* In press
69. Luyendyk, B. P., MacDonald, K. C., Bryan, W. B. 1973. Rifting history of the Woodlark Basin in the southwest Pacific. *Geol. Soc. Am. Bull.* 84:1125–34
70. McKee, E. H. 1971. Tertiary igneous chronology of the Great Basin of western United States—implications for tectonic models. *Geol. Soc. Am. Bull.* 82:3497–3502
71. McKenzie, D. P. 1969. Speculations on the consequences and causes of plate motions. *Geophys. J.* 18:1–32
72. Matsuda, T., Uyeda, S. 1971. On the Pacific type orogeny and its model-extension of the paired belts concept and possible origin of marginal seas. *Tectonophysics* 11:5–27
73. Menard, H. W. 1967. Transitional types of crust under small ocean basins. *J. Geophys. Res.* 72:3061–73
74. Meyerhoff, A. A., Meyerhoff, H. A., Briggs, R. S. Jr. 1972. Continental drift, V: Proposed hypothesis of earth tectonics. *J. Geol.* 80:663–92
75. Mitchell, A. H. G., Reading, H. G. 1971. Evolution of island arcs. *J. Geol.* 79: 253–84
76. Mitchell, A. H. G., Warden, A. J. 1971. Geological evolution of the New Hebrides island arc. *J. Geol. Soc.* 127: 501–29
77. Mitronovas, W., Isacks, B., Seeber, L.

1969. Earthquake locations and seismic wave propagation in the upper 250 km of the Tonga island arc. *Bull. Seismol. Soc. Am.* 59:1115–37
78. Miyashiro, A. 1972. Metamorphism and related magmatism in plate tectonics. *Am. J. Sci.* 272:629–56
79. Molnar, P., Oliver, J. 1969. Lateral variations of attenuation in the upper mantle and discontinuities in the lithosphere. *J. Geophys. Res.* 74:2648–82
80. Moore, J. C. 1972. Uplifted trench sediments: Southwestern Alaska-Bering shelf edge. *Science* 175:1103–4
81. Moore, J. C. 1973. Complex deformation of Cretaceous trench deposits, southwestern Alaska. *Geol. Soc. Am. Bull.* 84: 2005–20
82. Murauchi, S. 1971. The renewal of island arcs and the tectonics of marginal seas. *Island Arcs and Marginal Seas,* 39–56. Tokyo: Tokai Univ. Press
83. Murauchi, S. et al 1968. Crustal structure of the Philippine Sea. *J. Geophys. Res.* 73:3143–71
84. Noble, D. C. 1973. Tertiary pyroclastic rocks of the Peruvian Andes and their relation to lava vulcanism, batholith emplacement, and regional tectonism. *Geol. Soc. Am. Abstr. with Programs* 5:86
85. Oxburgh, E. R., Turcotte, D. L. 1971. Origin of paired metamorphic belts and crustal dilation in island arc regions. *J. Geophys. Res.* 76:1315–27
86. Packham, G. H., Falvey, D. A. 1971. An hypothesis for the formation of marginal seas in the western Pacific. *Tectonophysics* 11:79–110
87. Petelin, V. P. 1964. Hard rocks in the deep-water trenches of the south-western part of the Pacific Ocean. *Int. Geol. Congr. 22nd, Reports of Soviet Geologists* 16:78–86
88. Pichler, H., Ziel, W. 1969. Andesites of the Chilean Andes. *Proc. Andesite Conf. Oreg. Dep. Geol. Miner. Ind. Bull.* 65: 165–74
89. Plafker, G. 1972. Alaskan earthquake of 1964 and Chilean earthquake of 1960: Implications for arc tectonics. *J. Geophys. Res.* 77:901–25
90. Raitt, R. W. 1967. Marine seismic studies of the Indonesian island arc. *Trans. Am. Geophys. Union* 48:217
91. Ross, D. A., Shor, G. G. Jr. 1965. Reflection profiles across the Middle America Trench. *J. Geophys. Res.* 70: 5551–72
92. Rutland, R. W. R. 1971. Andean orogeny and ocean floor spreading. *Nature* 233: 252–55
93. Scholl, D. W. et al 1971. Deep Sea Drilling Project Leg 19. *Geotimes* 16: 12–15
94. Scholl, D. W., Buffington, E. C., Marlowe, M. S. Plate tectonics and the structural evolution of the Aleutian-Bering Sea region. *Geol. Soc. Am. Mem.* In press
95. Scholz, C. H., Barazangi, M., Sbar, M. L. 1971. Late Cenozoic evolution of the Great Basin, western United States, as an ensialic inter-arc basin. *Geol. Soc. Am. Bull.* 82:2979–90
96. Sclater, J. G. 1972. Heat flow and elevation of the marginal basins of the western Pacific. *J. Geophys. Res.* 77: 5705–19
97. Sclater, J. G., Hawkins, J. W. Jr., Mammerickx, J., Chase, C. B. 1972. Crustal extension between the Tonga and Lau ridges: petrologic and geophysical evidence. *Geol. Soc. Am. Bull.* 83:505–18
98. Shor, G. G. 1964. Structure of the Bering Sea and the Gulf of Alaska. *Mar. Geol.* 1:213–19
99. Silver, E. A. 1971. Transitional tectonics and late Cenozoic structure of the continental margin off northernmost California. *Geol. Soc. Am. Bull.* 82:1–22
100. Silver, E. A. 1972. Pleistocene tectonic accretion of the continental slope off Washington. *Mar. Geol.* 13:239–49
101. Sleep, N., Toksöz, M. N. 1971. Evolution of marginal basins. *Nature* 233:548–50
102. Suess, E. 1888. *The Face of the Earth (Das Antlitz der Erde),* transl. H. B. C. Sollas, published 1906. Oxford: Clarendon. 575 pp.
103. Talwani, M. 1970. Gravity. *The Sea* 4:251–98
104. Talwani, M., Worzel, J. L., Ewing, M. 1961. Gravity anomalies and crustal section across the Tonga Trench. *J. Geophys. Res.* 66:1265–68
105. Umbgrove, J. H. F. 1947. *The Pulse of the Earth.* The Hague: Nijhoff. 358 pp.
106. Uyeda, S., Ben Avraham, Z. 1972. Origin and development of the Philippine Sea. *Nature Phys. Sci.* 240: 176–78
107. Vening Meinesz, F. A. 1951. A third arc in many island arc areas. *Kon. Ned. Akad. Wetensch. Proc. Ser. B* 54: 432–42
108. Vening Meinesz, F. A. 1964. *The Earth's Crust and Mantle.* Amsterdam: Elsevier. 124 pp.
109. von Huene, R. 1972. Structure of the continental margin and tectonism of

the eastern Aleutian Trench. *Geol. Soc. Am. Bull.* 83:3613–26

110. von Huene, R. et al 1971. Deep Sea Drilling Project Leg. 18. *Geotimes* 16: 12–15
111. Wageman, J. M., Hilde, T. W. C., Emery, K. O. 1970. Structural framework of the East China Sea and Yellow Sea. *Am. Assoc. Petrol. Geol. Bull.* 54: 161–64
112. Westerveld, J. 1952. Quaternary vulcanism on Sumatra. *Geol. Soc. Am. Bull.* 63:561–94
113. Wilson, J. T., Burke, K. 1972. Two types of mountain building. *Nature* 239:448–49
114. Wyllie, P. J. 1971. Role of water in magma generation and initiation of diapiric uprise in the mantle. *J. Geophys. Res.* 76:1328–38

GROWTH LINES IN INVERTEBRATE SKELETONS

George R. Clark II[1]

Department of Geology, University of New Mexico, Albuquerque, New Mexico 87106

. . . here are the years of their growth,
numbered on their shells . . .

Leonardo da Vinci (34)

INTRODUCTION

As will presently be shown, growth lines are in some ways a record of the environment in which they form; such records have received only limited attention in biological studies, but now appear to have unique applications in geology.

Until recently, features of fossils which reflected environmental conditions were avoided rather than sought. Classical paleontology is primarily concerned with the classification of fossil organisms and the interpretation of their evolution, and only genetically determined features will serve these ends. In contrast, the relatively new field of paleoecology is directly concerned with the environments of the past, and growth line records could become invaluable tools in this research.

Growth lines can also be of service in the field of planetary dynamics. The periods most commonly preserved in growth line sequences—days, tides, lunar months, and years—are all expressions of planetary motions. Fossils with such growth lines have preserved a valuable astronomical record reaching far back into geologic history.

Despite these potential applications, growth lines are not yet well enough understood to be interpreted with complete confidence. They are complex phenomena, and research on their environmental relationships is still in its infancy. This, of course, is what makes growth line research an exciting and enjoyable occupation.

GROWTH LINES AS SKELETAL FEATURES

Definition and Expression

Growth lines can be briefly defined as abrupt or repetitive changes in the character of an accreting tissue.[2] Accreting tissues are tissues, commonly skeletal, forming in

[1] This article is based largely on research supported by the National Science Foundation, Grants GB-6275, GB-20692, and GA-33494.

[2] "Tissue" is used here in a broad sense and includes noncellular hard parts of organisms.

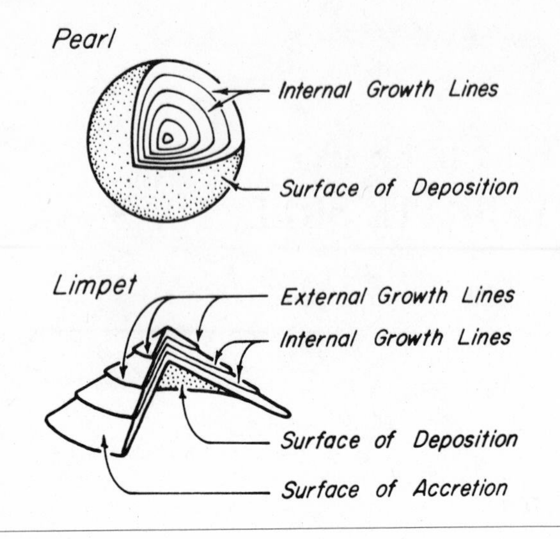

Figure 1 Two examples of accretive tissue illustrating the major features associated with their growth.

sequential layers (growth increments) along a *surface of deposition.* Two examples of accretive tissues are the woody tissues (xylem) of a tree and the molluscan pearl; in these examples the surface of deposition is essentially continuous, so that the last-formed layer conceals all earlier growth increments. In two other examples of accretive tissues, the skeleton of a coral and the shell of a limpet, the surface of deposition is not continuous, but is limited to the portion of the skeleton (or shell) in contact with the living tissues. Beyond this point the surface of the skeleton consist of a succession of margins of growth layers, and can be termed the *surface of accretion.* Figure 1 illustrates these concepts.

It follows that there are two ways to observe growth lines. Lines visible on a surface of accretion, called *external growth lines,* can be seen directly and can often be studied without disturbing the living organism. Lines visible only after sectioning or otherwise exposing a sequence of growth layers within the skeleton are called *internal growth lines,* and can normally be studied only after the death of the animal. These, too, are shown in Figure 1.

Resolution of Lines

We have defined growth lines above as abrupt or repetitive changes in the character of accreting tissue. There is no stipulation here that these changes be obvious, only that they be present. It is the task of the investigator to resolve them, by whatever means he may.

External growth lines are commonly found as differences in pigmentation, relief, or reflectivity, or as changes in direction of growth. They are usually readily apparent, but may require special treatment such as removing organic layers, or illuminating with certain wavelengths of light.

To distinguish internal growth lines usually requires some effort. In some cases nondestructive techniques can be used, such as holding a bivalve shell in front of a strong light; a more common approach, and the best for small-scale growth lines, is to section the skeleton and study thin sections or etched peels.

Other techniques which have received limited attention are electron microprobe and laser microprobe traverses, and isotopic analysis of sequential increments. In some cases these approaches can detect growth lines not visible in the ordinary sense.

Taxonomic Distribution

As suggested above, growth lines may be found in any accretive tissues. This independence of taxonomic position is reflected in the widespread distribution of growth lines in living organisms. Forms with robust accretive skeletons, such as mollusks, corals, and brachiopods have received the most attention, but growth lines are prominent features in many others, including echinoderms (56, 64), annelid worms (29, 44), and even those aberrant arthropods, the barnacles (54). Vertebrates commonly form growth lines on specialized skeletal elements, such as scales, otoliths, and teeth, and the higher plants are known for a variety of growth lines, including the familiar annual rings of trees. Even blue-green algae, single celled and lacking a skeleton, can form a permanent structure with prominent growth lines, the stromatolite.

GROWTH LINES AS RECORDS OF THE ENVIRONMENT

Influence of the Environment on Growth

It has been observed that it is the business of organisms to stay alive until they have reproduced themselves (3). A corollary to this is that an organism cannot achieve this goal without growth. A typical metazoan grows rapidly until it achieves maturity, whereupon reproduction becomes the dominant concern and growth slows. Aside from this, variations in growth rate are generally caused by environmental conditions.

For example, a physical injury to a growing shell will simply and effectively limit growth, for the organism must devote its energies to repair before continuing growth. A less direct effect is the limitation placed on metabolic activity, and therefore growth, by extremes in temperature; Gunter (24) has discussed this and more subtle effects of temperature in some detail. Similar effects can be traced to nearly every conceivable environmental variable; Hedgpeth (28), Moore (43), and Hallam (27) have excellent discussions on this topic.

The Environment as a Source of Stimuli

GENERAL Because growth is strongly influenced by environmental conditions, variations in growth, or growth lines, are likely to be the result of variations in environmental conditions. If so, certain characteristics of the growth lines should reflect these environmental variations.

Perhaps the most important features common to growth lines and their environmental stimuli are the pattern and frequency of their occurrence. Patterns of environmental stimuli can be broadly divided into the random and the periodic; growth lines caused by random events are called *disturbance lines,* while those resulting from periodic events can be called *periodic lines* (or by their frequency, such as annual or daily lines).

RANDOM EVENTS Perhaps the most common random event to produce disturbance lines is a storm. In a shallow marine environment a severe storm can cause major changes in light intensity, wave action, turbidity, substrate stability, temperature, and salinity, any one of which could interrupt growth sufficiently to cause a disturbance line.

A very different sort of random event, but one quite capable of stimulating the formation of a disturbance line, is an unsuccessful attack by a predator. The disturbance line can be directly caused by damage inflicted by the predator, or indirectly caused by an enforced period of inactivity, such as occurs when a bivalve keeps its shell closed in the presence of a predator.

A third sort of event capable of producing a disturbance line is a period of slightly abnormal environmental conditions, such as high summer temperatures. This may be enough to exceed the average tolerance of a population, and individually result in death or disturbance lines (or in some cases, no effects at all).

The disturbance lines produced in populations subjected to these sorts of events can be readily distinguished, even in some fossil assemblages. In the first example, correlative disturbance lines would be found on all specimens living in the shallow water area affected by the storm. In the second, disturbance lines would be found at random positions on the shells of random members of the population. In the third, correlative disturbance lines would be found on random members of the population.

Following Pannella & MacClintock (51), events producing simultaneous disturbance lines in a large part of a population are called *universal events,* while events producing disturbance lines at different times in individual members of a population are called *private events.*

PERIODIC EVENTS Periodic events seem to affect the growth process in more ways than random events. Some periodic lines, such as the sharp annual lines found on many bivalves, appear to be little more than a reaction to an environmental disturbance (such as winter cold) which happens to be periodic rather than random. Other periodic lines, however, have been shown to be associated with biological rhythms (10, 11, 41) which suggests that the environmental variations are much less directly involved; this means that virtually any periodicity which could be sensed by an organism could conceivably regulate the formation of growth lines.

It follows that an understanding of naturally occurring periodicities is important to the interpretation of growth lines. These periodicities will now be discussed in some detail.

The longest reasonable period commonly associated with growth lines is the *year*. In one possible mechanism, annual environmental extremes (such as cold temperatures) could slow or stop the growth and thus cause a line. In another, the line could be a reflection of slow growth during annual spawning which can be in turn triggered by the spring rise (or autumn decline) in temperature. It should be noted that in either case the mechanism is annual in the sense of happening once a year, and not in the sense of occurring at intervals of 365 solar days.

Aside from one of the possible tidal periodicities which will be discussed collectively later, the next longest reasonable period is that of the *synodic month*. This is the time between consecutive new moons, and can be measured as 29.53 solar days or 28.51 lunar days. The specific environmental variable involved would seem to be illumination from the full moon, which apparently triggers spawning in a wide variety of marine animals.

The sun is certainly a major factor in the marine environment, so we would expect to find the *solar day* to be an important period in the lives of marine organisms. There are numerous environmental variables associated with the sun: light, for photosynthesis and visual stimulation; temperature, especially in shallow waters; oxygen levels, varying with photosynthetic activity; and food availability, varying with plankton migration and the nocturnal habits of much of the shallow benthos.

Tides are significant factors in shallow waters. They are of course vitally important to the abundant forms of life in the intertidal zone, but also affect the subtidal environment, by creating currents, carrying food, varying the light intensity (by varying water depth), and in some extreme cases, such as at the mouth of estuaries, subjecting organisms to completely different water masses, ocean and estuary, with each reversal. Tides have a number of periodicities with widely ranging causes and effects. As some of these are quite complex, we will discuss them at some length.

Tides can be considered a byproduct of the dynamic forces involved in the relative motions of the earth, moon, and sun. Because these motions and their resultant tidal forces are essentially confined to a single plane (the ecliptic), the geometry of the situation is relatively simple (Figure 2). The Earth's oceans respond to the gravitational attraction of the sun and moon by forming bulges at the points nearest and farthest from the attracting body; the resulting shape of the oceans' surface approximates an ellipsoid. Both sun and moon produce tidal ellipsoids on the earth, but the moon's influence is so much greater than the sun's that only one is obvious. Since the sun and moon have apparent motions of different periods (thus producing the phases of the moon), their tidal attractions will at times reinforce and at times oppose one another, such that the resultant tidal ellipsoid will vary in shape.

A major complication in this system is that the earth also has a motion of its own, that of rotation, with its axis of rotation inclined about 67 degrees from the ecliptic. Thus an observer on the earth is concerned with two planes, a plane of rotation and the plane of the ecliptic. For an observer at the equator, these two planes intersect at a line which also passes through the center of the earth.

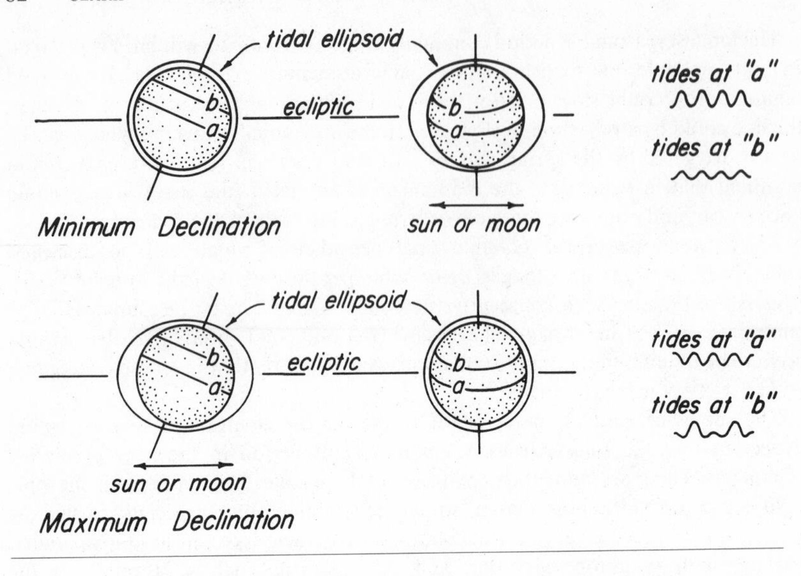

Figure 2 The influence of declination and latitude on tides.

When the sun or moon is on this line, it is said to be at minimum declination, and when it is on a line perpendicular to this line, it is at maximum declination (Figure 2).

Figure 2 also illustrates how both declination and latitude affect the tidal cycles. Latitude is of course a constant for a particular site, but declination is periodic. Thus, for a given site on the equator ("a" on Figure 2) the semidaily tides will tend to be of the same magnitude, but the magnitude will vary periodically with the declination. For a given site off the equator ("b" on Figure 2) the semidaily tides will vary periodically in both relative and absolute magnitude. These effects are further complicated by reinforcement and opposition of the declinations of the sun and moon.

In practice, of course, tidal motions are not so easily explained. The earth is not a uniform sphere covered with a thin layer of water, and the shape and size of individual bodies of water have a great effect on the tides they produce. For example, the Atlantic coast of North America is characterized by strong semidaily tides of nearly equal magnitude, the Pacific coast by strong unequal semidaily tides, and the Gulf coast by unequal semidaily or even daily (vanishing) tides. Thus it would not be a simple matter to estimate the paleolatitude of a fossil from the inequality of a semidaily periodicity observed in its growth lines.

An additional complication lies in the manner in which tidal motions act as stimuli. When considering abrupt growth lines, it should be noted that organisms living intertidally are dealing directly with the environment. They are likely to be forming lines in direct response to stimuli and are unlikely to record the times that a high tide *almost* reached them. In contrast, the long term periodicities involved in fortnightly or monthly variations are likely to leave a complete record.

There are a surprising number of periodicities associated with tides. The types discussed below do not make a complete list, but do include all those likely to be preserved in growth lines.

Semidaily tides are produced by the rotation of the earth past the tidal bulges (the ends of the tidal ellipsoid) produced by the moon and sun, with a period of 12.42 hr (1/2 lunar day), between successive high or low tides.

Daily tides are really better described as unequal semidaily tides, although if the inequality is severe enough there will indeed be only one high and low tide a day. The inequality is a function of latitude and the declination of the sun and moon. Unlike semidaily tides, the period may vary, for if the sun is at minimum declination and the moon at maximum, the moon's period of 24.84 hr will prevail, while during the opposite situation the sun's period (24.00 hr) will dominate. Also unlike semidaily tides, there will be times during which neither sun nor moon will have much effect; thus it is unlikely that long sequences of daily tidal lines will occur.

Fortnightly tides can occur in two ways. In one, the periodic coincidence and opposition of the sun and moon brings about an elongation and shortening of the tidal ellipsoid, causing variations in the heights of the semidaily tides. The largest tidal ranges are called spring tides and the smallest, neap tides. This cycle has a period of 14.8 solar or 14.3 lunar days (half a synodic month), although in practice the most extreme tides may be from 12 to 17 days apart. The second type of fortnightly tides, the tropic tides, involve the diurnal inequality of the moon. The inequality is greatest during maximum declination, and since the moon is at maximum declination twice during each siderial month the periodicity is 13.7 solar days or 13.2 lunar days.[3]

There are no strong *monthly* periodicities associated with tides, but Farrow (18) has proposed a very weak one as the cause of an observed series of growth lines.

One *semiannual* cycle with an important effect is the periodicity of the solar tropic tides, the diurnal inequality effect due to the declination of the sun. The sun is at maximum declination in June and December, and when this effect is combined with maximum declination of the moon, which of course also coincides with full or new moon and therefore spring tides, the tidal range exceeds any other. This can put a severe environmental stress on organisms living low in the tide zone.

Growth Line Investigations

GENERAL It is one thing to know that variations in the environment are capable of causing variations in a growing shell, but quite another to establish the relationship between the two.

For many purposes it is enough to identify the periodicity at which the growth lines are forming. This can usually be accomplished in the case of living organisms,

[3] This is really only half of a monthly cycle with respect to either of the semidaily tides, for each has a turn at being the largest; but as it is unlikely that this should be of concern to an intertidal organism, it is more realistic to consider it a fortnightly cycle.

and with reasonable care these results can be applied to related fossil groups. The most common difficulties encountered in studies of this sort are the random errors introduced by disturbance lines and the close similarities existing between some natural periodicities, such as solar and lunar days.

In some investigations it is necessary to identify the particular environmental variable causing the growth lines. This can be a difficult process, for the natural environment includes numerous variables, many with the same periodicity and some with complex interrelationships. For example, growth lines found to occur with 24-hr periodicity are almost certainly related to solar illumination, but elaborate experiments or fortuitous circumstances are usually necessary to determine whether this stimulus acts directly upon the organism, or indirectly through other stimuli such as temperature, plankton migration, or oxygen production by plants. An additional complication lies in the fact that some periodic growth lines have characteristics of biological rhythms, and continue to form in the absence of any stimulus; elaborate experimental controls are required to compensate for this phenomenon.

APPROACHES TO DETERMINING PERIODICITY The simplest but least convincing evidence for a particular periodicity is that growth lines occur with regular spacings and in numbers consistent with such subdivisions of the organism's lifespan (see Figure 3). This approach has its main value in the elimination of unlikely

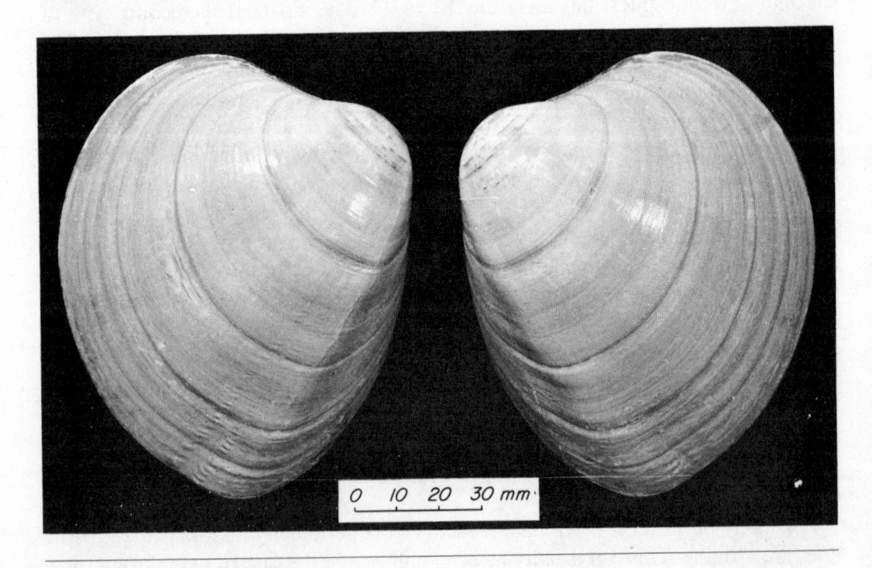

Figure 3 Shell of an adult *Megapitaria aurantiaca* showing four prominent growth lines. The number and spacing of the lines are suggestive of annual periodicity, with rapid growth during the first two years and a decline in growth rate with sexual maturity and age. The specimen is from the northern Gulf of California.

periodicities, and is usually applied intuitively in every study. In some studies, as in examining a single fossil specimen, it may be a significant fraction of the information available, but the weakness of the argument must be recognized.

Next in order of easily obtained evidence is that of numerologic relationships; an example is Wells' (66) observation of about 400 fine growth lines between adjoining major growth lines in Middle Devonian Corals. Such evidence may be strongly suggestive but lacks precision; in Wells' corals, for example, the major lines might represent the onset of winter, a variable phenomenon, and the fine lines might represent solar days (365 in a modern year) or lunar days (353 in a modern year).

The last type of evidence for periodicity which can be applied to fossil studies requires a population which experiences a universal event, causing catastrophic death or a distinctive disturbance line. If the individuals in the population are also forming periodic growth lines, the growth interval between the most recent periodic line and the record of the event should be the same in the entire population. This evidence, of course, does not identify the periodicity; it merely confirms that one is present.

Studies of living populations can utilize the types of evidence mentioned above, but much more convincing types of evidence are obtainable. One very simple approach is that of sampling the living population at intervals within the period under study, recording the amount of growth beyond the last growth line. If there is only one time during the period when no new growth is observed, then the periodicity *and* the time of formation is established. This has been successfully applied to many studies of annual growth lines (16, 25, 38, 42, 67) and to one study of monthly lines (8).

Another supposedly simple approach involves the observation of individual organisms over long periods (relative to the periodicity under study) by such means as tagging, marking, caging, or similar isolation. This has the advantage of requiring only two actions, tagging and recovering, rather than many samplings, but has the disadvantage that recovery of tagged specimens is sometimes difficult.[4] Another supposed advantage is the intuitive satisfaction with data from individuals rather than population averages; yet these data are not necessarily more reliable, as specimens which have been captured, marked, and released often grow poorly for a period thereafter. Moreover, this procedure commonly leaves a disturbance line, as will be discussed more fully later. Gibson (22) did not recognize this, and drew some unwarranted conclusions which he later corrected (23). Growth experiments conducted under laboratory conditions are prone to similar difficulties. In an experiment conducted with 12 juvenile *Pecten diegensis,* Clark (9) found only two specimens to form one growth line each day, with the others undergoing growth stoppages for as long as ten days; if there had not been alternative means of determining the growth rate, it would have been impossible to prove that the lines were daily. One final problem with this approach is that of resolution. An experiment in which an organism is marked, released, and recovered a year later

[4] Weymouth (67) released over 470 marked specimens of *Tivela,* but recovered none.

and found to have added one growth line does not prove the line to be annual. It is usually so construed because the evidence does rule out any other common periodicity. This is not the case with shorter periodicity growth lines. Although it is commonly recognized that evidence of one line being added in one day is insufficient to prove the line has a 24-hr periodicity, it is possible to overlook the fact that at least 29 days (a lunar month) are required to satisfactorily distinguish between solar (24.00 hr) and lunar (24.84 hr) daily periodicities. In spite of such problems, tagging experiments have enjoyed considerable success in demonstrating annual (12, 20, 38, 40, 42, 49), daily (2, 9–11, 14, 21, 51, 53, 57), and monthly (8) growth lines.

APPROACHES TO DETERMINING ENVIRONMENTAL STIMULI Much of the direct evidence for environmental stimuli of growth lines exists in observations of disturbance lines. Such evidence not only demonstrates this effect but also has considerable application in studies of annual growth lines, which usually are little more than periodic disturbance lines.

Probably the earliest observations on the causes of disturbance lines were those of shellfish biologists, who found that the subjects of their growth experiments formed disturbance lines each time they were handled for measurements, transplanting, or marking (4, 12, 31, 45, 48, 49, 67). It was a simple extrapolation to recognize that similar disturbances in nature, such as predation attempts, could cause such lines.

In a more widely applicable type of experiment, Posgay (55) produced disturbance lines in *Placopecten magellanicus* by abruptly varying water temperature to abnormal values; he also noted that the animals differed in the degree of their response, some forming a deep line and others only a faint trace. Similarly Shuster (61) produced disturbance lines in *Mya arenaria* by subjecting the animals to high levels of suspended silt, such as might be encounted in shallow water during storms. Unfortunately neither Posgay nor Shuster provided details of these experiments.

Observational evidence exists to support these experimental data. Shuster (61) correlated disturbance lines in the shells of juvenile *Mya arenaria* from Delaware Bay with summer storms in the area in 1951. House & Farrow (30) reported a similar effect in specimens of *Cardium edule* subjected to autumn gales in 1963, and Farrow showed minor growth disturbances to correlate with subzero air temperatures (17) and extreme low tides (18). Olsen (47) found sudden changes in color banding on shells of *Haliotis rufescens* to correlate with the seasonal availability of different algal food sources.

Less direct, but equally convincing evidence exists in the observation that the degree of exposure to environmental extremes can correlate with the prominence of growth lines (disturbance or periodic). A simple example is the observation of Orton (49) that the act of filing a notch in a shell causes a more prominent disturbance line than the mere handling of it. Of more environmental significance is his observation (48) that shells of *Cardium edule* picked up on beaches exposed to storms show numerous disturbance lines, while shells found at protected sites show none; Weymouth (67) used a similar approach to show that the annual growth rings of the *Tivela stultorum* are due to winter surf action rather than low temperatures. However, Newcombe (46) noted a correlation between annual ring prominence and latitude in *Mya arenaria,* and Davenport (14) found the same effect in species of *Argopecten* [this last is supported by a comparison of other studies of this species (4, 25, 58)]. More recently, Merrill et al (40) noted a negative correlation between annual line prominence and water depth, and inferred that the lines were formed by low temperatures.

Finally, we have the evidence that a particular periodic growth line is produced in conjunction with an entrained biological rhythm. This possibility was suggested by Millar (41) when he observed growth lines of possible daily frequency in larvae of oysters whether taken from plankton, taken from incubating females, or grown in the laboratory under constant conditions of temperature and illumination. This was later demonstrated experimentally by Clark (10, 11), who succeeded in altering the rhythm of formation of daily lines by subjecting specimens of *Pecten diegensis* to artificially shortened lighting regimes. Later experiments by Thompson & Barnwell (63) on *Mercenaria mercenaria* tended to confirm this effect.

Current Status of Growth Line Research

ANNUAL LINES AND MAJOR DISTURBANCE LINES Annual lines are the best established of all growth lines, and in connection with disturbance lines are the best understood. Fishermen have long regarded the prominent lines on commercial bivalves as forming annually—Leonardo da Vinci referred to them as familiar phenomena (34)—but it appears that their validity was not scientifically acceptable until relatively recently.

Belding, for example, who studied several species of commercial bivalves for the state of Massachusetts, demonstrated the validity of the annual growth line in *Argopecten irradians* (4), but recognized only disturbance lines in his studies of *Mactra solidissima* (5), *Mercenaria mercenaria* (6), and *Mya arenaria* (7).

Similarly, Isely (31) recognized that growth lines on freshwater mussels could be caused by a number of disturbances, winter cold among them. He considered them of little use in marking time intervals; in a later paper (32) he figured a mussel recovered alive 15 years after tagging, with prominent growth lines, yet did not mention them in the report.

The turning point of scientific opinion seems to have occurred in the twenties. First, Coker and others (12) showed that growth lines in freshwater mussels are caused by repetitions in the formation of the outer shell layers,[5] and suggested that annual (winter) lines could be distinguished from disturbance lines by their greater number of repetitions. This was followed by Weymouth's classic study on *Tivela stultorum* (67), in which he established the annual nature of the major growth lines by a number of independent techniques.

After this, nearly all investigations on shellfish biology (Weymouth's own papers were notable exceptions) included studies on the validity of growth lines, and most came to the conclusions that annual lines both existed and could be distinguished

[5] Isely also pointed this out in his 1913 paper (31), but credited Coker.

from disturbance lines. Two of the most interesting of these are Craig & Hallam (13) and Merrill et al (40). Later studies by Pannella & MacClintock (51) and Farrow (18), although concerned principally with daily and tidal periodicities, also provided considerable information on the formation of annual lines and disturbance lines.

DAILY GROWTH LINES Daily growth lines have been postulated for at least half a century (4, 67). They were first demonstrated by Davenport (14), who observed that marked specimens of *Argopecten irradians* added four lines in four days, and that young specimens collected about a month and a half after spatfall had about 50 lines.

Unfortunately evidence such as that presented by Davenport is not sufficient to demonstrate the precise periodicity of these lines. As discussed earlier, daily lines may follow either a solar (24.00 hr) or lunar (24.84 hr) periodicity, and under the best of conditions at least one lunar month's growth is required to distinguish them.

Similarly, the data presented by Petersen (53) could be used to support either sort of daily periodicity, for the number of lines counted falls slightly short of the number of either type of days passed. The circumstances suggest solar periodicity, for the animals (*Cardium*) were living subtidally and the tidal range was slight.

Similar observations by Clark (9) of daily growth lines in *Pecten diegensis* were more suggestive of solar periodicity, as some growth line counts closely approximated solar days and exceeded lunar days (Figure 4). As was pointed out in

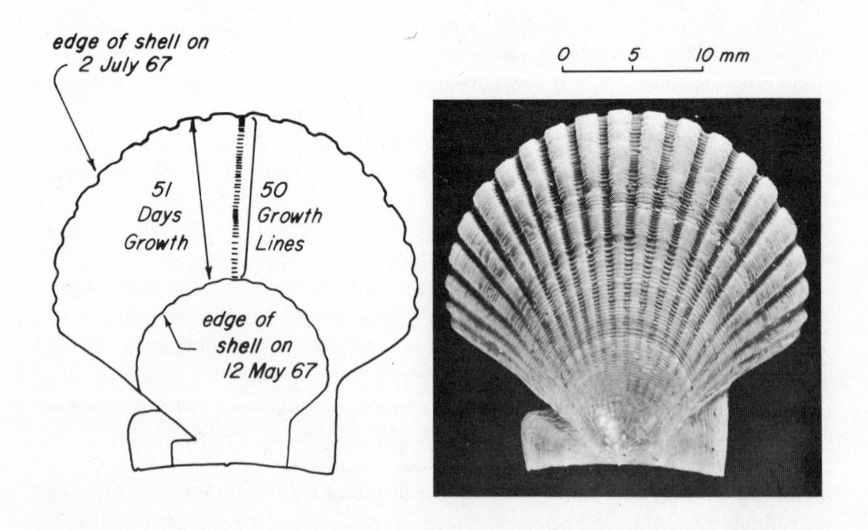

Figure 4 Right valve of *Pecten diegensis*, showing 50 growth lines formed in 51 days. Only two of the twelve subjects in this experiment added 51 lines.

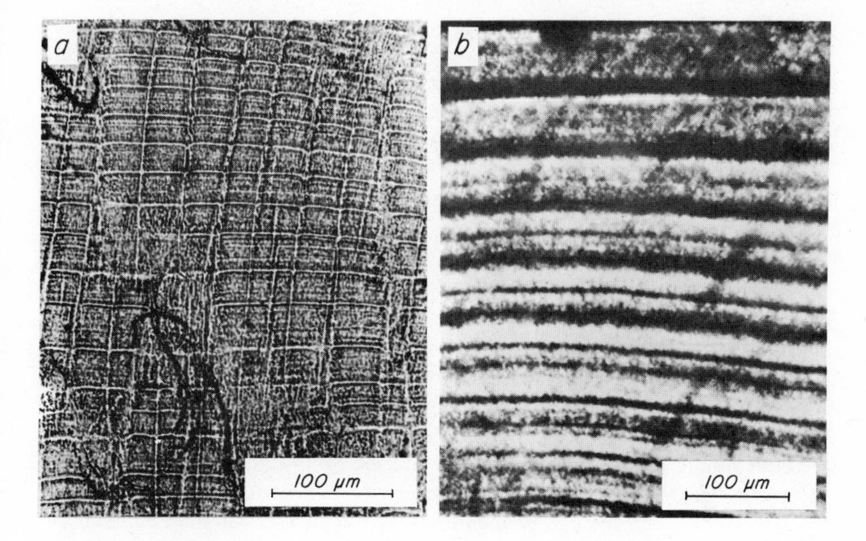

Figure 5 Daily growth lines in *Mercenaria mercenaria* (a; acetate peel) and *Tridacna squamosa* (b; thin section). The direction of growth is toward the top. Photographs of the same material have appeared in the *Journal of Paleontology* (51) and were generously provided by Giorgio Pannella and Copeland MacClintock.

that report, missing lines are more likely than extra lines. Later experiments (10, 11) showed that the growth lines were formed in reponse to an endogenous rhythm stimulated by changes in illumination; thus there can be no question but that these lines follow a solar periodicity.

Gebelein (21) demonstrated a daily periodicity in subtidal stromatolites. Again, his data involved short periods and cannot themselves distinguish between solar and lunar periodicity, but, because a major factor involved in stromatolite formation is algal growth, the solar stimulus appears more likely.

Barnes (2), working on coral skeletons, was able to demonstrate with vital stains that the fine growth lines in *Isophylia* are formed during the day, and are therefore of solar periodicity. This is a logical periodicity for reef-forming corals, for these organisms have symbiotic algae in their tissues whose photosynthetic activity appears to aid the coral's calcification process during daylight. This does not necessarily extend to non–reef-forming corals, which do not possess this symbiote.

Pannella & MacClintock (51), working with acetate peels of sectioned shells of *Mercenaria* and other mollusks, reported the presence of internal daily growth lines (Figure 5). Some sequences of lines were quite distinct and readily counted, but others were confused by secondary lines and further subdivisions. Pannella & MacClintock, in fact, recognized two types of increment, simple and complex, the latter consisting of two increments rather than one (Figure 5b). As might be imagined, the necessity of distinguishing between these two types of increments and of dealing with intermediate cases made the counting of these lines very difficult. In spite of this, they felt that they could count from 360 to 370 daily lines for a 368 day growth period, and from 720 to 725 lines for a 723 day growth period, in specimens notched and left in an intertidal environment. This was in remarkable agreement with the theory that the lines formed with a solar periodicity. However, the authors have since decided, on the basis of further research, that these lines are formed under the influence of tides, and that the daily increments must be of lunar periodicity (Pannella, 50). If so, this would affect the agreement of their counting, for there are about 356 lunar days in 368 solar days, and 699 lunar days in 723 solar days. Unfortunately the details of their further research have not appeared at the time of writing, and the matter remains unresolved.

In a recent report on the internal growth lines of an intertidal bivalve, Evans (15) made an excellent case for their being formed at times of exposure due to low tide (Figure 6). Here, as in the *Mercenaria* of Pannella & MacClintock (51), there are both simple and complex increments, but the complex increments can be seen to be areas of transition where the prominent boundaries of one series of simple increments fade out and the boundaries of the next series of simple increments first appear. Evans compared this phenomenon with the pattern of low tide

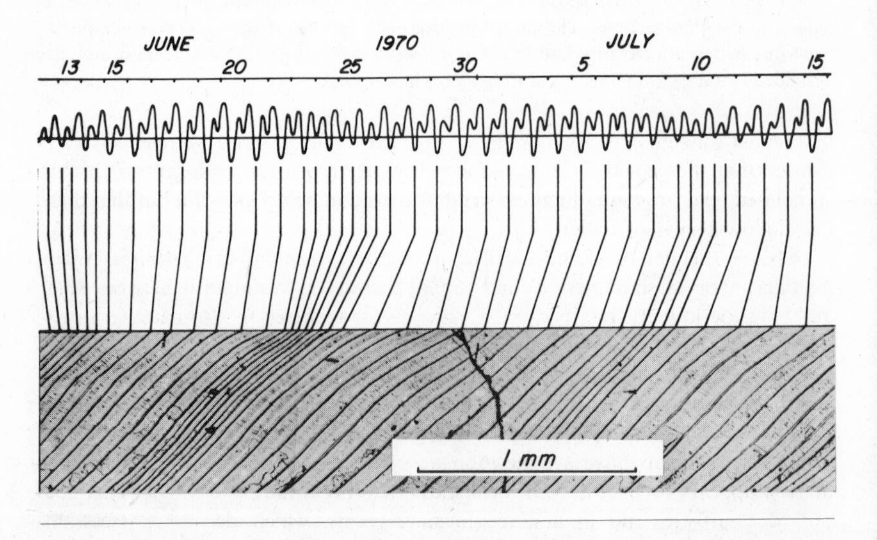

Figure 6 Internal growth lines in *Clinocardium nuttalli* collected at Charleston, Oregon, compared with tidal predictions for the same period for Empire, Oregon. The horizontal line drawn through the tidal curves marks the intertidal position at which the specimens were found. Daily, semidaily, and fortnightly tidal periodicities can be clearly seen and compared. Slightly modified from a figure which appeared in *Science* (176:716), April 28, 1972, copyright 1972 by the American Association for the Advancement of Science, and which was kindly made available by John W. Evans.

exposure characteristic of the mixed diurnal tides experienced in this area, and found excellent agreement between the tidal cycle and the growth line sequence of a portion of shell believed to have formed at the same time. Evans further pointed out that the daily lines figured by Pannella & MacClintock (51) in *Tridacna* showed this same pattern (Figure 5b), and suggested that these, too, are tidal in origin.

This elegant analysis cannot be directly applied to the question of the daily lines in Pannella & MacClintock's (51) *Mercenaria*. For one thing, there were no obvious shifts from one series of simple increments to another in the intervals of complex increments shown in their figures. Moreover, such shifts would not be expected; the tidal pattern in Massachusetts, where the *Mercenaria* grew, is one of nearly equal semidaily tides, rather than very unequal semidaily tides like that in Oregon, where Evans (15) collected his specimens. If there were to be a correspondence between periods of exposure and growth lines in *Mercenaria*, there should be two lines formed for every lunar day. As this does not appear to have happened, we must look to further explanations, possibly involving combinations of lunar and solar periodicities such as number of exposures during daylight.

Meanwhile, another detailed investigation involving daily growth lines was taking place in England. House & Farrow (30) first reported internal daily lines in *Cardium edule*. They did not report their line counts so it could not be determined whether the periodicity was solar or lunar; a statement that specimens collected late in the day all have the same part of the increment outermost is also ambiguous, for collecting was presumably accomplished at the same tidal stage as well. Later studies by Farrow (17, 18) added much to our understanding of other types of growth lines but did not materially improve on the evidence for daily periodicity.

OTHER PERIODIC LINES Wells (65) working on Devonian corals, noted prominent growth lines in numbers more appropriate to a monthly than to an annual periodicity. He noted further that there were other, less prominent growth lines present, and that their numbers agreed closely with the age in years calculated on the assumption that the prominent lines were lunar months.

Scrutton (59) examined several middle Devonian corals for evidence of periodic growth lines. He noted fine growth ridges similar to those described by Wells (66) as probable daily increments, but was unable to recognize any features strongly suggestive of annual lines. However, he did find that a series of constrictions around the epitheca divided the fine growth lines into regular groupings of about 31 lines each (Figure 7) and suggested that these represent lunar months. In this and a later paper (60) he discussed possible mechanisms involved, and favored the hypothesis that the fine lines represent solar days and the constrictions represent lunar breeding periodicities. Unfortunately, the numerologic relationships cannot help refine the hypothesis because the precise length of the day or month is not known for the middle Devonian (see discussion later).

Berry & Barker (8) studied the formation of growth ridges in *Chione* and interpreted them as fortnightly features related to tidal cycles; unfortunately they did not present their data. In the same paper they also noted that the Cretaceous

Figure 7 External growth lines on a Middle Devonian coral, ? *Heliophyllum* sp., showing fine ridges (a) divided into regular groupings (b) of about 31 lines each. Photographs of the same material have appeared in *Palaeontology* (59) and were graciously provided by Colin T. Scrutton.

bivalve *Idonearca* has three sorts of growth lines; they note from 8 to 12 major bands, each composed of from 24 to 25 ridges, which in turn are composed of clusters of fine growth lines. Here, too, they considered the ridges to be fortnightly features. However, in another Cretaceous bivalve, *Crassatella,* they found similar ridges to be composed of about 29 fine growth lines each, and suggested these to be monthly.

Pannella & MacClintock (51) figured one specimen of *Mercenaria* with prominent ridges consisting of approximately 30 lines each. They considered these to represent monthly periodicity.

Petersen (54) has called attention to the presence of various sorts of growth lines in barnacles, particularly on the scutae. In addition to annual lines, he noted distinct lines which he considered to represent molting.

In addition to growth lines recognizable as discrete skeletal features, there are those consisting only of repetitive sequences of growth lines. Such sequences were noted in the internal structures of corals by Ma (35–37) and Faul (19), and related to annual periodicity. More recently, the study of internal growth lines in bivalves by means of thin section and acetate peel has resulted in the description of a considerable number of periodicities, some with no satisfactory environmental equivalent. A few of these will be described here.

Apparently the first such study was conducted by Barker (1). Barker made thin sections of three genera of bivalves (*Mactra, Mercenaria,* and *Chione*) and examined their structure in considerable detail with light microscopy. As a result of this study, he proposed a hierarchy of five orders of cyclic growth layers; these consisted of: a fifth-order layer, formed of a pair of elemental layers; a fourth-order layer, formed of 4 fifth-order layers; a third-order layer, formed of 12 to 15 fourth-order layers; a second-order layer, formed of an alternation of zones rich in organic material and rich in carbonate; and a first-order layer, formed of a relatively thick and a relatively thin second-order layer. Largely from their numerological relationships, Barker proposed that the first-order layers express an annual temperature cycle, that the second-order layers mark the occurrence of equinoctial storms, that the third-order layers are the effect of differences in availability of food at different tidal extremes, that the fourth-order layers reflect differences in water temperature between night and day, that the fifth-order layers mark the opening and closing of the valves with the tides, and that the elemental layers reflect deposition of organic material during the process of opening or closing, and the deposition of carbonate while the shell is open or closed. To his credit, Barker does suggest that growth experiments be conducted to test these hypotheses.

Later workers (15, 30, 51, 57), using acetate peels rather than thin sections, found fortnightly periodicities to be widespread among intertidal bivalves; Figure 6, showing the cycle Evans (15) related to unequal diurnal tides, is an excellent example of this.

House & Farrow (30) and Farrow (18) demonstrated the existence of a monthly periodicity in intertidal bivalves, and Farrow (18) showed that this could be related to long periods of exposure due to very low high tides at every other neap for part of the year.

Reliability of Growth Line Records

GENERAL Unfortunately, the demonstration that a particular type of growth line is caused by a particular environmental variable does not always confirm its suitability for practical applications. For this, further examination of the evidence may be required.

There are two main factors in determining such suitability: the lines should form a precise and faithful record of the environmental variable, and they should be sufficiently distinctive to permit counting or other measurement without ambiguity. These requirements were noted as long ago as Weymouth's report on *Tivela* (67), but are still not too readily satisfied.

FIDELITY OF THE RECORDING PROCESS One of the first considerations in examining this factor is whether the variable known to cause a growth line is the only variable which can do so. In the case of the annual growth line, a number of both annual variables and random disturbances have been related to line formation (4, 16, 25,

31, 38, 62, 67); the possibility of similar effects being involved in other types of lines cannot be lightly dismissed.

A second consideration of some importance is whether there might be individual differences in response to the stimulus of the environmental variable. There are two ways in which this might occur. In one, slight differences in location of organisms might mean that some are more severely affected by the environmental variable than others; as an example of this, Farrow (18) found a monthly periodicity best expressed in cockles living high in the intertidal zone. The other involves genetic variability within a population, which may affect the degree of response by individuals to environmental variables; Posgay (55) reports such differences within a group of scallops.

A third consideration involves continuity of the record. Growth lines might be produced regularly with a daily periodicity so long as the shell grows, but if the growth stops for a few days the record will be missing some lines. This was shown by Clark (9) for specimens of *Pecten diegensis* grown in an aquarium; the sites of the missing lines could be shown by a comparison of growth rate curves constructed from the existing lines. Continuity of the record is especially important in those applications of growth lines which involve numerical relationships between different types of growth lines. For example, the number of daily lines between annual lines cannot be counted if the annual line is the result of some weeks or months of no growth; unfortunately some annual lines seem to form in just this way (4, 38).

A final question is whether aberrant behavior might occur under some circumstances. For example, Clark (in preparation) found that some specimens of *Argopecten gibbus* placed in an abnormally deep environment began to form semidaily growth lines on the left valve, while continuing their daily periodicity on the right valve. Such situations are probably exceedingly rare, but they show that growth line interpretation has no place for a complacent attitude.

PROBLEMS IN RESOLUTION The most common, and often the most difficult problem associated with growth lines is their definition; this involves the decision of where to place boundaries (especially important in measuring intervals), how to distinguish between poorly developed periodic lines and relatively prominent subunit boundaries, and how to handle short intervals where the record is obscured. Some of the decisions involved can be appreciated by attempting to define the growth lines in Figures 5a and 7a.

Perhaps the next most common problem is one of finding that part of a shell or skeleton with the best growth line record. For example, Pannella & MacClintock (51) note that winter growth is discontinuous at the outer surface of the shell of *Mercenaria* but continuous within the shell. Scrutton (59) points out that the fine ridges on coral epitheca often register preferential growth on one side, fading out and coalescing when traced around the circumference (Figure 7b). Clark (10) noted quite the opposite phenomenon in specimens of *Argopecten irradians* (Figure 8), where the growth lines are least well developed at the center, the area of fastest growth. In addition to such growth phenomena, most carbonate skeletons have

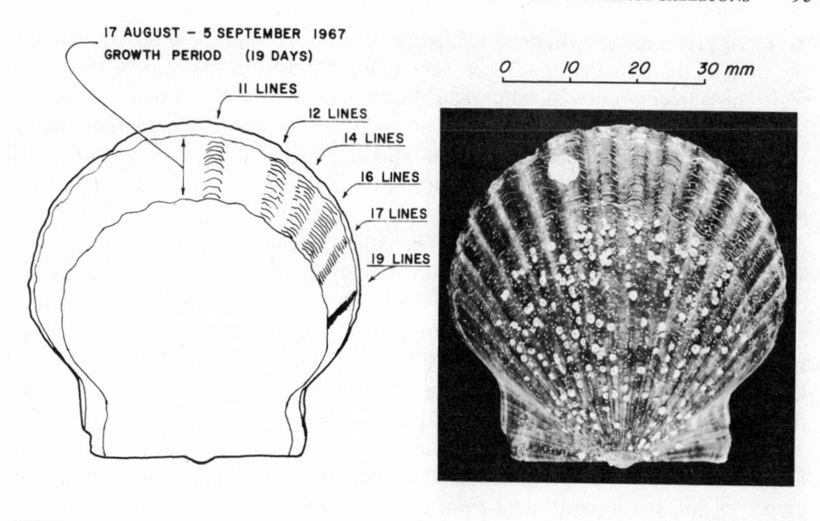

Figure 8 Left valve of *Argopecten irradians*, showing complete development of daily growth lines to occur preferentially at the posterior and anterior margins. The experimental growth period is bounded by strong disturbance lines.

suffered some degree of abrasion or epiphytic overgrowth which further obscures the record.

One final problem requires mention. This arises when the environmental variable is itself irregular or of uncertain periodicity. For example, annual lines are usually annual only in the sense of occurring once each year, not in occurring with a 365 day periodicity. Similar variation can accompany tidal cycles or breeding periodicities; such situations require considerably more data than situations where the record more closely reflects the environment.

APPLICATIONS OF GROWTH LINE RECORDS

Geochronology

In 1963 Wells (66) made the exciting suggestion that growth lines could be used to determine the absolute age of individual fossils. His reasoning was based on the widely accepted geophysical theory that the earth's rotation rate has been slowing throughout geologic time, such that a year in the geologic past would contain more (and shorter) days than a year today. If organisms could be found which form both annual and daily growth lines, then their fossil record could be used to determine the number of days in the geologic past, and thus test this theory. Moreover, if the theory were proven correct then each point in time should have its characteristic ratio of days to years, so that fossils of unknown age could be dated.

Wells (66) also undertook to test this elegant concept, using Devonian corals with well-preserved growth lines. Under the assumption that these lines were daily and annual, he estimated that there were about 400 days in the Middle Devonian year. This agreed quite well with predictions based on geophysical theory.

Soon after, Scrutton (59) reported that sets of lines which could represent daily and monthly periodicity occur on other Middle Devonian corals. Using these, he found about 30.6 days in a lunar month, another figure consistent with geophysical theory.

These reports were followed by a few other studies of growth line ratios in fossils (8, 39, 50, 52), which also depended heavily upon assumptions of the nature of the periodicities being studied. The overall contribution of this evidence toward establishing the history of the earth's rotation rate seems less than overwhelming; yet discussion papers appeared, in considerable numbers, offering further interpretations; in some cases (33) these even involved refinements beyond the accuracy of the original data.

It appears that widespread interest in the possible geophysical applications of fossil growth line data has stimulated the widespread acceptance of all available data; yet not one piece of data from these six studies can be proven absolutely to be valid. It is hoped it will not be long before further work on growth line formation will permit such proof, but until that time such data should be viewed with caution. The time has not yet come for Wells' dream of "every fossil a geochronometer" (66) to become a reality.

Paleoecology

Growth lines can be used as sources of information about the lives of fossil organisms. For example, periodic lines can provide evidence on the time of year an organism first appeared, its lifespan, rate of growth, breeding periodicity, and the season of its death. Applied to fossil populations, similarity in the patterns of disturbance lines (13, 55, 61) or of periodic lines (9, 16) could be used to determine which members of the population had lived (and died) at the same place and time. In some circumstances it could even be possible to overlap the records of individuals in a population and construct a chronology of events far exceeding any single lifespan.

Growth lines also have potential use in the interpretation of paleoenvironments. For example, the presence of disturbance lines, or of variation in the spacing of periodic lines, can be used as evidence for a variable environment; and variability argues for relatively shallow water. Similarly, the presence of an annual growth line suggests a climate with well-defined seasons, and tidal periodicity implies a habitat in or near the intertidal zone. Differences between growth line records noted in adjacent populations can further refine this sort of interpretation, and can lead to other applications, such as the identification of stunted faunas (18, 26).

In general, virtually any approach which can be used to demonstrate that growth lines form today as reflections of the environment can be turned about to make interpretations of past environments from fossil lines; the difficulty is that the interpretations thus derived are no better than the identification of the growth lines. It follows that extensive applications of growth lines in paleoecology must follow a discrete distance behind investigations of growth line formation in modern environments.

It is not surprising, then, to find relatively few reports (14, 26, 35–37) of studies using fossil growth lines for paleoecological interpretation. However, there is one report of special interest, for it pre-dates not only the study of paleoecology but most of the science of geology; this is the argument used by Leonardo da Vinci to refute the concept that fossils are created or originate within the rocks of mountains: "And if you were to say that these shells were created, and were continually being created in such places by the nature of the spot, and of the heavens which might have some influence there, such an opinion cannot exist in a brain of much reason; because here are the years of their growth, numbered on their shells, and there are large and small ones to be seen which could not have grown without food, and could not have fed without motion—and here they could not move" (34).

Literature Cited

1. Barker, R. M. 1964. Microtextural variation in pelecypod shells. *Malacologia* 2:69–86
2. Barnes, D. J. 1970. Coral skeletons: an explanation of their growth and structure. *Science* 170:1305–8
3. Barrington, E. J. W. 1967. *Invertebrate Structure and Function.* Boston: Houghton Mifflin. 549 pp.
4. Belding, D. L. 1910. *A Report upon the Scallop Fishery of Massachusetts.* Boston: Commonwealth of Massachusetts. 150 pp. 41 plates
5. Belding, D. L. 1910. The growth and habits of the sea clam (*Mactra solidissima*). *Rep. Comm. Fish. Game 1909,* 26–41, 6 plates. Boston: Commonwealth of Massachusetts
6. Belding, D. L. 1912. *A Report upon the Quahaug and Oyster Fishery of Massachusetts.* Boston: Commonwealth of Massachusetts. 134 pp. 36 plates
7. Belding, D. L. 1916. A report upon the clam fishery. *Rep. Comm. Fish. Game 1915,* 93–229, 6 plates. Boston: Commonwealth of Massachusetts
8. Berry, W. B. N., Barker, R. M. 1968. Fossil bivalve shells indicate longer month and year in Cretaceous than present. *Nature* 217:938–39
9. Clark, G. R. II. 1968. Mollusk shell: daily growth lines. *Science* 161:800–2
10. Clark, G. R. II. 1969. *Shell characteristics of the family pectinidae as environmental indicators.* PhD thesis. Calif. Inst. Technol., Pasadena. 101 pp.
11. Clark, G. R. II. 1969. Daily growth lines in the bivalve family pectinidae. *Geol. Soc. Am. Abstr. Programs,* Vol. 1, Pt. 7:34–35
12. Coker, R. E., Shira, A. F., Clark, H. W., Howard, A. D. 1921. Natural history and propagation of fresh-water mussels. *Fish. Bull. (US)* 37:75–181
13. Craig, G. Y., Hallam, A. 1963. Size-frequency and growth-ring analyses of *Mytilus edulis* and *Cardium edule,* and their palaeoecological significance. *Palaeontology* 6:731–50
14. Davenport, C. B. 1938. Growth lines in fossil pectens as indicators of past climates. *J. Paleontol.* 12:514–15
15. Evans, J. W. 1972. Tidal growth increments in the cockle *Clinocardium nuttalli. Science* 176:416–17
16. Fairbridge, W. S. 1953. A population study of the Tasmanian "commercial" scallop, *Notovola meridionalis* (Tate) (lamellibranchiata, pectinidae). *Aust. J. Mar. Freshwater Res.* 4:1–41
17. Farrow, G. E. 1971. Periodicity structures in the bivalve shell: experiments to establish growth controls in *Cerastoderma edule* from the Thames estuary. *Palaeontology* 14:571–88. 1 plate
18. Farrow, G. E. 1972. Periodicity structures in the bivalve shell: analysis of stunting in *Cerastoderma edule* from the Burry Inlet (South Wales). *Palaeontology* 15:61–72. 3 plates
19. Faul, H. 1943. Growth-rate of a Devonian reef-coral (*Prismatophyllum*). *Am. J. Sci.* 241:579–82
20. Forster, G. R. 1967. The growth of *Haliotis tuberculata*: results of tagging experiments in Guernsey 1963–65. *J.*

Mar. Biol. Assoc. UK 47:287–300. 1 plate
21. Gebelein, C. D. 1969. Distribution, morphology, and accretion rate of recent subtidal algal stromatolites, Bermuda. *J. Sediment. Petrology* 39:49–69
22. Gibson, F. A. 1953. Tagging of escallops (*Pecten maximus*, L.) in Ireland. *J. Cons., Cons. Perma. Int. Explor. Mer* 19:204–8
23. Gibson, F. A. 1956. Escallops (*Pecten maximus* L.) in Irish waters. *Sci. Proc. Roy. Dublin Soc.* 27:253–71
24. Gunter, G. 1957. Temperature. *Geol. Soc. Am., Mem. 67*, 1:159–84
25. Gutsell, J. S. 1931. Natural history of the bay scallop. *Fish. Bull.* (*US*) 46: 568–632. 6 plates
26. Hallam, A. 1963. Observations on the palaeoecology and ammonite sequence of the Frodingham Ironstone (Lower Jurassic). *Palaeontology* 6:554–74
27. Hallam, A. 1965. Environmental causes of stunting in living and fossil marine benthonic invertebrates. *Palaeontology* 8:132–55
28. Hedgpeth, J. W., Ed. 1957. *Treatise on Marine Ecology and Paleoecology, Vol. 1, Ecology. Geol. Soc. Am., Mem. 67, Vol. 1.* 1296 pp.
29. Hedley, R. H. 1958. Tube formation by *Pomatoceros triqueter* (polychaeta). *J. Mar. Biol. Assoc. UK* 37:315–22
30. House, M. R., Farrow, G. E. 1968. Daily growth banding in the shell of the cockle, *Cardium edule. Nature* 219: 1384–86
31. Isely, F. B. 1913. Experimental study of the growth and migration of fresh-water mussels. *Rep. US Comm. Fish. 1913*, Appendix 3. 24 pp. 3 plates
32. Isely, F. B. 1931. A fifteen year growth record in fresh-water mussels. *Ecology* 12:616–19
33. Lamar, D. L., Merifield, P. M. 1966. Length of Devonian day from Scrutton's coral data. *J. Geophys. Res.* 71:4429–30
34. Leonardo da Vinci. Quotation from *The Literary Works of Leonardo da Vinci*, compiled transl. J. P. Richter, 1883. Republ. 1970 as *The Notebooks of Leonardo da Vinci.* New York: Dover
35. Ma, T. Y. H. 1933. On the seasonal change of growth in some Palaeozoic corals. *Proc. Imp. Acad. Tokyo* 9:407–9
36. Ma, T. Y. H. 1934. On the seasonal change of growth in a reef coral, *Favia speciosa* (Dana), and the water-temperature of the Japanese seas during the latest geological times. *Proc. Imp. Acad. Tokyo* 10:353–56
37. Ma, T. Y. H. 1934. On the growth rate of reef corals and the sea water temperature in the Japanese islands during the latest geological times. *Sci. Rep. Tohoku Imp. Univ., Ser. 2* (*Geology*) 16:165–89. 4 plates
38. Mason, J. 1957. The age and growth of the scallop, *Pecten maximus* (L.), in Manx waters. *J. Mar. Biol. Assoc. UK* 36:473–92
39. Mazzullo, S. J. 1971. Length of the year during the Silurian and Devonian periods: new values. *Geol. Soc. Am. Bull.* 82:1085–86
40. Merrill, A. S., Posgay, J. A., Nichy, F. E. 1966. Annual marks on shell and ligament of sea scallop (*Placopecten magellanicus*). *Fish. Bull.* (*US*) 65:299–311
41. Millar, R. H. 1968. Growth lines in the larvae and adults of bivalve molluscs. *Nature* 217:683
42. Moore, H. B. 1934. On "ledging" in shells at Port Erin. *Proc. Malacol. Soc.* 21:213–17. 1 plate
43. Moore, H. B. 1958. *Marine Ecology.* New York: Wiley. 493 pp.
44. Nelson-Smith, A. 1967. *Catalogue of Main Marine Fouling Organisms, Vol. 3: Serpulids.* Paris: Organ. Econ. Coop. Develop. 79 pp.
45. Newcombe, C. L. 1935. Growth of *Mya arenaria* L. in the Bay of Fundy region. *Can. J. Res. Sect. D* 13:97–137
46. Newcombe, C. L. 1936. Validity of concentric rings of *Mya arenaria*, L. for determining age. *Nature* 137:191–92
47. Olsen, D. 1968. Banding patterns of *Haliotis rufescens* as indicators of botanical and animal succession. *Biol. Bull.* 134:139–47
48. Orton, J. H. 1923. On the significance of "rings" on the shells of *Cardium* and other molluscs. *Nature* 112:10
49. Orton, J. H. 1926. On the rate of growth of *Cardium edule.* Pt. I. Experimental observations. *J. Mar. Biol. Assoc. UK* 14:239–79
50. Pannella, G. 1972. Paleontological evidence on the earth's rotational history since early Precambrian. *Astrophys. Space Sci.* 16:212–37
51. Pannella, G., MacClintock, C. 1968. Biological and environmental rhythms reflected on molluscan shell growth. *J. Paleontol.* 42 (No. 5, pt. 2):64–80. 9 plates
52. Pannella, G., MacClintock, C., Thompson, M. N. 1968. Paleontological evidence of variations in length of synodic month since late Cambrian. *Science* 162:792–96
53. Petersen, G. H. 1958. Notes on the

growth and biology of the different *Cardium* species in Danish brackish water areas. *Meddr. Danm. Fisk.- og Havunders. N.S.* 2:1–31. 2 plates
54. Petersen, G. H. 1966. *Balanus Balanoides* (L.) (cirripedia): life cycle and growth in Greenland. *Meddr. Grønland* 159 (No. 12):1–114. 38 plates
55. Posgay, J. A. 1950. Investigations of the sea scallop, *Pecten grandis. Third Report on Investigations of Methods of Improving the Shellfish Resources of Massachusetts.* Chap. 4, 24–30. Boston: Commonwealth of Massachusetts
56. Raup, D. M. 1968. Theoretical morphology of echinoid growth. *J. Paleontol.* 42 (No. 5, Pt. 2): 50–63
57. Rhoads, D. C., Pannella, G. 1970. The use of molluscan shell growth patterns in ecology and paleoecology. *Lethaia* 3:143–61
58. Sastry, A. N. 1961. Studies on the bay scallop, *Aequipecten irradians concentricus* Say, in Alligator Harbor, Florida. PhD thesis, Florida State Univ. 118 pp.
59. Scrutton, C. T. 1965. Periodicity in Devonian coral growth. *Palaeontology* 7:552–58. 2 plates
60. Scrutton, C. T. 1970. Evidence for a monthly periodicity in the growth of some corals. *Palaeogeophysics,* ed. S. K. Runcorn, 11–16. London: Academic
61. Shuster, C. N. Jr. 1951. On the formation of mid-season checks in the shell of *Mya. Anat. Rec.* 111:127
62. Tang, S-F. 1941. The breeding of the escallop (*Pecten maximus* (L.)) with a note on the growth rate. *Proc. Trans. Liverpool Biol. Soc.* 54:9–28
63. Thompson, I. L., Barnwell, F. H. 1970. Biological clock control and shell growth in the bivalve *Mercenaria mercenaria. Geol. Soc. Am. Abstr. Programs* 2:704
64. Weber, J. N. 1969. Origin of concentric banding in the spines of the tropical echinoid *Heterocentrotus. Pac. Sci.* 23: 452–66
65. Wells, J. W. 1937. Individual variation in the rugose coral species *Heliophyllum halli* Edwards and Haime. *Paleontogr. Am.* 2:1–22
66. Wells, J. W. 1963. Coral growth and geochronometry. *Nature* 197:948–50
67. Weymouth, F. W. 1923. The life-history and growth of the pismo clam (*Tivela stultorum* Mawe). *Fish. Bull. (Calif.)* 7:1–120

THE PHYSICAL CHEMISTRY OF SEAWATER[1]

Frank J. Millero
Rosenstiel School of Marine and Atmospheric Science, University of Miami, Miami, Florida 33149

INTRODUCTION

This review is largely (but not exclusively) devoted to developments that have been made on the physical chemistry of seawater subsequent to the recent review of Pytkowicz & Kester (320). A number of other reviews have been written that deal with specific topics of the physical chemistry of seawater (53, 86, 114, 137, 160, 171, 173, 176, 210, 242, 257, 265, 271, 314, 343, 353, 380). Because the application of physical chemical techniques to ocean problems has been widespread in recent years, no attempt is made to cover all of these areas in this review. The author is presently writing a book on marine physical chemistry (273) that hopefully will cover these developments. We confine our discussion in this review to the physical chemistry of typical or average seawater and the interactions that occur in this medium. We discuss in detail the interactions of the major ionic components of seawater, how these interactions affect chemical processes occurring in the oceans, and how the physical chemical properties of seawater depend upon these interactions. Because of space requirements this review is limited in scope and is based on the author's major interests.

In recent years, there have been a number of excellent books written on marine chemistry that show how one can directly apply physical chemical principles to marine systems (16, 28, 104, 126, 140, 141, 172, 174, 354). Berner (16), for example, has written a book in which he applies physical chemical principles to low temperature geological problems. Stumm & Morgan (354) have written a book in which they discuss chemical equilibria in natural waters. Garrels & Christ's (126) book on the application of thermodynamics to mineral equilibria is still the classic text used by many workers as a source of information. Horne (172) has written a text on marine chemistry that stresses the importance of understanding the nature of water and a less descriptive science to one that attempts to probe the chemical processes occurring in the oceans, Horne's book is certainly needed. Although one might question

[1] Contribution Number 1668 from the School of Marine and Atmospheric Science University of Miami, Miami, Florida.

this approach, the author feels that his book represents an important first step toward the ultimate aim of marine physical chemistry, which is obtaining a molecular understanding of the chemical processes that occur in the marine environment.

In the coming year a number of books (139, 176, 273) will be available that discuss in detail the physical chemical properties of seawater (176) and the important contributions made by Sillen to marine physical chemistry (139).

THE PHYSICAL CHEMISTRY AND STRUCTURE OF WATER

To completely understand the physical chemistry of seawater on a molecular level it is necessary to know and understand the physical chemistry and structure of water and aqueous solutions. Although it is possible to treat the physical chemical properties and structure of seawater in terms of ion-water and ion-ion interactions without considering the structure of water (or water-water interactions), we must eventually understand pure water if we are to understand seawater (as well as other aqueous electrolyte solutions).

Most workers in oceanography (69, 357) are well aware (by reading *The Oceans*, 357) that all the unique physical chemical properties of seawater are due to the unique properties of water. Many, however, seem to forget too soon that water is the major constituent of seawater. In recent years, through the efforts of Horne (171–176) and others (71, 72), many marine chemists have been reminded that the understanding of water is basic to the understanding of seawater. Although one might take issue with some of the points raised by Horne (and others), it is the author's belief that this overall approach to studying the marine environment is necessary if we hope to eventually understand the mechanisms of the chemical processes occurring in the oceans.

Because the cause of all the unique properties of water and seawater are related to the structure of water, it is appropriate to examine briefly the present status on the structure of water and what the future might hold. The physical chemical properties and structure of pure water have been discussed by a number of workers in recent years (62, 71, 89, 115–117, 146, 174, 186, 190, 191, 339, 367). The books edited by Franks (117) and Horne (174) contain an up-to-date account of the structure and properties of water. The chapter of Kell (191) completely describes the thermodynamic and transport properties of water.

Structure of Water

In recent years the structure of water has caused considerable controversy (117). For example, the work on polywater has been greatly debated (117), as has the existence of thermal anomalies (71, 72, 202) (also referred to as kinks). The history of the polywater controversy is quite interesting (117) because it shows how sketchy evidence can lead to an explosion of claims and counterclaims. Few full length papers have been written on the subject; instead, the literature is confined to numerous short notes based on little or no experimental effort. Russian workers were the first to find and measure the properties of polywater. They found that water (and other liquids) confined in small capillaries after condensation had very interesting properties (it did not freeze at 0°C, it had a density of 1.3 g/cm^3, and had very high viscosity). After a group of British scientists substantiated the Russian workers' claims, the polywater bandwagon started to roll. Since then, many claims (most more advanced than initially made by the Russian workers) and counterclaims have been advanced, generally based on sketchy evidence. Speculation became as farfetched as a warning to scientists against experimenting with polywater because it might affect the oceans by turning them solid. Although the unlikelihood of this happening was quickly pointed out, several newspapers and popular science publications featured the story. At present, the existence of polywater as a unique form of water appears to be improbable and the observed effects are probably due to impurities.

Although polywater does not appear to be a new form of water, it is important to point out that all work on water near interfaces indicates that it is quite different from bulk water (72, 175). Thus, one would expect that water near interfaces in the oceans (for example, in pore waters or on the surfaces of suspended particles) has quite different properties from bulk water.

The current models of water structure can be divided into two major categories. The first includes the uniformist or continuum models (190). In the uniformist models no local domains of structure exist in water different from that of any other domain. The individual water molecules behave at any time in the same manner as any other water molecule. Kell (190) has recently reviewed the uniformist models. The second major category includes the mixture models (62). These models have received more attention than the uniformist models. The mixture models can be divided into the following: 1. broken-down ice lattice models (ice-like structural units in equilibrium with monomers); 2. cluster models (clusters in equilibrium with monomers); 3. interstitial models (clathrate-like cages in equilibrium with monomers); and 4. significant structure theory or polymer models (the bulky species is not necessarily a monomer). In each case, at least two species of water exist, namely, a bulky species representing some type of structured units and a dense species such as monomeric water molecules. The mixture models have normally been developed to express a special property of water (e.g. the maximum density, the minimum in compressibility or heat capacity, the anomalous acoustic absorption, etc). The ice lattice models and cluster models (146) do not appear to satisfy the X-ray scattering criterion for density fluctuations. With the exception of the interstitial model (339), none of the models provide an answer for the environment of the broken-down molecule.

The present best guess for the structure of water has recently been summarized by Frank (116). A brief summary of the bonding, structural patterns, equilibrium relations, and molecular motions is given below.

Both major models for water agree that water is a very structured liquid. The major differences concern the question: does water contain or not contain broken hydrogen bonds? At present, no unambiguous experiments have shown that broken hydrogen bonds exist in liquid water. Some workers feel that the Raman and IR data indicate that two distinguishable types of environment exist for OH stretching. If the Raman work is correct and broken bonds exist, one can rule out the uniformist models. Which of the mixture models is correct cannot be stated with certainty. The low-angle X-ray scattering data tends to support the interstitial model. The exact interstitial model is thought to be ice-like, not a pentagonal dodechedral framework (clathrate) as suggested by Pauling (116, 117). The interstitial model (116, 117) indicates that 20% of cold water is in the interstitial form. Because the other methods do not yield an absolute percentage of made or broken hydrogen bonds, it is not possible to compare this estimate with other estimates. All one can say at present is that the 20% value is reasonable compared to other model estimates. The interstitial model is consistent with the requirement for the concerted mechanism of the movement of water molecules (needed to explain the discrepancies that exist among the time scales found for molecular reorientations).

Although many workers have discussed the structure of water in seawater (72, 171–176) or aqueous electrolyte solutions, such discussions are only qualitative since it is normally not possible to separate water structure effects from ion-water and ion-ion interactions. Before we can make such discussion quantitative, we must understand the structure of water in bulk solutions as well as ion-water and ion-ion interactions. The electrostatic effects of ion-water and ion-ion interactions are normally considerably larger than the changes in the water structure by the ions. Thus it is difficult to learn much about water-water interactions in electrolyte solutions. Many workers have attempted to get around this problem by studying electrolytes in D_2O and H_2O (and the transfer from one medium to the other), or nonelectrolytes in H_2O. In the next section we briefly review ion-water interactions.

ION-WATER INTERACTIONS

The addition of an electrolyte to water multiplies the complexities of the system. Before we examine the interactions of ions in seawater, it is appropriate to briefly review the interactions that occur between ions and water molecules devoid of ion-ion interactions. Numerous books (21, 145, 174, 186) and papers (49, 50, 66, 67, 115, 165, 257, 258, 260, 339) have been written that review ion-water interactions (or the hydration of ions). For a full discussion of hydration and ion-water interactions, the reader should refer to these sources.

To examine the interactions of ions with water molecules, we will use the partial molal volume of ions, $\bar{V}^\circ$(ion) at infinite dilution (251, 253, 257, 258, 260). The partial molal volume of an ion can be visualized by considering the change in volume that occurs when one mole of ions is added to a large reservoir of water. The experimental methods of determining the $\bar{V}^\circ$ of ions (or electrolytes) have been discussed elsewhere (258, 260). By using a simple model (66, 67, 251, 258) for the interactions of ions and water molecules, we obtain

$$\bar{V}^\circ(\text{ion}) = \bar{V}^\circ(\text{int}) + \bar{V}^\circ(\text{elect}) \qquad 1.$$

where $\bar{V}^\circ$ (int) is the intrinsic size of the ion [$\bar{V}^\circ$ (cryst) $= 2.52r^3$, where r is the crystal radius in Å units, plus void space packing] and $\bar{V}^\circ$ (elect) is the decrease in volume or electrostriction due to ion-water interactions. For many ions (251, 257, 258, 260) one must add a third term to Equation 1: $\bar{V}^\circ$ (struct), the structure or caged partial molal volume which is due to changes in the water structure that cannot be explained by $\bar{V}^\circ$ (elect). Many workers in recent years have divided ions (75, 115) into two classes: 1. structure-breakers, which have a net effect of breaking down the structure of water; and 2. structure-makers, which have a net effect of making more structure. In general, the use of these terms is ambiguous (169) as we know little about the structure being made or broken. By confining our arguments to hydration effects (50, 66, 67, 75, 253, 257, 258, 260, 339) it is possible to discuss ion-water interactions using two regions around the ion: 1. positive hydrating ions, where $\bar{V}°$ (ion) $-$ $\bar{V}°$ (cryst) is negative due to electrostriction; and 2. negative hydrating ions where $\bar{V}°$ (ion) $-$ $\bar{V}°$ (cryst) is positive. In the electrostriction region the ionic charge on the ion predominates and the water dipoles are completely orientated toward the central ion. Although the water molecules in this region are rapidly exchanging, they are immobilized to a considerable extent compared with bulk water; the volume is decreased. For small ions of high charge this region predominates and $\bar{V}°$ (ion) $-$ $\bar{V}°$ (cryst) is negative. In the intermediate region between the electrostricted region and the bulk water, the effect of the ionic charge has diminished to such an extent that it can only partially orientate the water molecules; but the orientations are still large enough to interfere with the formation of the normal structure of water. Water in this region has a lower degree of hydrogen bonding and structure and, hence, it is called the disordered region. Some workers feel that the volume effect of this disordered region is negative; however, the author believes that the disordered region may reflect packing effects (251, 253, 257, 258, 260). For example, the ion with its electrostricted first-layer water molecules may not be able to fit into the water structure that exists before the ion is added. Thus, one can obtain a positive structural effect over and above what is normally expected for void space packing within the electrostricted zone.

Millero (251, 253) has successfully used the very simple model (given by Equation 1) for the $\bar{V}$ of ions in seawater (73, 74) that do not form ion pairs. He found the transfer of ions from pure water to seawater was given by

$$\Delta\bar{V}(\text{trans}) = a(Z^2/r)+b \qquad 2.$$

where a and b are salinity-dependent constants and Z is the charge on the ion. Some ions (SO_4^{2-}, HCO_3^-, and CO_3^{2-}) did not fit this general relationship and he interpreted the deviations in terms of ion pairing. In recent papers, he has used these methods to estimate the $\bar{V}$ of free ions (252) and paired ions in seawater (259, 261) and in aqueous salt solutions (244, 272).

Some workers have attempted to discuss ion-water interactions in terms of hydration numbers (6, 21, 66, 67, 75, 145, 165, 241, 242, 260, 295, 331). This hydration model equates the $\bar{V}°$ (elect) to $h(V^h_{H_2O} - V^b_{H_2O})$ where h is the hydration number (the number of waters attached to the ion), $V^h_{H_2O}$ is the molar volume of hydrated water, and $V^b_{H_2O}$ is the molar volume of bulk water. Estimates for $(V^h_{H_2O} - V^b_{H_2O})$ range from -2.0 to -7.0 cm^3 (258). Other thermodynamic and transport properties of solution can be discussed in a similar manner (21, 66, 67, 145).

THE COMPOSITION OF SEAWATER

Before we can discuss the physical chemistry of seawater in terms of the interactions of the major components, it is necessary to characterize the composition of seawater. In this section, we discuss the stoichiometry of seawater and the physical chemical models that have been used to arrive at the present composition.

Stoichiometry

As has been adequately demonstrated by numerous workers (60, 228, 240, 320, 330) the relative composition of the major (greater than 1 ppm by weight) dissolved inorganic components of seawater is nearly constant. Throughout most of the oceans Na/Cl, K/Cl, Mg/Cl, and Br/Cl ratios have been shown to be constant (60, 228, 240). Small variations may occur for Sr/Cl, HCO_3/Cl, $B(OH)_3$/Cl, and F/Cl and bottom waters are known to be higher in Ca/Cl (240). Fluctuations due to river run off, confinement, and freezing processes are locally significant; however, in this review we are concerned only with average seawater in the open ocean. The ratios of the mass of the major constituents of natural seawater, g_i, (g/kg of solution) to chlorinity, Cl (‰) (g/kg of solution) are given in column two of Table 1 (265). The results given for HCO_3^- are actually values of the carbonate alkalinity expressed as though it were all carbonate. The value given for Na^+ was determined by difference, by making the total cation equivalents equal to the total anion equivalents. The recent values obtained for Mg/Cl (‰) by Carpenter and Manella (39, 40) and the revised estimate for Na^+ are given in parentheses in

Table 1 The composition of seawater

Species	g_i/Cl (‰)[a]	Cl (‰) = 19.374[b] g_i, g/kg SW	e_i, eq/kg H_2O
Na^+	0.55566 (0.55668)	10.7653 (10.7851)	0.48534 (0.48623)
Mg^{2+}	0.06680 (0.06626)	1.2942 (1.2837)	0.11038 (0.10949)
Ca^{2+}	0.02125	0.4117	0.02129
K^+	0.02060	0.3991	0.01058
Sr^{2+}	0.00041	0.0079	0.00019
Cl^-	0.99894	19.3534	0.56579
SO_4^{2-}	0.14000	2.7124	0.05853
HCO_3^-	0.00735	0.1178	0.00200
CO_3^{2-}	—	0.0122	0.00042
Br^-	0.00348	0.0674	0.00087
F^-	0.00006_7	0.0013	0.00007
$B(OH)_3$	0.00132	0.0203	0.00034
$B(OH)_4^-$	—	0.0066	0.00009

[a] Taken from the compilation of Millero (265). The values for Mg^{2+} in parenthesis are taken from the recent work of Carpenter & Manella (39, 40). The values of Na^+ were obtained by difference by making the Σ of cation equivalents equal to the Σ of the anion equivalents.

[b] The grams and equivalent molality for average seawater at pH = 8.1 and 25°C. The values for HCO_3^-, CO_3^{2-}, $B(OH)_3$ and $B(OH)_4^-$ were calculated using the apparent constants of Lyman (227)—$[HCO_3^-]_T/[CO_3^{2-}]_T = 9.55$; $[B(OH)_3]_T/[B(OH)_4^-]_T = 3.98$.

column two of Table 1. At present, it is not possible to state with certainty which value is more reliable for Mg/Cl ($^{o}/_{oo}$).

The total grams per kilogram and equivalent molality for the components of average seawater of S ($^{o}/_{oo}$) = 35.000 or Cl ($^{o}/_{oo}$) = 19.374 and pH = 8.1 (on the N.B.S. scale) at 25°C are also given in Table 1. The total concentrations of HCO_3^-, CO_3^{2-}, $B(OH)_3$, and $B(OH)_4^-$ have been determined by using the apparent constants of Lyman (227). For many calculations (265) it is frequently convenient to assume that all the carbonate exists as HCO_3^- and all the boron exists as $B(OH)_3$. This assumption will have only a small effect when calculating the variations of the physical chemical properties of seawater; however, it will not be valid when discussing the effect of the medium of seawater on chemical processes occurring in the oceans.

The total molarity (c_T), molality (m_T), equivalent molarity or normality (N_T), equivalent molality (e_T), volume ionic strength (I_V), and weight ionic strength (I_m) for seawater solutions are given by

$$c_T = 0.0289047 \text{ Cl } (^{o}/_{oo}) \times d \qquad 3.$$

$$m_T = [28.9047 \text{ Cl } (^{o}/_{oo})]/[1000\text{–}1.81531 \text{ Cl } (^{o}/_{oo})] \qquad 4.$$

$$N_T = 0.0312803 \text{ Cl } (^{o}/_{oo}) \times d \qquad 5.$$

$$e_T = [31.2803 \text{ Cl } (^{o}/_{oo})]/[1000\text{–}1.81531 \text{ Cl } (^{o}/_{oo})] \qquad 6.$$

$$I_V = 0.0360145 \text{ Cl } (^{o}/_{oo}) \times d \qquad 7.$$

$$I_m = [36.0145 \text{ Cl } (^{o}/_{oo})]/[1000\text{–}1.81531 \text{ Cl } (^{o}/_{oo})] \qquad 8.$$

where d is the density. These equations can be redefined in terms of salinity by using the relation (368)

$$\text{S } (^{o}/_{oo}) = 1.80655 \text{ Cl } (^{o}/_{oo}) \qquad 9.$$

The mean molecular weight $M_T = 1/2\Sigma m_i M_i = 62.808$ whereas the mean equivalent weight $M'_T = 1/2\Sigma e_i M_i = 58.049$ for sea salt (265). It should be pointed out that the M_T and M'_T given above are in total grams of sea salt [$g_T = 1.81531$ Cl ($^{o}/_{oo}$)], not in salinity grams. There are 57.754 salinity g/equivalent of sea salt. Kester et al (193) have improved the recipe of Lyman & Fleming (228) for preparing artificial seawater. The general methods of preparation given by Kester et al can be used with the more reliable composition given in Table 1 to give an improved artificial mixture.

The concentrations of the minor elements of seawater are quite variable due to their high reactivity. Although average values are given by some workers (136–138, 330), in most instances reliable analytical data are not available.

In recent years, ion selective electrodes have been used to study the chemical composition of seawater. Warner (373) has recently reviewed their use in oceanography. A more detailed discussion of specific ion electrodes and reference electrodes is given by Durst (79), Eisenman (90), Rechnitz (324), and Ives & Janz (178). Warner's survey indicates that available electrodes for F^-, S^{2-}, Na^+, Cu^{2+},

Cl^- are promising for in situ measurements. In situ pH measurements in the ocean have been successfully accomplished by Ben-Yaakov & Kaplan (8–10). Recently, Whitfield (379) has described a compact potentiometric sensor that can be used to measure in situ pH, pS^{2-}, and pE in seawater.

Controlling Mechanism

The chemical reactions controlling the major composition of seawater have been discussed by a number of workers in recent years. Most of the models used are based on Sillen's physical chemical equilibrium models (343, 345, 346). Recent improvements have been made by Holland (167, 168), Mackenzie & Garrels (124–128, 230–235), Siever (341, 342), etc (25, 130, 161, 162, 217, 312). Mackenzie (235) has recently reviewed some of these models. In general, all of the equilibrium models have yielded reasonable suggestions of how seawater attained its present composition and how it is being maintained. Many questions still exist concerning the mechanisms and chemical reactions regulating the composition of seawater.

For example, we do not know where many of the controlling reactions take place: in sediments, in estuaries, in the surface layers, etc. In recent years, many workers have been concerned with the effect of the circulation of the world's oceans (27) and the effect of biochemical processes (41, 43, 365) on the composition. Recent workers (27, 210, 321) have also started to use kinetic models for the regulating mechanism rather than the steady state or equilibrium models used by earlier workers. Although these methods will probably lead to improved models for chemical mechanisms controlling the composition of seawater, the success of the simple equilibrium model proposed by Sillen strongly indicates that at least the major ions are in a near equilibrium or steady state.

IONIC INTERACTIONS IN SEAWATER

To understand the chemistry of seawater in terms of its major components and how the medium of seawater affects chemical processes that occur in the oceans it is necessary to understand ionic interactions in multicomponent electrolyte solutions. In recent years, there have been a number of major advances made on the interpretation of interactions in multicomponent electrolyte solutions. A number of workers (212–215, 265–270, 334, 380) have applied some of these techniques to the medium of seawater. Normally, the ionic interactions in seawater have been treated by using Bjerrum's (19) ion pairing model; we will discuss the ion pairing models in the next section.

Ion Pairing Models

Garrels & Thompson (123) were the first to completely apply this model to the major ionic components of seawater. In recent years, a number of other workers have attempted to improve the model (2, 158, 175, 206). Recently, the model has been applied to freshwaters (381). The basic assumption of this approach is that short-range interactions in seawater can be represented by the formation of cation-anion pairs and that there are standard solutions in which no association occurs (337). The procedure of using the association model to describe the ionic interactions in seawater has been described in detail elsewhere (2, 83, 123, 257, 317). Before we discuss these results, we will briefly review the theory of ion pair formation.

The extent of ion pair formation

$$M^{+} + A^{-} = MA^{0} \qquad 10.$$

is characterized by an association constant

$$K_A = [a_{MA}/(a_M + a_{A^-})] = [(MA^0)/(M^+)(A^-)][(\gamma_{MA^0})/(\gamma_{M^+}\gamma_{A^-})] \qquad 11.$$

where a_i, (i), and γ_i are, respectively, the activity, molal concentration, and activity coefficient of species i. There are four classes of ion pairs: 1. complexes—when the ions are held in contact by covalent bonds; 2. contact ion pairs—when the ions are in contact and linked electrostatically (with no covalent bonding); 3. solvent-shared ion pair—pairs of ions linked electrostatically, separated by a single water molecule; and 4. solvent-separated ion pairs—pairs of ions linked electrostatically but separated by more than one water molecule. Bjerrum (19) defined the distance between oppositely charged ions which can be classified as being associated by: $q = [(Z_+ Z_-)e^2/(2DkT)]$, where Z_i is the charge on the ion i; e is the electrostatic charge; D is the dielectric constant; k is the Boltzman constant; and T is the absolute temperature. In this treatment, two ions of opposite charge are considered to form an ion pair when they are between a, the ion size parameter, and q. This can include ion pairs of classes 2, 3, and 4.

The association constant of the Bjerrum method is given by

$$K_A = [(4\pi N)/1000][(Z_+ Z_-)e^2/(DkT)]^3 Q(b) \qquad 12.$$

where $b = |Z_+ Z_-| e^2/aDkT$ and the function of $Q(b)$ is given by Robinson & Stokes (331). The theory predicts greater ion pair formation the higher the valencies and the smaller the dielectric constant of the solvent, which is in agreement with experimental results.

Many workers have criticized the theory because of the arbitrary cutoff distance. It has now been superseded by other theories (63, 119, 275, 292).

For example, the model of Fuoss (119) considers only anions on the surface of a cation in volume, $v = 2.52\, a^3$ to be ion pairs. Fuoss obtained for K_A

$$K_A = [(4\pi N a^3)/3000] \exp [Z_+ Z_- e^2/DakT] \qquad 13.$$

where the first term is the excluded volume around the cation. Others (292) have made further elaborations on these methods and discussed the shortcomings of the model.

By differentiating Equations 12 and 13 with respect to pressure and temperature, one can determine the theoretical volume, enthalpy, and entropy change for the ion pairing process

$$-\Delta\bar{V}^\circ = RT(\partial \ln K_A/\partial P) \qquad 14.$$

$$\Delta\bar{H}^\circ = RT^2(\partial \ln K_A/\partial T) \qquad 15.$$

$$\Delta\bar{S}^\circ = RT(\partial \ln K_A/\partial T) \qquad 16.$$

Hemmes (163) has recently estimated the $\Delta\bar{V}^\circ$ for ion pair formation by using the Fuoss and Bjerrum constants and their pressure dependence. He obtained from the Fuoss equation

$$\Delta\bar{V}^\circ = [(Z_+ Z_- e^2 N)/aD](\partial \ln D/\partial P) - RT\beta \qquad 17.$$

where the second term involving the compressibility of water (β) is needed because both the Fuoss and Bjerrum association constants are based on the molarity scale. From the Bjerrum equation he obtained

$$\Delta\bar{V}^\circ = RT[3 + (\exp^b/Q(b)b^3)(\partial \ln D/\partial P) - \beta] \qquad 18.$$

For large values of b, the compressibility term is negligible ($RT\beta = 1.1$ cm^3/mol at 25°C) compared to the first term in both equations. Since $(\partial \ln D/\partial P)_T$ is positive for water (and other solvents), $\Delta\bar{V}^\circ$ is positive for the association of an ion pair in agreement with experimental results. For $a = 7.36$ Å, $\Delta\bar{V}^\circ = 8.98$ cm^3/mol from Equation 17, and for $a = 4.19$ Å, $\Delta\bar{V}^\circ = 6.89$ from Equation 18 (163). These equations can only be used for outer-sphere complexes. The ion size parameter needed in the calculation can be back calculated from Equations 12 and 13 by using the measured K_A. For example, Hemmes (163) calculated for the formation of $LaFe(CN)_6^0$, $a = 7.36$ Å and $\Delta\bar{V}^\circ = 8.98$ cm^3/mol from the Fuoss equations, and $a = 4.19$ Å and $\Delta\bar{V}^\circ = 6.89$ cm^3/mol from the Bjerrum equation, which is in reasonable agreement with the measured value of 8.0 cm^3/mol (148). For $MgSO_4^0$, he obtained $\Delta\bar{V}^\circ = 7.42$ and 4.86 cm^3/mol, respectively, from the Fuoss and Bjerrum equations, compared with the measured values of 7.3 cm^3/mol (108). For $MnSO_4^0$ he calculated $\Delta\bar{V}^\circ = 8.3$ and 5.0, respectively, from the Fuoss and Bjerrum equations, compared with the experimental value of 7.4 cm^3/mol (109).

For ion pairs that form inner-sphere complexes like $LaSO_4^+$ (112) and $EuSO_4^+$ (147), the predicted $\Delta\bar{V}$'s are much smaller than the directly measured values. It is thus possible to use the magnitude of the experimental $\Delta\bar{V}$'s to infer the structure of the ion pair.

The effect of temperature on ion pair formation calculated from the theory of Bjerrum and Fuoss has recently been examined by Prue (307). For the Fuoss equation he obtained

$$\Delta H^\circ = -RTb[1 + (\partial \ln D/\partial \ln T)] \qquad 19.$$

$$\Delta S^\circ = R \ln(4\pi N a^3/3000) - Rb(\partial \ln D/\partial \ln T) \qquad 20.$$

Because $(\partial \ln D/\partial \ln T)$ is negative for pure water, one would expect ΔH° to be positive (endothermic) or unfavorable to association. Negative values of ΔH° can be accounted for by adding a covalent contribution to K_A.

Because the second term of Equation 20 is normally larger than the first term, the Fuoss theory predicts that ΔS° should be positive for ion pair formation. In general, the $\Delta H°$, $\Delta S°$, and $\Delta V°$ for ion pair formation are more sensitive to changes of solvent than is $\Delta G°$ [similar to the behavior for the dissociation of acids and bases (254)].

In recent years, a number of workers have determined $\Delta H°$ by calorimetry (65, 102, 103, 179, 180, 292, 293). Various correlations between $\Delta H°$ and $\Delta S°$ have also been discussed (102, 103, 179, 180). Fay & Purdie (103) have pointed out that these correlations are not always the result of common structural properties for a given series. Linear correlations between $\Delta \bar{V}°$ and $\Delta \bar{K}°$ (98) have also been found for ion pairing processes [similar to that found for acid and base dissociation (226)].

It should be kept in mind that all of the theoretical models for ion pair formation are based on the use of the continuum models (treating the ions as hard spheres in a continuous dielectric medium).

APPLICATIONS TO SEAWATER The themodynamics of ion pair formation in binary electrolyte solutions has been studied experimentally by many workers. These results have been compiled and reviewed by many workers (20, 45, 63, 160, 275, 292, 293, 306, 344, 389). In this section, we will discuss the application of ion pairing methods to seawater speciation.

Although some discrepancies exist regarding the measurement, calculation, and

Table 2 The calculated and measured activity coefficients of the major ions in seawater at 25°C

Ion	Measured[a]	Ionic strength[b]	Ion pairing[c]	Guggenheim[d]
H^+	0.74	0.85	0.74	0.74
Na^+	0.68	0.71	0.70	0.68
Mg^{2+}	0.23	0.29	0.25	0.23
Ca^{2+}	0.21	0.26	0.22	0.21
K^+	0.64	0.63	0.62	0.63
Sr^{2+}	—	0.25	0.22	—
Cl^-	0.68	0.63	0.63	0.66
SO_4^{2-}	0.11	0.22	0.10	0.11
HCO_3^-	0.55	0.68	0.43	0.59
CO_3^{2-}	0.02	0.21	0.02	0.03
F^-	—	0.68	0.31	—
OH^-	—	0.65	0.11	0.56
$B(OH)_4^-$	0.35	0.68	0.38	—

[a] Taken from references 15, 59, 257, 265, 302–304, 311, 360, and 361 for 35‰ salinity seawater.

[b] Taken from references 2 and 265 at I = 0.7. The estimates were made by using the mean salt method.

[c] Calculated from the results given in Tables 5 and 6 and the text.

[d] Taken from references 214, 215, and 265 at I = 0.7. The values for HCO_3^- and CO_3^{2-} are for I = 0.51.

interpretation of activity coefficient, it is quite apparent that the activity coefficients calculated for free ions (Table 2) in seawater by the ionic strength principle or an extended Debye-Hückel equation for some ions (HCO_3^-, CO_3^{2-}, SO_4^{2-}, Mg^{2+}, and Ca^{2+}) are higher than the experimentally determined values. This difference can be interpreted by using the ion pairing model. Some of the major ion pairs thought to be important in seawater are the following: $NaSO_4^-$, KSO_4^-, $MgSO_4^0$, $CaSO_4^0$, $SrSO_4^0$, $NaCO_3^-$, $MgCO_3^0$, $CaCO_3^0$, $SrCO_3^0$, $NaHCO_3^0$, $MgHCO_3^+$, $CaHCO_3^+$, HCO_3^+, MgF^+, CaF^+, and $MgOH^+$. The relative concentration or speciation of these species in seawater can be determined by two methods: 1. using infinite dilution K_A data and estimated γs for the free ions; and 2. using stoichiometric association constants $K_A^* = K_A(\gamma_{M^+}\gamma_{A^-}/\gamma_{MA^0})$ determined in ionic media (132) at the same ionic strength of seawater (0.7 M). Many workers have pointed out the advantages (83, 317, 318, 343) and disadvantages (2, 22, 132) of using the ionic medium method. Both methods should yield the same results provided the experimental data is self-consistent. Garrels & Thompson (123) used the first

Table 3 Thermodynamic association constants, K_A, for the major sea salts at 25°C and 1 atm

	log K_A				
	SO_4^{2-}	HCO_3^-	CO_3^{2-}	OH^-	F^-
Na^+	0.72[a,b,c]	0.16[a,d]	0.55[a,d]	−0.57[a]	−0.26[e]
	0.65[c]	−0.08 to −0.30[f]	0.35[f]		
	0.70[g]	−0.25[b]	1.27[b]		
Mg^{2+}	2.21[a]	1.21[a]	3.28[a]	2.58[a]	1.82[a]
	2.23[h]	1.16[b]	3.24[i]	2.20[j]	
	2.25[k]	1.23[i]	3.4[b]		
	2.36[b,l]				
Ca^{2+}	2.31[a,b]	1.25[a]	3.92[a]	1.27[a]	1.04[a]
	2.28[g]	1.26[b]	3.2[b]		
	2.43[h]	1.23[m]	3.1[n]		
K^+	0.96[a,b,c]				
	1.0[g]				
	0.75[h]				
	0.85[o]				
Sr^{2+}	2.31[a]	1.25[a]	3.92[a]	0.82[a]	1.04[a]
H^+	1.98[h]	6.35[p]	10.33[p]	14.00[p]	

[a] Ref. 2 (the values for Sr^{2+} were assumed to be equal to that of calcium).
[b] Ref. 123.
[c] Ref. 184.
[d] Ref. 286.
[e] Ref. 76, 332.
[f] Ref. 36.
[g] Ref. 63.
[h] Ref. 179, 180.
[i] Ref. 287.
[j] Ref. 247.
[k] Ref. 284.
[l] Ref. 183.
[m] Ref. 7.
[n] Ref. 207.
[o] Ref. 366.
[p] Ref. 150.

method to estimate the speciation of the major components in seawater; more recently, Atkinson et al (2) have made similar calculations using more reliable thermodynamic data. Kester & Pytkowicz (195) have used the second method for sulfate ion pairs and arrived at a distribution that was slightly different from the results of Garrels & Thompson. The advantage of using the ionic medium is that it is not necessary to estimate single ion activity coefficients in the medium of seawater. If, however, ion-selective electrodes are used, one must normally estimate single ion activity coefficients in the standard solutions used in the calibration (since the electrodes measure activity and not total concentration). The ionic medium method assumes that single ion activity coefficients are only functions of ionic strength. Thus, in actual operation the two methods are not significantly different and should yield the same results provided the experimental results are self-consistent.

Because reliable stoichiometric constants are not available (except for the sulfate and carbonate systems), we are forced to use the second method to determine the

Table 4 Stoichiometric association constants, K_A^*, for sea salt ion pairs at 25°C and 1 atm

	SO_4^{2-}	HCO_3^-	CO_3^{2-}	OH^-	F^-	$B(OH)_4^-$
Na^+	1.21[a] 2.02[c] 0.70[d]	0.70[a] 0.26[c,d] 0.28[f] 0.32[g]	0.78[a] 4.16[c,d] 4.25[f] 1.57[g]	0.14[a]	0.25[b] 0.16[e]	0.57[m]
Mg^{2+}	10.3[a] 10.2[c] 6.1[d]	4.70[a] 5.22[c,d] 1.62[f]	116[a] 160[c,d] 112.3[f]	121[a]	19.2[a] 18.8[h] 20[i]	2–6[j] 8.03[m]
Ca^{2+}	11.7[a] 10.8[c] 8.8[d]	4.63[a] 5.10[c,d] 1.96[f]	454[a] 78.5[c,d] 162.3[f]	5.33[a]	2.9[a] 4.2[f] 3.2[g]	2–6[j] 13.01[m]
K^+	1.86[a] 1.03[c,d]					
Sr^{2+}	11.2[a]	4.45[a]	437[a]	182[a]	2.8[a]	
H^+	12.3[k] 16.6[j] 15.7[l]	7.2×10^5[j]	8.9×10^8[j]	1.6×10^{13}[j]	550[k]	4.1×10^8[j]

[a] Ref. 2.
[b] Calculated from the K_A given in Ref. 76, 332.
[c] Ref. 194, 195, 317; the values for $NaSO_4^-$, $MgSO_4^0$, and $CaSO_4^0$ were directly measured.
[d] Ref. 123.
[e] Ref. 35 measured at I = 1.0.
[f] Ref. 158, based on direct measurements.
[g] Ref. 36, linearly interpolated for I = 0.7 from measured values.
[h] Ref. 92.
[i] Ref. 47.
[j] Ref. 80, 83, 152 156.
[k] Ref. 59.
[l] Ref. 93.
[m] Ref. 37.

speciation of most ionic species in seawater (although the ionic medium method would be preferable). The methods that can be used to estimate activity coefficients for free ions at the ionic strength of seawater are discussed elsewhere (6, 63, 93, 126, 150, 331). The thermodynamic association constant for the major components of seawater are given in Table 3. The stoichiometric association constants calculated from these thermodynamic values and those measured directly by various workers (158, 194–196, 317) are given in Table 4.

In the calculation of K_A^* by Garrels & Thompson (123) they assumed $\gamma = 1.13$ for noncharged ion pairs, $\gamma = 0.68$ for singly charged ion pairs, and $\gamma = 0.20$ for double charged ion pairs. Atkinson et al (2) have assumed $\gamma = 1.0$ for noncharged ion pairs, $\gamma = 0.68$ for singly charged ion pairs, and $\gamma = 0.21$ for doubly charged ion pairs.

The speciation of the major cations and anions in seawater and 25°C are given in Tables 5 and 6. These results are discussed in the next section.

THE MAJOR ION SPECIATION The validity of the ion pairing model for the major components of seawater has been demonstrated by a number of methods. For example, the direct measurement of the major ions in seawater using ion selective

Table 5 The speciation of the major cations in 35‰ salinity seawater at 25°C and 1 atm

Ion	% Free	% MSO_4	% $MHCO_3$	% MCO_3	% MF	% MOH
Na^+	99[a]	1.2[a]	0.01[a]			
	97.7[b,e]	2.2[b,e]	0.03[b]			
	98.3[c]	1.6[c]	0.08[c]	0.002[c]		0.002[c]
	97.7[d]	2.3[d]	0.03[d]			
			0.01[e]			
Mg^{2+}	87.0[a]	11[a]	1[a]	0.3[a]		
	89.0[b]	10.3[b,e]	0.6[b]	0.13[b]		
	87.3[c]	11.9[c]	0.5[c]	0.2[c]	0.06[c]	0.02[c]
	88.9[d]	9.2[d]	0.4[d]	0.4[d]		
	89.2[e]		0.3[e]	0.1[e]	0.1[e]	
Ca^{2+}	91[a]	8[a]	1[a]	0.2[a]		
	88.5[b,e]	10.8[b,e]	0.6[b]	0.07[a]		
	85.4[c]	13.2[c]	0.5[c]	0.9[c]	0.01[c]	0.001[c]
	91.8[d]	7.3[d]	0.6[d]	0.2[d]		
			0.3[e]	0.3[e]		
K^+	99[a]	1[a]				
	98.8[b,e]	1.2[b,e]				
	97.6[c]	2.4[c]				
	98.8[d]	1.2[d]				
Sr^{2+}	86.4[c]	12.3[c]	0.4[c]	0.9[c]	0.01[c]	0.0003[c]

[a] Ref. 123.
[b] Ref. 195.
[c] Ref. 2.
[d] Ref. 206.
[e] Ref. 158.

electrodes (360, 361) agree with the model. The total activity coefficient of the ionic species (γ_T) can be calculated for the fraction of ions paired ($[M^+]_F/[M^+]_T$) from Tables 5 and 6

$$\gamma_T = ([M^+]_F/[M^+]_T)\gamma_F \qquad 21.$$

where γ_F is the activity coefficient of the free ion. The estimated γ_Ts for the major components of seawater calculated from Equation 21 are given in Table 2. As is quite apparent, the ion pairing model gives estimates for γ_T that are in good agreement with measured values.

The model has also been shown to be consistent with the solubility of minerals in seawater (121–123, 315), the difference chromatography of seawater (239), the

Table 6 The speciation of the major anions in 35‰ salinity seawater at 25°C and 1 atm

Ion	% Free	% NaA	% MgA	% CaA	% KA	% SrA
SO_4^{2-}	54[a]	21[a]	21.5[a]	3[a]	0.5[a]	
	39.0[b,e]	37.2[b]	19.4[b]	4.0[b,e]	0.4[b,e]	
	45.9[c]	25.9[c]	22.5[c]	4.8[c]	0.84[c]	0.04[c]
	39.8[d]	36.4[d]	20.6[d]	2.6[d]		
		37.1[e]	19.5[e]			
HCO_3^-	69[a]	8[a]	19.0[a]	4[a]		
	70.0[b]	8.6[b]	17.8[b]	3.3[b]		
	62.9[c]	20.5[c]	14.0[c]	2.6[c]		0.02[c]
	78.7[d]	7.9[d]	10.0[d]	3.3[d]		
	81.3[e]	10.7[e]	6.5[e]	1.5[e]		
CO_3^{2-}	9[a]	17[a]	67[a]	7[a]		
	9.1[b]	17.3[b]	67.3[b]	6.4[b]		
	9.2[c]	3.4[c]	49.6[c]	37.5[c]		0.3[c]
	10.6[d]	17.5[d]	67.1[d]	4.8[d]		
	8.0[e]	16.0[e]	43.9[e]	21.0[e]		
F^-	51.8[c]		46.8[c]	1.3[c]		0.01[c]
	45.4[f]	4.4[f]	47.0[e,f]	2.0[e,f]		
	51.0[e]					
OH^-	14.6[c]	0.6[c]	83.9[c]	0.9[c]		0.002[c]
$B(OH)_4^-$	76.0[g]	15.1[h]	24[g]	7.0[h]		
	56.1[h]		21.8[h]			
Cl^-, Br^-	100[a,b,c,d]					

[a] Ref. 123.
[b] Ref. 195.
[c] Ref. 2.
[d] Ref. 206.
[e] Ref. 158; this worker finds 7.4% $Mg_2CO_3^{+2}$ and 3.8% $MgCaCO_3^{+2}$.
[f] This paper.
[g] Ref. 83; the value for $MgB(OH)_4^+$ is actually for both Mg^{2+} and Ca^{2+}.
[h] Ref. 37.

large sonic absorption of seawater (88, 111, 216, 363, 364), and, recently, in explaining the Raman spectra of (acidified) solutions similar to seawater (61). For example, Daly et al (61) have examined the Raman spectra of $MgSO_4$, NaCl, and Na_2SO_4 solutions acidified with HCl. Since HSO_4^- exhibits a strong Raman band (besides the normal SO_4^{2-} band) they were able to examine the changes in this band upon the addition of Na^+ and Mg^{2+} ions. Because the Raman spectra of Na_2SO_4 and $MgSO_4$ do not reveal the presence of $NaSO_4^-$ and $MgSO_4^0$ (without the addition of acid), their results and others (164) indicate that these species are predominately solvent-separated ion pairs, which is in agreement with the ultrasonic measurements (88) and the partial molal volume measurements (252). The predicted ratios of the SO_4^{2-} band to the HSO_4^- band using the ion pairing model were found to be in good agreement with the measured values. Their results confirm the fact that Na^+ and Mg^{2+} ions can compete with H^+ for SO_4^{2-} and the ion pairing constants are reasonably accurate.

It should be emphasized that when one examines the speciation of ions in seawater using the ion pairing model, it is not certain that all the species actually exist as real entities. Additional experimental evidence (from other methods) is needed before we can be sure the species exist, even though we can successfully predict reasonable values of γ_T that are consistent with other direct measurements.

In the next section, we discuss the results of the speciation for the major components of seawater.

Sodium speciation All the results given in Table 4 indicate that most of the Na^+ ions are free in seawater. Although some workers have mathematically examined NaCl solutions in terms of ion pairing, as discussed later (in the Cl, Br speciation), most workers (including myself) do not believe ion pairing (i.e. the classical electrostatic type) is important for NaCl solutions. Some workers have also treated Na_2SO_4 solutions as if these are completely dissociated (131). Most workers, however (2, 3, 63, 93, 179, 184, 317, 329), feel that $NaSO_4^-$ does exist. The recent Raman work (61) gives added evidence for the existence of $NaSO_4^-$. The small differences that exist for the percentage of $NaSO_4^-$ in seawater are due to the small differences in the K_A^* used. A value of 2.0% $NaSO_4^-$ and 98.0% for Na^+ is probably a good compromise.

Magnesium speciation The major species found by all the workers for magnesium is free Mg^{2+}. The tendency of magnesium and sulfate ions to form ion pairs has been studied extensively by many workers using calorimetric methods (180, 284), conductivity methods (108, 329), viscosity methods (64), potentiometry (194, 196), solubility (315), ultrasonic relaxation (1, 88, 110, 111, 113), and Raman spectroscopy (61, 164). The thermodynamic and stoichiometric constants for $MgSO_4^0$ are in excellent agreement. The slight difference in the percentage of $MgSO_4^0$ (195) is the result of differences in the K_A^* values used for $NaSO_4^-$. The 10 to 12% magnesium obtained by the model calculations is in reasonable agreement with the value of 9.2% found by Fisher (111) by ultrasonic absorption measurements.

Potassium speciation The speciation of potassium is very similar to sodium in that 98 to 99% is in the free form. The small differences in the percentage of KSO_4^- found by various workers reflects the different K_A^*s used. Because all the work at infinite dilution, as well as the difference chromatography work (239), indicates that $K_A^*(KSO_4^-)$ should be larger than $K_A^*(NaSO_4^-)$, I prefer the results of Atkinson et al (2).

Calcium speciation All the results in Table 4 indicate that most of the calcium in seawater is in the free form, but the spread (85 to 91%) is quite large, largely because of the different values used for $K_A^*(CaSO_4^0)$. Although the thermodynamic values for $K_A(CaSO_4^0)$ are in reasonable agreement, the estimated K_A^*s do not agree with the directly measured value. One would normally prefer the directly measured value, but it may be in error because of possible problems with the Ca^{2+} electrode. For example, the Ca^{2+} electrode measurements of Thompson & Ross (361) yield 84% free Ca^{2+} in seawater, which is closer to the value obtained by Atkinson et al (2). Because the results of Atkinson et al were obtained with more reliable K_A^*s for the $CaSO_4^0$ ion pair, I prefer their results.

Strontium speciation The speciation of strontium in seawater is very similar to the results obtained for Mg^{2+} and Ca^{2+}, which is not surprising inasmuch as Atkinson et al estimated all the K_As for Sr^{2+} by assuming these were equal to the appropriate Ca^{2+} values.

Chloride and bromide speciation Although some workers have examined NaCl and KCl solutions in terms of ion pairing, most workers feel that Cl as well as Br does not form ion pairs with the major cations in seawater. For example, partial molal volume data (256), conductance data (44, 77), dielectric constant data (W. Ho and W. F. Hall, personal communication), and activity coefficient data (87) for NaCl [as well as KCl (78)] solutions between 0.2 and 1 M can mathematically be treated in terms of ion pairing.

As has been pointed out by many workers in recent years, the K_A determined from conductance data (as well as other methods) is quite dependent upon the value assigned to the ion size parameter (149, 245). For example, Hanna et al (149) have recently reanalyzed the conductivity data for some 1:1 and 2:2 salts. They found that $K_A = 0.025$ and 0.026 for NaCl and KCl using $a = 3.53$ and 3.60 Å, respectively. In other words, no NaCl and KCl ion pairs exist if the proper ion size parameter is used. In summary, there is little or no reliable evidence for the existence of chloride and bromide (electrostatic ion pairs) with the major cations in seawater.

Sulfate speciation Although the speciation of sulfate in seawater has been studied by many workers, the speciation results are widely divergent. Estimates for the free sulfate range from 39 to 54%. This large range of values reflects the use of different K_A^*s by the various workers. The infinite dilution K_As (Table 3) for $NaSO_4^-$,

$MgSO_4^0$, KSO_4^-, and $CaSO_4^0$ obtained by various workers are in reasonable agreement; however, the calculated K_A^*s for $NaSO_4^-$ and the directly measured values are not in good agreement. At present, it is not possible to state with certainty which value for the K_A^* of $NaSO_4^-$ is more reliable. Since most experimental evidence indicates that the K_A^* of KSO_4^- is larger than $NaSO_4^-$, I prefer the estimated values of Atkinson et al (2). The recent estimates of the speciation by Dyrssen & Wedborg (83) are in error because of the use of an unreliable value for the K_A^* of $MgSO_4^0$. In summary, the estimates for speciation for sulfate by Atkinson et al are preferred.

Bicarbonate speciation Most of the bicarbonate is in the free form (63 to 81%). The wide spread of values obtained by various workers reflects the different values used for the K_A^*s. The variation of the values used for the K_A^* for $NaHCO_3^0$ cause the major differences. The estimated K_A^* ($NaHCO_3^0$) determined by Atkinson et al (from 286) is twice as large as the directly measured values (158), as well as the earlier estimates of Garrels & Thompson (123). The lower K_A^* ($NaHCO_3^0$) is probably more reliable. The estimated values for K_A^* ($MgHCO_3^+$) and K_A^* ($CaHCO_3^+$) are also larger than the directly measured values (158). Because the K_As used to estimate K_A^* ($MgHCO_3^+$) and K_A^* ($CaHCO_3^+$) are in good agreement with recent measurements (7, 287), the errors are the result of either errors in the direct measurements of K_A^*, or errors in the methods used to estimate the free ion activity coefficients. Because it is hard to believe that such a large error could be made in estimating activity coefficients, I prefer the estimated K_A^*s for $MgHCO_3^+$ and $CaHCO_3^+$. It is thus my opinion that the original estimates made by Garrels and Thompson are the most reliable.

Carbonate speciation Unlike the other major anions in seawater, the free CO_3^{2-} is very small (~9%). Although the estimates for the free CO_3^{2-} are in good agreement, the percentage of CO_3^{2-} paired to the various cations is quite variant. These differences are due to the various K_A^*s used by the various workers. The K_A^* ($NaCO_3^-$) estimated by Atkinson et al (from 286) is in reasonable agreement with the direct measurements of Butler & Huston (36), whereas the estimates of Garrels & Thompson (123) are in excellent agreement with the direct measurements of Hawley (158). Which values are correct is not certain. The estimated K_A^* for $MgCO_3^0$ is in excellent agreement with the direct measurements of Hawley. This may be fortuitous, however, because Hawley has also postulated the existence of $Mg_2CO_3^{2+}$ ($K_A^* = 386.6$), $MgCaCO_3^{2+}$ ($K_A^* = 1040.3$), and $Ca_2CO_3^{2+}$ ($K_A^* = 2597$). The estimates for the K_A^* of $CaCO_3^0$ are quite variable. The value estimated by Atkinson et al (2) appears to be too high compared to the experimentally determined value as well as the earlier estimate of Garrels & Thompson (123). It is not possible to state with certainty which values are correct. Before one can believe the new species postulated by Hawley (158) exist, new measurements should be made on the $CaCO_3$ system, perhaps using the techniques used by Nakayama (287) to study $MgCO_3^0$ pairing. If the new species do exist, they will be very important species to consider when discussing the dissolution and precipitation of

$CaCO_3$. At present, the complete speciation of CO_3^{2-} is in doubt and the truth is probably somewhere between the estimates of Hawley and Garrels and Thompson.

Borate speciation Dyrssen & Wedborg (83) have determined the speciation of borate anion in seawater at 25°C using the stoichiometric constants of Dyrssen & Hansson (80) for the dissociation of boric acid and the formation constant of $MgB(OH)_4^+$. At a pH = 8.0 and 25°C, they found $B(OH)_3 = 80.2\%$, $B(OH)_4^- = 15.1\%$, and $MB(OH)_4^+ = 4.77$ ($M = Ca^{+2} + Mg^{2+}$) of the total boric acid in seawater (i.e. 76% of $B(OH)_4^-$ is free).

The authors have neglected to consider the formation of $NaB(OH)_4^0$ ion pair, however, from an examination of the activity coefficient data of NaCl and $NaB(OH)_4$ solutions (371), it appears that $B(OH)_4^-$ is about 30% paired with Na^+ at $I = 0.7$. We presently are examining the Na^+ activity in $NaB(OH)_4$ and NaCl solution by using a specific ion electrode to determine a reliable stoichiometric constant for the formation of $NaB(OH)_4^0$. The recent results (which I prefer) of Byrne & Kester (37) for the speciation of $B(OH)_4^-$ are given in Table 6.

Fluoride speciation Elgquist (92) recently determined the speciation of fluoride in seawater at 25°C. He determined the stability constants of MgF^+ and CaF^+ in 0.1, 0.4, 0.7, and 1.0 *M* NaCl solutions using a fluoride ion selective electrode. His results (Table 4) are in reasonable agreement with the earlier work of Connick & Tsao (47). Elgquist (92) found 48.4% free F^-, 49.4% MgF^+, and $2.1_5\%$ CaF^+ in seawater at 25°C using his constants. Dyrssen & Wedborg (83) made a similar estimate; they obtained $F^- = 51.1\%$, $MgF^+ = 46.9\%$, and $CaF^+ = 2.0\%$ of the total fluoride, which is in excellent agreement with the estimates of Atkinson et al (2). These estimates are probably in error because these workers neglected the formation of NaF^0.

Duer, Robinson, and Bates (76, 332) have recently determined K_A for the formation of NaF^0 from conductance and activity coefficient data. They found from their conductance data $K_A = 0.29$ to 1.41 using ion size parameters of 3.0 to 7.0 Å. By using $a = 3.98$ Å (the hydrated radius), they obtained $K_A = 0.58$ from conductance data, which is consistent with the value of $K_A = 0.53$ obtained from activity coefficient data. The stoichiometric K_A^* calculated from the results of Duer et al (76) are in reasonable agreement with the value of K_A^* obtained by Butler & Huston (35) in 1 *M* NaCl solutions from potentiometric data. If $a = 2.31$ Å (the sum of the crystal radii) is used, the Pitts conductance equation yields $K_A = 0$, whereas the sum of the hydrated ions $a = 3.9$ Å in the Pitts equation gives $K_A = 0.55$. Thus it is important to know what ion size parameter is used when determining K_A from conductance data as well as other physical chemical data. It is necessary to make this point clear because conductivity data (or other physical chemical data) can not be used to unequivocally prove or disprove that an ion pair exists in solution. Supporting evidence from a number of physical chemical experiments is needed before one can be sure that ion pairing actually exists and is not just a convenient mathematical method of fitting the data.

Since NaF^0 appears to exist, Robinson et al (332) suggest the use of KF solutions to calibrate the F^- specific ion electrode (frequently used in marine chemistry).

MINOR ELEMENT SPECIATION Sillen (343) calculated the main dissolved species of some minor elements in seawater. His calculations were based on tabulated stability constants using pH = 8.1 and pE = 12.5. These principal species were listed by Goldberg et al (135–138), who also commented on the speciation of minor constituents in seawater. In recent years, a number of workers have considered the speciation of minor elements in seawater. In this section, we will briefly review a portion of their results. It should be pointed out that the use of ion pairing methods for minor elements is difficult because of: 1. the lack of reliable concentrations of the elements; 2. the lack of reliable thermodynamic data; 3. the lack of direct experimental proof that the complexes exist; and 4. the lack of knowledge of the concentrations of organic ligands that normally form stronger complexes than inorganic ion pairs.

Although the state of the minor elements is controlled by interactions with biological and geochemical matter, it is still important to know the chemical state of these minor elements to understand different separation procedures and the chemical reactions of minor elements.

Proton speciation The speciation of H^+ in seawater has been examined by a number of workers. Most of the H^+ is free and the complexed H^+ is mainly in the form of HSO_4^- whereas a small amount is in the form of HF^0. If we use the apparent association constants of HSO_4^-, HF, and total activity coefficient for H^+, $\gamma_T(H^+) = 0.74$ obtained by Culberson et al (59), one obtains 67.2% of the H^+ in the free form, 32.5% as HSO_4^-, and 0.3% as HF^0. Because the K_A^* as determined by Dyrssen & Hansson (80) are in agreement with the values of Culberson et al, similar results will be obtained from their work. Elgquist & Wedborg (93) recently determined the stoichiometric formation constant $K_A^*(HSO_4^-) = 17.0$, using a UV spectrophotometric method. Their value yields 66.7% free H^+ and 33.0 HSO_4^- and 0.3‰ HF. I prefer the estimates made with the K_A^*s of Culberson et al (59) and Dyrssen & Hansson (80), although the differences are small. Using the $K_A = 95.5$ for the formation of HSO_4^- and the total activity coefficients (γ_T) given in Table 1 and assuming $\gamma_T(HSO_4^-) = 0.68$, we obtain $K_A^* = 14.1$, which is in good agreement with the measured values. The activity of H^+ (a'_H) (and OH^-, a'_{OH}), determined from pH measurements (5) using infinite dilution buffers, does not yield true activities due to liquid junction potentials (158). The true activity of H^+, $a_H = a'_H/1.13$, has been estimated by Hawley (158).

Hydroxide speciation Atkinson et al (2) have estimated the speciation of OH^- in seawater. Their results are given in Table 5. Most of the OH^- (84%) is ion paired with Mg^{2+} and only ~15% of the OH^- is free in seawater.

Phosphate speciation Dyrssen & Wedborg (83) have recently determined the speciation of phosphate in seawater at 25°C. Using the apparent constants of

Kester & Pytkowicz (192), they determined (for a pH = 8.0) $[H_2PO_4^-]_T$ was 0.66%, $[HPO_4^{2-}]_T$ was 84.9%, and $[PO_4^{3-}]_T$ was 14.4% of the total dissolved phosphate. Using estimated constants for the divalent metal ions ($M = Ca^{2+} + Mg^{2+}$), they obtained the following speciation: $MHPO_4^0 = 68.4\%$, $MPO_4^{2-} = 16.5\%$, $MPO_4^- = 14.4\%$, $H_2PO_4^+ = 0.18\%$, and $PO_4^{3-} = 0.023\%$ of the total phosphate. Most of the phosphate is in the uncharged form which may be an important finding for membrane transport and adsorption of phosphate in the marine environment.

Recently, there have been two studies on formation of calcium complexes with phosphate ions. Chughtai et al (46) have determined the formation constants of $CaH_2PO_4^+$, $CaHPO_4^0$, and $CaPO_4^-$ at 25 and 37°C; while McDowell et al (246) have determined the K_A for $CaH_2PO_4^+$ and $CaHPO_4^0$.

Iron speciation Kester & Byrne (198) have recently estimated the chemical forms of dissolved iron in seawater. In the pH range of seawater the most significant forms are $Fe(OH)_2^+$ and $Fe(OH)_4^-$. They found that at the pH of seawater organic ligands cannot successfully compete with OH^- for Fe^{3+}. By relating the chemical forms of iron to the solubility of ferric hydroxide, they concluded that 10 to 50% of the observable iron in seawater may be in solution.

Heavy metal speciation (Zn, Cd, Hg, Pb) Sillen (343) and Goldberg (137) have listed the principal species of the heavy metals in seawater. Recently, Dyrssen & Wedborg (83) and Zirino & Yamamoto (394) have examined the complexing of heavy metals in seawater. Zirino & Yamamoto developed a pH dependent model for the complexing of Cu, Zn, Cd, Pb with Cl^-, SO_4^{2-}, HCO_3^-, CO_3^{2-}, and OH^-. With the exception of Cd, the distribution of all the metal species varies greatly with changes in pH. They neglected complexes with Br^- and F^-, which should not cause a serious error because of the low concentrations. The most serious limitation of their model is the omission of mixed liquid and polynuclear complexes. Dyrssen and Wedborg also considered the complexing of heavy metals with the same ligands considered by Zirino and Yamamoto; however, they considered mixed polynuclear complexes, as well as Br^-, F^-, and glycine. A comparison of results obtained in these studies at pH = 8.1 is made below.

Copper Zirino & Yamamoto (394) found 90% as $Cu(OH_2)^0$, 87% as $CuCO_3^0$, 1% as $CuOH^+$, and 1% as Cu^{2+}. Dyrssen & Wedborg (83) found 65.2% as CuOHCl, 21.6% as $CuCO_3^0$, 5.8% as $CuCl^+$, 3.7% as $CuOH^+$, 1.6% as $CuCl^+$, 3.7% as $CuOH^+$, 1.6% as $CuCl_2^0$, 0.7% as $ClCl_3^-$, 0.7% as Cu^{2+}, and 0.1% as Cu-glycine and 0.5% as $CuCl_4^2$. Both studies indicate that very little free copper exists in seawater, and most of the complexed copper is in the form of neutral ion pairs with OH^- and CO_3^{2-}.

Zinc Zirino & Yamamoto (394) found 62% as $Zn(OH)_2^0$, 17% as Zn^{2+}, 6.4% as $ZnCl^+$, 5.8% as $ZnCO_3^0$, 4% as $ZnSO_4^0$, and 4% as $ZnCl_2^0$. Dyrssen & Wedborg (83) found 44.3% as $ZnCl^+$, 16.1% as Zn^{2+}, 15.4% as $ZnCl_2^0$, 12.5% as $ZnOHCl^0$,

3.3% as $ZnCO_3^0$, 2.3% as $ZnOH^+$, 2.3% as $ZnCl_4^{2-}$, 1.9% as $ZnSO_4^0$, 1.7% as $ZnCl_3^-$, and 0.3% as $ZnHCO_3^+$. Both workers yield similar results for free Zn, but quite different results for the complexed forms. Both studies indicate that a large percentage of the complexed Zn is in a neutral complex.

Cadmium Zirino & Yamamoto (394) found 51% as $CdCl_2^0$, 39% as $CdCl^+$, 6% as $CdCl_3^-$, 2.5% as Cd^{2+}, and ~1.0% as $CdCO_3^0$. Dyrssen & Wedborg found 35.1% as $CdCl_2^0$, 32.7% as $CdCl_3^-$, 27.2% as $CdCl^+$, 2.7% as $CdOHCl^0$, 1.7% as Cd^{2+}, 0.2% as $CdCO_3^0$, 0.2% as $CdSO_4^0$, and 0.1% as $CdBr^+$. Both workers agree that the major ion pairs of Cd are with Cl and that very little free Cd exists, which is in general agreement with Goldberg's early estimates (83% as $CdCl^+$, 16.7% as Cd^{2+}, and 1% as $CdSO_4^0$).

Lead Zirino & Yamamoto (394) found 80% as $PbCO_3^0$, ~11% as $PbCl^+$, ~3% as $PbCl_2^0$, and 2% as Pb^{2+}. Dyrssen & Wedborg (83) found 42% as $PbCl_2^0$, 18.9% as $PbCl^+$, 10.2% as $PbOH^+$, 9.2% as $PbCl_3$, 8.8% as $PbOHCl^0$, 4.5% as Pb^{2+}, 3.6% as $PbCl_4^{2-}$, 1.4% as $PbHCO_3^+$, 0.5% as $PbSO_4^0$, 0.4% as $PbCO_3^0$, and 0.1% as $PbBr^+$. Both studies indicate that most of the Pb is complexed in seawater, but differ significantly concerning which is the predominant ion pair.

The significance of these models must await further experimental measurements. The pH dependency of the complexes may be important when the particles sink and are incorporated into the sediments (where the pH is lower) and may be released back into solution. The Cu and Zn complexes with OH^- might cause a significant fraction of these species to be colloidal or absorbed on particulate material in seawater.

Recently, polarographic studies have been made (4, 26, 391–393) on some of the heavy metals. Baric & Branica (4) have examined the complexes of Zn and Cd in seawater. They postulated that Zn exists mainly as the free ion and $ZnOH^+$, but Cd is present as $CdCl^+$. Zirino & Healy (391–393), however, have shown that the neutral complexes $ZnCO_3^0$ (4%) and $Zn(OH)_2^0$ (75%) may be present rather than $ZnOH^+$. They also found 80% of the cadmium to be $CdCl^+$ and $CdCl_2^0$ whereas lead was predominately $PbCO_3^0$. For zinc they found no evidence of a chelating influence of natural organic compounds in seawater.

Maljkovic & Branica (237) have also examined the formation of cadmium complexes with EDTA in seawater and NaCl solutions using polarography. Their results indicate that the degree of complex formation of Cd with EDTA is diminished in seawater by the formation of $CaCl^+$ and by the competition between Ca^{2+} and Cd^{2+} for EDTA.

Minor element organic complexes Few studies have been made on the complexing of minor metal elements with organic ligands. Although reliable association constants are available for metal organic ligand complexes, reliable estimates of organic ligand speciation cannot normally be made because we do not know which organic components exist in seawater and in what concentrations they are present. Goldberg (137) has discussed the relative complexing of some metal ions with organic ligands. Mangel (238) has recently calculated the percentage of complexed ion concentrations in seawater for 26 metals. By examining the ratio of stability constants (K^*_{MB} and K^*_{MA}) and the ligand concentrations ($[A^-]$ and $[B^-]$), he has calculated the relative percentage of the complexed species, $[MA]/[MB] = [A^-]K^*_{MB}/[B^-]K^*_{MA}$. He apparently has used infinite dilution stability constants without making the appropriate activity coefficient corrections; thus, his results are not reliable.

Dyrssen & Wedborg (83) have examined the complexing of divalent cations with acetic acid and glycine in seawater. They found for CH_3COOH that 81.6% exists as CH_3COO^-, 15.9% as $MgO_2CCH_3^+$, 2.5% as $CaO_2CCH_3^+$. For NH_2CH_2COOH, they found 40.7% as NH_2CH_2COOH or $NH_3^+CH_2COO^-$, 1.2% as $NH_2CH_2COO^-$, 57.3% as $MgO_2CCH_2NH_2^+$, 0.7% as $CuO_2CCH_2NH_2^+$, and 0.1% as $CaO_2CCH_2NH_2^+$. The significance of these results must await further experimental studies.

Effect of Temperature and Pressure on Speciation

In recent years, a number of workers (196, 206, 259, 261, 322) have attempted to estimate the effect of temperature and pressure on the speciation of seawater. The temperature (ΔH^0) and pressure ($\Delta \bar{V}^0$) dependence for some of the sea salts at infinite dilution have been measured by a number of workers (3, 7, 102, 108, 179, 180, 183, 284, 366) at 25°C. The ΔH^0 results, however, are not in very good agreement. For example, the ΔH^0 for $NaSO_4^-$ varies from -0.49 to 1.1 kcal/mol (3, 179); the ΔH^0 for $MgSO_4^0$ varies from 0.51 to 5.7 kcal/mol (180, 183, 208, 284); and the ΔH^0 for $CaSO_4^0$ varies from 0.81 to 1.50 kcal/mol (180, 208). The only infinite dilution $\Delta \bar{V}^0$ results available for a sea salt is for $MgSO_4^0$ (108). To use these thermodynamic values in seawater, one must estimate the temperature and pressure dependencies of activity coefficients in seawater (which are normally not known). Lafon (206) has extended the ion pairing model for seawater from 0 to 60°C and 1 to 1000 bars by assuming the temperature and pressure dependence of activity coefficients are negligible. His results are shown in Tables 7 and 8. As is quite apparent, there appears to be no regular temperature dependencies. Also given in Table 7 are the recent results of Kester and Pytkowicz based on direct measurements of ΔH^* (in the medium of seawater) for $NaSO_4^-$ and $MgSO_4^0$ and $\Delta \bar{V}^*$ for $NaSO_4^-$. The cause of the different temperature effects are the result of differences in the signs of ΔH^0 in pure water (positive) and the ΔH^* in seawater (negative). We presently are making titration calorimetric measurements (45) and heat of dilution measurements (179, 180, 208) on these and other ion pairing systems in pure water and seawater in an attempt to resolve these differences.

The estimates for the effect of pressure on the speciation made by various workers (Table 8) are in general agreement, except for $MgSO_4^0$ and $CaSO_4^0$. The percentage of free ions all increases whereas the percentage of the ion pairs decreases with increasing pressure as one would expect (108, 109, 112, 147, 148, 163). The differences in the effect of pressure on $MgSO_4^0$ and $CaSO_4^0$ reflect the choice of $\Delta \bar{V}^*$ for $NaSO_4^-$, $MgSO_4^0$, and $CaSO_4^0$ used in the calculations. As discussed elsewhere (259), we feel that our estimates are more reasonable. Recently,

Table 7 Effect of decreasing the temperature from 25 to 0°C on the speciation of the major ions in 35‰ seawater at 1 atm

Ion	Δ% Free	Δ% SO_4^{2-} pair	Δ% HCO_3^- pair	Δ% CO_3^{2-} pair
Ca^{2+}	−0.4[a]	+0.9[a], 0[b]	−0.4[a]	−0.11[a]
Mg^{2+}	+3.0[a]	−2.6[a], +2[b]	−0.3[a]	−0.1[a]
Na^+	+0.6[a]	−0.6[a], +9[b]	+0.01[a]	—
K^+	+0.2[a]	−0.2[a]	—	—

Ion	Δ% Free	Δ% Ca^{2+} pair	Δ% Mg^{2+} pair	Δ% Na^+ pair	Δ% K^+ pair
SO_4^{2-}	+14.9[a], −11[b]	+0.3[a]	−5.2[a]	−9.9[a]	−0.07[a]
HCO_3^-	+8.9[a]	−1.8[a]	−7.5[a]	+0.4[a]	—
CO_3^-	+1.2[a]	−1.3[a]	+12.0[a]	−11.8[a]	—

[a] Ref. 206.
[b] Ref. 196.

Table 8 Effect of increasing the pressure from 1 to 1000 atm on the speciation of the major ions in 35‰ seawater at 25°C

Ion	Δ% Free	Δ% SO_4^{2-} pair	Δ% HCO_3^- pair	Δ% CO_3^{2-} pair
Ca^{2+}	+3.1[a]	−2.8[a], +1[b], 0[c]	−0.4[a]	0.1[a]
Mg^{2+}	0.0[a]	+0.2[a], +3[b], −2[c]	−0.2[a]	0.0[a]
Na^+	+0.8[a]	−0.8[a], −15[b], −12[c]	−0.01[a]	—
K^+	+0.4[a]	−0.4[a]	—	—

Ion	Δ% Free	Δ% Ca^{2+} pair	Δ% Mg^{2+} pair	Δ% Na^+ pair	Δ% K^+ pair
SO_4^{2-}	+14.9[a], +11[b], +14[c]	−1.0[a]	+0.4[a]	−34.2[a]	−0.14[a]
HCO_3^-	+10.6[a]	−1.6[a]	−5.1[a]	−3.9[a]	—
CO_3^{2-}	+10.5[a]	−0.4[a]	−8.1[a]	−2.0[a]	—

[a] Ref. 206.
[b] Ref. 196.
[c] Ref. 259.

we have analyzed the effect of pressure on ion pair formation by using the relationship (98)

$$RT\ln(K^P/K) = -\Delta\bar{V}^0(P-1) + 1/2\Delta\bar{K}^0(P-1)^2 \qquad 22.$$

This relationship has been used (226, 263, 372) to represent the pressure effect of acid and base ionization more reliably than the equations derived by Owen & Brinkley (296). We (98) have also found that the same linear correlation between $\Delta \bar{V}^0$ and $\Delta \bar{K}^0$ that has been shown to exist for acid-base ionization (226) is also valid for ion pairing systems, which is consistent with the hydration models of North (295) and Marshall (241). Further experimental volume and pressure work on the formation of ion pairs is needed, both in pure water and seawater. Such results could be used to study the structure of the ion pair as well as be useful in estimating the effect of pressure on their speciation. One might expect the ΔV and ΔH of transfer for ion pairing processes to be similar to acid-base equilibria; however, this is not necessarily the case, because the ion pairs may lose water molecules when transferred into seawater (inasmuch as they are solvent separated), whereas the thermodynamic transfer functions of acids and bases will not change very much.

Although the ion pairing model for ionic interactions in seawater has been quite successful in predicting activity coefficient data, there are a number of other methods that can be used to study ionic interactions in seawater. These methods are reviewed in the next section.

Non-Ion Pairing Methods

In recent years, there has been a number of major advances made on the interpretation of interactions in multicomponent electrolyte solutions (6, 151, 212–215, 257, 258, 265, 273). In this section we briefly review these recent developments and discuss the application of some of these methods to seawater solutions. These new methods should provide another approach to study the physical chemistry of seawater (i.e. one that differs from the ion pairing model).

SPECIFIC INTERACTION MODELS An alternate approach to the ion pairing method to study short-range interactions is the specific interaction model of Brønsted (33). Brønsted's theory assumes that like-charged ions do not have specific interactions. Although this assumption is similar to Bjerrum's (19) ion pairing model, in the Brønsted theory the specific interaction can be electrostatic and nonelectrostatic (whereas the ion pairing model assumes that the interaction between plus and minus ions is entirely electrostatic). Guggenheim (143, 144) and others (150, 211, 257, 301, 340) have extended the Brønsted theory while keeping the assumption that like-charged ions have little or no specific interactions. Equations have been derived (331) which include both oppositely charged pair and triplet interactions. The early Guggenheim equations (143) include only oppositely charged ion interactions, whereas his more recent extensions (144) include plus-plus and minus-minus terms. These earlier methods of treating the interactions in multicomponent electrolyte solutions are reviewed in detail in the classical texts on electrolyte solution chemistry (150, 211, 331). Leyendekkers (214, 215) has recently estimated the total activity coefficient γ_T for a number of salts and ions in seawater using these equations (Table 2) (212–215). As is quite apparent, the Guggenheim equations yield excellent estimates for the γ_T of ions in the medium of seawater. Robinson & Wood (334) have also estimated γ_T for salts in seawater, using a similar method (327, 328). Since both methods yield excellent results, it is impossible, at present, to say which method is more reliable. The difficulty in using this method for minor species in seawater is in the lack of reliable data. Bates, Staples & Robinson (6) have recently developed a hydration method to determine single ion activity coefficients for ions. Elgquist & Wedborg (93) have recently used these methods to estimate the γ_F for free ions in seawater, in agreement with those determined by the mean salt method.

One of the basic findings of the specific interaction model is the empirical relationship (150, 211, 331) called Harned's Rule

$$\log \gamma_2 = \log \gamma_2(0) - \alpha_{23} m_3 \qquad 23.$$

where γ_2 is the mean activity coefficients of electrolyte 2 in the presence of electrolytes 3 (of molality m_3), $\gamma_2(0)$ is the mean activity coefficient of electrolyte 2 in itself at the ionic strength of the mixture, the coefficient α_{23} is the specific interaction term. This relationship has been shown to hold for a number of two-salt systems, including many of the sea salts (131).

The difficulty of separating the ion pairing model from the specific interaction model is exemplified by the recent work of Butler & Huston (36). They (36) have determined the activity coefficient of NaCl in $NaHCO_3$ and Na_2CO_3 solutions using potentiometric methods. They found that Harned's Rule is obeyed over the entire ionic strength range (0.5 to 3.0) with a coefficient $\alpha = 0.047 \pm 0.03$ for either CO_3^{2-} or HCO_3^-. They also found that the results could be interpreted in terms of an ionic pairing model.

STATISTICAL MECHANICAL TREATMENTS Although the statistical mechanical treatments are complicated, in recent years major developments have been made in the application of these methods to multicomponent electrolyte solutions (257, 265, 387). The major advancement was made by developing simple methods to calculate osmotic coefficients in terms of repulsive and attractive forces of clusters of molecules. The radial distribution function (which gives the probability of finding a molecule at a specified distance from an arbitrary central molecule) has been directly related to the properties of ionic solutions. These recent applications (118) of the virial cluster expansion to electrolyte solutions appear to be the most important theoretical developments made since the Debye-Hückel theory was first proposed. It should be pointed out that when one applies the cluster expansion theory to ionic solutions, no attempt is made to separate the ionic interactions into electrostatic and nonelectrostatic effects.

Although the coefficients of the cluster expansions are complex functions of ionic strength, for mixing processes at constant ionic strength the coefficients are constant and can be related to specific type ion-ion interactions. The excess enthalpy ($\Delta_m H$) and volume ($\Delta_m V$) of mixing NX with MX at constant ionic strength is given by

$$\Delta_m H = y(1-y)RT\,\mathrm{I}^2 h_{\mathrm{N,M^X}} \qquad 24.$$

$$\Delta_m V = y(1-y)RT\,\mathrm{I}^2 v_{\mathrm{N,M^X}} \qquad 25.$$

where h_{N,M^X} and v_{N,M^X} are the coefficients for the interactions of N and M in the presence of X. Both the ion pairing model and the specific interaction model assume that like-charged ions do not have specific interactions. Thus, $\Delta_m G$, $\Delta_m H$, and $\Delta_m V$ should be zero for the mixing of NX and MX at constant ionic strength. The cluster theory, however, predicts that like-charged ions should have specific interactions and that these interactions should be more important than triplet interactions. To understand the interactions that occur in seawater, it is important to study all of the interactions (cation-cation, anion-anion, and cation-cation). For example, to understand the interactions of the major cations (Na^+ and Mg^{2+}) and anions (Cl^- and SO_4^{2-}) in seawater, one must examine all the possible interactions (Na^+–Na^+, Na^+–Mg^{2+}, Mg^{2+}–Mg^{2+}, Na^+–Cl^-, Mg^{2+}–Cl^-), which can be represented by

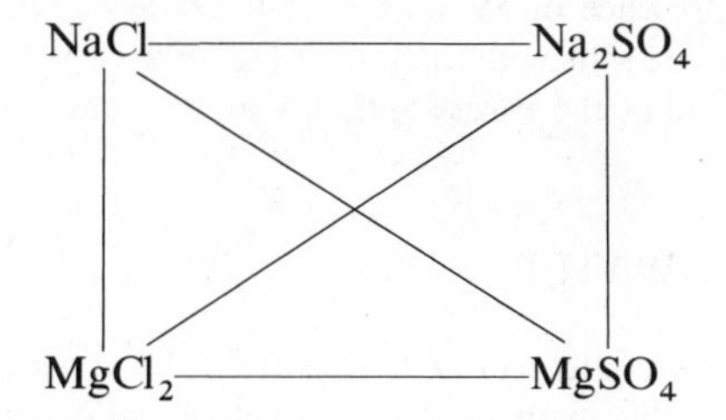

The excess properties around the side of this diagram can be used to study plus-plus and minus-minus interactions. To study plus-minus interactions (both coulombic and noncoulombic), which are the more dominant terms, one must understand the interactions in the single salt solutions as well as the mixtures. Since the plus-minus interactions are quite large for $MgSO_4$ solutions many workers have compared its properties to the sum obtained from $MgCl_2 + Na_2SO_4 - 2\,NaCl$ (259, 272).

YOUNG'S RULE One of the most useful generalizations to be developed for multicomponent electrolyte solutions is Young's Rule (257, 265, 390) and the modifications made by Wood and co-workers (327, 328, 386, 387). Young's Rule is given by

$$\Phi = \sum E_i \phi_i \qquad 26.$$

where Φ is any apparent equivalent property (such as volume, expansibility, compressibility, heat capacity, etc) of seawater, E_i is the equivalent ionic fraction of species i (equal to e_i/e_T, where e_i is the equivalent molality of species i and $e_T = \sum e_i$ is the total equivalent molality of the solution) and ϕ_i is the apparent equivalent property of species i at the total ionic strength corresponding to a given salinity. This equation essentially states that as a first approximation the excess properties of mixing electrolyte solutions at constant ionic strength could be neglected. This relationship is a very useful first approximation and has been shown to be applicable to many multicomponent electrolyte solutions including seawater solutions (257, 265).

Wood & Anderson (386) proposed an improved equation for Young's Rule that incorporates a number of other experimental findings (the cross square rule). The equation for the mean equivalent volume of a multicomponent electrolyte solution is given by

$$\Phi_V = \sum_{MX} E_M E_X \phi_V(MX) + (RT I^2/m) \sum_{M<N,X} E_M E_N E_X v_{M,N^X}$$

$$+ (RT I^2/m) \sum_{X<Y,M} E_X E_Y E_M v_{X,Y^M} \qquad 27.$$

where E_M is the equivalent fraction of M, $\phi_V(MX)$ is the apparent equivalent volume of the pure electrolyte MX at ionic strength I and v_{M,N^X} and v_{X,Y^M} are the interaction parameters, respectively, for the mixing of N and M in the presence of X and the mixing of X and Y in the presence of M. Except in the case of free energies, the excess properties for mixing sea salts with each other are unknown. In my laboratory, we are presently determining the excess enthalpy and volume of mixing sea salts with one another.

CHEMICAL PROCESSES IN SEAWATER

The ultimate aim of understanding the ionic interactions in seawater is being able to understand the mechanisms and rates of the many chemical processes that occur in the oceans. The estimation of total activity coefficients or total activities of solutes in seawater are of great importance because the data are very useful in interpreting various geochemical and biochemical equilibria. In recent years, many workers have studied the chemical equilibria and kinetics of processes of importance to the marine environments. Although it will not be possible to discuss all these studies, we will briefly outline some of the recent results in this section.

Acid Ionization

The thermodynamics of acid and base ionization in seawater has been studied by a number of workers (86, 218, 314, 348). Most of the recent studies have been made on the carbonic acid (336, 348) and boric acid systems. Edmond & Gieskes (86) have reviewed the thermodynamics of the CO_2 system in seawater. Hansson & Dyrssen (152–156) have redetermined the ionization constants of carbonic acid, water, and boric acid using the ionic medium method. They also developed a new ionic medium pH scale (based on a TRIS buffer) for seawater work that gives reproducible pH measurements, regardless of what reference electrode is used. Although many workers (310) will continue to use the infinite dilution scale for pH measurements, the author believes that the new pH scale should be used for thermodynamic measurements. The methods used to study the carbonate system have been improved in recent years (81, 82, 85, 156, 264, 297). By using the Gran titration technique (81, 82, 85, 142, 156) the total alkalinity (A_T) and total CO_2($\sum CO_2$) can be determined with great precision. Titration calorimetry (45) has also been used to determine A_T and $\sum CO_2$ with a comparable precision (264). The effect of pressure on the carbonate equilibria in seawater has been examined by direct measurements (57, 58, 70, 97, 157) and by using partial molal volume data (24, 94–96, 371, 372). For the effect of pressure on ionization of carbonic acid and boric acid, the directly measured values and estimated values from partial molal volume are in excellent agreement (261). Wangersky (370) has commented on the control of the pH of seawater by ion pairing processes. As pointed out by others (197, 323), the effect of ion pairing in controlling the pH is small compared to the effect of geochemical and biochemical processes (298).

Ben-Yaakov & Goldhaber (13) have used the ion pairing model to estimate how changes in the composition of seawater influence the carbonic acid apparent constants. The model predicts apparent constants that are in reasonable agreement with the measured values. The estimated value for K_2 is $\sim 6\%$ lower than the measured value (227). This discrepancy is probably the result of the incompleteness of the model; although some workers (17, 359) feel that Lyman's values (227) may be too low.

Gas Solubility

A number of recent studies have been made on the solubility of gases in seawater (101, 129, 221, 279–283, 352). Kester (200) has recently reviewed the dissolved gases other than CO_2 in seawater; a number of recent studies have been made on the solubility of CO_2 in seawater (221, 279, 283). These results are in good agreement with earlier estimates (86). Murray et al (280–282) have redetermined the solubility of Ar, N_2, and O_2 in seawater. Weiss (374–377) has analyzed these and other results by using the Setchenow equation

$$\log(\alpha/\alpha^0) = kS(^\circ/_{oo}) \qquad 28.$$

where α and α^0 are the Bunsen coefficients in seawater and pure water and k is the salting out coefficient (225, 243). The effect of pressure on the solubility of gases in water and seawater have been determined by a number of workers (101, 129, 352). Bradshaw (24) has directly determined the $\bar{V}$ of CO_2 in seawater, while Millero & Berner (261) have estimated the $\bar{V}$ by using data for the effect of pressure on the dissociation of carbonic acid (57, 58, 70). A number of studies have also been made on rate of dissolution of gases (170, 200, 378).

Mineral Solubility

A number of studies have been made on the solubility of minerals in recent years (17, 18, 34, 42, 120, 157, 177, 181, 285, 299, 313, 316, 336). Most of these studies have been concerned with the compensation depth (11, 86, 157, 159, 218, 305, 309, 319) and lysocline (14, 18, 276–278, 300) of $CaCO_3$ minerals in seawater. Most of the recent work has been concerned with the kinetics of dissolution (14, 17, 18, 222, 259, 276–278, 290, 291, 299, 300, 305, 309) and crystallization (222–224, 250, 288, 289, 291, 308, 325, 326, 349). Berner & Morse (17, 18, 276), for example, have examined the kinetics of dissolution of $CaCO_3$ in seawater and the formation of the lysocline. Nancollas and co-workers (222–224, 250, 288–291, 325, 326) have made a number of excellent studies on the rates of crystallization and dissolution of minerals in water. The inhibition effects of ionic solutes (185, 278) and organic solutes (42, 181, 308, 356) on the solubility of minerals has also been investigated. Ben Yaakov & Goldhaber (13) have estimated that the solubility product presently being used (86) to calculate the solubility of $CaCO_3$ in seawater may be 6% too high. This difference may be responsible in part to the differences in the saturation depth determined by the saturometer method (11) and by calculations using apparent constants (157).

The effect of pressure on the equilibrium solubility of $CaCO_3$ has been determined by direct measurements (157, 313) and by partial molal volume measurements (23, 73, 261). The estimated effects of pressure on the solubility are not in agreement with the measured values. At present, it is not possible to state with certainty what causes this difference. It could be caused by nonequilibrium in the direct measurements (because of inhibition effects) or to errors in the estimation of the volume change. For example, the $\bar{V}$ used for the mineral phase is the crystal molar volume. If the equilibrium exists between a layer of $CaCO_3$ molecules on the surface of the mineral and the solution, this volume would be incorrect (because it includes crystal void space). Further studies are needed on both the thermodynamics and kinetics of mineral solubilities, both in pure water and seawater.

PHYSICAL PROPERTIES OF SEAWATER

As mentioned in a previous section, a number of major advances have been made in the interpretation of the thermodynamic properties of multicomponent electrolyte solutions. In this section, we will review some of the recent measurements on the physical chemical properties of seawater and demonstrate how the properties of seawater can be examined in terms of its major components. A more thorough review is given elsewhere (265).

Application of Young's Rule to Seawater

As mentioned previously, except for excess free energies, the excess properties for sea salts are not known. Thus we are forced to use only the first term of Wood and Anderson's revised form of Young's Rule. For seawater we have (265)

$$\Phi = \sum_{MX} E_M E_X \phi_{MX} + E_B \phi_B \qquad 29.$$

where ϕ_B is the apparent equivalent property for boric acid and E_B is the equivalent fraction of boric acid. The extension of Young's Rule to electrolyte and non-electrolyte solutions [i.e. the addition of $E_B \phi_B$ to Equation 29 has recently been demonstrated to be reliable for the volumes of boric acid—NaCl solutions over a wide concentration range (371)]. By making the summation for each cation over all its possible salts, one obtains the equivalent weighted cation contribution ($\phi[M \sum X_i]$)

$$\phi(M \sum X_i) = E_M E_{Cl} \phi(MCl) + E_M E_{SO_4} \phi(MSO_4) + \ldots \qquad 30.$$

The total Φ is given by

$$\Phi = \phi(\mathrm{Na}\sum \mathrm{X}_i)+\phi(\mathrm{Mg}\sum \mathrm{X}_i)+\phi(\mathrm{Ca}\sum \mathrm{X}_i)+\ldots+\phi(\mathrm{B}) \qquad 31.$$

where $\phi(\mathrm{B}) = \mathrm{E_B}\phi_\mathrm{B}$ is the boric acid contribution. It should be pointed out that by determining Φ by this method one essentially eliminates excess cation-anion interactions (possibly related to ion pair formation).

The difficulty of using these equations to calculate the apparent properties of seawater rests in the paucity of reliable physical chemical data for all the major sea salts over the concentration (0.1 to 0.8 molal ionic strength) and temperature (0 to 40°C) range of interest. Because salts such as $CaSO_4$ are not soluble in pure water at high ionic strengths, one must estimate its properties by some additivity method, e.g. $\phi(CaSO_4) = \phi(CaCl_2)+\phi(Na_2SO_4)-2\phi(NaCl)$. By dividing the apparent property into an infinite dilution term (Φ^0) and one or more concentration terms (S′ or b)

$$\Phi = \Phi^0 + \mathrm{S'I}_V^{1/2} \qquad 32.$$

$$\Phi = \Phi^0 + \mathrm{S\,I}_V^{1/2} + b\,\mathrm{I}_V \qquad 33.$$

it is possible to simplify the use of Young's Rule. Because the infinite dilution properties are always additive, the $\phi^0(i)$s can be estimated from data on soluble salts. Thus, Young's Rule need only be applied to the concentration dependent terms (S′ and b – S is the theoretical Debye-Hückel term and can be estimated for any system).

The Φ for seawater solutions are directly related to measured physical properties by

$$\Phi = (P_{\mathrm{SW}} - P_{\mathrm{H_2O}})/n_T \qquad 34.$$

where P_{SW} and $P_{\mathrm{H_2O}}$ are, respectively, the physical property of seawater and pure water and n_T is the total equivalents. By combining Equation 34 with either 32 or 33, and noting that n_T and I_V are proportional to Cl_V (the volume chlorinity, $\mathrm{Cl}_V = \mathrm{Cl}(‰) \times d$, the density), we have

$$P_{\mathrm{SW}} = P_{\mathrm{H_2O}} + \mathrm{A'\,Cl}_V + \mathrm{B'\,Cl}_V^{3/2} \qquad 35.$$

$$P_{\mathrm{SW}} = P_{\mathrm{H_2O}} + \mathrm{A\,Cl}_V + \mathrm{B\,Cl}_V^{3/2} + \mathrm{C\,Cl}_V \qquad 36.$$

where A and A′ = const × Φ^0, B′ = const × S′, B = const × S and C = const × b. The constants A and A′ are related to ion water interactions and the constants B′, B, and C are related to ion-ion interactions. The B term is related to the theoretical Debye-Hückel ion-ion interaction and C is related to deviations from the limiting law. For the density of seawater, Equation 35 is equivalent to the Root equation (335) developed by Wirth (385) over thirty years ago.

At a given concentration any physical chemical property of seawater can be visualized as being equal to

$$P_{\mathrm{SW}} = P_{\mathrm{H_2O}} + \sum \text{ion-water interactions} + \sum \text{ion-ion interactions} \qquad 37.$$

The second term is due to the weighted ion-water interactions of the major sea salts at infinite dilution and the third term is due to the weighted ion-ion interactions of the major sea salts. The ion-ion interaction term can be split up into a theoretical Debye-Hückel limiting law term and term due to deviations from the limiting law

$$\sum \text{ion-ion interaction} = \text{Debye-Hückel term} + \sum \text{deviations from Debye-Hückel} \qquad 38.$$

The general approach of examining the physical chemical properties serves two purposes: 1. it provides the theoretical concentration dependence of the physical chemical properties of seawater; and 2. it emphasizes the importance of ion-water and ion-ion interactions of the major components of seawater.

The use of the systematic approach, as exemplified above for physical chemical properties, is preferred for the study of the physical chemical properties of seawater because: 1. it provides an integrated theory which considers all possible interactions; 2. it treats analogous properties in an analogous fashion, thus allowing for maximum systemization of knowledge in this complex area; and 3. it describes the properties of seawater as being perturbations of the properties of pure water and aqueous sea salt solutions.

Recently, it has been shown that these equations are valid for predicting and representing the density or apparent equivalent volume (265, 266, 268); expansibility or apparent equivalent expansibility (265, 266, 268); compressibility or apparent equivalent compressibility (209, 265–267); enthalpy or apparent equivalent enthalpy (265, 269); specific heat or apparent equivalent heat capacity (265, 270); and osmotic coefficients (265, 333, 334) and the viscosity (68, 265, 274) of seawater. For a more exact determination of the physical properties of seawater (Equation 37) from its major components, it is necessary to know the excess properties of all the possible combinations of the major ions. Some of the recent experimental results on the physical properties of seawater are reviewed in the next section.

Recent Experimental Results

In the last few years a number of studies have been made on the physical chemical properties of seawater. Pytkowicz & Kester (320) and others (53, 171, 172, 273, 357) have reviewed most of these early studies. In the book the author is presently writing (273) and the book edited by Horne (176) compilations on the physical properties of seawater will be given. In this section, we will briefly review some of the recent results.

VOLUME PROPERTIES OF SEAWATER

Density In recent years a number of workers have made measurements on the *P-V-T* properties of seawater (23, 38, 51, 54–56, 84, 99, 100, 105–107, 204, 205, 219, 268, 294, 355, 358, 369, 384). Presently the 1 atm densities of seawater tabulated by Knudsen (201, 358) and the compressibilities of Ekman (91) make up the tables that are currently being used in oceanography. Although many redeterminations have been made on the volume properties of seawater in recent years, many of these studies (54, 294, 384) were not made with a precision comparable to the original studies (± 1 ppm in density and ± 0.05 ppm in compressibility).

Most of the density measurements have been made in terms of the 1901 liter

(1 liter = 1.000028 dm^3) where water is assigned a density of 1 g/ml at its maximum density (187). Because the original density measurements on pure water were made on water of unknown isotopic composition, Cox et al (51) have suggested the use of water of known isotopic composition (248) as a density standard. The water selected by Cox et al was distilled deep Mediterranean ocean water (DMOW), while Menach & Girard (133, 134, 249) have suggested the use of distilled standard mean ocean water (SMOW). Recently, the relative densities of various standards waters have been measured by Emmet (100). The maximum differences between DMOW and SMOW and Miami ion exchanged water was found to be 2 ppm. Menach & Girard (249) have suggested that the absolute density of distilled SMOW be assigned the value of 999.975 kg m^{-3} at 4°C (i.e. until direct measurements are made).

Most recent density measurements (54, 204, 205) on seawater indicate that the original density measurements of Knudsen (201) are, on the average, too low by 6 to 13 ppm, which is in agreement with earlier findings (362). Cox et al measured the densities of numerous seawater samples (9 to 39‰ S and 0 to 25°C) from various world oceans. They found the densities of Knudsen are, on the average, too low by 6 ppm, which is within their quoted standard deviation of 10 ppm. Kremling (204, 205) more recently made density measurements on nine of the same samples (0 to 25°C) used by Cox et al. The average difference between his results and Cox et al is ± 8 ppm. Kremling made his measurements with a vibrating densimeter (203) that measured densities to ± 2 ppm. He found the densities of Knudsen (201) to be too low by an average of 13 ppm. From the recent work of Bradshaw (24), the combined effect of changes in pH and total alkalinity (A_T) in the ocean may be of the order of 13 ppm in density. It is, therefore, important to characterize the CO_2 system (i.e. by pH and A_T) when defining densities to within ± 10 ppm.

Expansibility Although recent density measurements in our laboratory also indicate that the densities of Knudsen are too low by ~ 10 ppm, the expansibilities or effect of temperature on the densities determined from Knudsen's tables [or the mathematical reformulations (358)] appear to be in excellent agreement with recent measurements (23, 268). It appears at present that although the absolute densities of Knudsen may be too low, the internal temperature precision appears to be excellent. In two recent studies the high pressure (to 1000 bars) expansibility of seawater (23, 38) has been directly measured. Both studies are in good agreement with each other and the values determined from Ekman's work and the sound speed equation of state (56, 105, 106, 369).

Compressibilities Many workers have questioned the reliability of Ekman's compressibility measurements in seawater (55, 84, 384). Direct measurements on the isothermal compressibility of seawater at 1 atm have been made by Lepple & Millero (209) by using a piezometric technique (255). Their results are in excellent agreement with the values determined from sound speeds (55, 219, 369) and the directly determined values of A. Bradshaw and K. E. Schleicher (personal communication). Although Ekman's absolute compressibility values are too high by ~0.25 ppm, his internal precision was very good and his results for the difference between the compressibilities of seawater and pure water (188, 189) are in excellent agreement with the recent direct measurements (209), and values calculated from sound speeds. These findings prove conclusively that the internal precision of Ekman's original work was excellent and that his absolute measurements are in error because of the manner he used to calibrate his apparatus with pure water.

Recently, two separate sets of measurements have been made on the high pressure compressibility of seawater and both agree very well with values determined from sound speeds (55, 105–107, 369). Emmet & Millero (99, 100) have directly measured the density of seawater to 1000 bars using a high pressure magnetic float system (262) to a precision within ±5 ppm. Bradshaw and Schleicher (personal communication) recently combined their high pressure expansibility measurements (23) with compressibility measurements at 10°C to 1000 bars. Both these recent compressibility studies are in excellent agreement with the values determined from sound speeds (55, 105, 106, 369).

Equation of state The ultimate aim of obtaining reliable *P-V-T* data for seawater is to define a reliable equation of state of seawater. As discussed in earlier sections, the present equation of state based on Knudsen's tables (201) and Ekman's (91) compressibilities is in error. During the last few years we have been attempting to refine the equation of state of water and seawater (99, 100, 105, 106, 369). Although other workers have derived equations of state for seawater (55, 84, 219, 384), most have been based on unreliable *P-V-T* data. Recently, we have derived an equation of state for water and seawater (105, 107, 369) by an iterative computer technique (369) from the sound speed data of Wilson (382, 383) and Del Grosso & Mader (65). The equation of state (369) is a second degree secant bulk modulus equation (229)

$$(V^{\circ}-V^{P})/P = D/(B+A_1P+A_2P^2) \qquad 39.$$

where V° and V^{P} are the specific volumes at zero (1 atm) and P is applied pressure, D is the specific volume of pure water at 1 atm (187), B is the reciprocal of the 1 atm compressibility (β), and A_1, A_2 are temperature dependent constants. This equation is more accurate (369) than others derived from sound data (55, 219) because of the manner in which the constants are derived (i.e. by directly equating sound speeds to compressibility). *P-V-T* data derived from this equation of state is in excellent agreement with the direct volume measurements of Emmet & Millero (99, 100) and A. Bradshaw and K. E. Schleicher (personal communication); and the direct expansibility measurements of Bradshaw & Schleicher (23) (which agree with the expansibilities calculated from Ekman's work). A complete comparison of these results is given elsewhere (107, 369). Since the maximum error in the compressibilities derived from sound data is within ±0.02 ppm, our equation of state based on sound speed data is more reliable than any direct measurements made to date. Because of the discrepancies (106) in the high pressure sound data (as much as 0.9 m/sec) of various workers (65, 382, 383) it is not possible, at present, to make any further improvements in the high pressure compressibilities (i.e. to better than ±0.02 ppm). Further work is needed on the high pressure sound speed in seawater.

OSMOTIC COEFFICIENT OF SEAWATER The osmotic coefficient of seawater, as well as other colligative properties such as the activity of sea salt and seawater, can be determined from vapor pressure measurements and freezing point measurements (211, 271, 331). The most reliable vapor pressure measurements of seawater are those determined by Robinson (333). The activity of sea salt has been calculated recently by Millero (265) and Leyendekkers (214). Leyendekkers (214) and Robinson & Wood (334) have recently estimated the osmotic pressure of seawater from sea salt data. Both estimates are in good agreement with the measured values. Because the colligative properties are determined from the osmotic coefficient, these results mean that one can probably estimate all the colligative properties to nearly within the experimental error of direct measurements.

As pointed out by Pytkowicz & Kester (320), the earlier workers are in disagreement concerning the freezing point of seawater. Recently, there have been two measurements that are in good agreement with each other and the earlier work of Knudsen (53, 357). Kester & Doherty (199) measured the freezing point of seawater using a number of methods and found their results agreed with the earlier work of Knudsen (53, 357). Mayneord & Murray's recent results (accurate to ±0.004°C) are quoted in Riley & Chester's book (330).

ENTHALPY OF SEAWATER The enthalpy of seawater can be used to determine the effect of temperature on the activity of sea salt and water in seawater. Bromley et al (31, 347) have measured the enthalpies of mixing and dilution of seawater from 0 to 70°C and 0 to 120‰. Connors (48) derived an empirical equation for the enthalpy over the oceanographic range (0 to 30°C, 10 to 40‰S) using the heat capacity data of Cox & Smith (52) and his enthalpies of mixing. More recently, Millero et al (269) have made heats of dilution and mixing measurements for seawater in the range of 0.4 to 40‰ S and from 0 to 30°C, using a microcalorimeter. Their enthalpy results and its temperature dependence are in good agreement with the results of Bromley et al (31, 347) and Connors (48). Millero (265) has estimated the enthalpy of seawater from sea salt data using Young's Rule. His results are in good agreement at low salinities, however, large errors (10%) occur at high salinities. These results indicate that the excess heat of mixing sea salts contribute significantly to the enthalpy unlike most of the other physical chemical properties studied. We presently are determining $\Delta_m H$ for sea salts to examine these excess terms.

HEAT CAPACITY OF SEAWATER The heat capacity of seawater is of particular interest in oceanography because it is used for various thermodynamic calculations (114). Bromley et al (29, 30, 32) have measured the heat capacity of seawater from 2 to 200°C and from 10 to 120‰ salinity. Over the oceanographic range, their results are in general agreement (±0.002 J/g deg) with the previous measurements of Cox & Smith (52) except near 0°C. Millero et al (270) have recently measured the heat capacity (to ±0.0005 J/g deg) of seawater as a function of temperature (5 to 35°C) and chlorinity (0.5 to 22‰). The relative heat capacity calculated from these results is in agreement with the temperature dependency of the enthalpy data of Millero et al (269). Their results agree with the earlier workers to within the earlier workers' experimental error. They used Young's Rule to estimate the heat capacity of seawater to within experimental error using pure water, sea salt data.

THERMAL CONDUCTIVITY The thermal conductivity, although not normally valid for calculations in the open oceans (due to turbulent motion), is needed for various engineering calculations in desalinization. Jamieson & Tudhope (182) have measured the thermal conductivity of seawater for 35‰ salinity between 0 to 40°C.

DIFFUSION COEFFICIENT The diffusion of ions in geochemical mass transport processes has received a lot of attention in recent years (12, 220, 388). Recent analysis of deep-sea cores indicates large-scale mass transport of solutes from sediments to the oceans (236). Ben-Yaakov (12) has recently derived equations to explain ionic diffusion from seawater to dilute solutions. He found the calculated rates for Cl^-, K^+, Mg^{2+}, and Ca^{2+} were compatible with his experimental measurements. His results suggested that ionic interactions may modify the fluxes of diffusing ions (affecting the relative distribution). Thus, a change in ion/Cl does not necessarily mean that ion is involved in a chemical reaction or that it migrates. Li & Gregory (220) have made measurements on the diffusion of ions in seawater and sediments (red clay). They have also discussed the effect of temperature, salinity, and pressure on the diffusion of ions. They found that tracer-diffusion coefficients of ions in seawater and pure water differ by no more than 8%. Their results are in general agreement with the findings of Ben-Yaakov (12). Wollast & Garrels (388) have made direct measurements of the diffusion of H_4SiO_4 in seawater.

DIELECTRIC PROPERTIES Besides being of interest in studying the structure of seawater, the dielectric constant of seawater is useful in determining the brightness temperature needed for the determination of the ocean surface temperature by radiometry. Ho & Hall (166) have recently measured the dielectric constant and dielectric loss of seawater from 15 to 36‰ S and 5 to 25°C at a frequency of 2.653 GH_z. To explain the concentration dependence they treat the solution as two phases: 1. hydrated water molecules having negligible response, and 2. bulk water.

VISCOSITY Dexter (68) has made measurements on the viscosity of seawater at 1 atm from 0 to 30°C and 0 to 40‰ S, using a precision automatic viscometer (202). His results are more reliable than the earlier results of Miyake & Koizumi (274). Millero (265) has shown that the viscosity of seawater can be estimated within experimental error by using Young's Rule and single salt data. Stanley & Butler (350) have measured the viscosity of seawater as a function of pressure.

REFRACTIVE INDEX The refractive index of seawater is frequently used to determine the salinity of seawater. Although the salinities determined from refractive index are normally not as accurate as the conductivity method, with improved accuracy (338) it may yield salinities to ± 0.006‰ and be more convenient to use. Rusby (338) has made measurements of from 17 to 30°C and 30.9 to 38.8‰ S of the refractive index of seawater relative to Copenhagen standard water using an interferometer (with a standard deviation of 0.006%). Stanley (351) has measured the absolute refractive index of 35‰ S seawater from 0 to 30°C and 0 to 1000 bars applied pressure using an interferometer (SD 0.00002).

ACKNOWLEDGMENTS

The author would like first to acknowledge the Office of Naval Research (Contract N00014-67-A-0201-0013) and the Oceanographic Branch of the National Science Foundation (GA-17386) for their support of this study. I wish to thank my graduate students: Arthur Chen, Rana Fine, Wing Leung, Sheldon Schrager, and Gary Ward, who aided in the literature research and Augustin Gonzalez for checking the references.

Literature Cited

1. Atkinson, G., Petrucci, S. 1966. Ion association of magnesium sulfate in water at 25°. *J. Phys. Chem.* 70:3122–28
2. Atkinson, G., Dayhoff, M. O., Ebdon, D. W. 1972. *Computer Modeling of Inorganic Equilibria in Seawater.* Paper presented at Electrochemistry meeting, Miami Beach, to be published in the Proceedings
3. Austin, J. M., Mair, A. D. 1962. The standard enthalpy of formation of complex sulfate ions in water. I. HSO_4^-, $LiSO_4^-$, $NaSO_4^-$. *J. Phys. Chem.* 66:519–21
4. Baric, A., Branica, M. 1967. Polarography of seawater. I. Ionic state of cadmium and zinc in seawater. *J. Polarogr. Soc.* 13:4–8
5. Bates, R. G. 1964. *Determination of pH.* New York: Wiley. 435 pp.
6. Bates, R. G., Staples, B. R., Robinson, R. A. 1970. Ionic hydration and single ion activities in unassociated chlorides at high ionic strengths. *Anal. Chem.* 42:867–71
7. Bauman, J. E. Jr., Hostetler, P. B., Almon, W. R. 1973. *Bicarbonate Complex of Calcium.* Presented at 165th Am. Chem. Soc. Nat. Meet., Dallas, Texas
8. Ben-Yaakov, S., Kaplan, I. R. 1968. pH-temperature profiles in ocean and lakes using an in-situ probe. *Limnol. Oceanogr.* 13:688–93
9. Ben-Yaakov, S., Kaplan, I. R. 1968. A versatile probe for in-situ oceanographic measurements. *J. Ocean Technol.* 3:25–29
10. Ben-Yaakov, S., Kaplan, I. R. 1971. An oceanographic instrumentation system for in-situ application. *Mar. Technol. Soc. J.* 5:41–46
11. Ben-Yaakov, S., Kaplan, I. R. 1971. Deep-sea in-situ calcium carbonate saturometry. *J. Geophys. Res.* 76:722–31
12. Ben-Yaakov, S. 1972. Diffusion of sea water ions-I. Diffusion of sea water into a dilute solution. *Geochim. Cosmochim. Acta* 36:1395–1406
13. Ben-Yaakov, S., Goldhaber, M. B. 1973. The influence of sea water composition on the apparent constants of the carbonate system. *Deep Sea Res.* 20:87–99
14. Berger, W. H. 1967. Foraminiferal ooze: solution at depth. *Science* 156:383–85
15. Berner, R. A. 1965. Activity coefficients of bicarbonate, carbonate, and calcium ions in sea water. *Geochim. Cosmochim. Acta* 29:947–65
16. Berner, R. A. 1971. *Principles of Chemical Sedimentology.* New York: McGraw-Hill. 240 pp.
17. Berner, R. A., Wilde, P. 1972. Dissolution kinetics of calcium carbonate in sea water I. Saturation state parameters for kinetic calculations. *Am. J. Sci.* 272:826–39
18. Berner, R. A., Morse, J. W. Dissolution kinetics of calcium carbonate in sea water IV. Theory of calcite dissolution. *Am. J. Sci.* In press
19. Bjerrum, N. 1926. Ionic association I. Influence of ionic association on the activity of ion at moderate degree of association. *Kgl. Danske Vidensk. Selsk. Mat.- Fys. Medd.* 7:1–48
20. Bjerrum, J., Schwarzenbach, G., Sillen, L. G. 1958. *Stability Const. Chem. Soc. Spec. Publ.* 7
21. Bockris, J. O'M., Reddy, A. K. N. 1971. *Modern Electrochemistry.* New York: Plenum
22. Bond, A. M. 1970. Some considerations

of the electrolyte used to maintain constant ionic strength in studies on concentration stability constants in aqueous solutions. Application to the polarographic evaluation of Thallium (I) complexes. *J. Phys. Chem.* 74:331–38
23. Bradshaw, A., Schleicher, K. E. 1970. Direct measurement of thermal expansion of sea water under pressure. *Deep Sea Res.* 17:691–706
24. Bradshaw, A. 1973. The effect of carbon dioxide on the specific volume of sea water. *Limnol. Oceanogr.* 18:95–105
25. Brancazio, P. J., Cameron, A. G. W. 1964. *The Origin and Evolution of Atmospheres and Oceans.* New York: Wiley
26. Branica, M., Petek, M., Baric, A., Jeftic, L. 1969. Polarographic characterization of some trace elements in sea water. *Rapp. P. V. Renn. Comm. Int. Explor. Sci. Med. Mediterr.* 19:929–33
27. Broecker, W. S. 1971. A kinetic model for the chemical composition of sea water. *Quart. Res.* 1:188–207
28. Broecker, W. S., Oversby, V. M. 1971. *Chemical Equilibria in the Earth.* New York: McGraw-Hill. 318 pp.
29. Bromley, L. A., Desaussure, V. A., Clipp, J. C., Wright, J. S. 1967. Heat capacities of sea water solutions at salinities of 1 to 12% and temperatures of 2 to 80°C. *J. Chem. Eng. Data* 12:202–6
30. Bromley, L. A. 1968. Heat capacity of sea water solutions: partial and apparent values for salts and water. *J. Chem. Eng. Data* 13:60–62
31. Bromley, L. A. 1968. Relative enthalpies of sea salt solutions at 25°C. *J. Chem. Eng. Data* 13:399–402
32. Bromley, L. A., Diamond, A. E., Salami, E., Wilkins, D. G. 1970. Heat capacities and enthalpies of sea salt solutions to 200°C. *J. Chem. Eng. Data* 15:246–53
33. Brønsted, J. N. 1922. Studies on solubility IV. Principle of the specific interaction of ions. *J. Am. Chem. Soc.* 44: 877–98
34. Burton, J. D., Marshall, N. J., Phillips, A. J. 1968. Solubility of barium sulfate in sea water. *Nature* 217:834–35
35. Butler, J. N., Huston, R. 1970. Potentiometric studies of multi-component activity coefficients using the lanthanum fluoride membrane electrode. *Anal. Chem.* 42:1308–11
36. Butler, J. N., Huston, R. 1970. Activity coefficients and ion pairs in the system sodium chloride-sodium bicarbonate-water and sodium chloride-sodium carbonate-water. *J. Phys. Chem.* 74: 2976–83
37. Byrne, R. H. Jr., Kester, D. R. Inorganic speciation of boron in sea-water. *J. Mar. Res.* In press
38. Caldwell, D. R., Tucker, B. E. 1970. Determination of thermal expansion of sea water, by observing onset of convection. *Deep Sea Res.* 17:707–19
39. Carpenter, J. H. 1972. Problems in application of analytical chemistry to oceanography. In *Analytical Chemistry: Key to Progress on National Problems,* 393–419. Washington, DC: Bur. Standards, Spec. Publ. 351
40. Carpenter, J. H., Manella, M. E. 1973. Magnesium to chlorinity ratios in sea water. *J. Geophys. Res.* 78:3621–26
41. Chave, K. E., Smith, S. V. 1972. *Inorganic Cycles of Calcium and Carbon in the Ocean.* Symposium on the biogeochemistry of the oceans, Joint Oceanogr. Assembly, Tokyo, 1970
42. Chave, K. E., Suess, E. 1970. Calcium carbonate saturation in sea water effects of dissolved organic matter. *Limnol. Oceanogr.* 15:633–37
43. Chave, K. E. 1971. Chemical reactions and the composition of sea water. *J. Chem. Educ.* 48:148–51
44. Chin, Y-C, Fuoss, R. M. 1968. Conductance of the Alkali Halides XII. Sodium and potassium chlorides in water at 25°. *J. Phys. Chem.* 72:4123–29
45. Christensen, J. J., Izatt, R. M. 1968. Thermochemistry in inorganic solution chemistry. In *Physical Methods in Advanced Inorganic Chemistry,* 539–98. New York: Interscience
46. Chughtai, A., Marshall, R., Nancollas, G. H. 1968. Complexes in calcium phosphate solutions. *J. Phys. Chem.* 72: 208–11
47. Connick, R. E., Tsao, M. S. 1954. Complexing of magnesium ion by fluoride ion. *J. Am. Chem. Soc.* 76: 5311–14
48. Connors, D. N. 1970. On the enthalpy of sea water. *Limnol. Oceanogr.* 15: 587–94
49. Conway, B. E., Desnoyers, J. E., Smith, A. C. 1964. On the hydration of simple ions and polyions. *Phil. Trans. Roy Soc. London A* 256:389–437
50. Conway, B. E. 1966. Electrolyte solutions: solvation and structural aspects. *Ann. Rev. Phys. Chem.* 17:481–528
51. Cox, R. A., McCartney, M. J., Culkin, F. 1968. Pure water for relative density standard. *Deep Sea Res.* 15:319–25
52. Cox, R. A., Smith, N. D. 1959. The specific heat of sea water. *Proc. Roy Soc. A* 252:51–62
53. Cox, R. A. 1965. Physical properties of

sea water. *Chemical Oceanography,* ed. J. P. Riley, G. Skirrow, 1:73–120. New York: Academic

54. Cox, R. A., McCartney, M. J., Culkin, F. 1970. The specific gravity/salinity/temperature relationship in natural sea water. *Deep Sea Res.* 17:679–89
55. Crease, J. 1962. The specific volume of sea water under pressure as determined by recent measurements of sound velocity. *Deep Sea Res.* 9:209–13
56. Crease, J. 1971. Determination of the density of seawater. *Nature* 233:329
57. Culberson, C., Kester, D. R., Pytkowicz, R. M. 1967. High pressure dissociation of carbonic and boric acids in sea water. *Science* 157:59–61
58. Culberson, C., Pytkowicz, R. M. 1968. Effect of pressure on carbonic acid and the pH of sea water. *Limnol. Oceanogr.* 13:403–17
59. Culberson, C., Pytkowicz, R. M., Hawley, J. E. 1970. Sea water alkalinity determination by the pH method. *J. Mar. Res.* 28:15–21
60. Culkin, F. 1965. The major constituents of sea water. In *Chemical Oceanography,* ed. J. P. Riley, G. Skirrow, 1:121–61. London: Academic
61. Daly, F. P., Brown, C. W., Kester, D. R. 1972. Sodium and magnesium sulfate ion pairing: evidence from Raman Spectroscopy. *J. Phys. Chem.* 76:3664–68
62. Davis, C. M., Jarzynski, J. 1972. Mixture models of water. In *Water and Aqueous Solutions,* ed. R. A. Horne, 377–424. New York: Wiley
63. Davies, C. W. 1962. *Ion Association.* London: Butterworth. 190 pp.
64. Davies, C. W., Mapass, V. E. 1964. Ion association and the viscosity of dilute electrolyte solutions part I-aqueous inorganic salt solutions. *Trans. Faraday Soc.* 2075–78
65. Del Grosso, V. A., Mader, C. W. 1972. Speed of sound in sea water samples. *J. Acoust. Soc. Am.* 52:961–74
66. Desnoyers, J. E. *Ionic Solute Hydration.* Presented at XV Conseil International de Chemie Solvay, Brussels, June 1972, Proceeding. In press
67. Desnoyers, J. E., Jolicoeur, C. 1969. Hydration effects and thermodynamic properties of ions. In *Modern Aspects of Electrochemistry,* ed. J. O'M. Bockris, B. E. Conway, 5:1–89. New York: Plenum
68. Dexter, R. 1971. *I. Viscosity of heavy water and artificial sea water. II. In situ measurements of some oceanographically important chemical parameters.* MS Thesis. Univ. Miami, Miami, Fla.
69. Dietrick, G. 1963. *General Oceanography.* New York: Wiley
70. Disteche, A., Disteche, S. 1967. The effect of pressure on the dissociation of carbonic acid from measurements with buffered glass electrode cells; the effects of NaCl, KCl, Mg^{++}, Ca^{++}, $SO_4^=$ and of boric acid with special reference to sea water. *J. Electrochem. Soc.* 114: 330–40
71. Drost-Hansen, W. 1967. *Equilibrium Concepts in Natural Water Systems,* 67:70–120, Am. Chem. Soc. Publ.
72. Drost-Hansen, W. 1972. Molecular aspects of aqueous interfacial structures. *J. Geophys. Res.* 77:5132–46
73. Duedall, I. W. 1972. The partial molal volume of calcium in sea water. *Geochim. Cosmochim. Acta* 36:729–34
74. Duedall, I. W., Weyl, P. K. 1967. The partial equivalent volumes of salts in sea water. *Limnol. Oceanogr.* 12:52–59
75. Duedall, I. W. 1973. *Seawater: An explanation of hydration and ion-water interactions in terms of differential isothermal compressibility measurements.* PhD Thesis. Dalhousie Univ., Halifax, Nova Scotia
76. Duer, W. C., Robinson, R. A., Bates, R. G. 1972. Molar conductivity of sodium fluoride in aqueous solution at 25°C. *J. Chem. Soc. Faraday Trans.* 68:716–22
77. Dunn, L. A., Marshall, W. L. 1969. Electrical conductances and ionization behavior of sodium chloride in dioxane-water mixtures at 100°. *J. Phys. Chem.* 73:2619–22
78. Dunsmore, H. S., Jalota, S. K., Paterson, P. 1972. Ion association of rubidium chloride in aqueous solutions at 25°C. *J. Chem. Soc. Faraday Trans. I* 6: 1583–85
79. Durst, R. A. 1969. *Ion Selective Electrodes.* Washington, DC: Nat. Bur. Standards, Spec. Publ. 314. 452 pp.
80. Dyrssen, D., Hansson, I. Ionic medium effects in sea water. A comparison of acidity constants of carbonic acid and boric acid in sodium chloride and synthetic sea water. *Mar. Chem.* In press
81. Dyrssen, D., Jagner, D., Wengelin, F. 1968. *Computer Calculation of Ionic Equilibria and Titration Procedures.* Stockholm: Almqirst & Wiksell. 250 pp.
82. Dyrssen, D., Sillen, L. G. 1967. Alkalinity and total carbonate in sea water; a plea for P-T-independent data. *Tellus* 19:113–21

83. Dyrssen, D., Wedborg, M. Equilibrium calculations of the speciation of elements in sea water. In *The Sea,* ed. E. Goldberg, Vol. 5. New York: Wiley. In press
84. Eckart, C. 1958. Properties of water-II. The equations of state of water and sea water at low temperatures and pressures. *Am. J. Sci.* 256:225–40
85. Edmond, J. M. 1970. High precision determination of titration alkalinity and total carbon dioxide content of sea water by potentiometric titration. *Deep Sea Res.* 17:737–50
86. Edmond, J. M., Gieskes, J. M. T. M. 1970. On the calculation of the degree of saturation of sea water with respect to calcium carbonate under in-situ conditions. *Geochim. Cosmochim. Acta* 34:1261–91
87. Eigen, M., Wicke, E. 1954. The thermodynamics of electrolytes at higher concentration. *J. Phys. Chem.* 58:702–14
88. Eigen, M., Tamm, K. 1962. Sound absorption in electrolytes as a consequence of chemical relaxation. I. Relaxation theory of stepwise dissociation. *Z. Elektrochem.* 66:93–107, 107–21
89. Eisenberg, D., Kauzmann, W. 1969. *The Structure and Properties of Liquid Water.* Oxford: Oxford Univ. Press
90. Eisenman, G. 1967. *Glass Electrodes for Hydrogen and Other Cations: Principles and Practice.* New York: Dekker. 582 pp.
91. Ekman, V. W. 1908. Die Zusammendrückbarkeit des Meerwasser nebst einigen Werten fur Wasser und Quecksilber. *Pubs. Circonst. Cons. Perm. Int. Explor. Mer.* 43:1–47
92. Elgquist, E. 1970. Determination of the stability constants of MgF^+ and CaF^+ using a fluoride ion selective electrode. *J. Inorg. Nucl. Chem.* 32:937–44
93. Elgquist, B., Wedborg, M. Sulphate complexation in sea water. A relation between the stability constants of sodium sulphate and magnesium sulphate in sea water from a determination of the stability constant of hydrogen sulphate and a calculation of single ion activity coefficient of sulphate. *Mar. Chem.* In press
94. Ellis, A. J. 1970. Quantitative interpretation of chemical characteristics of hydrothermal systems. *Geothermics* 2: 516–28
95. Ellis, A. J., Giggenbach, W. 1971. Hydrogen sulphide ionization and sulphur hydrolysis in high temperature solution. *Geochim. Cosmochim. Acta* 35: 247–70
96. Ellis, A. J., McFadden, I. M. 1972. Partial molal volumes of ions in hydrothermal solutions. *Geochim. Cosmochim. Acta* 36:413–26
97. Ellis, A. J. 1959. The effect of pressure on the first dissociation constant of "carbonic acid". *J. Chem. Soc.* 3689–99
98. Emmet, R. T., Millero, F. J. 1971. *The effect of pressure on ion pair formation. Abstr. FLACS, Am. Chem. Soc. Meet. Miniature, Gainesville, Florida, No. 52, Vol. XXIV*
99. Emmet, R. T., Millero, F. J. 1973. *High pressure density measurements of sea water—Preliminary results. Abstr. 09, EOS, Trans.* 54:(4)300
100. Emmet, R. T. 1973. *Density studies in aqueous solutions and sea water at various temperatures and pressures.* PhD thesis. Univ. Miami, Miami, Fla.
101. Enns, T., Scholander, P. F., Bradstreet, E. D. 1965. Effect of hydrostatic pressure on gases dissolved in water. *J. Phys. Chem.* 69:389–91
102. Fay, D. P., Purdie, N. 1969. Calorimetric determination of the heats of complexation of the lanthanide monosulfates $LnSO_4^+$. *J. Phys. Chem.* 73: 3462–67
103. Fay, D. P., Purdie, N. 1970. The linear correlation of ΔH and ΔS of complexation. *Inorg. Chem.* 9:195–96
104. Faust, S. D., Hunter, J. V. 1967. *Principles and Applications of Water Chemistry.* New York: Wiley. 643 pp.
105. Fine, R. A., Wang, D. P., Millero, F. J. 1972. *The equation of state of water and sea water as determined from sound velocity data. Abstr. Ann. Meet., Acoust. Soc. Am., Miami Beach, Fla.*
106. Fine, R. A., Wang, D. P., Millero, F. J. 1973. *The equation of state of sea water. Abstr. 013, EOS, Trans.* 54: (4)301
107. Fine, R. A., Millero, F. J. The compressibility of water as a function of temperature and pressure. *J. Chem. Phys.* Submitted for publication
108. Fisher, F. H. 1962. The effect of pressure on the equilibrium of magnesium sulfate. *J. Phys. Chem.* 66: 1607–11
109. Fisher, F. H., Davis, D. F. 1965. The effect of pressure on the dissociation of manganese sulfate ion pairs in water. *J. Phys. Chem.* 69:2595–98
110. Fisher, F. H. 1965b. Ultrasonic absorption in $MgSO_4$ solutions as a function of pressure and dielectric constant. *J. Acoust. Soc. Am.* 38: 805–12
111. Fisher, F. H. 1967. Ion pairing of magnesium sulfate in sea water: deter-

mined by ultrasonic absorption. *Science* 157:823
112. Fisher, F. H., Davis, D. F. 1967. Effect of pressure on the dissociation of the ($LaSO_4$) complex ion. *J. Phys. Chem.* 71:819–22
113. Fisher, F. H. 1972. Effect of pressure on sulfate ion association and ultrasonic absorption in sea water. *Geochim. Cosmochim. Acta* 36:99–101
114. Fofonoff, N. P. 1962. Physical properties of sea water. In *The Sea,* ed. M. N. Hill, 1:3–30. New York: Interscience
115. Frank, H. S., Wen, W. Y. 1957. Structural aspects of ion-solvent interaction in aqueous solutions. *Discuss. Faraday Soc.* 24:113–40
116. Frank, H. S. 1972. Structural Models. In *Water, A Comprehensive Treatise,* ed. F. Franks, Vol. 1. New York: Plenum
117. Franks, F. 1972. The physics and physical chemistry of water. In *Water, a Comprehensive Treatise,* Vol. 1. New York: Plenum. 596 pp.
118. Friedman, H. L. 1960. Mayer's ionic solution theory applied to electrolyte mixtures. *J. Chem. Phys.* 32:1134–44
119. Fuoss, R. M. 1958. Ionic association III, The equilibrium between ion pairs and free ions. *J. Am. Chem. Soc.* 80:5059–61
120. Gardner, A. W., Glueckauf, E. 1970. Thermodynamic data of the calcium sulphate solution process between 0 and 200°C. *Trans. Faraday Soc.* 66: 1081–86
121. Garrels, R. M., Thompson, M. E., Siever, R. 1960. Stability of some carbonates at 25°C and one atmosphere total pressure. *Am. J. Sci.* 258: 402–18
122. Garrels, R. M., Siever, R., Thompson, M. E. 1961. Control of carbonate solubility by carbonate complexes. *Am. J. Sci.* 259:24–45
123. Garrels, R. M., Thompson, M. E. 1962. A chemical model for sea water at 25°C and one atmosphere total pressure. *Am. J. Sci.* 260:57–66
124. Garrels, R. M. 1965. The role of silica in the buffering of natural waters. *Science* 148:69
125. Garrels, R. M., Mackenzie, F. T. 1967. Origin of the chemical compositions of some springs and lakes. In *Equilibrium Concepts in Natural Water Systems,* Advan. Chem. Ser. No. 67, 222–42. Washington, DC: Am. Chem. Soc.
126. Garrels, R. M., Christ, C. L. 1965. *Solutions, Minerals, and Equilibria.* New York: Harper & Rowe. 450 pp.
127. Garrels, R. M., Mackenzie, F. T., Siever, R. 1972. Sedimentary cycling in relation to the history of the continents and oceans. In *The Nature of the Solid Earth,* ed. E. C. Robertson, Chap. 5. New York: McGraw-Hill
128. Garrels, R. M., Mackenzie, G. T. Chemical history of the oceans deduced from post-depositional charges in sedimentary rocks. In *Studies in Paleo-oceanography.* SEPM Spec. Publ. In press
129. Gibbs, R. E., Van Ness, H. C. 1971. Solubility of gases in liquids in relation to the partial molar volumes of the solute. Carbon dioxide-water. *Ind. Eng. Chem. Fundam.* 10:312–15
130. Gibbs, R. J. 1970. Mechanism controlling world water chemistry. *Science* 170:1088–90
131. Gieskes, J. M. T. 1966. The activity coefficient of sodium chloride in mixed electrolyte solutions at 25°C. *Z. Phys. Chem. Frankfurt am Main.* 50:78–79
132. Ginstrup, O. 1970. Scope of the ionic medium method. On measurement on cells with hydrogen electrode-silver-silver halogenide electrodes in 3M media. *Acta Chem. Scand.* 24:875–88
133. Girard, G., Menaché, M. 1971. Variation de la masse volumique de l'eau en fonction de sa composition isotopique. *Metrologia* 7:83–87
134. Girard, G., Menaché, M. 1972. Sur le calcul de la masse volumique de l'eau. *C. R. Acad. Sci.* 274:377–79
135. Goldberg, E. D., Arrhenius, G. O. S. 1958. Chemistry of Pacific pelagic sediments. *Geochim. Cosmochim. Acta* 13:153–212
136. Goldberg, E. D. 1963. The oceans as a chemical system. In *The Sea,* 2:3–25. New York: Interscience
137. Goldberg, E. D. 1965. Minor elements in sea water. In *Chemical Oceanography,* 1:163–96. New York: Academic
138. Goldberg, E. D., Broecker, W. S., Gross, M. G., Turekian, K. K. 1971. Marine chemistry. In *Radioactivity in the Marine Environment,* 137–46. Washington, DC: Nat. Acad. Sci.
139. Goldberg, E. D. 1973. *The Sea,* Vol. 5. New York: Interscience
140. Gould, R. F. 1967. *Equilibria Concepts in Natural Water Systems, Advan. Chem. Ser.* 67. Washington, DC: Am. Chem. Soc. 344 pp.
141. Gould, R. F. 1968. *Trace Inorganics in Water, Advan. Chem. Ser.* 73.

Washington, DC: Am. Chem. Soc. 396 pp.
142. Gran, G. 1952. Determination of the equivalence point in potentiometric titrations. Part II. *Analyst* 77:661–71
143. Guggenheim, E. A. 1935. Thermodynamic properties of aqueous solutions of strong electrolytes. *Phil. Mag.* 19:588–643
144. Guggenheim, E. A. 1966. Mixture of 1:1 electrolytes. *Trans. Faraday Soc.* 62:3446–50
145. Gurney, R. W. 1953. *Ionic Processes in Solution.* New York: Dover
146. Hagler, A. T., Scheraga, H. A., Nemethy, G. 1972. Structure of liquid water. Statistical thermodynamic theory. *J. Phys. Chem.* 76:3229–43
147. Hale, C. F., Spedding, F. H. 1972. Effect of high pressure on the formation of aqueous $EuSO_4^+$ at 25°C. *J. Phys. Chem.* 2925–29
148. Hamman, S. D., Pearce, P. J., Strauss, W. 1964. The effect of pressure on the dissociation of lanthanum ferricyanide ion pairs in water. *J. Phys. Chem.* 68:375–80
149. Hanna, E. M., Pethylridge, A. D., Prue, J. E. 1971. Ion association and the analysis of precise conductimetric data. *Electrochim. Acta* 16:677–86
150. Harned, H. S., Owen, B. B. 1958. *The Physical Chemistry of Electrolytic Solutions,* A.C.S. Monogr. Ser., No. 137. New York: Reinhold. 803 pp.
151. Harned, H. S., Robinson, R. A. 1968. *Multicomponent Electrolyte Solutions.* Oxford: Pergamon
152. Hansson, I. 1973. The determination of dissociation constants of carbonic acid in synthetic sea water media in the salinity range of 20–40‰ and the temperature range of 5 to 30°C. *Acta Chem. Scand.* 27:931–44
153. Hansson, I. 1973. Determination of the acidity constant of boric acid in synthetic sea water media. *Acta Chem. Scand.* 27:924–30
154. Hansson, I. 1973. A new set of acidity constants for carbonic acid and boric acid in sea water. *Deep Sea Res.* 20:
155. Hansson, I. 1973. A new pH-scale and set of standard buffers for sea water. *Deep Sea Res.* 20:479–91
156. Hansson, I., Jagner, D. 1973. Evaluation of accuracy in Gran plots by means of computer calculations. Application to the potentiometric titration of total alkalinity and carbonate in sea water. *Anal. Chim. Acta* 65:363–73
157. Hawley, J., Pytkowicz, R. M. 1969. Solubility of calcium in sea water at high pressures and 2°C. *Geochim. Cosmochim. Acta* 33:1557–61
158. Hawley, J. E. 1973. *Bicarbonate and carbonate ion association with sodium, magnesium and calcium at 25°C and 0.72 ionic strength.* PhD Thesis. Oregon State Univ., Corvallis
159. Heath, G. R., Culberson, C. 1970. Calcite: degree of saturation rate of dissolution, and the compensation depth in the deep oceans. *Geol. Soc. Am. Bull.* 81:3157–60
160. Helgeson, H. C. 1969. Thermodynamics of hydrothermal systems at elevated temperatures and pressures. *Am. J. Sci.* 267:729–804
161. Helgeson, H. C., Garrels, R. M., Mackenzie, F. T. 1969. Evaluation of irreversible reactions in geochemical processes involving minerals and aqueous solutions. II applications. *Geochim. Cosmochim. Acta* 33:455–81
162. Helgeson, H. C., Mackenzie, F. T. 1970. Silicate—sea water equilibria in the ocean system. *Deep Sea Res.* 17:877–92
163. Hemmes, P. 1972. The volume changes of ionic association reactions. *J. Phys. Chem.* 76:895–900
164. Hester, R. E., Plane, R. A. 1964. Solvation of metal ions in aqueous solutions: the metal-oxygen bond. *Inorg. Chem.* 3:768–69
165. Hinton, J. F., Amis, E. S. 1971. Solvation numbers of ions. *Chem. Rev.* 71:627–74
166. Ho, W., Hall, W. F. Measurements of the dielectric properties of sea water and NaCl solutions at 2.65 GH_z. *J. Geophys. Res.* In press
167. Holland, H. D. 1965. The history of ocean water and its effect on the chemistry of the atmosphere. *Proc. Nat. Acad. Sci. USA* 53:1173–83
168. Holland, H. D. 1972. The geologic history of sea water—an attempt to solve the problem. *Geochim. Cosmochim. Acta* 36:637–51
169. Holtzer, A., Emerson, M. F. 1969. On the utility of the concept of water structure in the rationalization of the properties of aqueous solutions of proteins and small molecules. *J. Phys. Chem.* 73:26–33
170. Hoover, T. E., Berkshire, D. C. 1969. Effects of hydration on carbon dioxide exchange across an air-water interface. *J. Geophys. Res.* 74:456–64
171. Horne, R. A. 1965. The physical chemistry and structure of sea water. *Water Resour. Res.* 1:263–76
172. Horne, R. A. 1969. *Marine Chemistry.*

New York: Wiley-Interscience. 568 pp.
173. Horne, R. A. 1970. *Sea water.* In *Advan. Hydrosc.*, 6: 107–40. New York: Academic
174. Horne, R. A. 1972. *Water and Aqueous Solutions.* New York: Wiley-Interscience. 837 pp.
175. Horne, R. A. 1972. Structure of sea water and its role in chemical mass transport between the sea and the atmosphere. *J. Geophys. Res.* 77: 5170–75
176. Horne, R. A. 1974. *The Oceans Handbook.* New York: Dekker. In press
177. Hurd, D. C. 1972. *Interactions of biogenic opal, sediment, and sea water in the central equatorial Pacific.* PhD Thesis. Univ. Hawaii, Honolulu, Hawaii
178. Ives, D. J. G., Janz, G. J. 1961. *Reference Electrodes, Theory and Practice.* New York: Academic. 651 pp.
179. Izatt, R. M., Eatough, D., Christensen, J. J., Bartholomew, C. H. 1969. Calorimetrically determined log K, ΔH^0 and ΔS^0 values for the interaction of sulphate ion with H^+, Na^+, and K^+ in the presence of tetra-*n*-alkylammonium ions. *J. Chem. Soc. A*, 45–47
180. Izatt, R. M., Eatough, D., Christensen, J. J., Bartholomew, C. H. 1969. Calorimetrically determined log K, ΔH^0, and ΔS^0 values for the interaction of sulphate ion with several Bi- and Trivalent metal ions. *J. Chem. Soc. A*: 47–53
181. Jackson, T. A., Bischoff, J. L. 1971. The influence of amino acids on the kinetics of the recrystallization of aragonite to calcite. *J. Geol.* 79: 493–97
182. Jamieson, D. T., Tudhope, J. S. 1970. Physical properties of sea water solutions: thermal conductivity. *Desalination* 8: 393–401
183. Jones, H. W., Monk, C. B. 1952. E. M. F. studies of electrolytic dissociation. Part 2—magnesium and lanthanum sulphates in water. *Trans. Faraday Soc.* 48: 929–33
184. Jenkins, I. L., Monk, C. B. 1950. The conductances of sodium, potassium and lanthanum sulfates at 25°. *J. Am. Chem. Soc.* 72: 2695–98
185. Katz, A. 1973. The interaction of magnesium with calcite crystal growth: an experimental study at 25°–90°C and one atmosphere. *Geochim. Cosmochim. Acta* 37: 1563–86
186. Kavanau, J. L. 1964. *Water and Solute-Water Interactions.* San Francisco: Holden-Day
187. Kell, G. S. 1967. Precise representation of volume properties of water at one atmosphere. *J. Chem. Eng. Data* 12: 66–69
188. Kell, G. S., Whalley, E. 1965. P-V-T properties of water. Part I. Liquid water at 0 to 150 degrees and at pressures to 1 kilobar. *Phil. Trans. Roy. Soc. A* 258: 565–614
189. Kell, G. S. 1970. Isothermal compressibility of liquid water at one atmosphere. *J. Chem. Eng. Data* 15: 119–22
190. Kell, G. S. 1972. Continuum theories of liquid water. *Water and Aqueous Solutions,* ed. R. A. Horne, 331–76. New York: Wiley
191. Kell, G. S. 1972. Thermodynamic and transport properties of fluid water. *Water: A Comprehensive Treatise,* ed. F. Franks, Vol. 1. New York: Plenum
192. Kester, D. R., Pytkowicz, R. M. 1967. Determination of the apparent dissociation constants of phosphoric acid in sea water. *Limnol. Oceanogr.* 12: 243–52
193. Kester, D. R., Duedall, I. W., Connors, D. N., Pytkowicz, R. M. 1967. Preparation of artificial sea water. *Limnol. Oceanogr.* 12: 176–79
194. Kester, D. R., Pytkowicz, R. M. 1968. Magnesium sulfate association at 25°C in synthetic sea water. *Limnol. Oceanogr.* 13: 670–74
195. Kester, D. R., Pytkowicz, R. M. 1968. Sodium, magnesium and calcium sulfate ion-pairs in sea water at 25°C. *Limnol. Oceanogr.* 14: 686–92
196. Kester, D. R., Pytkowicz, R. M. 1970. Effect of temperature and pressure on sulfate ion association in sea water. *Geochim. Cosmochim. Acta* 34: 1039–51
197. Kester, D. R. 1972. Effect of ion pairing on the pH of sea water. *Limnol. Oceanogr.* 17: 959–60
198. Kester, D. R., Byrne, R. H. Jr. 1972. Chemical Forms of iron in sea water. In *Ferromanganese Deposits on the Ocean Floor,* ed. D. R. Horn, 107–16. Palisades, New York: Lamont-Doherty Geol. Observ. Columbia Univ.
199. Kester, D. R., Doherty, B. T. 1972. Freezing point of sea water. *Abstr. EOS, Trans.* 53(4): 404
200. Kester, D. R. Dissolved gases other than CO_2. In *Chemical Oceanography,* ed. J. P. Riley, G. Skirrow, Vol. 1. 2nd ed. New York: Academic. 2nd ed. In press
201. Knudsen, M. 1901. *Hydrographical tables according to the measurings of Carl Forch, P. Jacobsen, Martin*

Knudsen, S. P. L. Sorensen. Copenhagen: G.E.C. Gad; London: Williams Norgate. 63 pp.

202. Korson, L., Drost-Hansen, W., Millero, F. J. 1968. Viscosity of water at various temperatures. *J. Phys. Chem.* 73:34–39
203. Kratky, O., Leopold, H., Stabinger, H. 1969. Dichtemessungen an Flüssigkeiten und Gasen auf 10^{-6} g/cm^3 bei 0.6 cm^3 Präparatvolumen. *Z. Angew. Phys.* 27:273–77
204. Kremling, K. 1971. Measurement of sea water density by a new laboratory method. *Nature* 229:109–10
205. Kremling, K. 1972. Comparison of specific gravity in natural sea water from hydrographical tables and measurements by a new density instrument. *Deep Sea Res.* 19:377–83
206. Lafon, G. M. 1969. *Some quantitative aspects of the chemical evolution of the oceans.* PhD Thesis. Northwestern Univ., Evanston, Illinois
207. Lafon, G. M. 1970. Calcium complexing with carbonate ion in aqueous solutions at 25°C and 1 atmosphere. *Geochim. Cosmochim. Acta* 34:935–40
208. Larson, J. W. 1970. Thermodynamics of divalent metal sulfate dissociation and the structure of the solvated metal sulfate ion pair. *J. Phys. Chem.* 74:3392–96
209. Lepple, F. K., Millero, F. J. 1971. The isothermal compressibility of sea water near one atmosphere. *Deep Sea Res.* 18:1233–54
210. Lerman, A. 1971. Time to chemical steady-states in lakes and oceans, chap. 2, *Nonequilibrium Systems in Natural Water Chemistry.* Advan. Chem. Ser. (106), 30–76. Washington, DC: Am. Chem. Soc.
211. Lewis, G. N., Randall, M. 1961. *Thermodynamics.* New York: McGraw-Hill. 2nd ed. (Revised by K. S. Pitzer, L. Brewer)
212. Leyendekkers, J. V. 1971a. Single ion activities in multicomponent systems. *Anal. Chem.* 43:1835–43
213. Leyendekkers, J. V. 1971b. Thermodynamics of mixed electrolyte solutions. Ionic entropy correlations and volume fraction statistics. *J. Phys. Chem.* 75:946–56
214. Leyendekkers, J. V. The chemical potentials of sea water components. *Mar. Chem.* In press
215. Leyendekkers, J. V. The ionic activity function of water and the activity coefficient of the hydrogen ion in sea water. *Limnol. Oceanogr.* In press
216. Liebermann, L. N. 1948. The origin of sound absorption in water and in sea water. *J. Acoust. Soc. Am.* 20:868–73
217. Li, Y-H. 1972. Geochemical mass balance among lithosphere, hydrosphere, and atmosphere. *Am. J. Sci.* 272:119–37
218. Li, Y-H., Takahashi, T., Broecker, W. S. 1969. The degree of saturation of $CaCO_3$ in the oceans. *J. Geophys. Res.* 74:5507–25
219. Li, Y-H. 1967. Equation of state of water and sea water. *J. Geophys. Res.* 72:2665–78
220. Li, Y-H., Gregory, S. Diffusion of ions in sea water and in deep sea sediments. *J. Geophys. Res.* Submitted for publication
221. Li, Y-H., Tsui, T-F. 1971. The solubility of CO_2 in water and sea water. *J. Geophys. Res.* 76:4203–7
222. Little, D. M. S., Nancollas, G. H. 1970. Kinetics of crystallization and dissolution of lead sulphate in aqueous solutions. *Trans. Faraday Soc.* 66:3103–12
223. Liu, S. T., Nancollas, G. H. 1973. Linear crystallization and induction-period studies of the growth of calcium sulphate dihydrate crystals. *Talanta* 20:211–16
224. Liu, S. T., Nancollas, G. H. 1973. The crystallization of magnesium hydroxide. *Desalination* 12:75–84
225. Long, F. A., McDevit, W. F. 1952. Activity coefficients of nonelectrolyte solutes in aqueous salt solutions. *Chem. Rev.* 51:119–69
226. Lown, D. A., Thirsk, H. R., Wynne-Jones, L. 1968. Effect of pressure on ionization equilibria in water 25°C. *Trans. Faraday Soc.* 64:2073–80
227. Lyman, J. 1957. *Buffer mechanism of sea water.* PhD Thesis. Univ. Calif., Los Angeles, Calif.
228. Lyman, J., Fleming, R. H. 1940. Composition of sea water. *J. Mar. Res.* 3:134–46
229. Macdonald, J. R. 1969. Review of some experimental and analytical equations of state. *Rev. Mod. Phys.* 41:316–49
230. Mackenzie, F. T., Garrels, R. M. 1965. Silicates: reactivity with sea water. *Science* 150:57–58
231. Mackenzie, F. T., Garrels, R. M. 1966. Chemical mass balance between rivers and oceans. *Am. J. Sci.* 264:507–25
232. Mackenzie, F. T., Garrels, R. M. 1966. Silica-bicarbonate balance in the ocean and early diagenesis. *J. Sed. Petrol.* 36:1075–84

233. Mackenzie, F. T., Garrels, R. M., Bricker, O. P., Bickley, F. 1967. Silica in sea water: control by silica minerals. *Science* 155:1404–5
234. Mackenzie, F. T. 1969. Chemistry of sea water. In *Encyclopedia of Marine Resources,* ed. F. E. Firth, 106–12. New York: Van Nostrand-Reinhold
235. Mackenzie, F. T. Sedimentary cycling and the evolution of sea water. *Chem. Oceanogr.* Submitted for publication
236. Mainheim, F. T. 1970. The diffusion of ions in unconsolidated sediments. *Earth Planet. Sci. Lett.* 9:307–9
237. Malijkovic, D., Branica, M. 1971. Polarography of sea water. II. Complex formation of cadmium with EDTA. *Limnol. Oceanogr.* 16:779–85
238. Mangel, M. S. 1971. A treatment of complex ions in sea water. *Mar. Geol.* 11:M24–M26
239. Mangelsdorf, P. C. Jr., Wilson, T. R. S. 1971. Difference chromatography of sea water. *J. Phys. Chem.* 75:1418–25
240. Mangelsdorf, P. C. Jr., Wilson, T. R. S. 1972. Constancy of ionic proportions in the Pacific ocean. *Abstr. EOS Trans.* 53:(4)402
241. Marshall, W. L. 1972. A further description of complete equilibrium constants. *J. Phys. Chem.* 76:720–31
242. Marshall, W. L. 1972. Predictions of the geochemical behavior of aqueous electrolytes at high temperatures and pressures. *Chem. Geol.* 10:59–68
243. Masterton, W. L., Lee, T. P. 1970. Salting coefficients from scaled particle theory. *J. Phys. Chem.* 74:1776–82
244. Masterton, W. L., Welles, H., Knox, J., Millero, F. J. 1973. *Volume change of ion pair formation; Rubidium nitrate and potassium nitrate in water. Abstr. 47. Phys. Chem. Sec. Am. Chem. Soc. Nat. Meet., 165th, Dallas, Texas, April 8–13*
245. Matheson, R. A. 1968. Formation constants of some 2:2 and 3:3 ion pairs. *J. Phys. Chem.* 72:3330–32
246. McDowell, H., Brown, W. E., Sutter, J. R. 1971. Solubility study of calcium hydrogen phosphate. Ion-pair formation. *Inorg. Chem.* 10:1638–43
247. McGee, K. A., Hostetler, P. B. 1973. Stability constants for $MgOH^+$ and Brucite below 100°C. *Abstr. V47, EOS, Trans.* 54:(4)487.
248. Menaché, M. 1971. Verification, par analyse isotopique de la validite de la methode de Cox, McCartney et Culkin tendant à l'obtention d'un étalon de masse volumique. *Deep Sea Res.* 18: 449–56
249. Menaché, M., Girard, G. 1972. *Concerning the Different Tables of Thermal Expansion of Water Between 0 and 40°C.* France: Inst. Oceanogr. Bur. Inter. des Poids et Mesures
250. Meyer, J. L., Nancollas, G. H. 1972. The effect of pH and temperature on the crystal growth of hydroxyapatite. *Arch. Oral Biol.* 17:1623–27
251. Millero, F. J. 1968. Apparent molal expansibility of some divalent chlorides in aqueous solution at 25°. *J. Phys. Chem.* 72:4589–93
252. Millero, F. J. 1969. The partial molal volumes of ions in sea water. *Limnol. Oceanogr.* 14:376–85
253. Millero, F. J. 1969. The partial molal volumes of ions in various solvents. *J. Phys. Chem.* 73:2417–20
254. Millero, F. J., Wu, C. H., Hepler, L. G. 1969. Thermodynamics of ionization of acetic and chloroacetic acid in ethanol-water mixtures. *J. Phys. Chem.* 73:2453–55
255. Millero, F. J., Curry, R. W., Drost-Hansen, W. 1969. Isothermal compressibility of water at various temperatures. *J. Chem. Eng. Data* 14:422–25
256. Millero, F. J. 1970. The apparent and partial molal volume of aqueous sodium chloride solutions at various temperatures. *J. Phys. Chem.* 74:356–62
257. Millero, F. J. 1971. The physical chemistry of multicomponent salt solutions. In *Biophysical Properties of Skin,* ed. H. R. Elden, 329–76. New York: Wiley
258. Millero, F. J. 1971. The molal volume of electrolytes. *Chem. Rev.* 71:147–76
259. Millero, F. J. 1971. Effect of pressure on sulfate ion association in sea water. *Geochim. Cosmochim. Acta* 35:1089–98
260. Millero, F. J. 1972. The partial molal volumes of electrolytes in aqueous solutions. In *Water and Aqueous Solution,* ed. R. A. Horne, 519–95. New York: Wiley-Interscience
261. Millero, F. J., Berner, R. A. 1972. Effect of pressure on carbonate equilibria in sea water. *Geochim. Cosmochim. Acta* 36:92–98
262. Millero, F. J., Knox, J. H., Emmet, R. T. 1972. A high precision variable pressure magnetic float densimeter. *J. Solution Chem.* 1:173–86
263. Millero, F. J., Hoff, E. V., Kahn, L. 1972. The effect of pressure on the ionization of water at various temperatures from molal volume data. *J. Solution Chem.* 1:309–27
264. Millero, F. J., Schrager, S. R., Hansen,

L. D. 1972. *Titration calorimetry of sea water. Abstr. 1972 Fall Meet. Am. Geophys. Union, San Francisco, Calif.*
265. Millero, F. J. Sea water as a multicomponent electrolyte solution. *The Sea,* ed. E. D. Goldberg, Vol. 5. New York: Interscience. In press
266. Millero, F. J. 1973. Sea water—a test for multicomponent electrolyte solution theories I. The apparent equivalent volume, expansibility and compressibility of artificial sea water. *J. Solution Chem.* 2:1–22
267. Millero, F. J. 1973. Theoretical estimates of the isothermal compressibility of sea water. *Deep Sea Res.* 20:101–5
268. Millero, F. J., Lepple, F. K. 1973. The density and expansibility of artificial sea water solutions from 0 to 40°C and 0 to 21‰ chlorinity. *Mar. Chem.* 1: 89–104
269. Millero, F. J., Hansen, L. D., Hoff, E. V. 1973. The enthalpy of sea water from 0 to 30°C and from 0 to 40‰ salinity. *J. Mar. Res.* 31:21–39
270. Millero, F. J., Perron, G., Desnoyers, J. E. 1973. The heat capacity of sea water solutions from 5 to 35°C and 0.5 to 22‰ chlorinity. *J. Geophys. Res.* 78:4499–4507
271. Millero, F. J. Thermodynamic properties of sea water. In *The Oceans Handbook,* ed. R. A. Horne. New York: Dekker. In preparation
272. Millero, F. J., Masterton, W. L. The volume change for the formation of magnesium sulfate ion pairs at various temperatures. *J. Phys. Chem.* Submitted for publication
273. Millero, F. J. 1974. *Marine Physical Chemistry.* New York: Dekker. In preparation
274. Miyake, Y., Koizumi, M. 1948. The measurement of the viscosity coefficient of sea water. *J. Mar. Res.* 7:63–66
275. Monk, C. B. 1961. *Electrolytic Dissociation.* London: Academic. 1st ed. 320 pp.
276. Morse, J. W., Berner, R. A. 1972. Dissolution kinetics of calcium carbonate in sea water (II): a kinetic origin of the lysocline. *Am. J. Sci.* 272:840–56
277. Morse, J. W. Dissolution kinetics of calcium carbonate in sea water III. A new method for the study of carbonate reaction kinetics. *Am. J. Sci.* 273. In press
278. Morse, J. W. Dissolution kinetics of calcium carbonate in sea water VI. Effects of natural inhibitors and position of the chemical lysocline. *Am. J. Sci.* 273: In press
279. Munjal, P., Stewart, P. B. 1971. Correlation equation for solubility of carbon dioxide in water, sea water and sea water concentrates. *J. Chem. Eng. Data* 16:170–72
280. Murray, C. N., Riley, J. P., Wilson, T. R. S. 1969. The solubility of gases in distilled water and sea water I. Nitrogen. *Deep Sea Res.* 16:297–310
281. Murray, D. N., Riley, J. P. 1969. The solubility of gases in distilled water and sea water II. Oxygen. *Deep Sea Res.* 16:311–20
282. Murray, C. N., Riley, J. P. 1970. The solubility of gases in distilled water and sea water III. Argon. *Deep Sea Res.* 17:203–9
283. Murray, C. N., Riley, J. P. 1971. The solubility of gases in distilled water and sea water IV. Carbon Dioxide. *Deep Sea Res.* 18:533–41
284. Nair, V. S. K., Nancollas, G. H. 1958. Thermodynamics of ion association Part IV. Magnesium and zinc sulfates. *J. Chem. Soc.* 3706–10
285. Nakayama, F. S., Rasnick, B. A. 1967. Calcium electrode method for measuring dissociation and solubility of calcium sulfate dihydrate. *Anal. Chem.* 39:1022–23
286. Nakayama, F. S. 1970. Sodium bicarbonate and carbonate ion pairs and their relation to the estimation of the first and second dissociation constants of carbonic acid. *J. Phys. Chem.* 74: 2726–28
287. Nakayama, F. S. 1971. Magnesium complex and ion-pairs in $MgCO_3$-CO_2 solution system. *J. Chem. Eng. Data* 16:178–81
288. Nancollas, G. H., Mohan, M. S. 1970. The growth of hydroxyapatite crystals. *Arch. Oral Biol.* 15:731–45
289. Nancollas, G. H., Reddy, M. M. 1971. The crystallization of calcium carbonate II. Calcite growth mechanism. *J. Colloid Interface Sci.* 37:824–30
290. Nancollas, G. H., Marshall, R. W. 1971. Kinetics of dissolution of dicalcium phosphate dihydrate crystals. *J. Dental. Res.* 50:1268–72
291. Nancollas, G. H., Reddy, M. M., Tsai, F. 1972. An autoclave for the study of crystal growth and dissolution in aqueous solution at high temperature. *J. Phys. Sci. Instrum.* 5:1186–88
292. Nancollas, G. H. 1966. *Interactions in Electrolyte Solutions.* New York: Elsevier
293. Nancollas, G. H. 1970. The thermodynamics of metal-complex and ion-pair formation. *Coord. Chem. Rev.* 5: 379–415

294. Newton, M. S., Kennedy, G. C. 1965. An experimental study of the P-V-T-S relations of sea water. *J. Mar. Res.* 23:88–103
295. North, N. A. 1973. Pressure dependence of equilibrium constants in aqueous solutions. *J. Phys. Chem.* 77:931–34
296. Owen, B. B., Brinkley, S. R. Jr. 1941. Calculation of the effect of pressure upon ionic equilibria in pure water and in salt solutions. *Chem. Rev.* 29:461–74
297. Park, K. P. 1969. Oceanic CO_2 system; an evaluation of ten methods of investigation. *Limnol. Oceanogr.* 14:179–86
298. Park, K. 1966. Deep-sea pH. *Science* 154:1540–42
299. Pesret, F. 1972. *Kinetics of carbonate-sea water interactions.* MS Thesis. Univ. Hawaii, Honolulu, Hawaii. 49 pp.
300. Peterson, M. N. A. 1966. Calcite: rates of dissolution in a vertical profile in the central Pacific. *Science* 154: 1542–44
301. Pitzer, K. S. 1973. Thermodynamics of electrolytes. 1. Theoretical basis and general equations. *J. Phys. Chem.* 77: 268–77
302. Platford, R. F., Dafoe, T. 1965. The activity coefficient of sodium sulfate in sea water. *J. Mar. Res.* 23:63–68
303. Platford, R. F. 1965. The activity coefficient of sodium chloride in sea water. *J. Mar. Res.* 23:55–62
304. Platford, R. F. 1965. Activity coefficient of the magnesium ion in sea water. *J. Fish. Res. Bd. Can.* 22:113–16
305. Pond, S., Pytkowicz, R. M., Hawley, J. E. 1971. Particle dissolution during settling in the oceans. *Deep Sea Res.* 18:1135–39
306. Prue, J. E. 1966. *Ionic Equilibria.* Oxford: Pergamon. 115 pp.
307. Prue, J. E. 1969. Ion pairs and complexes: free energies, enthalpies and entropies. *J. Chem. Educ.* 46:12–16
308. Pytkowicz, R. M. 1965. Rate of inorganic calcium carbonate precipitation. *J. Geol.* 73:196–99
309. Pytkowicz, R. M. 1965. Calcium carbonate saturation in the ocean. *Limnol. Oceanogr.* 10:220–25
310. Pytkowicz, R. M., Kester, D. R., Burgener, B. C. 1966. Reproducibility of pH measurements in sea water. *Limnol. Oceanogr.* 11:417–19
311. Pytkowicz, R. M., Duedall, I. W., Connors, D. N. 1966. Magnesium ions: activity in sea water. *Science* 152:640–42
312. Pytkowicz, R. M. 1967. Carbonate cycle and the buffer mechanism of recent oceans. *Geochim. Cosmochim. Acta* 31:63–73
313. Pytkowicz, R. M., Fowler, G. A. 1967. Solubility of foraminifera in sea water at high pressures. *Geochim. J.* 1:169–82
314. Pytkowicz, R. M. 1968. The carbon dioxide—carbonate system at high pressures in the oceans. *Oceanogr. Mar. Biol. Ann. Rev.* 6:83–135
315. Pytkowicz, R. M., Gates, R. 1968. Magnesium sulfate interactions in sea water from solubility measurements. *Science* 161:690–91
316. Pytkowicz, R. M. 1969. Chemical solution of calcium carbonate in sea water. *Am. Zool.* 9:673–79
317. Pytkowicz, R. M., Kester, D. R. 1969. Harned's rule behavior of NaCl-Na_2SO_4 solutions explained by an ion association model. *Am. J. Sci.* 217: 217–29
318. Pytkowicz, R. M. 1969. Use of apparent equilibrium constants in chemical oceanography, geochemistry, and biochemistry. *Geochim. J.* 3:181–84
319. Pytkowicz, R. M. 1970. On the carbonate compensation depth in the Pacific Ocean. *Geochim. Cosmochim. Acta* 34:836–39
320. Pytkowicz, R. M., Kester, D. R. 1971. The physical chemistry of sea water. *Oceanogr. Mar. Biol. Ann. Rev.* 9:11–60
321. Pytkowicz, R. M. 1972. The chemical stability of the oceans and the CO_2 system. In *The Changing Chemistry of the Oceans,* ed. D. Dyrssen, D. Jagner. Stockholm: Almqvist & Wiksell
322. Pytkowicz, R. M. 1972. The status of our knowledge of sulfate association in sea water. *Geochim. Cosmochim. Acta* 36:631–33
323. Pytkowicz, R. M. 1972. Comments on "the control of sea water pH by ion pairing" (P. J. Wangersky). *Limnol. Oceanogr.* 17:958–59
324. Rechnitz, G. A. 1965. Chemical measurements with cation-sensitive glass electrodes. *Rec. Chem. Progr.* 26:241–55
325. Reddy, M. N., Nancollas, G. H. 1971. The crystallization of calcium carbonate 1. isotopic exchange and kinetics. *J. Colloid Interface Sci.* 36: 166–72
326. Reddy, M. M., Nancollas, G. H. 1973. Calcite crystal growth inhibition by phosphonates. *Desalination* 12:61–74
327. Reilly, P. J., Wood, R. H. 1969. The prediction of the properties of mixed electrolytes from measurements on common ion mixtures. *J. Phys. Chem.* 73:4292–97

328. Reilly, P. J., Wood, R. H., Robinson, R. A. 1971. The prediction of osmotic and activity coefficients in mixed electrolyte solutions. *J. Phys. Chem.* 75:1305–15
329. Righellato, E. C., Davies, C. W. 1930. The extent of dissociation of salts in water II. Uni-bivalent salts. *Trans. Faraday Soc.* 26:592–600
330. Riley, J. P., Chester, R. 1971. *Introduction to Marine Chemistry.* New York: Academic. 465 pp.
331. Robinson, R. A., Stokes, R. H. 1959. *Electrolyte Solutions.* London: Butterworth. 571 pp.
332. Robinson, R. A., Duer, W. C., Bates, R. G. 1971. Potassium fluoride—a reference standard for fluoride ion activity. *Anal. Chem.* 43:1862–65
333. Robinson, R. A. 1954. The vapor pressure and osmotic equivalence of sea water. *J. Mar. Biol. Assoc. UK* 33:449–55
334. Robinson, R. A., Wood, R. H. 1972. Calculations of the osmotic and activity coefficients of sea water at 25°C. *J. Solution Chem.* 1:481–88
335. Root, W. C. 1933. An equation relating density and concentration. *J. Am. Chem. Soc.* 55:850
336. Roques, H. 1969. A review of present-day problems in the physical chemistry of carbonates in solution. *Trans. Cave Res. Group Gt. Brit.* 11:139–63
337. Rossotti, F. J. C., Rossotti, H. 1961. *Determination of Stability Constants.* New York: McGraw-Hill. 425 pp.
338. Rusby, J. S. M. 1967. Measurements of the refractive index of sea water relative to Copenhagen standard sea water. *Deep Sea Res.* 14:427–39
339. Samoilov, O. Ya. 1965. *Structure of Aqueous Electrolyte Solutions and the Hydration of Ions,* transl. D. J. G. Ives. New York: Consultant Bureau
340. Scatchard, G., Rush, R. M., Johnson, J. S. 1970. Osmotic and activity coefficients for binary mixtures of sodium chloride, sodium sulfate, magnesium sulfate and magnesium chloride in water at 25°C III. Treatment with the ions as components. *J. Phys. Chem.* 74:3786–96
341. Siever, R. 1968. Sedimentological consequences of a steady-state ocean-atmosphere. *Sedimentology* 11:5–29
342. Siever, R. 1969. Establishment of equilibrium between clays and sea water. *Earth Planet. Sci. Lett.* 5:106–10
343. Sillen, L. G. 1961. The physical chemistry of sea water. In *Oceanography,* ed. M. Sears, 549–81. Washington, DC: Am. Assoc. Advan. Sci., Publ. no. 67
344. Sillen, L. G., Martell, A. E. 1964. *Stability Constants of Metal-Ion Complexes,* Spec. Publ. No. 17. London: Chem. Soc. 754 pp. (Suppl. 25, 1971)
345. Sillen, L. G. 1967. The ocean as a chemical system. *Science* 156:1189–96
346. Sillen, L. G. 1967. How have sea water and air got their present composition? *Chem. Brit.* 3:291–97
347. Singh, D., Bromley, L. A. 1973. Relative enthalpies of sea salt solutions at 0 to 70°C. *J. Chem. Eng. Data* 18:174–81
348. Skirrow, G. 1965. The dissolved gases —CO_2. In *Chemical Oceanography,* ed. J. P. Riley, G. Skirrow, Vol. 1. New York: Academic. 227 pp.
349. Smith, B. R., Sweett, F. 1971. The crystallization of calcium sulfate dihydrate. *J. Colloid Interface Sci.* 37:612–18
350. Stanley, E. M., Batten, R. C. 1969. Viscosity of sea water at moderate temperatures and pressures. *J. Geophys. Res.* 74:3415–20
351. Stanley, E. M. 1971. The refractive index of sea water as a function of temperature, pressure and two wavelengths. *Deep Sea Res.* 18:833–40
352. Stewart, P. B., Munjal, P. 1970. Solubility of carbon dioxide in pure water, synthetic sea water, and synthetic sea water concentrates at $-5°$ to 25°C and to 45 atm. Pressure. *J. Chem. Eng. Data* 15:67–71
353. Stumm, W. 1973. Chemical Speciation. In *Chemical Oceanography,* ed. J. P. Riley, G. Skirrow, Vol. 1. New York: Academic. 2nd ed.
354. Stumm, W., Morgan, J. J. 1970. *Aquatic Chemistry, An Introduction Emphasizing Chemical Equilibria in Natural Waters.* New York: Wiley-Interscience. 583 pp.
355. Sturges, W. 1970. On the thermal expansion of sea water. *Deep Sea Res.* 17:637–40
356. Suess, E. 1970. Interactions of organic compounds with calcium carbonate-I. Association phenomena and geochemical implications. *Geochim. Cosmochim. Acta* 34:157–68
357. Sverdrup, H. W., Johnson, M. W., Fleming, R. H. 1942. *The Oceans.* Englewood Cliffs, NJ: Prentice-Hall. 1087 pp.
358. Sweers, H. E. 1971. Sigma-t, specific volume anomaly and dynamic height. *Mar. Technol. Sci. J.* 5:7–25
359. Takahashi, T. et al 1970. A carbonate

chemistry profile at the 1969 Geosecs intercalibration station in the eastern Pacific Ocean. *J. Geophys. Res.* 75: 7648–66

360. Thompson, M. E. 1966. Magnesium in sea water: an electrode measurement. *Science* 153:866–67
361. Thompson, M. E., Ross, J. W. Jr. 1966. Calcium in sea water by electrode measurement. *Science* 154:1643–44
362. Thompson, T. G., Wirth, H. E. 1931. The specific gravity of sea water at zero degrees in relation to chlorinity. *J. Cons. Cons. Perma. Int. Explor. Mer.* 6:232–40
363. Thorp, W. H. 1965. Deep-ocean sound attenuation in the sub and low-kilocycle-per-second region. *J. Acoust. Soc. Am.* 38:648–54
364. Thorp, W. H. 1967. Analytic description of the low-frequency attenuation coefficient. *J. Acoust. Soc. Am.* 42:270
365. Thorstenson, D. C., Mackenzie, F. T. 1971. Experimental decomposition of algae in sea water and early diagenesis. *Nature* 234:543–45
366. Truesdell, A. H., Hostetler, P. B. 1968. Dissociation constants of KSO_4^- from 10°–50°C. *Geochim. Cosmochim. Acta* 32:1019–22
367. Turner, E. S. 1968. *Structure and Properties of Liquid Water,* 124. Technical Information Libraries, Bell Telephone Labs. 154 pp.
368. UNESCO 1966. *Second report of the Joint Panel on Oceanographic Tables and Standards. UNESCO Tech. Pap. Mar. Sci.,* No. 4, 9 pp. (Mimeographed)
369. Wang, D.-P., Millero, F. J. Precise representation of the P-V-T properties of water and sea water determined from sound speeds. *J. Geophys. Res.* In press
370. Wangersky, P. J. 1972. The control of sea water pH by ion pairing. *Limnol. Oceanogr.* 17:1–6
371. Ward, G. K., Millero, F. J. 1972. The apparent molal volume of boric acid and sodium borate in pure water and sodium chloride solutions. *Abstr. No. 9, FLACS, Vol. XXV. Am. Chem. Soc. Meet., Miniature, Key Biscayne, Fla.*
372. Ward, G. K., Millero, F. J. The effect of pressure on the ionization of boric acid from molal volume data. *J. Solution Chem.* Submitted for publication
373. Warner, T. B. 1972. Ion selective electrodes—properties and uses in sea water. *Mar. Technol. Soc. J.* 6:24–33
374. Weiss, R. F. 1970. The solubility of nitrogen, oxygen and argon in water and sea water. *Deep Sea Res.* 17: 721–35
375. Weiss, R. F. 1970. Helium isotope effect in solution in water and sea water. *Science* 168:247–48
376. Weiss, R. F. 1971. The effect of salinity on the solubility of argon in sea water. *Deep Sea Res.* 18:225–30
377. Weiss, R. F. 1971. Solubility of helium and neon in water and sea water. *J. Chem. Eng. Data* 16:235–41
378. Welch, M. J., Lifton, J. F., Seck, J. A. 1969. Tracer studies with radioactive oxygen—15. Exchange between carbon dioxide and water. *J. Phys. Chem.* 73: 3351–56
379. Whitfield, M. 1971. A compact potentiometric sensor of novel design. In situ determination of pH, pS^{2-} and Eh. *Limnol. Oceanogr.* 16:829–37
380. Whitfield, M. 1973. Sea water as an electrolyte solution. In *Chemical Oceanography,* ed. J. P. Riley, G. Skirrow, Vol. 1. New York: Academic. 2nd ed.
381. Wigley, T. M. L. 1971. Ion pairing and water quality measurements. *Can. J. Earth Sci.* 8:468–76
382. Wilson, W. 1960. The speed of sound in sea water as a function of temperature, pressure and salinity. *J. Acoust. Soc. Am.* 32:641–44
383. Wilson, W. D. 1960. Equation for the speed of sound in sea water. *J. Acoust. Soc. Am.* 32:(10)1357
384. Wilson, W., Bradley, D. 1968. Specific volume of sea water as a function of temperature, pressure and salinity. *Deep Sea Res.* 15:355–63
385. Wirth, H. E. 1940. The problem of the density of sea water. *J. Mar. Res.* 3: 230–47
386. Wood, R. H., Anderson, H. L. 1966. Heats of mixing of aqueous electrolytes III. A test of the general equations with quarternary mixtures. *J. Phys. Chem.* 70:1877–79
387. Wood, R. H., Reilly, P. J. 1970. Electrolytes. *Ann. Rev. Phys. Chem.* 21:287–406
388. Wollast, R., Garrels, R. M. 1971. Diffusion coefficients of silica in sea water. *Nature Phys. Sci.* 239:94
389. Yatsimirskii, K. B., Vasilév, V. P. 1960. *Instability Constants of Complex Compounds,* New York: Consultants Bur. 120 pp.
390. Young, T. F., Smith, M. B. 1954. Thermodynamic properties of mixtures of electrolytes in aqueous solutions. *J. Phys. Chem.* 58:716–24
391. Zirino, A., Healy, M. L. 1970. Inorganic

zinc complexes in sea water. *Limnol. Oceanogr.* 15:956–58
392. Zirino, A., Healy, M. L. 1971. Voltammetric measurement of zinc in the northeastern tropical Pacific Ocean. *Limnol. Oceanogr.* 16:773–78
393. Zirino, A., Healy, M. L. 1972. pH-controlled differential voltammetry of certain trace transition elements in natural waters. *Environ. Sci. Technol.* 6:243–49
394. Zirino, A., Yamamoto, S. 1972. A pH-dependent model for the chemical speciation of copper, zinc, cadmium and lead in sea water. *Limnol. Oceanogr.* 17:661–71

GEOPHYSICAL DATA AND THE INTERIOR OF THE MOON

M. Nafi Toksöz
Department of Earth and Planetary Sciences, Massachusetts Institute of Technology, Cambridge, Massachusetts 02139

I. INTRODUCTION

The past decade has seen an enormous increase in the amount of scientific data pertaining to the moon. With significant returns from the Ranger, Surveyor, Orbiter, and Luna missions, and most importantly from the Apollo flights, our understanding of the moon has improved greatly. Data from orbital measurements, geophysical stations on the moon, and returned lunar samples have led to major discoveries and conclusions with respect to lunar composition, properties, and evolution. In this article some of this data is summarized with particular emphasis on the interior and the evolution of the moon.

After the earth, the moon is the only planetary body which has been explored in some depth. Although the moon is much smaller than the earth (2% of the earth by volume and 1.2% by mass), in many respects it is a more suitable example for the study of the origin and nature of the terrestrial planets. First, the moon seems to have evolved rapidly and then preserved a record of the early history of solar and planetary evolution in greater detail than the earth because of the absence of major volcanic activity and erosional processes in the past three billion years. Second, the relatively low pressure and temperature (less than about 50 kbar and 1500°C, respectively) in the lunar interior are generally reachable in the laboratory. Thus the lunar data can be interpreted in the light of laboratory measurement without excessive extrapolation.

In spite of the above advantages, knowledge of the lunar interior is still sketchy. So far, only a very limited area of the moon has been directly sampled and explored (Figure 1). In the polar regions and the back side no landed measurements have been made. Some geophysical networks have now reached their full configuration and more useful data will continue to come from these arrays. Further study and analysis of returned lunar samples and data will extend our knowledge of lunar structure and properties to greater depths.

In this article we present a review of current geophysical data and knowledge of the lunar interior with emphasis on physical properties. In Section II geophysical data and constraints most relevant to the general properties of the lunar interior are summarized. In Section III, detailed models of the lunar interior based on seismic and other data are presented. Implications of these models with respect to the thermal state and evolution of the moon are discussed briefly in Section IV.

Figure 1 Lunar earthside and the landing sites of the US Apollo (*A*) and the USSR Luna (*L*) missions which returned lunar samples.

II. GEOPHYSICAL DATA AND PHYSICAL PROPERTIES OF THE LUNAR INTERIOR

A large amount of data has been obtained from study of the moon by earth based, orbital, and lunar surface observations and from returned lunar samples. A substantial cross section of this data can be found in the many hundreds of papers published recently (see 1970–1973 *Proceedings of the Lunar Science Conferences, Geochimica et Cosmochimica Acta,* Suppl. 1–4). In this section we review a limited portion of the data which is directly relevant to the lunar interior. This includes the figure, density, and gravity field of the moon, the viscosity, strength, and electrical and magnetic properties. Other data, such as seismic and heat flow, will be discussed in Sections III and IV, respectively.

II.1 *Figure, Density, and Gravity Field*

The mean density of the moon is well determined as $\bar{\rho} = 3.344 \pm 0.004$ g/cm^3 (36). That this value is lower than that of the earth, even when both values are corrected for pressure effect, implies a difference in bulk composition between the earth and the moon. The mean radius is $a = 1738.09$ km. The mass product for the moon is $GM = 4902.8$ km^3 s^{-2}, where M = mass and G = gravitational constant (48). With newer data, the moment of inertia factor has been revised extensively, and the latest value given by Kaula (38) is $C/Ma^2 = 0.395\,(+0.005$ or $-0.010)$. This is slightly lower than the value of $C/Ma^2 = 0.402\,(+0.004$ or $-0.002)$ of Michael & Blackshear (48) which was used extensively for density model calculations. Within the uncertainties, it is clear that C/Ma^2 is slightly less than that of a uniform density moon (0.400). This implies that the lunar interior could be of nearly uniform density below the lunar crust. Density models calculated for the lunar interior depend strongly on temperature and other constraints (26, 70). The simplest model would be a 60 km thick crustal layer with $\rho = 3.0$ g/cc, and a mantle density varying between $\rho_2 = 3.38$ and 3.40 g/cc.

The pressure profile inside the moon depends on the details of the density model. Generally for the outer layers the pressure gradient is approximately 1 kbar for 18 km depth. The pressure at the center of the moon is about 46 ± 6 kbar, the exact value depending on the density model.

So far we have discussed the general characteristics of the spherical and laterally homogeneous moon models. There are large scale and regional variations from lateral homogeneity and sphericity. Most comprehensive data on the subject come from orbital gravity, laser altimetry, and Apollo landmark tracking investigations (35, 37, 38, 69, 96). On the figure of the moon, the orbital results (37, 96) have shown that the center of mass is displaced toward the earth by about 2 km relative to the center of figure. The mean altitude of terrae is about 3 km above maria. The altitude profile relative to a sphere of 1738 km radius is generally positive on the farside (except for a 1400 km wide basin) and negative on the nearside, as shown in Figure 2, taken from Kaula et al (37). It should also be noted that the terrae are quite rough while the maria are generally smooth and level.

The Ranger, Orbiter, and Apollo missions provided important information on both the long-wavelength and short-wavelength gravity anomalies. In terms of spherical harmonics, the gravitational potential U can be expressed as:

$$U = \frac{GM}{r}\left[C_{0,0} + \sum_{n=2}^{\infty}\sum_{m=0}^{n}\left(\frac{a}{r}\right)^{n} P_{n,m}(\sin\phi)(C_{n,m}\cos m\lambda + S_{n,m}\sin m\lambda)\right] \quad 1.$$

where G is the gravitational constant, M is the lunar mass, a is the mean lunar radius, r is the distance from center of mass, $P_{n,m}$ is the associated Legendre polynomial of degree n and order m, ϕ is the selenographic latitude, and λ is the selenographic longitude. Spherical harmonic coefficients (up to degree and order 13)

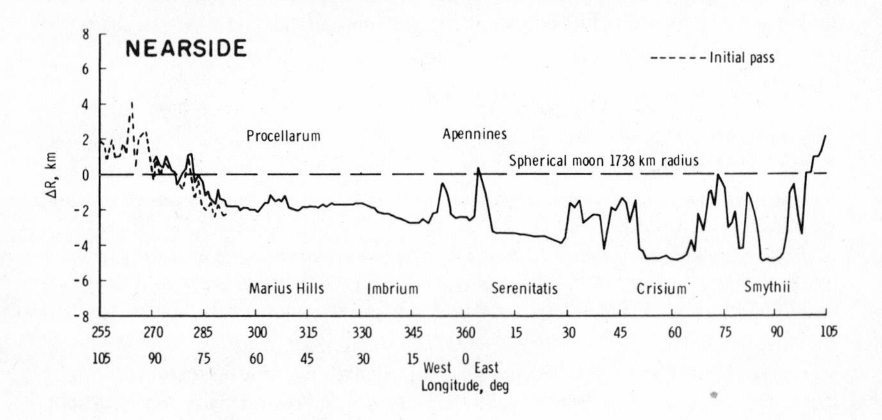

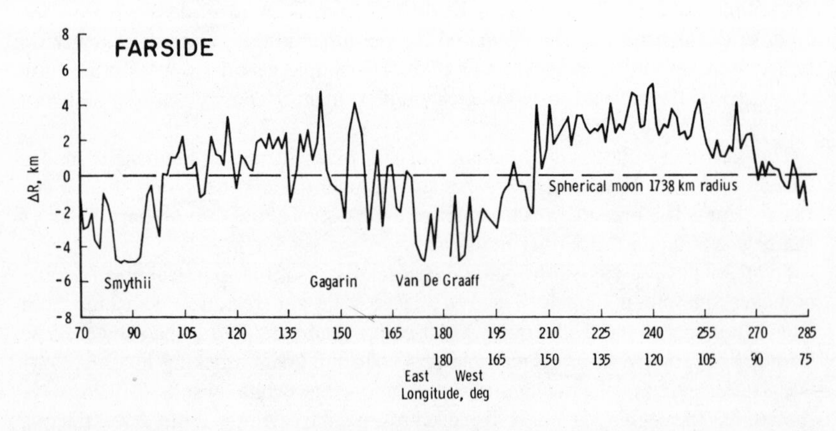

Figure 2 Altitude profiles (ΔR) of the lunar surface relative to a spherical moon of 1738 km radius taken from Kaula et al (37). Ground tracks to which altitudes correspond can be identified by the named features. Note the relatively smooth maria, rough terrae, and the great depression on the far side.

Table 1 Lower spherical harmonic coefficients for the lunar gravitational potential [data from Michael & Blackshear (48), thirteenth degree and order solution]

n	m	$C_{n,m}$	$S_{n,m}$
0	0	1.000054	—
2	0	-2.038×10^{-4}	0
2	1	1.105×10^{-5}	1.301×10^{-5}
2	2	2.485×10^{-5}	-0.001×10^{-5}

are given by Michael & Blackshear (48). The five lowest order coefficients are listed in Table 1. From two of these coefficients and the physical libration parameters β or γ, the three moments of inertia (A, B, C) can be determined by

$$C_{2,0} = \frac{1}{Ma^2}\left[\frac{A+B}{2} - C\right]$$

$$C_{2,2} = \frac{1}{4Ma^2}[B-A]$$

$$\beta = \frac{C-A}{B} \qquad \gamma = \frac{B-A}{C} \tag{2.}$$

where A is the moment about the mean moon-earth line in the lunar equatorial plane (x-axis), B is the moment about y-axis normal to x-axis in the equatorial plane, and C is the moment about the axis of rotation (z-axis). The libration parameters have been determined by the lunar laser ranging to be $\beta = 630.6\times10^{-6}$ and $\gamma = 226.0\times10^{-6}$ (13a).

Without reference to β or γ, we can determine $(B-A)/Ma^2 \simeq 1\times10^{-4}$. This indicates that there is excess mass in the moon along the x-axis. The cross moments show that the excess mass is in the earthside in agreement with the earthward displacement of the center of mass relative to center of figure, discussed earlier.

Compared to the earth and Mars, the gravitational field of the moon is relatively smooth, indicating that isostatic compensation has taken place, most likely relatively early in the lunar history. There are some well defined gravity anomalies, however, which are correlated with surface features indicating that isostatic compensation has not taken place over these features.

The most interesting of these gravity anomalies are the "mascons"—positive anomalies associated with some major ringed basins [such as Nectaris, Imbrium, Crisium, and Serenitatis (52)]. The gravity anomalies correspond to an excess mass density of about 800 kg/cm^2 over these flooded basins (69). They are disk-shaped, shallow features and are attributed to the high density of the mare basalts filling the basins. If one takes a value of $\Delta\rho = 0.4$ g/cc for the density difference between mare basalts and terrae material and does not allow for elevation changes and isostatic compensation, then a 20 km thick high-density filler is required in the mascon basins. Although the surface mass anomaly is unique, basalt thickness is model dependent. Unflooded craters of less than 100 km diameter generally have negative gravity anomalies and are uncompensated. Some higher features such as the Apennines (+85 mgal) and Marius Hills (+62 mgal) are also uncompensated. More complicated features (such as the ringed structured Orientale) are sometimes reflected in the gravity picture (67). The traverse gravimeter carried on Apollo 17 showed a well-defined, positive gravity anomaly over the valley floor. At that site the thickness of the higher density mare basalts is estimated at about 1.5–2.0 km (75).

II.2 *Viscosity and Strength of the Lunar Interior*

The gravitational anomalies together with the surface topography can be utilized to estimate the shear stresses and viscosity in the lunar interior. Such calculations are strongly dependent, however, on the structure and the rheology assumed for the moon. In the case of viscosity estimates, an evolution model or a relaxation time needs to be specified.

Stress models that have been calculated for the moon on the basis of surface topography and gravity anomalies indicate maximum stress differences of about 100 bars or less (10, 11). These are relatively small stresses and are probably below the strength of lunar materials at temperatures in the outer 500 km of the moon. The viscosity estimates for the outer layers of the moon down to a depth of several hundred kilometers range between 10^{26} and 10^{27} poise (8, 9, 12). Since these estimates are based primarily on the depths of the mascon basins and the magnitude of the anomalies, they apply to present-day conditions and probably to the past 3.0 billion years.

The relatively low stress differences, the high viscosity, and high strength of lunar materials in the absence of water and volatiles imply that faulting, convection, or other tectonic motions probably could not take place in the outer layers of the moon at the present time. The general aseismicity of the moon and the absence of shallow moonquakes, discussed in Section III, support the estimated high viscosity and high strength for the lunar mantle.

II.3 *Electrical Conductivity*

Three magnetometers with continuous recording capability have been deployed on the lunar surface, one each with the Apollo 12, 15, and 16 missions. These instruments, in addition to recording permanent lunar magnetic fields, have recorded the magnetic fields induced in the moon by large scale extended transient events. In addition to these, the ambient and time-dependent magnetic fields in the lunar environment are measured by the Explorer 35 satellite magnetometer that is orbiting the moon.

The analysis of the transient fields recorded simultaneously by the Explorer 35 magnetometer and the surface magnetometer has made it possible to calculate the electrical conductivity inside the moon (23, 24, 72, 73). The moon responds to transient magnetic events; these transients induce eddy currents inside the moon. The response or transfer function is computed from the induced field at the moon's surface and the forcing (source) function recorded by the Explorer 35 magnetometer.

The experimental results are interpreted by calculating the theoretical response of a spherical moon with radially varying conductivity. Although there are some limitations to the theoretical calculations and some difficulty in matching all the available data, the results definitely show an increase of electrical conductivity with depth into the moon. To a depth of about 1000 km the conductivity profile can be characterized by three layers (23) or by a greater variety of models including current layers (73). The resolving power of the data is not sufficient to detail the conductivity below this depth.

Conductivity depth profiles given by Dyal et al (23) and the three layer model (a good representation for the average models) of Sonett et al (73) are shown in Figure 3. Basically the models require very low conductivity ($\sigma_1 \leqq 10^{-9}$ mho/m)

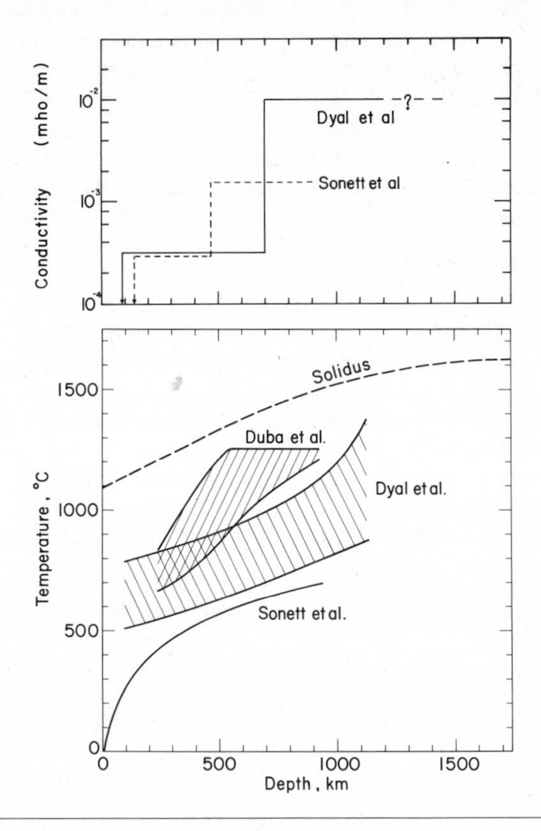

Figure 3 Electrical conductivity profiles (top) and some temperature inferences (bottom) that can be made from these. Conductivity profiles shown are from Dyal et al (23) and the three-layer model of Sonett et al (73). The temperature profiles shown include: the lowest estimate of Sonett et al (72), and the bounds based on the same electrical conductivity profile, but the temperature-conductivity relationship of Duba et al (21) for olivine with no Fe^{3+}. Temperature bounds given by Dyal et al (23) correspond to the bounds on their conductivity profiles and assumed olivine and peridotite compositions. "Solidus" is for anhydrous lunar basalt (62).

for the outer 50 km of the moon. For the intermediate layer ($1100 < R \leqq 1700$ km; R is radius) the conductivity is in the range of $\sigma_2 = 1$ to 7×10^{-4} mho/m. For the deeper interior there is a greater discrepancy. While Sonett et al (73) gave a value of about $\sigma_3 \approx 10^{-3}$ mho/m for $R < 1200$ km from the sunlit side data, Dyal et al (23) give $\sigma_3 \simeq 10^{-2}$ for $R \leqq 1000$ km.

From the conductivity profiles one can estimate the temperatures inside the moon if the composition and temperature dependence of conductivity are known. For a given composition, conductivity, σ_T, increases with temperature: $\sigma_T = \sigma_0 \exp(-A/kT)$ where A is the activation energy, k is the Boltzmann constant, T is the absolute temperature, and σ_0 is a constant dependent on the material. If olivine were assumed for a composition, and a lower bound on the conductivity of olivine is used, a temperature of about 1000°C is estimated for the moon at about 700 km depth (23, 73). If newer laboratory data (21, 32, 56) are used, higher temperatures (i.e. 1000°C at 400 km depth) are estimated (5, 86). Some bounds on temperature models are shown in Figure 3; in all cases temperatures are below the solidus to a depth of about 600–700 km. This point will be discussed further in conjunction with the thermal models of the lunar interior.

II.4 *Magnetic Properties of the Moon*

Magnetic properties of the moon and the magnetization of the lunar crust have been studied extensively by the orbital and surface magnetometers and by the remanent magnetic properties of lunar samples. Both the Explorer 35 (55) and the Apollo 15 and 16 subsatellite magnetometers (18, 65) have indicated the absence of a significant dipole field associated with the moon. The upper limit for the magnetic dipole moment is $M_M \leqq 8 \times 10^{19}$ cgs units (18) and most likely the limit is $M_M \leqq 3.6 \times 10^{18}$ cgs units, a value limited by the resolution of the data and its analysis (65). Even if the higher value is taken, this is about 9×10^{-7} that of the earth. A centered dipole with $M_M = 8 \times 10^{19}$ cgs units would produce surface

Table 2 Steady magnetic fields measured on the lunar surface by Apollo magnetometers [data from Dyal et al (24)]

Site	Field Magnitude (Gammas)	Components (Gammas)		
		Up	East	North
Apollo 12—ALSEP	38 ± 2	-25.8 ± 1	$+11.9 \pm 0.9$	-25.8 ± 0.4
Apollo 14—1	103 ± 5	-93 ± 4	$+38 \pm 5$	-24 ± 5
—2	43 ± 6	-15 ± 4	-36 ± 5	-19 ± 8
Apollo 15—ALSEP	3.4 ± 2.9	$+3.3 \pm 1.5$	$+0.9 \pm 2$	-0.2 ± 1.5
Apollo 16—ALSEP	235 ± 4	-186 ± 4	-48 ± 3	$+135 \pm 3$
—2	189 ± 5	-189 ± 5	$+3 \pm 6$	$+10 \pm 3$
—5	112 ± 5	$+104 \pm 5$	-5 ± 4	-40 ± 3
—13	327 ± 7	-159 ± 6	-190 ± 8	-214 ± 6
—LRV	113 ± 4	-66 ± 4	-76 ± 4	$+52 \pm 2$

magnetic field strengths between 1.5 and 3.0 gammas (1 gamma = 10^{-5} oersted). For all practical purposes, it is reasonable to assume that the moon does not have a global dipole magnetic field at the present time.

Permanent magnetic fields have been measured at four landing sites. The magnitudes of the fields range from about 6 gammas at the Apollo 15 site to a high of 327 gammas at the Apollo 16 site (23, 24). Measured fields are listed in Table 2; both the magnitudes and the directions of the fields are highly variable. At the Apollo 14 site, two measurements are within 1.1 km of each other, and at the Apollo 16 site all 5 measurements are over a distance of 7.1 km. These rapid variations indicate that the sources of the measured fields are local rather than global. Some of these variations may be due to edge effects of broken magnetized layers and to craters. Generally the front side highlands have stronger magnetic anomalies than the mare.

The orbital magnetometer showed great variability of magnetic field on the farside with local minima strongly correlated with crater positions (18). Although these anomalies are very small (with maximum amplitudes less than about one gamma) the shapes are well-defined, especially those correlated with craters Van de Graaff, Korolev, and Hertzsprung on the backside, and Balmer and Lagrenus on the nearside [see Figure 4, taken from Coleman et al (18)]. On the nearside, the field

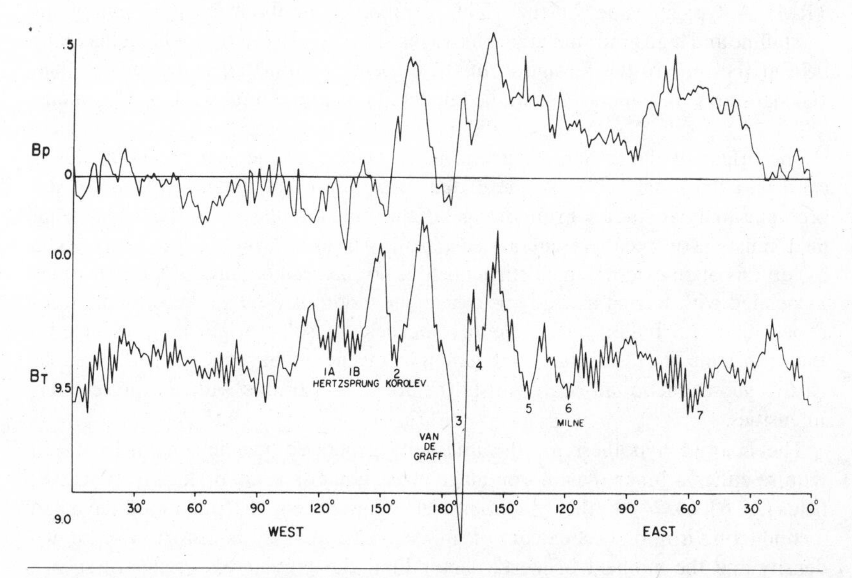

Figure 4 Components of the magnetic field measured by the Apollo 15 subsatellite magnetometer while the moon was in the relatively quiet geomagnetic tail of the earth [after Coleman et al (18)]. B_P is the vector component parallel to the spin axis of the spacecraft (normal to the ecliptic) and positive northward. B_T is the absolute value of the component transverse to spin axis. Units are in gammas (1 gamma = 10^{-5} oersted). The plotted values are averages over 17 successive orbits, at altitudes between 80 and 160 km.

over the maria is generally smooth. The variations associated with backside craters may be due to shock demagnetization of a magnetized crust by the crater forming impacts. Another possibility is that these craters are filled with highly magnetized breccias.

Studies of the remanent magnetic properties of returned lunar samples have provided additional data for the analysis of crustal magnetization and the nature of the magnetizing field. All samples (soils, breccias, crystalline rocks) have natural remanent magnetization (19, 22, 27, 53, 59, 64, 74). The initial magnetic intensity of the soils (fines) is the largest (about 8×10^{-3} emu/g). Breccias have highly variable initial intensities ($0.5–14.0 \times 10^{-4}$ emu/g). The typical values for lunar basalts are about $1.0–2.0 \times 10^{-6}$/emu/g (59). Both alternating field (AF) demagnetization and thermal hysteresis studies have determined the most important magnetic carrier to be metallic iron. Contributions of other magnetic phases (nickel, cobalt), if any, appear to be minor. The variation of the magnetic characteristics of samples is primarily due to the abundance and grain size distribution of the iron present.

Typically most rocks have a soft component of magnetization which may be due to magnetic contamination (inside the spacecraft or on the earth), and a stable component most likely acquired on the moon as the rocks cooled through the Curie point of iron in the presence of a magnetic field (Thermoremanent magnetization—TRM). A typical value for this stable component is about 2×10^{-6} emu/g for crystalline and high grade metamorphic rocks. The intensity of the ambient magnetic field at the time of the formation of these rocks is estimated to be greater than 1000 gammas, and in one case studied in detail it is found to be 2100 ± 80 gammas (27).

The origin of the ancient magnetic fields, which magnetized the initial lunar crust and the lunar rocks, is one of the most important and as yet unresolved problems in lunar studies. From the ages of the crust and the rocks, the magnetizing field must have been present at least from 4.6 to 3.2 billion years ago. The hypothesis of an external magnetizing field is not favored because of the difficulties associated with having a steady field near the moon of 1000 gammas or more for a period of 1.5 billion years. Neither the field of solar origin (at present even the transients are less than 100 gammas in the lunar environment) nor the earth's geomagnetic tail could satisfy the above constraints with the present day intensities.

The favored hypothesis of the internally generated magnetizing field would require either a lunar dynamo or magnetized lunar interior or locally generated fields (24, 63, 64, 74). In the dynamo model, it is necessary that the moon have had a conducting iron or Fe/FeS core of sufficient size early in its history. The known density and the moment of inertia factor limit the present day radius of such a core to less than about 500 km. Whether such a small core could have sustained a self-exciting dynamo for more than a billion years is an open question.

In the case of a highly magnetized model of the lunar interior (63, 74) it is required that temperatures below a few hundred kilometer depth remain below the Curie temperature (about 760°C) while all the magmatic activity and differentiation take place at shallower depths. Such a model would put severe requirements on the thermal evolution models and also require a high magnetizing field at the time of the lunar accretion.

In summary, the magnetization of the lunar crust is a fact, but the source of the magnetizing field is not yet resolved.

III. SEISMIC DATA AND STRUCTURE OF THE LUNAR INTERIOR

A seismic network which consists of four stations at the present (Apollo 12, 14, 15, and 16 sites) is operating on the moon. Seismograms from moonquakes, meteoroid impacts, and man-made impacts (Saturn SIV-B stage and LM ascent stage) have been recorded and analyzed. In Figure 5, typical seismograms from moonquakes and impacts are shown. They are very long and reverberating and are dissimilar to terrestrial seismograms. These reverberating characteristics have been explained by strong scattering of seismic waves in the upper 10–15 km of the lunar crust (20, 42, 43). Topographic features, lunar regolith, compositional boundaries, and especially cracks and joints in the crust become very efficient scatterers in the absence of water and the absence of damping. The Q-value for the lunar crustal material is about 3000–5000, an order of magnitude greater than that of the earth's crust. This high Q is primarily because of the absence of water and volatiles.

III.1 *Moonquakes and Lunar Tectonism*

The moon is very aseismic compared to the earth. The seismic energy release is about 10^{15} ergs/year, ten orders of magnitude less than that in the earth (44). Although each one of the seismic stations detects, on the average, between 600 and 3000 moonquakes per year, all moonquakes are very small (Richter magnitude 2 or less). There are many more very small micromoonquakes correlated with lunar sunset and sunrise at a given station. These may be thermally activated near surface events (45).

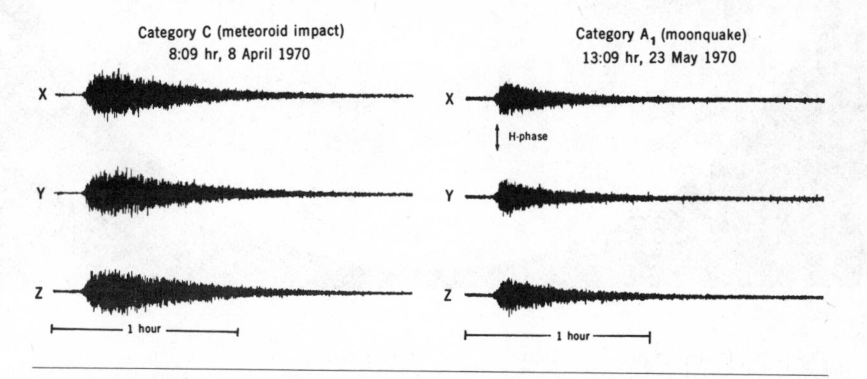

Figure 5 Typical seismograms of impacts and moonquakes recorded by Apollo Passive Seismic Experiment instruments.

The moonquakes can be grouped into sets with almost identical seismograms. Each set probably comes from a well-defined source region with dimensions of a few kilometers. Forty-one categories of matching events have been identified (45). The epicenters of these larger events recorded by the network are shown in Figure 6. The focal depths of these events (only for half of the events can the focal

Figure 6 Distributions of moonquake epicenters. Each epicenter represents a swarm of moonquakes belonging to a given category. The solid circles are epicenters where focal depths are determined. Open circles are epicenters where depth was not determined independently but constrained to 800 km [after Latham et al (45)].

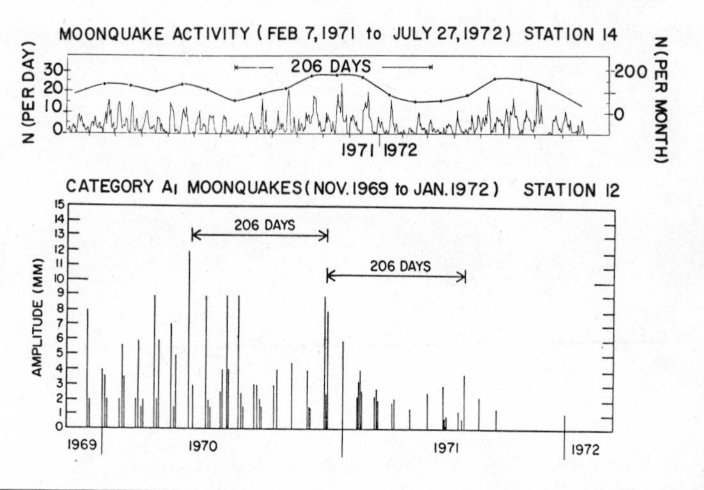

Figure 7 Time history of the moonquake activity recorded at two Apollo stations as given by Latham et al (45). The upper figure gives the number of events per day (the jagged curve) and per month (smooth curve) at station 14. The lower diagram shows the maximum peak-to-peak trace amplitude of moonquakes recorded at station 12.

depths be determined independently) are between 700 and 1200 km with the greatest majority occurring in the 800–1000 km depth range (42, 43, 44, 45). An important characteristic of moonquake activity is the biweekly periodicity (Figure 7). The peaks in the activity are correlated with the physical (latitudinal) librations of the moon (45). In addition, there is a 206-day variation in the activity, as shown in Figure 7, probably due to the solar perturbation of the lunar orbit. Longer periodicities or trends may also exist. These correlations suggest that the moonquakes are triggered by tidal forces, and at the present, tidal stresses play a more important role in lunar dynamics than tectonic forces.

III.2 *Velocity Structure*

The seismic velocity structure for the outer 150 km of the moon has been determined from the analysis of man-made impact seismograms recorded by the Lunar Seismic Network. Saturn SIV-B booster and LM ascent stage impacts have provided 22 seismograms covering a source-to-station distance range of 9 to 1750 km. Travel times, amplitudes, and seismogram shapes of all events with identifiable P-waves were used to derive a compressional wave velocity structure for the outer 150 km of the moon (81, 82, 83, 85). These are combined with those derived from distant meteorite impacts and moonquakes (45, 54) to infer the structure and physical conditions in the lunar interior.

The station network and the distribution of impact sites are shown in Figure 8. The travel time–distance plots for the compressional waves and distinct later arrivals

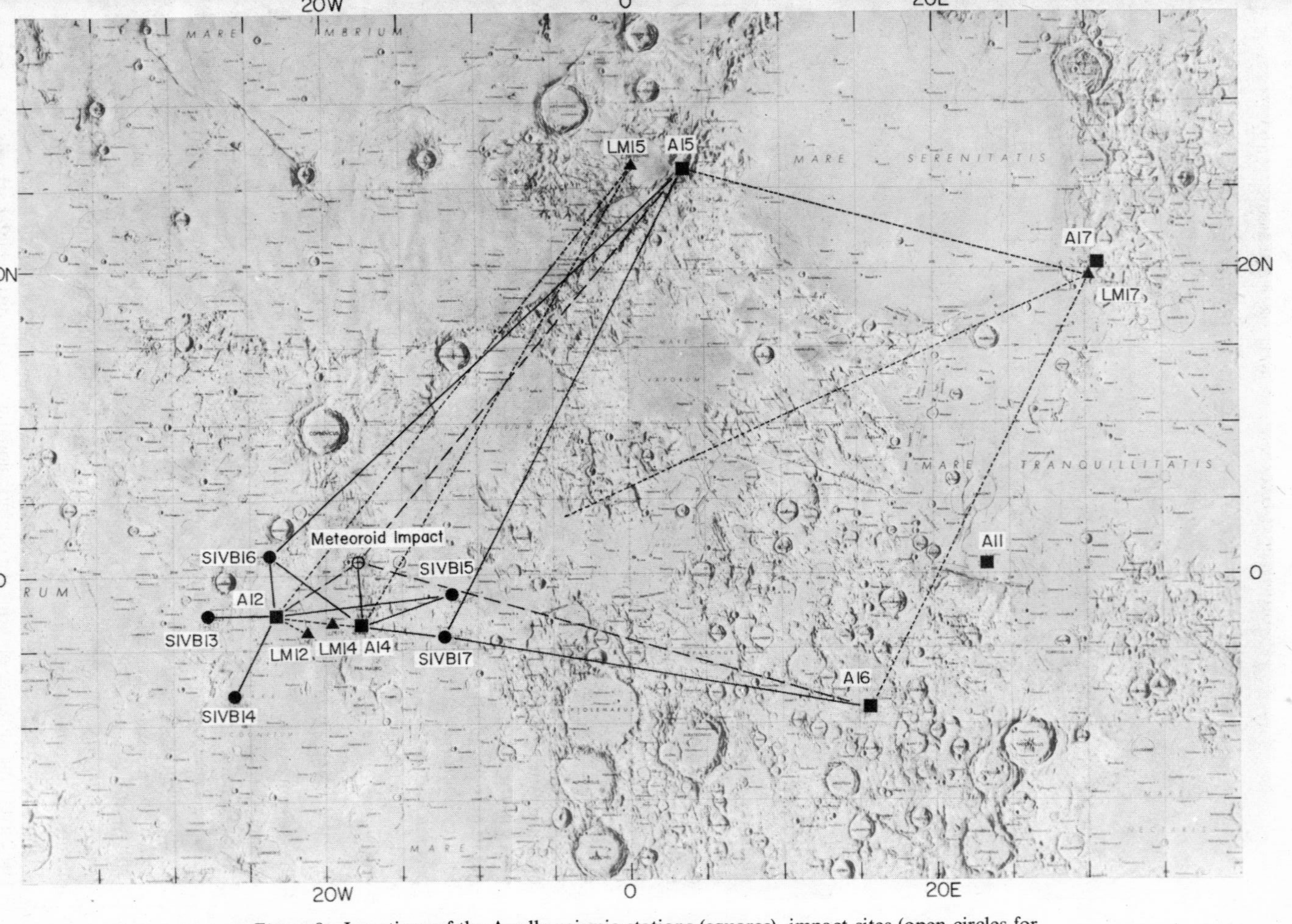

Figure 8 Locations of the Apollo seismic stations (squares), impact sites (open circles for

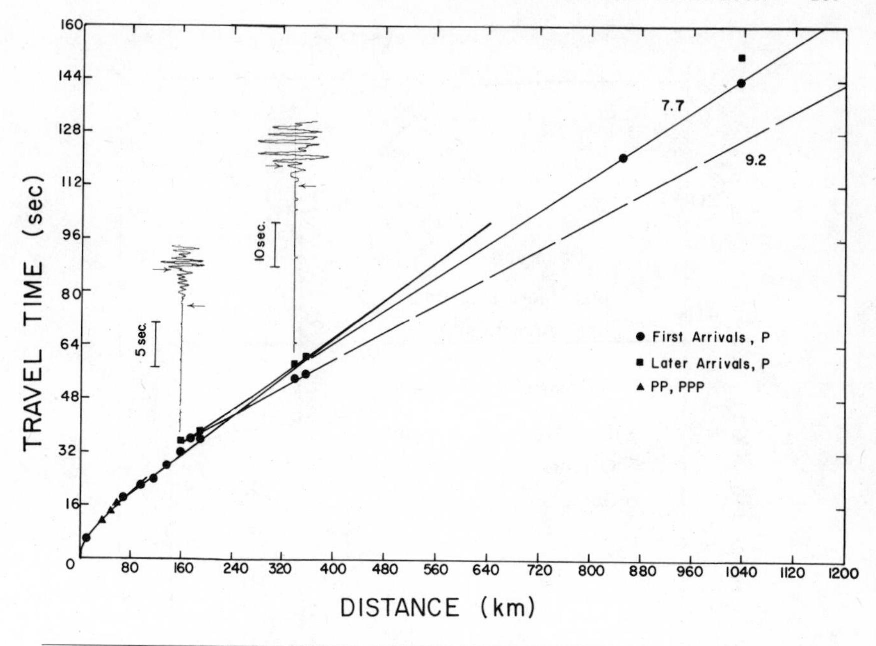

Figure 9 Travel time versus distance plot for seismic compressional waves generated by man-made impacts. Observed points are shown by circles (first arrivals), squares (later arrivals) and triangles (surface reflected phases). The two theoretical curves "7.7" and "9.2" km/sec indicate the range of possible upper mantle velocities. Two seismograms show the first and large second arrivals due to the large velocity discontinuity at 60 km depth—the base of the lunar crust.

are shown in Figure 9. Also shown in Figure 9 are the theoretical travel times for models with different upper mantle velocities. To a distance of 360 km, the travel time curve is well-constrained by the first as well as later arrival phases. The triplication of the travel time curve between 170 and 650 km (Figure 9) is due to the velocity jump at the base of the lunar crust, at a depth of about 60 km.

At far distances the amplitudes of the first recognizable phases are very small. The two arrivals from SIVB-17 at far distances (850 and 1040 km) correspond to an average velocity of 7.7 km/sec. Whether these are first arrivals or surface reflected later arrivals has not been determined. The velocity models constructed for the lunar mantle incorporate this uncertainty.

The compressional velocity model of Toksöz et al (85) is shown in Figure 10. It includes a two-layered lunar crust with a total thickness of about 60 km. In the upper few kilometers the velocity increases rapidly because of the transition from regolith, to fractured rocks, to competent crustal materials. The shallow layering for the Apollo 17 site, determined by Kovach et al (38) is also shown on the figure.

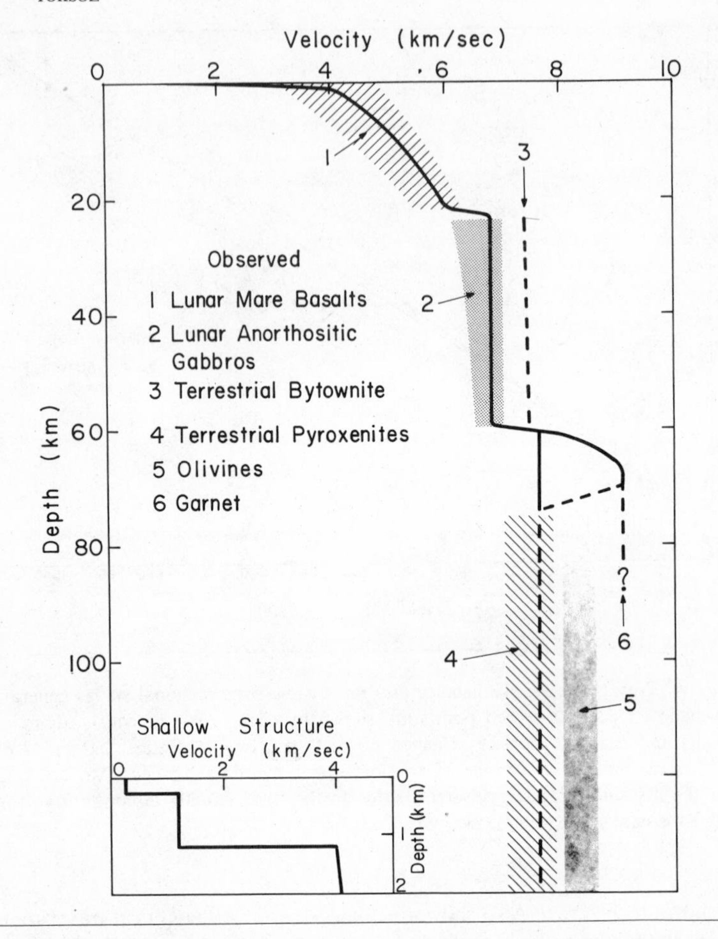

Figure 10 Compressional velocity profile for the lunar crust and upper mantle and laboratory data on velocities of lunar and terrestrial rocks. The velocity model is shown by a heavy line (or dashed heavy line where the model is uncertain). Inset shows shallow structure at Apollo 17 landing site and is based on data of Kovach et al (39). The laboratory data for lunar mare basalts and anorthositic gabbros include samples from different missions. Terrestrial bytownite velocities are from Wang et al (91), and olivines are taken from Chung (15). Pyroxenites cover a wide range of compositions and garnet velocity is a theoretical value for the high pressure phase of anorthite-rich rocks.

Below the crust there are two possible velocity models. These have velocities of 7.7 and 9.0 km/sec, respectively, immediately below the crust.

The nature of the 9.0 km/sec velocity requires some discussion. We see that evidence for such a high velocity layer comes from seismograms near station 12. If such a high velocity layer indeed exists, either it is localized and not a moon-wide phenomenon or it is of relatively small thickness (less than about 40 km) so that refracted arrivals cannot be observed at distant stations.

It should be re-emphasized that the crustal model discussed above applies primarily to the general region of the Apollo 12 and 14 landing sites. From gravity and laser altimetry data discussed earlier, one would expect the crustal thickness to change laterally. Also, sufficient data does not exist for an independent determination of the crustal structure under highlands.

Extension of the structural model to the deep lunar interior was done with seismic data from moonquakes and meteorite impacts. The recording of well-defined S-waves from the deep moonquakes implies that the lunar mantle must be sufficiently rigid to 1000 km depth to prevent appreciable S-wave attenuation. However, the attenuation of S-waves that have penetrated deeper than about 1000 km into the moon (45, 54) indicates a "softening" of the material in the central region below 1000 km. (These S-waves were generated by a farside impact and farside moonquakes.) Any softening which would reduce Q to less than about 500 would explain the observations. This could be achieved by temperatures approaching the solidus, by a very small amount of partial melt or by other mechanisms (such as perhaps an increase in the amount of volatiles in the deep lunar interior or a different bulk composition). Total melting is not a preferred model for 1000 km depth, but cannot be ruled out at the center of the moon.

By analogy to the earth, the lunar interior can be divided into a "lithosphere" and an "asthenosphere." The 1000 km thick outer shell would be the relatively rigid (viscosity $\geqq 10^{26}$ poise) "lunar lithosphere." The velocity in the "lithosphere" is estimated to be around 8 km/sec with possibly a slightly negative gradient with depth. The relatively soft deep interior of the moon ($r \leqq 700$ km) may resemble the earth's asthenosphere in its rheological properties.

The major units of lunar structure are shown schematically in Figure 11. From the displacement of the center of mass from the center of figure and the differences between the principal moments of inertia, a case can be made for possible thickening of the crust to about 100–150 km on the backside of the moon (98). Due to an absence of mare basalts, the crust on the backside is modeled as a single layer. The lunar "lithosphere" and "asthenosphere" are defined in a broad sense on the basis of preliminary results. A detailed characterization of the properties of these regions requires additional data. The presence or absence of a small (radius < 500 km) iron-rich core cannot be resolved with present data.

III.3 *Compositional Implications of Velocity Structure*

From studies of returned lunar samples, in situ chemical analysis (such as Surveyor data) and orbital measurements (1), vast amounts of information have been gathered about the chemistry and the mineralogy of the lunar surface. To review these is not within the scope of this article. As an overly simplified generalization, one may state that the mare material is primarily basaltic and terrae material is more anorthositic and may be characterized as anorthositic gabbro. Within each class there are wide variations both in composition and mineralogy (13).

With this background, the compositional implications of the lunar velocity models can be explored with the aid of high-pressure laboratory measurements on lunar and terrestrial rocks. Velocity measurements have been made on lunar soils, breccias, and igneous rocks from six Apollo missions (7, 16, 17, 34, 49, 51, 78, 79, 80, 90, 91, 93, 94). Regardless of composition, these rocks are characterized by very low velocities at low pressures relative to terrestrial rocks. This can be attributed to the absence of water in the lunar rocks combined with the effects of porosity and microcracks.

The measured velocities of appropriate lunar samples are shown in a generalized form along with the observed compressional velocity profile in Figure 10. Compressional velocities of terrestrial pyroxenites and olivines (6, 15, 60) shown on the figure specify general bounds between which most values fall.

From the comparison of the laboratory data and the lunar velocity profile, the following units can be identified:

(*a*) Near the surface the extremely low seismic velocities (about 100 m/sec near the surface) correspond to those of lunar regolith.

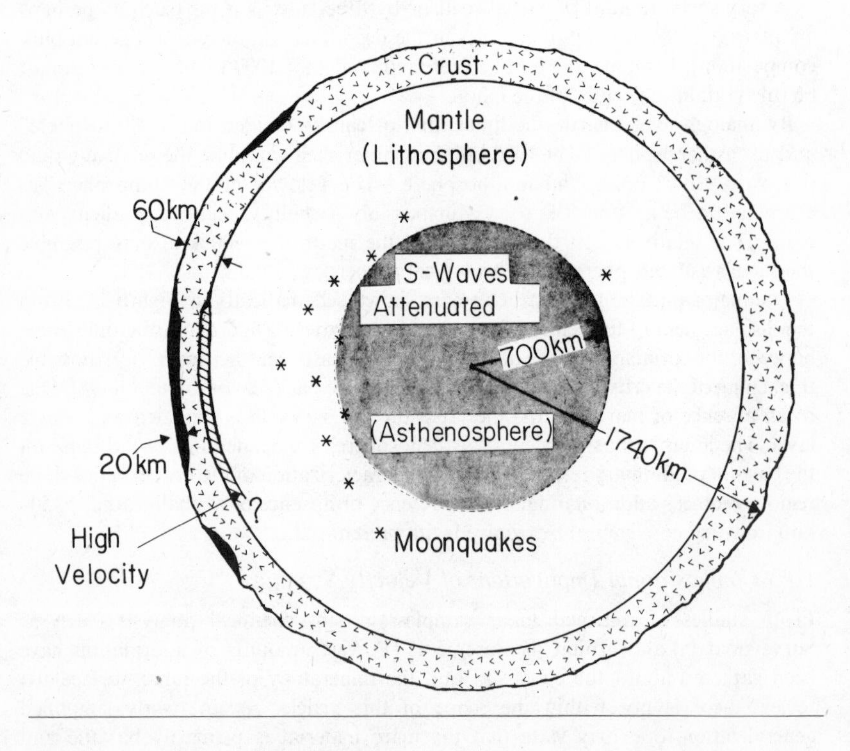

Figure 11 Schematic diagram of lunar structure with the thickness of the crust exaggerated. The earthside is to the left of the figure. Basalt filling under maria is shown in black. The possible limited extent of the high-velocity zone is indicated by stippling.

(*b*) Below a depth of about a kilometer, the measured velocities of lunar basaltic rocks fit the seismic velocity profile to a depth of about 20 km in the general vicinity of the Apollo 12 site. This is consistent with other data indicating similar values for the thickness of mare basalts. The elevation difference between Oceanus Procellarum (particularly Mare Cognitum) and the adjacent highlands is about 4 km. A model based on isostatic compensation between basaltic mare and feldspathic highland requires approximately 20–25 km of basalt in Oceanus Procellarum to explain the elevation difference.

(*c*) The second layer of the lunar crust (20 to 60 km) appears to be made of competent rock. The petrological interpretation of the velocity curve is not very simple. The velocities of lunar "anorthositic gabbros" defined by the laboratory data corrected for microfracturing effects, are close to the observed velocity profile. While the observed velocity profile is compatible with anorthositic gabbros, it does not limit the composition. Other models, especially some basalts, or a fixed composition with changes in physical state and microcrack densities, cannot be ruled out.

(*d*) Below the 60 km discontinuity neither the velocities nor the composition of the lunar mantle are specified uniquely. Even without the problem of the high velocity layer, a mantle with velocities of 7.7 or 8.0 km/sec could be consistent with a pyroxene-olivine composition. The velocity density systematics of this composition imply density $\rho = 3.4$ g/cm^3 consistent with moment of inertia considerations (70, 71). Such a composition is favored by Ringwood & Essene (62) and Green et al (28) as the source of mare basalts. Some plagioclase-rich compositions can also be accommodated within the constraints of the data with appropriate temperature models (3).

IV. THERMAL STATE AND THE EVOLUTION OF THE MOON

The structural, electrical, and elastic properties of the lunar interior described in the preceding sections indicate a differentiated moon and impose constraints on the thermal state of the lunar interior. Magmatization, differentiation, and evolution of the moon are controlled by its thermal history.

Thermal evolution models for the moon have been calculated by a number of investigators (30, 31, 46, 57, 61, 71, 84, 86, 87, 89, 97). Here some results are given without going into extensive detail. The computations on which these models are based are described by Toksöz et al (84) and Toksöz & Solomon (86).

The constraints that must be satisfied by the thermal models include: (*a*) the differentiation history and chronology of lunar igneous activity, (*b*) development early in lunar history of a lithosphere strong enough to support mascons, (*c*) lunar heat flow values at the Apollo 15 and 17 sites, (*d*) the electrical conductivity distribution inside the moon, (*e*) very low tectonic and moonquake activity at the present time, and (*f*) the relatively greater attenuation of S-waves below 1000 km. The initial conditions, heat sources, and mechanisms of heat transfer must be specified in the calculations. These must be adjusted within the limits of the data as required by the constraints.

The chronology of lunar igneous activity starts from the formation of the original crust about 4.6 billion years ago (67, 76, 95). The crust probably differentiated within a relatively short time (100 million years or so). Although the record of pre-mare volcanism has been obscured by subsequent events, there is evidence of magmatic activity in the time between the appearance of the original crust and the filling of the mare basins (33, 58, 66). The filling of mare basins spans a time interval between 3.16 and 3.80 billion years (57, 58, 77). No crystalline rocks younger than 3.16 billion years have been found.

The above chronology indicates some activity in the moon from the time of formation of the original crust to the time of the emplacement of mare basalts, although there may have been a cataclysmic jump in activity at about 3.95 billion years (77). The span of igneous activity is shown in Figure 12.

Direct measurements of the surface heat flux of the moon were carried out at the Apollo 15 and 17 sites (40, 41). The value at the 15 site is about 30 ergs/cm^2 sec and the preliminary result at the 17 site is about 28 ergs/cm^2 sec. Both of these values are still being revised with additional corrections and the uncertainties are probably about ± 5 ergs/cm^2 sec. Heat flow values place narrow limits on the total radioactivity, and to a lesser extent, on the temperatures in the moon, and they were utilized as a strong constraint in our thermal models. Other constraints were discussed in detail by Toksöz & Solomon (86).

To calculate the thermal evolution it is necessary to specify the initial conditions. The initial temperature distribution in the moon depends on the process of lunar formation. The evidence favoring an early episode of extensive near-surface melting in the moon has been outlined above. Probably the most important energy source which was available to heat the moon to melting temperatures was the gravitational energy liberated during accretion. Tidal dissipation, solar wind flux, short-lived

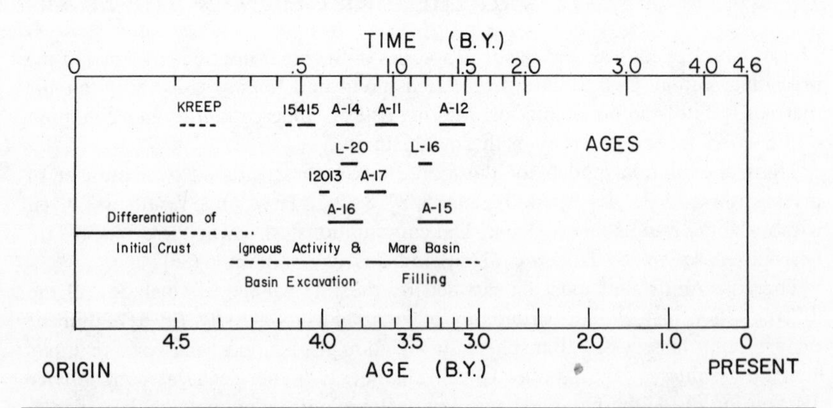

Figure 12 Ages of lunar samples in billion years (b.y.) and a summary of lunar igneous activity at the lunar surface. Samples from different missions are designated by *A* for Apollo and *L* for Luna. Special rocks are designated by their lunar catalog sample numbers. Reference sources for the summary are discussed in the text.

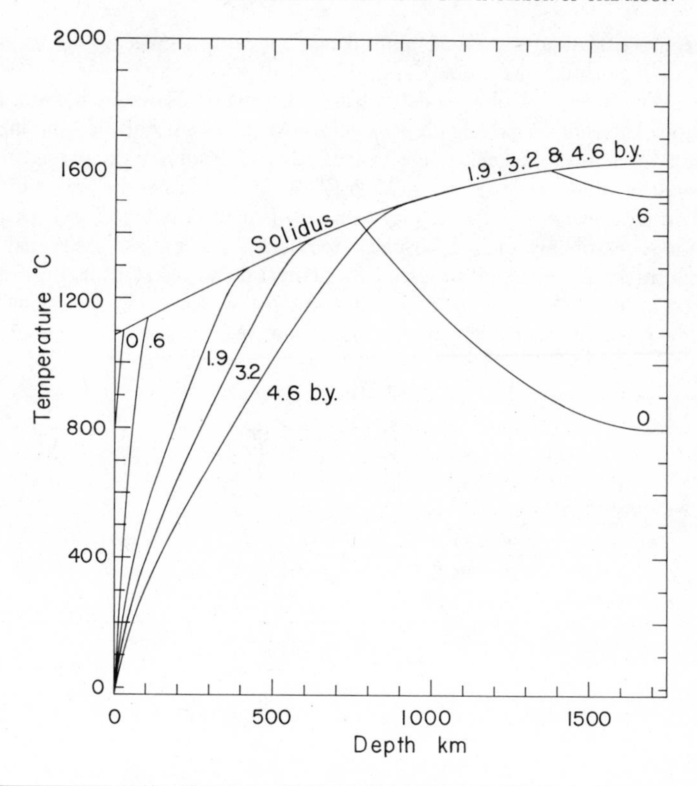

Figure 13 Calculated temperatures inside the moon as a function of time since its formation 4.6 billion years ago. The numbers on the curves indicate the time after lunar origin. The initial temperature profile is for a moon accreted in 100 years from material at 800°C. The radiogenic heat sources are differentiated during the first 1.6 b.y. The present day concentration of heat sources averaged for the whole moon are: U = 60 ppb, Th = 240 ppb, K = 120 ppm.

radioactivity and adiabatic compression are lesser sources. Taking a time-dependent accretion rate (29, 50), the accretion time may be adjusted so as to give an initial temperature profile in excess of the solidus in the outer portions of the moon.

A number of thermal history models were calculated with different initial conditions and heat sources by Toksöz & Solomon (86). In Figure 13, the temperature curves for a typical model, which satisfies most of the constraints, are shown. The initial temperature profile is from an accretion model with a starting temperature of 500°C and a total accretion time of 100 years. Radioactive heat sources are differentiated upward, and after 1.0 billion years, their concentration decreases exponentially with depth. Ratios of Th/U = 4 and K/U = 2000 are used. The present day uranium concentration averages 60 ppb, for the whole moon. The surface concentration is about 1 ppm, consistent with average values for lunar rocks and the orbital measurements (47).

It may be seen that the model satisfies the major constraints upon lunar evolution. The zone of melting deepens with time. The lithosphere during the first 2 billion years (b.y.) thickens at the rate of about 120 km/b.y. In particular, the shallowest melting progresses from 120 to 170 km depth during the period of mare filling, in agreement with the depth of origin of mare basalts (62) and with the need for a reasonably thick lithosphere to sustain the stresses associated with mascon gravity anomalies. At present, the temperatures below a depth of about 1000 km reach that of the basalt solidus, but the whole moon is cooling. The heat flow value for this model is about 30 ergs/cm^2 sec.

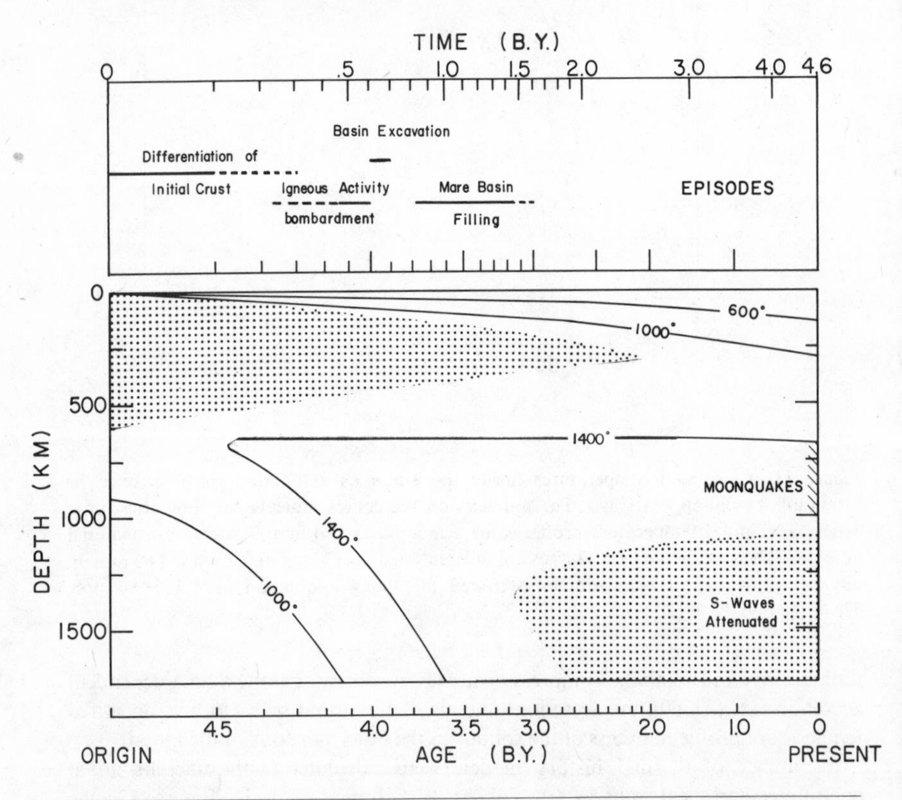

Figure 14 Thermal evolution of the lunar interior as a function of time (in billion years) based on the temperature calculations shown in Figure 13. Major episodes of differentiation and igneous activity are shown on top and are based on lunar sample ages. Isotherms are in centigrade degrees. Fine and coarse shading denote regions of partial and extensive melting, respectively. Zones of moonquake activity and high-attenuation of S-waves shown on the figure apply to the present.

The thermal evolution of the lunar interior as a function of time is shown in Figure 14. During the first two billion years of lunar history, the lunar upper mantle underwent sufficient melting to account for the differentiation of the crust and the subsequent lunar volcanism and mare filling. The disappearance of this melting coincides roughly with the termination of magmatic events. The deep interior is hot enough at the present time to be partially molten. Near-melting temperatures or a very small amount (less than one percent) of molten material, or both, could account for the attenuation of S-waves in this zone as described in Section III.2. If the deep lunar interior is indeed partially molten, its rheologic properties may be similar to the earth's asthenosphere and it may have convective motions. Such convection may exert small stresses at the bottom of the lithosphere where the moonquakes occur, but it could not induce sufficient stresses in the 1000 km thick lunar "lithosphere" to cause large moonquakes or active tectonic motions.

Present day temperatures shown in Figures 13 and 14 exceed the melting temperatures of iron or iron–iron sulfide combinations in the deep lunar interior. Thus if there were a concentration of Fe or Fe/FeS in the center these would be molten, and the moon would have a molten core.

V. SUMMARY AND CONCLUSIONS

The data and the models described above characterize the moon as a differentiated body which evolved relatively early in its history. At the present the moon has a crust, a relatively rigid and thick upper mantle—"lithosphere," and an "asthenosphere." It is not known whether it has a small iron rich core. The temperatures in the deep interior probably reach the melting temperature of basalts and possibly that of olivines. There is no evidence for tectonic activity in the moon at the present. This is consistent with results and calculations which indicate that convective motions could take place only very deep (below 1000 km) in the lunar interior. Such motions would not have sufficient energy to induce tectonic motions in the lunar "lithosphere." It appears that the moon expended its thermal energy early in its evolutionary history and completed its differentiation more than 3 billion years ago.

Large impacts have played an important role in shaping the lunar surface. Lateral variations in composition of surface rocks and the physical properties reflected in geophysical data are related to impact cratering and subsequent basalt filling.

The bulk chemistry of the moon as can be specified from the composition of the surface, the density and moment of inertia factor, and the seismic velocity profile is still uncertain. As a whole, the moon is enriched in refractory and rare earth elements and depleted in Na, K, Fe, and in volatiles relative to the earth and relative to chondrites (3, 25, 88, 92). The relative abundance of elements in the lunar samples, and the total amount of uranium determined from the heat flow data can be explained by a bulk composition derived by mixing two meteoritic components: approximately two thirds eucrites and one third ordinary chondrites (92).

Such chemical characteristics impose strong constraints on the formation of the moon. The moon could not have condensed from the solar nebulae in the same general region as the earth, and at the same time, have such a different chemical composition from the earth. Thus if the moon accreted in the solar system among the inner planets, it acquired its present orbit shortly after accretion (3, 4, 14, 61a). An alternate hypothesis is that the moon was formed outside the solar system and captured by the earth (2, 68).

ACKNOWLEDGMENTS

I would like to express my appreciation to Drs. A. Dainty and S. Solomon for their most valuable help and cooperative efforts in this paper. The article was written while the author was a visiting associate at the Division of Geological and Planetary Sciences of the California Institute of Technology. Many helpful discussions were held with Dr. D. L. Anderson.

This work was supported by NASA under Grant NGL 22-009-187.

Literature Cited

1. Adler, I. et al 1972. The Apollo 15 x-ray fluorescence experiment. In *Proceedings of the Third Lunar Science Conference,* ed. D. R. Criswell, 3:2157–78. Cambridge: MIT
2. Alfvén, H., Arrhenius, G. 1972. Origin and evolution of the earth-moon system. *The Moon* 5:210–30
3. Anderson, D. L. 1973. The formation of the moon. In *Lunar Science IV,* ed. J. W. Chamberlain, C. Watkins, 40–42. Houston: Lunar Sci. Inst.
4. Anderson, D. L. 1973. The composition and origin of the moon. *Earth Planet. Sci. Lett.* 18:301–16
5. Anderson, D. L., Hanks, T. C. 1972. Is the moon hot or cold? *Science* 178: 1245–49
6. Anderson, O. L., Liebermann, R. C. 1966. Sound velocities in rocks and minerals. *VESIAC State-of-the-Art Report* #7887-4-X. Willow Run Labs., Univ. Michigan. 189 pp.
7. Anderson, O. L., Scholz, C., Soga, N., Warren, N., Schreiber, E. 1970. Elastic properties of a micro-breccia, igneous rock and lunar fines from Apollo 11 mission. *Proceedings of the Apollo 11 Lunar Science Conference,* ed. A. A. Levinson, 3:1959–73. New York: Pergamon
8. Arkani-Hamed, J. 1973. Viscosity of the moon. I: After mare formation. *The Moon* 6:100–11
9. Arkani-Hamed, J. 1973. Viscosity of the moon. II: During mare formation. *The Moon* 6:112–26
10. Arkani-Hamed, J. 1973. Stress differences in the moon as evidence for a cold moon. *The Moon* 6:135–63
11. Baldwin, R. B. 1968. A determination of the elastic limit of the outer layer of the moon. *Icarus* 9:401–4
12. Baldwin, R. B. 1970. Absolute ages of the lunar maria and large craters. II. The viscosity of the moon's outer layers. *Icarus* 13:215–25
13. Bansal, B. M. et al 1973. Lunar rock types. In *Lunar Science IV,* ed. J. W. Chamberlain, C. Watkins, 48–50. Houston: Lunar Sci. Inst.
13a. Bender, P. L. et al 1973. The lunar laser ranging experiment. *Science* 182: 229–38
14. Cameron, A. G. W. 1973. Properties of the solar nebula and the origin of the Moon. *The Moon* 7:377–83
15. Chung, D. H. 1971. Elasticity and equations of state of olivines in the Mg_2SiO_4-Fe_2SiO_4 system. *Geophys. J. Roy. Astron. Soc.* 25:511–38
16. Chung, D. H. 1972. Laboratory studies on seismic and electrical properties of the moon. *The Moon* 4:356–72
17. Chung, D. H. 1973. Elastic wave velocities in anorthosite and anorthositic gabbros from Apollo 15 and 16 landing sites. In *Lunar Science IV,* ed. J. W. Chamberlain, C. Watkins, 141–42. Houston: Lunar Sci. Inst.
18. Coleman, P. J. Jr., Lichtenstein, B. R., Russell, C. T., Sharp, L. R., Schubert, G. 1972. Magnetic fields near the moon. See Ref. 1, 3:2271–86
19. Collinson, D. W., Runcorn, S. K., Stephenson, A., Manson, A. J. 1972. Magnetic properties of Apollo 14 rocks and fines. See Ref. 1, 3:2343–61
20. Dainty, A. M. et al Seismic scattering and shallow structure of the moon in

Oceanus Procellarum. *The Moon*. In press

21. Duba, A., Heard, H. C., Schock, R. N. 1972. The lunar temperature profile. *Earth Planet. Sci. Lett.* 15:301–4
22. Dunn, J. R., Fuller, M. 1972. On the remanent magnetism of lunar samples with special reference to 10048,55 and 14053,48. See Ref. 1, 3:2363–86
23. Dyal, P., Parkin, C. W., Cassen, P. 1972. Surface magnetometer experiments: Internal lunar properties and lunar field interactions with the solar plasma. See Ref. 1, 3:2287–2307
24. Dyal, P., Parkin, C. W., Daily, W. D. 1973. Surface magnetometer experiments: Internal lunar properties. *NASA Tech. Memorandum, NASA TM X-62, 278,* 41 pp.
25. Gast, P. W. 1972. The chemical composition and structure of the moon. *The Moon* 5:121–48
26. Gast, P. W., Giuli, R. T. 1972. Density of the lunar interior. *Earth Planet. Sci. Lett.* 16:299–305
27. Gose, W. A. 1973. A determination of the intensity of the ancient lunar magnetic field. *The Moon* 7:196–201
28. Green, D. H., Ringwood, A. E., Ware, N. G., Hibberson, W. O. 1972. Experimental petrology and petrogenesis of Apollo 14 basalts. See Ref. 1, 1:197–206
29. Hanks, T. C., Anderson, D. L. 1969. The early thermal history of the earth. *Phys. Earth Planet. Interiors* 2:19, 29
30. Hanks, T. C., Anderson, D. L. 1972. Origin, evolution and present thermal state of the moon. *Phys. Earth Planet. Interiors* 5:409–25
31. Hays, J. F. 1972. Radioactive heat sources in the lunar interior. *Phys. Earth Planet. Interiors* 5:77–84
32. Housley, R. M., Morin, F. J. 1972. Electrical conductivity of olivine and the lunar temperature profile. *The Moon* 4:35–38
33. Husain, L., Schaeffer, O. A., Sutter, J. F. 1972. Age of a lunar anorthosite. *Science* 175:428–30
34. Kanamori, H., Mizutani, H., Hamano, Y. 1971. Elastic wave velocities of Apollo 12 rocks at high pressures. *Proceedings of the Second Lunar Science Conference,* ed. A. A. Levinson, 3: 2323–26. Cambridge: MIT
35. Kaula, W. M. 1971. Interpretation of the lunar gravitational field. *Phys. Earth Planet. Interiors* 4:185–92
36. Kaula, W. M. 1971. Selenodesy and planetary geodesy. *Trans. Am. Geophys. Union* 52: IUGG1–IUGG4
37. Kaula, W. M., Schubert, G., Lingenfelter, R. E., Sjogren, W. L., Wollenhaupt, W. R. 1972. Analysis and interpretation of lunar laser altimetry. See Ref. 1, 3:2189–2204
38. Kaula, W. M., Lingenfelter, R. E., Schubert, G. 1973. Lunar topography from Apollo 15 and 16 laser altimetry. See Ref. 13, pp. 432–34
39. Kovach, R. L., Watkins, J. S., Nur, A., Talwani, P. 1973. The properties of the shallow lunar crust: An overview from Apollo 14, 16 and 17. See Ref. 13, pp. 444–45
40. Langseth, M. G. Jr., Clark, S. P. Jr., Chute, J. L. Jr., Keihm, S. J., Wechsler, A. E. 1972. The Apollo 15 lunar heat-flow measurement. *The Moon* 4:390–410
41. Langseth, M. G., Chute, J. L., Keihm, S. 1973. Direct measurements of heat flow from the moon. See Ref. 13, pp. 455–56
42. Latham, G. et al 1971. Moonquakes. *Science* 174:687–92
43. Latham, G. et al 1972. Moonquakes and lunar tectonism. *The Moon* 4: 373–82
44. Latham, G. et al 1972. Moonquakes and lunar tectonism results from the Apollo passive seismic experiment. See Ref. 1, 3:2519–26
45. Latham, G. et al 1973. Moonquakes, meteoroids and the state of the lunar interior. See Ref. 13, pp. 457–59
46. McConnell, R. K. Jr., Gast, P. W. 1972. Lunar thermal history revisited. *The Moon* 5:41–51
47. Metzger, A. E., Trombka, J. I., Peterson, L. E., Reedy, R. C., Arnold, J. R. 1972. A first look at the lunar orbital gamma-ray data. See Ref. 1, Vol. 3: frontispiece
48. Michael, W. H. Jr., Blackshear, W. T. 1972. Recent results on the mass, gravitational field and moments of inertia of the moon. *The Moon* 3:388–402
49. Mizutani, H., Fujii, N., Hamano, Y., Osako, M. 1972. Elastic wave velocities and thermal diffusivities of Apollo 14 rocks. See Ref. 1, 3:2557–64
50. Mizutani, H., Matsui, T., Takeuchi, H. 1972. Accretion process of the moon. *The Moon* 4:476–89
51. Mizutani, H., Newbigging, D. F. 1973. Elastic-wave velocities of Apollo 14, 15 and 16 rocks and thermal conductivity profile of the lunar crust. See Ref. 13, pp. 528–30
52. Mueller, P. M., Sjogren, W. L. 1968. Mascons: Lunar mass concentrations. *Science* 161:680–84
53. Nagata, T., Fisher, R. M., Schwerer, F. C., Fuller, M. D., Dunn, J. R. 1973.

Rock magnetism of Apollo 14 and 15 materials. See Ref. 1, 3:2423–47
54. Nakamura, Y. et al 1973. Missing shear waves from far-side meteoroid impact and moonquakes. *Science* 181:49–51
55. Ness, N. F., Behannon, K. W., Scearce, C. S., Cantarano, S. C. 1967. Early results from the magnetic field experiment on lunar Explorer 35. *J. Geophys. Res.* 72:5769–78
56. Olhoeft, G. R., Frisillo, A. L., Strangway, D. W., Sharpe, H. N. 1973. See Ref. 13, 575–77
57. Papanastassiou, D. A., Wasserburg, G. J. 1971. Lunar chronology and evolution from Rb-Sr studies of Apollo 11 and 12 samples. *Earth Planet. Sci. Lett.* 11: 37–62
58. Papanastassiou, D. A., Wasserburg, G. J. 1972. The Rb-Sr age of a crystalline rock from Apollo 16. *Earth Planet. Sci. Lett.* 16:289–98
59. Pearce, G. W., Strangway, D. W., Gose, W. A. 1972. Remanent magnetization of the lunar surface. See Ref. 1, 3:2449–64
60. Press, F. 1966. Seismic velocities. In *Handbook of Physical Constants,* ed. S. P. Clark. *Geol. Soc. Am. Mem.* 97: 195–218
61. Reynolds, R. T., Fricker, P. E., Summers, A. L. 1972. Thermal history of the moon. In *Thermal Characteristics of the Moon,* ed. J. W. Lucas, 303–37. Cambridge: MIT
61a. Ringwood, A. E. 1972. Some comparative aspects of lunar origin. *Phys. Earth Planet. Interiors* 6:366–76
62. Ringwood, A. E., Essene, E. 1970. Petrogenesis of Apollo 11 basalts, internal constitution and origin of the moon. See Ref. 7, 1:769–99
63. Runcorn, S. K., Urey, H. C. 1973. A new theory of lunar magnetism. *Science* 180:636–38
64. Runcorn, S. K. et al 1971. Magnetic properties of Apollo 12 lunar samples. *Proc. Roy. Soc. A* 325:157–74
65. Russell, C. T., Coleman, P. J. Jr., Lichtenstein, B. R., Schubert, G., Sharp, L. R. 1973. Apollo 15 and 16 subsatellite measurements of the lunar magnetic field. See Ref. 13, pp. 645–46
66. Shonfeld, E., Meyer, C. Jr. 1972. The abundances of components of the lunar soils by a least-squares mixing model and the formation age of KREEP. See Ref. 1, 2:1397–1420
67. Silver, L. T. 1970. Uranium-thorium-lead isotopes in some Tranquillity Base samples and their implications for lunar history. See Ref. 7, 2:1533–74
68. Singer, S. F. 1972. Origin of the moon by tidal capture and some geophysical consequences. *The Moon* 5:206–9
69. Sjogren, W. L., Muller, P. M., Wollenhaupt, W. R. 1972. Apollo 15 gravity analysis from the S-band transponder experiment. *The Moon* 4:411–18
70. Solomon, S. C. Density within the moon and implications for lunar composition. *The Moon.* In press
71. Solomon, S. C., Toksöz, M. N. 1973. Internal constitution and evolution of the moon. *Phys. Earth Planet. Interiors* 7:15–38
72. Sonett, C. P. et al 1971. Lunar electrical conductivity profile. *Nature* 230:359–62
73. Sonett, C. P., Smith, B. F., Colburn, D. S., Schubert, G., Schwartz, K. 1972. The induced magnetic field of the moon: Conductivity profiles and inferred temperature. See Ref. 1, 3:2309–36
74. Strangway, D. W., Sharpe, H. N., Gose, W. A., Pearce, G. W. 1973. Magnetism and the early history of the moon. See Ref. 13, pp. 697–99
75. Talwani, M., Thompson, G., Dent, B., Kahle, H. G., Buck, S. 1973. Traverse gravimeter results on Apollo 17. See Ref. 13, p. 704
76. Tatsumoto, M. 1970. Age of the moon: an isotopic study of U-Th-Pb systematics of Apollo 11 lunar samples—II. See Ref. 7, 2:1595–1612
77. Tera, F., Papanastassiou, D. A., Wasserburg, G. J. 1973. A lunar cataclysm at ~3.95 AE and the structure of the lunar crust. See Ref. 13, pp. 723–25
78. Tittman, B. R., Abdel-Gawad, M., Housley, R. M. 1972. Elastic velocity and Q factor measurements on Apollo 12, 14, and 15 rocks. See Ref. 1, 3: 2565–75
79. Todd, T., Wang, H., Baldridge, W. S., Simmons, G. 1972. Elastic properties of Apollo 14 and 15 rocks. See Ref. 1, 3:2577–86
80. Todd, T., Wang, H., Richter, D., Simmons, G. 1973. Unique characterization of lunar samples by physical properties. See Ref. 13, pp. 731–33
81. Toksöz, M. N. et al 1972. Velocity structure and properties of the lunar crust. *The Moon* 4:490–504
82. Toksöz, M. N. et al 1972. Lunar crust: structure and composition. *Science* 176:1012–16
83. Toksöz, M. N. et al 1972. Structure, composition and properties of lunar crust. See Ref. 1, 3:2527–44
84. Toksöz, M. N., Solomon, S. C., Minear, J. W., Johnston, D. H. 1972. Thermal evolution of the moon. *The Moon* 4:190–213, 5:249–50

85. Toksöz, M. N., Dainty, A. M., Solomon, S. C., Anderson, K. R. 1973. Velocity structure and evolution of the moon. See Ref. 13, pp. 734–36
86. Toksöz, M. N., Solomon, S. C. 1973. Thermal history and evolution of the moon. *The Moon* 7:251–78
87. Tozer, D. C. 1972. The moon's thermal state and an interpretation of the lunar electrical conductivity distribution. *The Moon* 5:90–105
88. Turkevich, A. L. 1971. Comparison of the analytical results from the Surveyor, Apollo and Luna missions. See Ref. 34, 2:1209–15
89. Urey, H. C., MacDonald, G. J. F. 1971. Origin and history of the moon. In *Physics and Astronomy of the Moon,* ed. Z. Kopal, 213–89. New York : Academic. 2nd ed.
90. Wang, H., Todd, T., Weidner, D., Simmons, G. 1971. Elastic properties of Apollo 12 rocks. See Ref. 34, 3: 2327–36
91. Wang, H., Todd, T., Richter, D., Simmons, G. 1973. Elastic properties of plagioclase aggregates and seismic velocities in the moon. See Ref. 13, pp. 758–60
92. Wänke, H. et al 1973. Multielement analysis of Apollo 16 samples and about the composition of the whole moon. See Ref. 13, pp. 761–63
93. Warren, N. et al 1971. Elastic and thermal properties of Apollo 11 and Apollo 12 rocks. See Ref. 34, 3:2345–60
94. Warren, N., Anderson, O. L., Soga, N. 1972. Applications to lunar geophysical models of the velocity-density properties of lunar rocks, glasses and artificial lunar glasses. See Ref. 1, 3:2587–98
95. Wasserburg, G. J., Papanastassiou, D. A. 1971. Age of an Apollo 15 mare basalt; lunar crust and mantle evolution. *Earth Planet. Sci. Lett.* 13:97–104
96. Wollenhaupt, W. R., Osburn, R. K., Ransford, G. A. 1972. Comments on the figure of the moon from Apollo landmark tracking. *The Moon* 5:149–57
97. Wood, J. A. 1972. Thermal history and early magmatism in the moon. *Icarus* 16:229–40
98. Wood, J. A. 1973. Asymmetry of the Moon. See Ref. 13, pp. 790–92

LOW GRADE REGIONAL METAMORPHISM: MINERAL EQUILIBRIUM RELATIONS[1]

E-an Zen
U.S. Geological Survey, Reston, Virginia 22092

Alan B. Thompson[2]
Department of Geology, University of Manchester, Manchester, England

INTRODUCTION

In 1961 Coombs (29) made an excellent survey of our state of knowledge of low grade rock metamorphism. Since then, there has been a blossoming of studies of such rocks, spurred by advances in determinative methods, in experimental techniques that have extended the range of study to lower temperatures and pressures, and in the theory and techniques permitting study of phase equilibria involving mixed volatile components to be made under calibrated conditions. Thus a new review seems justified.

In this paper, "low grade regional metamorphism" encompasses a range of geologic conditions and processes interpreted or known to be not too far removed from those prevailing at the earth's surface; low grade metamorphism grades into diagenesis as the conditions approach those at the earth's surface, and many mineralogic transformations usually considered diagenetic will be discussed here. For pelitic rocks, the upper cutoff of our discussion will be the biotite grade; for impure carbonate rocks, the cutoff will be tremolite-bearing marbles; for mafic volcanic rocks, the cutoff will be actinolite-epidote-hornblende–bearing greenstones. Blueschists are not considered here because they have recently been discussed elsewhere (Ernst et al, 38).

Some of the means used to estimate temperature and pressure will be reviewed. The principal method is by comparison with hydrothermal phase-equilibrium experiments, which furnish calibration points for natural processes. Many low grade

[1] Publication authorized by the Director, U.S. Geological Survey.

[2] Address as of September 1973: Department of Geological Sciences, Harvard University, Cambridge, Massachusetts 02138.

reactions involve one or more volatile components. Thus laboratory results can be applied only if the relation between total pressure and activity of the components is known. For ancient metamorphic rocks—in contrast to recent hydrothermal alterations—such relations can be ascertained only with difficulty. Fluid inclusion data could help to define the composition of the fluid phase. The observed lack of a unique relation between total pressure and fluid pressure in deep wells (see Hubbert & Rubey, 89; Burst, 19) should caution us from trying to use any simple rule concerning these partly independent variables. To be sure, given enough mineralogic data, unique values for all of the variables should be obtainable, and one might also devise probes—presumably electrochemical probes—to measure the chemical potentials of some components independently, at least in the laboratory, as is being done for oxygen (Sato, 156); however, we are not there yet.

METAMORPHIC ZONES AND FACIES

The term metamorphic zone was first introduced informally in 1893 by Barrow (5) apparently as a purely descriptive term. Tilley (183) equated his concept of isograds with Eskola's idea of metamorphic facies, and gave them physical significance as defining lines of equal T or P or both. Tilley fully appreciated the importance of compositional control on the appearance of minerals. Later, however, he (184) seemed to have equated Barrow's zones with isograds. In practice it is easier to describe the first appearance of a given mineral in a rock sequence than to pin down the specific reaction leading to the mineral. Metamorphic zones in the Barrovian sense are thus easy to delimit, but the facies boundaries (*sensu lato*) or isograds (synonyms here) in the physical-chemical sense [see also Thompson (180) for a more modern definition] are difficult to define. It would be impractical to insist that lines on geologic maps must await full understanding of the chemical reactions; we need a descriptive or petrographic scheme to show the appearance or disappearance of minerals. We will use the term "zone marker" in this descriptive sense. In contrast, a "facies boundary" or isograd refers to a specified first order heterogeneous reaction leading to an abrupt change in the mineral assemblage and a change in the topology of the particular phase diagram (180, p. 856).

It is clear that many facies changes correspond to new zones, though not all; most certainly the converse is not true. The reaction kaolinite + quartz = pyrophyllite is both a facies boundary and a zone marker, as is the conversion of kyanite to sillimanite, but the beginning of the biotite zone results from different reactions even for pelitic schists. Many experimentally determined reactions are possible facies boundaries, but because of compositional restrictions may not be directly applicable to rocks.

REVIEW OF SELECTED AREAS

This section summarizes the petrographic data for some low grade metamorphic terranes, chosen for the types of rocks involved and the types of reactions that have occurred. Three geothermal areas of contemporary active metamorphism subjected to a spectrum of physical-chemical conditions are particularly instructive because they permit calibration of the mineral assemblages with inhole measurements of physical and chemical variables. We have deliberately excluded several important areas of low grade metamorphic rocks because they are widely known through recent summaries in the English language; examples are the Sanbagawa district, Japan (Banno, 4); the Franciscan terrane, California [see Ernst et al (38) for data and references to both areas]; and the Otago Schist terrane, New Zealand (Brown, 16; Bishop, 9). There is much Soviet literature on low grade metamorphic and diagenetic rocks (e.g. Logvinenko & Osipova, 119; Lebedev & Bondarenko, 109; Lebedev et al, 110; see also Marakushev, 121); unfortunately space does not permit us to include these areas, for which we apologize to our Soviet colleagues. Metamorphism of ultramafic rocks was recently studied by Trommsdorff & Evans (189) and by Dungan & Vance (34). The metamorphism of saline deposits, though important and widely reported in the literature (e.g. Bradley & Eugster, 14; Jones, 97; Milton & Eugster, 133; Sheppard & Gude, 163) is omitted as it deals with unusual rocks.

Salton Sea, California

At Salton Sea in Imperial Valley, drilling in Pliocene and Pleistocene sand, silt, and clay deposits of the Colorado River has revealed a sequence of diagenetic-metamorphic assemblages (Muffler & White, 140). Penetration by drilling into hot zones is about 2 km (6000–8000 ft). The mineral assemblages in the cores are directly related to the temperature profiles in the holes, but not to the depth (140). The high temperatures encountered at fairly shallow depths (Helgeson, 68, p. 379) presumably promoted the mineralogic reactions.

Detrital montmorillonite is converted to illite (K-mica) below 100°C, and mixed layer montmorillonite-illite becomes K-mica at about 210°C. Dolomite-ankerite reacts with kaolinite to form chlorite + calcite at 180°C or lower (Muffler & White, 140). This reaction is a good isograd (Zen, 200). At about 300°C, epidote having about 1/3 of the Al substituted by ferric iron appears as a new mineral (Muffler & White, 140; Keith et al, 98), and the assemblage quartz-epidote-chlorite-K feldspar-albite-K mica essentially corresponds to the greenschist facies. Zeolites were not found.

From additional wells, Helgeson (68, 69) reported small quantities of newly formed diopside and tremolite, as well as high iron biotite, at a depth greater than 1.2 km (4000 ft) and temperature around 300°C. Helgeson studied the compositions of the aqueous solutions obtained from one of the wells. The typical solution is a concentrated Na-K-Ca chloride brine for which, at 300°C, the fugacity of S_2 is about 10^{-10} and that of O_2 about 10^{-30} bar. At 300°C the oxygen fugacity value is near the magnetite-hematite boundary at 1 bar total pressure (Eugster & Wones, 40, p. 92); Muffler & White (140, p. 168) reported that hematite is common in the most highly metamorphosed rocks, but pointed out (p. 178) that oxygen fugacity probably varies from well to well. The relatively high oxygen fugacity is consistent with the high ferric iron content of epidote, but it seems surprising to find "high iron biotite" and ankerite in this environment. The silica content of the solution seems to indicate equilibration with quartz.

The Wairakei and Broadlands Areas, North Island, New Zealand

The Wairakei and Broadlands geothermal areas show alteration of upper Cenozoic felsic volcanic rocks by aqueous fluids. The mineralogy and chemistry of Wairakei was reported by Steiner (167, 168), Coombs et al (30), and Ellis (36). The mineralogy and chemistry of Broadlands was reported by Browne & Ellis (17), Mahon & Finlayson (120), and Eslinger & Savin (39).

At both places, the volcanic rocks and interbedded sediments have been penetrated by drilling to depths of about 1.2 to 1.5 km (4000–5000 ft). Maximum temperatures of about 260°C are reported for Wairakei (Steiner, 168; Coombs et al, 30) with only minor reversals. At Broadlands, temperatures reach about 300°C and closely follow the boiling curve for water (Browne & Ellis, 17).

At Wairakei, the clay mineral sequence of any given drill hole is montmorillonite near the top (kaolinite is rare, and is confined to near-surface samples), grading downward to potassium-fixed illite-montmorillonite mixed-layer clay, and to illite at about 130°C. Illite is accompanied by chlorite. Epidote appears at about 220°C, and albite and adularia at about 240°C. "Ptilolite" (mordenite; see Coombs et al, 30, p. 72), accompanied by heulandite, occurs in the upper, cooler parts of the cores, about 100°C. Laumontite appears at about 160°C. Wairakite first appears at about the same temperature but becomes persistent only at about 230°C. Steiner (168) has reported that montmorillonite overlaps laumontite in its range of occurrence, but the K-mica–chlorite assemblage overlaps the ranges of wairakite, adularia, and albite. Quartz seems to be in equilibrium with the hydrothermal solution everywhere in the cores. Prehnite was reported (Steiner, 167, p. 7) but no details were given.

Calcite is found in the zone of albite and adularia (Coombs et al, 30, p. 72; see also Steiner, 168). Available information suggests that calcite and the calcium zeolites are, in general, not found together here.

The situation at Broadlands is similar as far as the micaceous minerals go. Calcite and the alkali feldspars occur mainly in the deeper samples. Quartz occurs as a product of alteration and devitrification, suggesting that the phase is equilibrated with the hydrothermal solution; this is compatible w.th the chemical data on the solutions (Mahon & Finlayson, 120, p. 53, 54). Epidote forms scantily at the deep parts of the drill holes, at temperatures not lower than 260°C (Browne & Ellis, 17, p. 260).

Wairakite, heulandite, laumontite, and mordenite are important and common at Wairakei, but are much less common at Broadlands. "Ptilolite" (mordenite again?) occurs at temperatures between 60° and 170°C in several cores at Broadlands, and wairakite occurs in a few cores only between about 230° and 280°C. These temperatures are similar to those for Wairakei.

The different abundance of the calcium zeolites puts Broadlands intermediate between Wairakei and the Salton Sea area. This is consistent with the available data on the CO_2 fugacity of the circulating fluids. At Salton Sea, the measured

CO_2 pressure and calculated CO_2 fugacity are a few bars (Helgeson, 68, p. 385; Rook & Williams, 154). Mahon & Finlayson (120), quoting and adding to Ellis (36), have indicated that the partial pressure of CO_2 is about 9 bars at Broadlands and 0.9 bars at Wairakei. Thus there seems to be a broad correlation between the occurrence of calcium zeolites and low fugacity of CO_2, supporting the argument for the stability relations of zeolites and other Ca-Al silicates versus carbonate minerals (Zen, 203; Thompson, 176).

The Wairakei and Broadlands mineral assemblages show that temperature, not lithostatic pressure, is the predominant factor. This accords with the observations at Salton Sea (Muffler & White, 140). One must remember, however, that the Wairakei and Broadlands geothermal drill holes penetrated only about 1.5 km, corresponding to a pressure of about 400 bars, and even the deeper Salton Sea drill hole for which mineralogical data are available did not exceed 1 kbar of rock pressure.

West-Central Vermont and Vicinity

Metamorphic mineral assemblages are developed in Eocambrian and lower Paleozoic sedimentary rocks of west-central Vermont. The assemblages tabulated by Zen (201, 202) at first sight suggest simple metamorphic zonation (Thompson & Norton, 182), but in reality the relations may be more complex because of polyphase metamorphism in the eastern, higher grade areas (Zen, 207).

The appearance of chloritoid is the only documented change in the topology of the phase relations for the pelitic rocks; Zen (201) suggested the reaction to be: paragonite + chlorite + quartz = albite + chloritoid + H_2O. Thompson (181) suggested the reaction ferrous chlorite + hematite = chloritoid + magnetite; a change from purple to green rock color would accompany the reaction. Frey (49, p. 112) provided chemical data from the Keuper Formation of central Switzerland to show that the change from purple to green color reflected a decrease of ferric-ferrous ratio. Thus a real difference in oxygen fugacity values probably existed between these alternative assemblages, a conclusion supported by the fact that in the eastern, higher grade area of west Vermont the chloritoid-bearing strata are stratigraphically controlled. Rocks carrying hematite-chlorite in lower grades become magnetite-chlorite rocks, without chloritoid, in the higher grade areas.

Stilpnomelane-bearing assemblages do not seem to have a low grade cutoff. Within large areas, stilpnomelane and chloritoid both occur in adjacent rock units, controlled by the bulk composition; the muscovite-chlorite ($\pm$iron oxide phases) join separates them in the AKFM diagram.

The impure limestones of the pelitic sub-biotite zone show the assemblage muscovite-chlorite-calcite-dolomite-quartz; talc was not found. Tremolite (slightly ferrous)$\pm$zoisite appear in calcite marbles. The occurrence of calcite-dolomite-quartz assemblage nearby indicates that the formation of tremolite may initially involve changing composition of dolomite or involve Al in tremolite.

No zeolite was found in west-central Vermont and this was attributed (Zen, 203) to the control of chemical potential of H_2O versus CO_2. Recently, analcime assemblages (Mossman & Bachinski, 139; Zen, 209) and prehnite-pumpellyite rocks

(Coombs et al, 32; Zen, 209) were found in several localities in the northern Appalachians, mainly in the same structural-metamorphic belt as western Vermont, but in altered mafic igneous rocks. These assemblages allow subdivision of the chlorite zone into several isograds, separated by abrupt changes of topology.

The Keuper Formation of Switzerland

The Upper Triassic Keuper Formation of central and eastern Switzerland originally consisted of sequences of interbedded shale, sandstone, and subsidiary carbonate (mainly dolostone) and calcareous shale. Volcanic rocks are minor [some "volcanic tuff" in the Quartenschiefer, part of the Keuper, was shown to be nonvolcanic (Frey, 49, p. 90)]. Samples from the Jura tableland and the Swiss plain (a drill-hole sample) are unmetamorphosed, those of the Glarus region begin to show metamorphism ("anchimetamorphism"; see Kubler, 104, 105; Dunoyer de Segonzac, 35), those of the Aarmassiv are above the stilpnomelane isograd, those of the Gotthardmassiv are above the chloritoid isograd, and those from the Lukmanierpass region are in the staurolite zone (Frey, 49).

The metamorphism of the Keuper Formation is everywhere a late Alpine event, mainly of early Tertiary age (Niggli, 144; Jäger, 90; M. Frey, 1972, personal communication). The diagenesis of the Keuper in the nonmetamorphosed areas, of course, could have proceeded ever since deposition. Polymetamorphism does not seem to have been likely here.

Frey showed (49, p. 98 ff, 51; Frey & Niggli, 53) that the first reaction observed was the formation of Al-rich chlorite and illite-muscovite from mixed layer montmorillonite-illite, as in the Salton Sea region. Before the first appearance of paragonite and chloritoid, the 1M polymorphs of muscovite are replaced by 2M polymorphs. Frey (49, p. 99) used the crystallinity of the K-mica to delineate metamorphic zones: the unmetamorphosed zone, the anchi-zone, and the epizone. The 1M–2M muscovite change is probably rate controlled as the 1M polymorph is now considered metastable (Velde, 194, p. 436), thus these changes may define metamorphic zones but they do not define isograds.

Frey (50, 51) showed that in the "greenschist facies" or his epizone, paragonite occurs, and in the anchimetamorphic zone mixed-layer phengite-paragonite appears, possibly formed simultaneously from a Na-bearing illite (Frey, 49, p. 115). The intimate textural relation of paragonite and phengite in low grade metamorphic rocks has also been reported by Laduron & Martin (106) and Black (10). Interpretation of the significance of phengite will be discussed in a later section.

Pyrophyllite occurs in the Glarus region, and could be caused by the reaction of kaolinite plus quartz (Frey, 49, p. 107), an association found in the drill hole in the Swiss plain. Because the phases belong to a simple chemical system, the Glarus region clearly is at a higher metamorphic grade.

Puerto Rico

The regional geology of the Cretaceous rocks, including greywacke, sandstone, mudstone, and volcanic rocks, was reviewed by Berryhill et al (6). The metamorphism, especially of the andesitic to basaltic rocks of the Robles

Formation, was described by Otalora (149) and Jolly (95). Jolly found that the mineral associations depend on whether they formed in the greywacke, or in the olivine or plagioclase phenocrysts, as shown below:

	Greywacke	*Olivine phenocrysts*	*Plagioclase phenocrysts*
Zone 1	Sericite, albite, analcime, celadonite, laumontite	Chlorite, analcime	Sericite
Zone 2	Laumontite, albite, prehnite, pumpellyite	Prehnite, pumpellyite, chlorite	Albite prehnite, pumpellyite
Zone 3	Albite, pumpellyite, prehnite, epidote	Pumpellyite, epidote, quartz, hematite	Albite, pumpellyite, epidote, hematite

The dependence of local assemblages on local bulk composition is not surprising, and has been reported also elsewhere (Brown, 15).

Mordenite and heulandite are not reported. Analcime, although common, is not found with prehnite-pumpellyite; instead, albite is present. Laumontite coexists with prehnite-pumpellyite at least at lower grades; this is true also in Chile (Levi, 111), in the Taveyanne Formation of the French and Swiss Alps (Martini & Vaugnat, 124, 125; Martini, 122, 123), and in the Keweenawan basalt of Michigan (Jolly & Smith, 96), among others. Whether the laumontite is or ever was fully hydrated, however, remains unknown.

Jolly (95, p. 208) reported the association of prehnite and sericite; at Wairakei, Steiner (168) reported wairakite and adularia. These assemblages are approximately related by the reaction

$$\text{muscovite} + \text{prehnite} + 5\ \text{quartz} + 3H_2O = 2\ \text{wairakite} + \text{adularia}$$

Increasing total pressure and temperature, and decreasing chemical activity of H_2O would favor the prehnite-muscovite association.

Jolly's data show clearly that material has migrated megascopic distances. Noteworthy is the substantial transport of aluminum into former olivine phenocrysts. Intergranular and grain-boundary movement, possibly with a fluid phase as carrier, presumably was important; study of fluid inclusions in such low grade rocks, if feasible, could add significantly to our understanding of the metamorphic process for these rocks. The movement of aluminum in these rocks is in sharp contrast to the idea of immobility of aluminum in pelitic rocks during regional metamorphism, as proposed by Carmichael (22) and Fisher (45); whether the contradiction is real remains to be seen.

Stilpnomelane is not reported from these Puerto Rico rocks. At lower grades, celadonite could be the substitute phase. Wise & Eugster (199, p. 1074) suggested this possibility, Frey et al (55) reported prograde change of glauconite-celadonite to stilpnomelane in some Alpine rocks, and Zen (209) found similar relations in the northern Appalachians. The absence of stilpnomelane in the Puerto Rican rocks might reflect a difference in oxygen fugacity.

Australia

Smith (165, 166) described "burial metamorphic" rocks from an Ordovician sequence in west-central New South Wales. The sequence consists of mafic lavas and sedimentary and pyroclastic rocks locally altered to prehnite-pumpellyite and actinote-epidote assemblages, described in terms of color domains. The "yellow-green domains" are aggregates of epidote + quartz ± albite, calcite. The associated amygdules contain epidote or quartz, ± coarse pumpellyite. The most common domain consists of albite + chlorite, and minor prehnite, pumpellyite (after chlorite), and epidote. Where amygdaloidal, the large amygdules (1 to 5 mm) are filled with calcite-chlorite and smaller ones are filled with chlorite + prehnite + epidote or pumpellyite.

Smith suggested that chemical segregation occurred during the alteration, leading to Ca-enriched epidotic domains and to the albite-chlorite-rich domains having a spilitic lithology (see also Vallance, 192, 193). The domains might reflect segregations that occurred at a previous, lower grade. In any event, they imply selective dissolution and transport of material during alteration, particularly CaO, Al_2O_3, and SiO_2 away from plagioclase and pyroxene to form calcium-aluminum silicates in regions of high permeability. The redistribution of material, occurring on a large scale, could lead to spilitic bulk compositions, as also suggested by Reed & Morgan (151) for the greenstones of the Catoctin Formation of Virginia.

Chile

The Andean geosyncline of northern and central Chile consists of Jurassic and Lower Cretaceous volcanic and sedimentary rocks, estimated to be 15 to 28 km thick. In this Mesozoic sequence, the basaltic-andesitic lavas, felsic ignimbrites and volcanogenic sediments are altered by "burial metamorphism" to zeolite, prehnite-pumpellyite, and actinolite-epidote assemblages.

The alteration minerals are most abundant in the brecciated tops of lava flows. The middle parts of flows are only slightly altered and the lower parts are unaltered. The amygdules are larger towards tops of flows; they are filled by extremely variable mineral associations in adjacent amygdules (A. B. Thompson, unpublished data). This variability may reflect strictly local, stagnant fluid-rock interaction or complex flow networks of the fluid causing the alteration. Studies should be made on spatial and temporal variations in amygdule fillings and their possible dependence upon the sizes of cavities.

In general, the metamorphic grade increases with depth in the volcanic pile. The uppermost zeolite bearing amygdules contain chabazite, stilbite, mordenite, and analcime. Lower down, heulandite (with analcime + quartz) appears, followed by laumontite + albite + celadonite (B. Levi, personal communication, 1970; A. B. Thompson, unpublished data). Epidote and prehnite then appear, followed by pumpellyite produced by reaction of chlorite with epidote, laumontite, or prehnite. At the lowest level of the volcanic pile actinolite joins the epidote-albite assemblage, whereas prehnite and pumpellyite gradually disappear. Most of the assemblages also include calcite and quartz; sphene is found throughout, but wairakite has not been identified.

The downward increase in grade appears to be independent of rock type. The metamorphic zones (some are isograds) generally parallel bedding planes of individual flows and, as indicated previously, reflect alteration in the most porous parts of the flows.

At Bustamante Hill west of Santiago the predominantly andesitic strata show alteration ranging from laumontite through prehnite-pumpellyite to actinolite-epidote assemblages in a 10-km section. (Thomas, 173; Levi, 111; A. B. Thompson, unpublished data.) Levi (111) noted that the assemblages above the "Peralillo unconformity" were of higher metamorphic grade (epidote-actinolite) than those below it (laumontite). Levi (112) reported discovering breaks in the progressive mineralogical sequences at stratigraphic and unconformable boundaries as well; four such sequences are found in the Coast Ranges and three in the Andean Range. Levi (112) considered that the separate series represent "episodic" alterations, and each alteration episode effectively sealed the rocks to subsequent episodes of alteration. However, because rocks with zeolite and calcite cement are still porous and are readily susceptible to alteration, the seal may not have been effective against polyphase metamorphism. In fact, the actual nature of the "unconformities" is not demonstrated; some might actually be thrust faults whereby higher grade rocks were transported over lower grade rocks. This interpretation would reduce the estimated cumulative stratigraphic thickness from about 28 to nearer 15 km.

Metamorphism on the Mid-Oceanic Ridges

The low grade metamorphic rocks dredged from mid-oceanic ridges and vicinity have been described by numerous authors; typical are the results reported by Melson & Van Andel (127), Cann (21), and Miyashiro et al (137, 138). They come mainly from lower parts of steep fault scarps and the depth of metamorphism is unknown (141, p. 60). Alteration minerals include zeolites (natrolite, thomsonite, analcime, chabazite, laumontite, stilbite; Miyashiro et al, 138), nontronite, pumpellyite, epidote, actinolite, and albite (Melson & Van Andel, 127).

If the metamorphism resulted from reaction with heated seawater, then the relations would be analogous to that in the Andean geosyncline of Chile (A. B. Thompson, unpublished data). Whether the observed low grade metamorphism is pervasive throughout ocean-floor basalts or is restricted to fracture zones should be investigated, as the conclusions would have profound implications on the petrology of the ocean floor and its possible thermal, mechanical, and mineralogical behavior in subduction zones.

SUMMARY: OBSERVED ZONE MARKERS, INCLUDING SOME ISOGRADS

The information summarized in the preceding section allows some generalization on mineralogic changes that can be useful zone markers for various rock types.

Lime-Poor Pelitic Rocks

For these rocks, the first notable changes are in the dioctahedral micaceous minerals: montmorillonite to mixed layer illite-montmorillonite to 1M muscovite, finally to 2M muscovite. The same general sequence is found in the various areas reviewed; Burst (19) found similar relations in the sediments of the Mississippi delta. The changes are mappable and, whatever the causes, can be used to mark zones. As one effect is progressive dehydration in the sequence, temperature may be expected to be a significant control, in agreement with the geologic relations of the areas and with Burst's observations.

Several changes occur after 2M muscovite appears. These include the reaction kaolinite + quartz = pyrophyllite [the Liassic shales and the Keuper beds; also the underclays of the Pennsylvanian anthracite beds (Hosterman et al, 87)], the disappearance of stilpnomelane and formation of biotite, and the consumption of hematite either to form other ferric-iron bearing minerals (Thompson, 181) or by reduction (Frey, 49). The reaction paragonite + chlorite + quartz = chloritoid + albite + H_2O, though non–first order, is a useful zone marker for aluminous rocks.

The top of the biotite zone is commonly given by the appearance of garnet in a Barrovian sequence. This is a good zone marker. In lime-free systems, the pair biotite + chloritoid is stable rather than chlorite + muscovite + almandine (Zen, 206; also Zen, unpublished data). For specified rock composition the reaction may be: muscovite + chlorite = chloritoid + biotite + quartz + H_2O and is a good zone marker.

Calcareous Rocks

Tilley (185) suggested that talc is the first metamorphic mineral to appear in impure quartz-bearing dolostones. This mineral is not commonly observed. If the rock contains any appreciable feldspar or clay mineral, then prehnite, laumontite (D. S. Coombs, personal communication, 1973), chlorite, and/or phlogopite forms instead, for instance: dolomite + muscovite + quartz + H_2O = calcite + Mg-chlorite + phlogopite + CO_2. This "biotite" zone marker, of course, may not be compared with the biotite zone marker in pelitic rocks.

Details of the reactions leading to the actinolite formation, another early zone marker, are obscure. In Vermont, adjacent beds contain dolomite-calcite-quartz and calcite-actinolite-quartz, so calcite-dolomite-actinolite-quartz must be stable (Zen, 201); thus according to the phase rule, the reaction dolomite + quartz = actinolite + calcite (Bowen, 13) is at least bivariant.

Frey & Niggli (54) reported various margarite assemblages and suggested that the appearance of this mineral is by the reaction pyrophyllite + calcite = margarite + quartz + H_2O + CO_2.

Mafic and Intermediate Volcanic Rocks

Metamorphism of mafic and intermediate volcanic rocks begins with the formation of zeolites, followed by prehnite-pumpellyite; many greywackes show similar sequence of mineral changes. These appearances of new phases are useful zone markers; some are also isograds (Zen, 209).

Mineral assemblages in metamorphosed pillow lavas, many of which are spilitic, commonly show different "grades" between pillow assemblages and interpillow assemblages, presumably related to the greater permeability of interpillow material, different cooling history, and possible addition or removal of material in the interpillow samples during sedimentation or later. The vesicular versus compact parts of individual flows from Chile show such relations. These different assemblages need not mean disequilibrium, but could simply indicate compositional controls or local equilibrium in an activity gradient or fracture system and are thus worthy of special study.

Felsic Volcanic Rocks

Low grade metamorphism of felsic volcanic rocks in many ways parallels that of the more mafic rocks. In calc-alkaline rocks, the alkali and calcium zeolites are the first products of devitrification of relict glass (Wairakei, Broadlands), provided the activity of CO_2 is not too high. In lime-poor rocks analcime is the dominant zeolite (Mossman & Bachinski, 139). Mordenite and clinoptilolite are reported from hydrothermal areas (Yellowstone National Park: Fenner, 44; Honda & Muffler, 85). The kinetics of nucleation and formation of zeolites and their dependence on the environment and thermal history of the glass need further laboratory study.

Because the bulk compositions of many felsic volcanic rocks correspond approximately to a mixture of quartz, alkali feldspars, and some subsidiary mafic minerals—stable together from the biotite zone up—monitoring of their progressive metamorphism beyond the zeolite grade may be difficult. Details of the feldspar composition and structural state or both may be useful, but will not furnish facies boundaries.

EXPERIMENTAL DATA AND CORRELATION WITH MINERAL ASSEMBLAGES

It is beyond the scope of this review to survey and critically comment on all the pertinent experimental data now available. In this section, then, we merely list the important data sources and the variables and equilibrium conditions used for the closely reversed experiments which may have direct application to low grade assemblages, with only occasional comments; the many syntheses and nonreversed studies are not considered. Equilibria in some simple systems (Al_2O_3-SiO_2-H_2O; CaO-Al_2O_3-SiO_2-H_2O-CO_2; Na_2O-K_2O-Al_2O_3-SiO_2-H_2O), in particular, have direct application as zone markers.

The Data

ALUMINUM SILICATES The reaction

$$\text{kaolinite} + 2\,\text{quartz} = \text{pyrophyllite} + H_2O \qquad 1.$$

was reversed by Reed & Hemley (150) at 300°C, 1 kbar; by Althaus (3) at 390–405°C, 2–7 kbar; and by Thompson (174) at 325–375°C, 1–4 kbar.

The upper stability limit of pyrophyllite given by the reaction

$$\text{pyrophyllite} = \text{andalusite} + 3\ \text{quartz} + H_2O \qquad 2.$$

was reversed by Althaus (3) at 490–525°C, 2–7 kbar; by Hemley (76) at 400°C, 1 kbar; and by Kerrick (99) at 410–430°C, 1.8–3.9 kbar. Both reactions reflect equilibrium for the condition $P_{H_2O} = P_{total}$.

ALKALI ALUMINUM SILICATES Preliminary results on the reaction

$$\text{K-feldspar} + \text{kaolinite} = \text{muscovite} + 2\ \text{quartz} + H_2O \qquad 3.$$

by Thompson (178) suggest that 2M muscovite may have a lower stability limit in this system near 100±25°C (see, however, Hemley & Jones, 77, p. 561).

Paragonite plus quartz may have a lower stability limit with respect to albite and kaolinite, analogous to reaction 3 (Zen, 201); the data of Kisch (102) lend support to this idea. The upper stability limit of paragonite with and without quartz was studied by Chatterjee (23, 26). The stability range is generally 470–650°C, 1–5 kbar, so for rocks of the proper bulk compositions it is stable over the whole range of low grade metamorphic conditions.

K-Na muscovite solid solutions are stable over a wide range of metamorphic conditions and are reported by several groups; see Eugster et al (42) for the most recent data and summary. The equilibrium decomposition of phengite and of celadonite can be determined from Velde (195, 197), and from Wise & Eugster (199). The influence of celadonitic substitution on the stability of paragonite is not known. The data of Black (10), Frey (50), and Laduron & Martin (106) suggest the importance of phengite-paragonite relations in nature.

The works of Hemley and co-workers (75, 77, 132) on the phase equilibria of Na-K micas, feldspars, kaolinite, pyrophyllite and the Al_2SiO_5 polymorphs, with the additional component HCl, demonstrate the effect of fluid compositional control, in addition to P_{total}, P_{H_2O}, and T, on mineral equilibria. In addition, Hemley et al (78) considered the system K_2O-Al_2O_3-SiO_2-H_2O-H_2SO_4, with alunite as an added phase.

The Na-zeolite analcime attracted attention because of its occurrence in sedimentary and lower grade metamorphic rocks. For systems having excess quartz, the reaction

$$\text{analcime} + \text{quartz} = \text{albite} + H_2O \qquad 4.$$

was reversed by Campbell & Fyfe (20) at about 190°C, 12 bars, by Thompson (177) at 150°–190°C, 4.5–2 kbar, and by Liou (114) at 183°–200°C, 5–2 kbar. The data confirm the negative P–T slope for the condition $P_{H_2O} = P_{total}$. Problems in the direct application of the data to natural parageneses include the stoichiometry of analcime, the structural state of albite, and the deviations of the activity of silica from that of quartz and of the activity of H_2O from that of pure water in nature. See Saha (155). Coombs & Whetten (31) and Senderov (161, 162).

The CaAl-NaSi substitution in wairakite-analcime (Seki, 159) and margarite-paragonite (Frey, 52) is probably significant at the conditions of low grade metamorphism (Thompson, 179). However, experimental data are scanty for intermediate compositions (Hinrichsen & Schürmann, 82).

CALCIUM ALUMINUM SILICATES Low grade metamorphic minerals in the system $CaO\text{-}Al_2O_3\text{-}SiO_2\text{-}H_2O$ include the calcium zeolites, prehnite, margarite, zoisite, grossularite, and lawsonite. Phase relations in this system have been summarized by Boettcher (11) and Newton (143) for the high $P\text{–}T$ region and by Thompson (179) for the low $T\text{–}P$ region. A summary of the earlier, mainly synthesis studies was presented by Coombs et al (30).

The dehydration of wairakite to anorthite according to the reaction

$$\text{wairakite} = \text{anorthite} + 2\ \text{quartz} + 2H_2O \qquad 5.$$

was reversed by Liou (113) at 330–385°C, 0.5–3 kbar; the metastable dehydration of laumontite to anorthite according to the reaction

$$\text{laumontite} = \text{anorthite} + 2\ \text{quartz} + 4H_2O \qquad 6.$$

was reversed by Thompson (175) at 310–347°C, 1–6 kbar. The reaction

$$\text{laumontite} = \text{wairakite} + 2H_2O \qquad 7.$$

was reversed by Liou (115) at 235–327°C, 0.5–6 kbar, and the reaction

$$\text{stilbite} = \text{laumontite} + 3\ \text{quartz} + 3H_2O \qquad 8.$$

was reversed at 170–185°C, 2–5 kbar (Liou, 116).

The conversions of wairakite and laumontite to lawsonite were investigated by Nitsch (145), Thompson (175), and Liou (115). The dehydrations of laumontite and wairakite to anorthite are metastable relative to their dehydration to prehnite and zoisite; attempts to determine the breakdown of Ca-zeolites to prehnite showed that Ca-montmorillonite and not kaolinite formed as the excess alumina phase (Liou, 117; Thompson, 176).

The dehydration of prehnite to anorthite plus wollastonite was shown by Liou (117) to be metastable relative to the breakdown to zoisite + grossularite according to the reaction

$$5\ \text{prehnite} = 2\ \text{zoisite} + 2\ \text{grossularite} + 3\ \text{quartz} + 4H_2O \qquad 9.$$

Liou reversed the reaction at 393–403°C, 5–3 kbar, but Strens (171) suggested that this reaction occurs at 380°C, 2 kbar.

All of the above equilibria were determined for the condition $P_{H_2O} = P_{total}$.

The upper stability limit of margarite relative to anorthite + corundum was found by Chatterjee (24; 1973, personal communication) to be at 485–630°C, 1–7 kbar, see Velde (196), Tu (190); but the lower stability limit or the effect of quartz is unknown. Hemley et al (79) could not positively identify margarite in their study of the system $CaO\text{-}Al_2O_3\text{-}SiO_2\text{-}H_2O\text{-}HCl$; they found that Ca-montmorillonite breaks down directly to andalusite, pyrophyllite, kaolinite, or anorthite, depending on the temperature and the ratio $a_{Ca^{2+}}/(a_{H^+})^2$ at constant fluid pressure. They do, however, report evidence for a mixed-layer phase (Ca-montmorillonite + margarite?) at higher temperatures in the "Ca-montmorillonite" field.

The substitution of ferric iron-Al in natural epidote has recently been studied by Holdaway (84) and by Liou (118) under controlled f_{O_2} conditions. The complex reaction relations in this system require careful interpretation for application to natural assemblages.

Additional hydrothermal synthesis data on zeolite minerals may be found in several articles in Flanigen & Sand (46).

IRON AND MAGNESIUM SILICATES Several experimental studies on end-member Fe-Mg bearing minerals provide limiting data on their stability.

The breakdown of celadonite, composition near $K(Mg, Fe^{2+})(Al, Fe^{3+})Si_4O_{10}(OH)_2$ to ferriphlogopite + ferrisanidine + quartz or ferri-biotite + quartz was investigated by Wise & Eugster (199) who found equilibrium in the range 400–425°C at $P_{total} = 2$ kbar over a wide range of f_{O_2} (10^{-15} to 10^{-29} bars for the former and 10^{-29} to 10^{-35} bars for the latter reaction). Velde (197) determined the compositions and breakdown of celadonite solid solutions at the same total pressure and range of values of f_{O_2}; the breakdown temperatures are in the range 300–430°C, depending on the composition.

Experimental data on chlorite are scanty. The most recent reversed data on clinochlore are by Chernosky (27). Other data on chlorites are presented by Turnock (191), Hellner, Hinrichsen & Seifert (74), Fawcett & Yoder (43), and Bird & Fawcett (8).

The decomposition of Fe-chlorite to almandine garnet was studied by Hsu (88) under controlled f_{O_2} conditions, at 500–525°C and 0.5–3 kbar. Ganguly (56) and Hoschek (86) investigated several reactions relating to the disappearance of Fe-chloritoid. Theoretical studies of the phase relations involving chloritoid were also made by others, e.g. Albee (1, 2) and Thompson & Norton (182).

Experimental studies on the phase relations of pumpellyite have been made by Hinrichsen & Schürmann (81), Landis & Rogers (108), and Nitsch (146). These studies did not have buffered oxygen fugacity or reversal of runs; natural compositions were used and full information on the identities of the phases was lacking. The results can be applied only with great uncertainty (Zen, 209).

MIXED VOLATILE EQUILIBRIA Experimental work in the system CaO-MgO-SiO_2-H_2O-CO_2 is not only important for correlating reactions in calcareous rocks during low grade metamorphism but for providing important end-member models for many other equilibria.

The equilibria in the MgO-SiO_2-H_2O-CO_2 subsystem were summarized by Greenwood (61) and Johannes (92). Greenwood found that at $X_{CO_2} = 0.5$, magnesite + quartz react to form talc at about 270°C and 2 kbar total pressure; Johannes observed the same reaction at 400°C. The work of Metz and co-workers in the CaO-MgO-SiO_2-H_2O-CO_2 system relates to important equilibria involving dolomite and tremolite in addition to those mentioned above (see Metz & Winkler, 128; Metz, Puhan & Winkler, 129; and Metz & Puhan, 131).

Dolomite-quartz-calcite-talc relations were also studied by Gordon & Greenwood (58). The study by Skippen (164) showed that many equilibria in related multi-systems may be evaluated through the precise formulation of equilibrium constants for selected reactions. The papers by Metz & Trommsdorff (130) and Skippen (164) summarize the equilibrium conditions for phases in this quinary system. Undoubtedly, future studies will expand this system to include other components. For example, addition of K_2O or of iron at controlled f_{O_2} would make more realistic model systems.

Several experimental studies exist for the system CaO-Al_2O_3-SiO_2-H_2O-CO_2. The early theoretical and experimental studies of Greenwood with H_2O-CO_2 fluids (60, 61) enabled calculations of many important equilibria in this quinary system, for instance by Storre (169) and Thompson (176). Gordon & Greenwood (59) investigated equilibria involving grossularite. Storre & Nitsch (170) studied zoisite equilibria relative to anorthite and calcite. Experimental studies of low temperature equilibria (A. B. Thompson, unpublished data) for the reactions

$$\text{laumontite} + CO_2 = \text{calcite} + \text{kaolinite} + 2\ \text{quartz} + 2H_2O \qquad 10.$$

and

$$\text{laumontite} + \text{calcite} = \text{prehnite} + \text{quartz} + 3H_2O + CO_2 \qquad 11.$$

confirm the calculated temperatures and equilibrium values of X_{CO_2} at a total pressure of 2 kbar (Thompson, 176, p. 150).

Nitsch (147) determined the temperature (310–390°C), X_{CO_2} (0.01–0.04), total pressure (4–7 kbar) for the equilibrium

$$\text{calcite} + \text{pyrophyllite} + H_2O = \text{lawsonite} + 2\ \text{quartz} + CO_2 \qquad 12.$$

these values correspond well with those calculated by Thompson (176, Reaction 28). Nitsch & Storre (148) investigated the reaction

$$\text{calcite} + 2\ \text{andalusite} + H_2O = \text{margarite} + CO_2 \qquad 13.$$

Reactions among prehnite, calcite, and zoisite in mixed H_2O-CO_2 gas have not been made to date, nor have the effects of Fe^{3+} substitution for aluminum and complex substitutions of the type NaSi-CaAl in phases prepared in such media.

Hewitt (80) and Johannes & Orville (94) determined some equilibria in the system K_2O-CaO-Al_2O_3-SiO_2-H_2O-CO_2, notably the reaction

$$\text{muscovite} + \text{calcite} + 2\ \text{quartz} = \text{orthoclase} + \text{anorthite} + CO_2 + H_2O \qquad 14.$$

All these studies using the H_2O-CO_2 medium were made under the condition of calcite as the stable $CaCO_3$ polymorph. This is true despite the recent study by Johannes & Puhan (93) which moved the calcite-aragonite boundary to lower pressures.

The effect of other volatile components in addition to H_2O and CO_2 has been considered theoretically by Eugster & Skippen (41) and experimentally by Skippen (164). The possible importance of graphite in carbonate sediments in controlling the fugacity of gaseous species was emphasized by these workers.

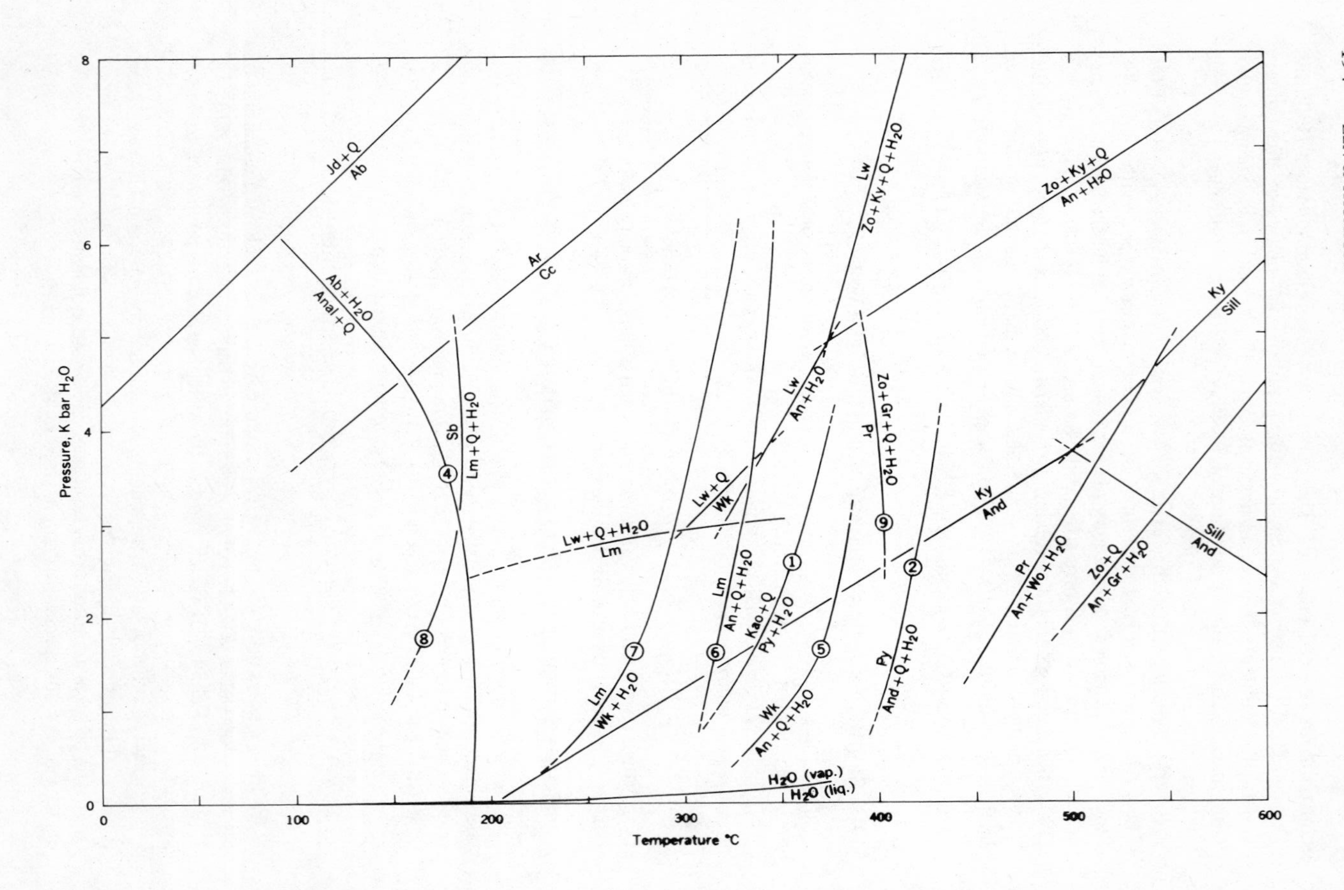

Jd+Q
Ab
Ab+H2O
Anal+Q
Ar
Cc
Sb
Lm+Q+H2O
Lw
Zo+Ky+Q+H2O
Zo+Ky+Q
An+H2O
Lw
An+H2O
Zo+Gr+Q+H2O
Pr
Lw+Q
Wk
Lw+Q+H2O
Lm
Lm
An+Q+H2O
Kao+Q
Py+H2O
Ky
And
Ky
Sill
Sill
And
Pr
An+Wo+H2O
Zo+Q
An+Gr+H2O
Py
And+Q+H2O
Lm
Wk+H2O
Wk
An+Q+H2O
H2O (vap.)
H2O (liq.)
1
2
4
5
6
7
8
9
Pressure, K bar H2O
0
2
4
6
8
0
100
200
300
400
500
600
Temperature °C

Figure 1 Temperature—total pressure plot of experimental equilibrium data for some low grade metamorphic reactions. Abbreviations used: *Ab*, Albite; *An*, Anorthite; *Anal*, Analcime; *And*, Andalusite; *Ar*, Aragonite; *Cc*, Calcite; *Gr*, Grossularite; *Jd*, Jadeite; *Kao*, Kaolinite; *Ky*, Kyanite; *Lm*, Laumontite; *Lw*, Lawsonite; *Pr*, Prehnite; *Py*, Pyrophyllite; *Q*, Quartz; *Sb*, Stilbite; *Sill*, Sillimanite; *Wk*, Wairakite; *Wo*, Wollastonite; *Zo*, Zoisite. $P_{H_2O} = P_{total}$. Curves *1* through *9* are referred to in the text by their respective reaction numbers. A few curves have been extrapolated.

1 kaolinite + 2 quartz = pyrophyllite + H_2O (Thompson, 174)
2 pyrophyllite = andalusite + 3 quartz + H_2O (Hemley, 76; Kerrick, 99)
4 analcime + quartz = albite + H_2O (Thompson, 177)
5 wairakite = anorthite + 2 quartz + $2H_2O$ (Liou, 113)
6 laumontite = anorthite + 2 quartz + $4H_2O$ (Thompson, 175)
7 laumontite = wairakite + $2H_2O$ (Liou, 115)
8 stilbite = laumontite + 3 quartz + $3H_2O$ (Liou, 116)
9 5 prehnite = 2 zoisite + 2 grossularite + 3 quartz + $4H_2O$ (Liou, 117)

In addition, the following equilibria are shown:

albite = jadeite + quartz (Boettcher & Wyllie, 12)
calcite = aragonite (Johannes & Puhan, 93)
laumontite = lawsonite + 2 quartz + $2H_2O$ (Nitsch, 145; Thompson, 175)
wairakite = lawsonite + 2 quartz (Liou, 115)
lawsonite = anorthite + $2H_2O$ (Crawford & Fyfe, 33)
4 lawsonite = 2 zoisite + kyanite + quartz + $7H_2O$ (Newton, 143)
2 zoisite + kyanite + quartz = 4 anorthite + H_2O (Newton, 143)
4 zoisite + quartz = grossularite + 5 anorthite + $2H_2O$ (Newton, 143)
Al_2SiO_5 polymorph relations (Holdaway, 83)
prehnite = anorthite + wollastonite + H_2O (Liou, 117)

Correlation of Experimental Data with Low Grade Mineral Assemblages

Figure 1 summarizes the reversed univariant curves reviewed above; we did not attempt to construct correct multisystems. It can be seen that the results are generally consistent with the notion that petrographically defined progressive assemblages do represent increasing metamorphic grade. If field data can be reasonably correlated with *P–T* diagrams, then progressive low grade metamorphism may be calibrated. Application of the experimental data is not without difficulty, however. One obvious problem is the lack of knowledge concerning the "equilibrium" fluid phases, now absent from most of the rock systems. Petrographic information increasingly suggests that low grade assemblages could buffer a_{CO_2}. If we could be sure that the fluid was simply mixed H_2O-CO_2, at least approximate adjustments could be made to the experiments where $P_{H_2O} = P_{total}$; however, fluid-inclusion studies suggest that dissolved salts and other gaseous species in the system C-H-O-N-S-Cl may be significant. The concentration of dissolved salts in inclusions is much greater than in samples of natural hydrothermal brines, and suggests that many natural fluids had a_{H_2O} values significantly below those encountered in hydrothermal experiments. The quantitative effect of other gaseous species on a_{H_2O} during low grade metamorphism is still largely conjectural.

The problem of attainment of equilibrium in both experimental and natural assemblages deserves mention. Many of the equilibria studies delineate synthesis fields rather than equilibrium boundaries. Locations of equilibrium boundaries for low *P–T* conditions require much more time than for medium and high grade equilibria. Even if we were sure of equilibrium in experiments, the direct application of the results to natural assemblages is still difficult. The patchiness of many low grade alteration assemblages suggests local reaction and local equilibrium, especially in unfractured rocks of low bulk permeability; for such rocks the activities of many components (dissolved and volatile components) may be locally buffered.

Application of experimental data to natural assemblages, where redox reactions are involved, requires that the f_{O_2} value be carefully controlled in the experiments. Different f_{O_2} values not only could affect the composition of individual phases produced (e.g. epidote versus zoisite), but could lead to different assemblages. As mentioned at the beginning of this paper, direct laboratory calibration of Fe^{3+}/Fe^{2+} values in minerals in specified assemblages as a function of *P*, *T*, and f_{O_2} promises the formulation of new limitations to metamorphic grades of given rocks. The fact that Fe^{3+}-rich and poor varieties of the same phase occur near one another in some rocks suggests significant variation of f_{O_2} over small distances.

Much of the relevant experimental data on iron-bearing mineral stability relations are difficult to interpret or apply to natural parageneses. In addition to lack of f_{O_2} buffering, these often are defective in the nonreversal of runs, random bulk compositions, and lack of information on the compositions of phases which must be in partial exchange equilibrium, presumably proceeding at rates different from the simultaneous phase equilibrium.

Under low-grade metamorphic conditions, material transport over megascopic distances probably depends heavily on a fluid phase. The permeability of rocks and molecular or ionic diffusion of dissolved species through the fluid are probably important kinetic factors (Fisher, 45), as is the possibility of membrane equilibria (see Hanshaw & Zen, 64, and references therein).

In all likelihood the natural assemblages record intermediate steps towards various equilibrium states, depending on the local conditions. The relationships envisioned reflect the effects of a rapidly moving fluid having sufficient time or reactive capacity to cause alteration only along the more permeable parts of massive rocks. Experimental calibration of these factors is needed, as well as of such factors as the buffering capacities of natural assemblages, rate studies on the interaction of nonequilibrium fluids with whole rocks or selected minerals, perhaps under situations where rate of fluid flow is rapid relative to reaction rate and where a steady state is not achieved.

The mass-transfer approach of Helgeson and co-workers (70–73) is a promising start at model building. Many of the necessary parameters, such as various diffusion rates, should be determined or tested experimentally before the models are accepted.

EVIDENCE OF PHASE EQUILIBRIUM IN LOW GRADE METAMORPHIC ROCKS

Evidence for the attainment of phase equilibrium in mineral assemblages is intrinsically difficult to evaluate, even for relatively coarse-grained rocks: much of the petrographic data supply necessary, but not sufficient, evidence for equilibrium (Zen, 204). For fine-grained metamorphic rocks, where petrographic observation of relations is itself difficult, conclusions regarding the establishment of chemical equilibrium can at best be made with a large grain of salt (a well-known flux).

The types of evidence used for equilibrium in rocks are: (*a*) lack of phases known to be incompatible; (*b*) apparent textural reorganization of the rock, reducing interface energy; (*c*) establishment of isotopic equilibrium of elements between mineral pairs; (*d*) nonviolation of the phase rule; (*e*) regularity in the partition of elements among phases; and (*f*) ascertainment of phase equilibrium in chemically simple subsystems.

Most low grade metamorphic rocks tend to be fine-grained. The domain of equilibrium among the minerals may be considerably smaller than that represented by a typical thin section. Therefore, attainment of equilibrium at this scale must not be assumed. Indeed, if the scale of equilibrium is not larger than the characteristic dimensions of the crystals, then even the concept of "mosaic" or local equilibrium ceases to be meaningful, even though the values of the chemical potentials of components may be continuous across grain boundaries.

Under these circumstances, perhaps the only justifiable statement is that the concept of phase equilibrium is a helpful limiting model, whereby the actual mineral assemblages may be compared and understood. This would be analogous to the use of the ideal solution law as a model to compare real solutions.

The fact that generalizations regarding the mineral assemblages of low grade rocks can be made at all argues for at least a tendency for such rocks to approach, if not attain, phase equilibrium. The limited microprobe data we have on some fine-grained rocks carrying the prehnite-pumpellyite assemblage show that within areas of a few millimeters local equilibrium is closely approached. The general conformity of megascopic mineral assemblages (on samples the size of thin sections or even hand specimens) to the phase rule indicates the same; as does the general similarity of phase assemblages for rocks of comparable bulk composition but of different ages and geographic locations, when due consideration is given to the variability to such parameters as the fugacity of the volatile components.

Since the various yardsticks by which we estimate the approach of a given rock system to phase equilibrium may not be mutually consistent, due care must be given in comparing results. For instance, does the attainment of oxygen isotopic exchange equilibrium between minerals and fluid (James & Clayton, 91; Eslinger & Savin, 39) necessarily imply phase equilibrium? Similarly, the existence of textural equilibrium in a rock indicates that material movement had occurred, which presumably is coupled with the reduction of bulk free energy of the system, but does not imply phase equilibrium for volumes larger than the scale of textural equilibrium (grain contacts).

Perhaps the most persuasive evidence for equilibrium among the low grade rocks comes from contemporary geothermal areas. As reviewed earlier, at Salton Sea and the New Zealand geothermal areas newly formed minerals occur in boreholes at definite temperatures [see Muffler & White (140), Figure 8, for a striking demonstration of this feature], and also depend on the f_{CO_2}/f_{H_2O} ratios. Muffler & White (140, p. 176) suggested that the mineral assemblages found at Salton Sea reflect the present thermal regime rather than past thermal history. R. O. Fournier (personal communication, 1973) has suggested that the present temperature distribution is close to the maximum temperature possible under the existing groundwater regime. The regularity of mineralogy indicates some systematic response of the rocks to the physical chemical environment, and therefore, at least approach to equilibrium—metastable or stable.

HEAT IN LOW GRADE METAMORPHISM

The source of heat for low grade alteration poses a problem. The patchiness of alteration assemblages (e.g. Smith, 165, 166) and the fact that the more porous parts of rocks are generally more altered (except where pervasive deformation had occurred) suggest that fluids were important agents in the alteration. Data from geothermal terranes suggest that hydrothermal fluids were heat carriers as well as media for material transport. The fluids may be circulating groundwater (Burst, 19), seawater, connate water, or even primary magmatic fluid. The fact that many regional metamorphic zones conform to structural patterns, however, indicates an intimate relationship between alteration and geometry of deformation of rock bodies on a large scale, possibly related to depth of burial.

Another possibly important source of heat is reactions involving hydration of igneous rocks or pyroclastics. Consider the formation of laumontite from a rock containing 10% anorthite. The reaction may be written: anorthite + 2 quartz + $4H_2O$ = laumontite. Using the thermochemical data of Robie & Waldbaum (152), Burnham et al (18), and Zen (208), and extrapolating from the experimental data of Thompson (175), we get for the equilibrium ΔH at 280°C, 500 bars P_{H_2O} a value of −33 kcal/mole anorthite. If the rock has 5% of pores filled with steam at T, P (density about 0.8), and a bulk density of 2.7, this leads to 30 cal/cm^3 of rock. Using a dry-rock heat capacity of 1 j/g (Goranson, 57) and steam (T, P) heat capacity of 4 j/g (Burnham et al, 18) the heat, if all conserved, would raise the temperature by about 45°C. If the reaction occurred in 10^3 years, the heat generation would be about 1×10^{-9} cal/cm^3/sec, which is to be compared with radioactive heat generation of about 1 to 10×10^{-13} cal/cm^3/sec (Birch et al, 7) for some igneous rocks.

The hydration of a forsterite-anorthite rich rock could be by the reaction

$$5 \text{ forsterite} + 4 \text{ anorthite} + 10H_2O = 2 \text{ clinochlore} + 2 \text{ prehnite} + \text{quartz}$$

At 25°C and 1 bar, with liquid water, the heat of reaction (Robie & Waldbaum, 152; Zen, 208) turns out to be about 164 cal/g forsterite. If a rock of bulk density 2.9 (5% water-filled pores) has 15 wt % forsterite and 30 wt % anorthite, complete reaction would lead to about 72 cal/cm^3 of rock. If all the heat were conserved, this could raise the temperature by 100°C.

Crude as these calculations are, they do suggest that the heat effect of such "retrograde" reactions may be very important (see also 141, p. 63). Exothermic reactions tend to be self-accelerating, and once initiated will proceed provided material can be supplied and removed with progress of the reaction. The rate-controlling step in many low grade reactions of igneous rocks may well be the rate of fluid flow, and some of the lowest grade assemblages may be bypassed by the heating effect. By the same token, sedimentary rocks that are intimately intercalated or mixed with pyroclastics might attain equilibrium more readily than those without intermixtures.

VARIABLES AFFECTING METAMORPHIC ASSEMBLAGES AND GRADES

For low grade metamorphic rocks, temperature is clearly a major variable. Temperatures up to 300°C have been measured (geothermal areas or oil fields; see Burst, 19; Kisch, 103) and the corresponding mineral assemblages belong to the lower "greenschist facies." Temperatures may be deduced from oxygen isotope thermometry. The data of Schwarcz et al (157) and James & Clayton (91) suggest that the chlorite zone is about 250°C or lower, and the biotite zone about 350°C or lower. Finally, temperatures have been estimated from experimental phase equilibrium data and are in the same general range.

A second variable is total rock pressure, which varies from one bar to several kilobars at the base of thick piles of sediments. By association with zeolite minerals which occur at low pressures, some prehnite-pumpellyite assemblages must also have formed at low to moderate pressures despite the high mineral densities (Zen, 209). In contrast to the blueschists, pumpellyite-bearing rocks do not necessarily betoken high-pressure metamorphism (Miyashiro, 134).

High phengitic content of muscovite has been suggested to be favored by high pressure (Ernst, 37; Ernst et al, 38). Phengitic mica is not confined to blueschist metamorphism (see Maxwell & Hower, 126, p. 850), so all low grade metamorphic muscovites must be regarded with wary eyes. The coexistence of phengitic muscovite and paragonite restricts the use of the usual X-ray determinative method for compositions of coexisting paragonite and muscovite (Zen & Albee, 210; Guidotti & Crawford, 63; Eugster et al, 42) to define metamorphic grades (see, for instance, Chatterjee, 25, p. 194).

In recent years the concept of "burial metamorphism" has been widely invoked to explain mineral assemblages ranging from zeolite bearing to prehnite-pumpellyite bearing. Coombs (28) tentatively suggested a general relation of mineralogy with depth for rocks of the Taringatura, New Zealand area, but recognized that other variables are also at work. Later Coombs (29, p. 214) formally proposed "burial metamorphism" to describe metamorphism without deformation. Walker (198) showed zeolite zonation according to the depth in volcanic piles in Iceland. Otalora (149) showed parallel trends between Coombs' rocks and those of Puerto Rico, but the pressure ranges differ markedly (for summary see Hay, 66, p. 71). This lack of one-to-one correlation between depth and mineralogy is supported by data from other areas (e.g. Levi, 111, 112; Seki et al, 160; Smith, 166).

Levi (111) used burial metamorphism as a key in reconstructing the metamorphic history in the Chilean Andes. The data from the northern Appalachians (Mossman & Bachinski, 139) and from the Taveyanne Formation of the Alps (Martini, 122) show that similar mineral assemblages do occur in deformed areas, so "burial metamorphism" probably is not a genre apart from regional metamorphism. The concentration of reports of zeolitic assemblages from nondeformed areas may mean that suitable rocks are not usually preserved at the portions of foldbelts where appropriate P–T conditions prevailed; the shallow-burial parts of foldbelts are commonly not rich in volcanics.

The importance of H_2O as an independent variable needs no apology. In near-surface conditions, the activity of H_2O may be affected by the salinity of connate water, by permeability to groundwater/magmatic water percolation, and possibly by clay-membrane phenomena. Admixture of other components such as CO_2, N_2, CH_4 could also affect the activity of H_2O. CH_4 in particular could be important in graphitic rocks (French, 47; Eugster & Skippen, 41; Skippen, 164); its presence is documented by fluid-inclusion studies (Touray & Jauzein, 187; Touray & Sagon, 188; Roedder, 153). It will both lower the activities of other components and provide more reducing environments, and could have promoted the appearances of zoisite and margarite in low grade graphitic rocks (Frey, 52). Nitrogen may be important in the graphitization of organic matter, generating NH_3 and producing a nonoxidizing environment. Volatiles that include sulfur and the halogens may also be important not only as inert pressure media but as additional components in phases such as the sulfide minerals or biotite.

Carbon dioxide is another important volatile component. Our review of geothermal areas showed the apparently dominant effect of H_2O/CO_2 ratio on the alteration mineralogy. High CO_2 activity could mask out assemblages that would otherwise result for the given bulk rock composition (Zen, 203; Browne & Ellis, 17). Detailed correlation of mineralogy with CO_2 fugacity rests on the assumption that the chemical regimes of the areas have remained steady; documentation of this assumption is clearly needed.

Discussions of phase equilibria in the literature commonly assume that both CO_2 and H_2O are boundary-value components (Zen, 204, p. 930). Although these discussions appear to apply directly to real assemblages, for some rocks at least equal success might be found if CO_2 were taken to be a buffered or initial-value component (Zen, 209). An environment of metamorphism having a CO_2-rich gas phase (such as at Salton Sea), or a system buffered by a carbonate phase that is not part of the observed assemblage, could be rationalized by either assumption regarding CO_2. Clearly, the size of the system chosen for observation, the dynamics of the flow regime, and the source and cause of generation of CO_2 all will affect our concept of the status of this component.

In rocks containing iron, manganese, and other elements of variable valence, redox reactions may be important and the oxygen activity is another relevant variable. Whether this is an "externally controlled" or boundary-value variable in given metamorphic rocks, however, is a much debated point (e.g. Zen, 204; Miyashiro, 135; Thompson, 181). Recent technical development suggests that we may soon be able to measure directly the oxygen activities in minerals or aggregates by means of doped zirconia electrodes (Sato, 156); this method could help to settle the question of the thermodynamic role of oxygen for specific rocks, as well as furnishing direct estimates of scales of redox equilibrium.

Correlation of mineral assemblages with physical and chemical parameters (temperature, pressure, and chemical activities of boundary-value components) by measurement in situ at active geothermal areas, or correlation of depth of mineral formation with stratigraphic or tectonic overburden assumes that these parameters actually existed at the time of mineral formation. This is a geologic assumption that must be evaluated individually in each case.

Finally, the possibility of large scale metastability must be remembered. Some apparently anomalous mineral assemblages reported, such as biotite + pyrophyllite (Tobschall, 186; Hosterman et al, 87) or albite + pyrophyllite (186) might conceivably be examples. Some of these anomalies could be the result of a choice of a system too large for local equilibrium. Such local relations have been suggested to be controlled by ionic equilibria in the fluid phase (Carmichael, 22; Fisher, 45; Surdam, 172); this possibility needs wider recognition and documentation and should be related also to the method of mass-transfer calculations (Helgeson and co-workers, 70–73).

Another possible manifestation of metastability is the apparent dependence of the genesis of zeolite assemblages on the nature of the protolith. Many reported zeolitic assemblages in regional metamorphism were derived from volcanic glasses (for additional references, see articles by Aiello, Colella, & Sersale, Minatu &

Utada, Sheppard, and Iljama, all in Flanigen & Sand, 46). Does the tetrahedral coordination of Al and Si in glasses favor the nucleation of zeolites, so that zeolites might actually form metastably under suitably low CO_2 fugacity (Zen, 205)? Senderov (161), Hay (66), and Honda & Muffler (85) all pointed out that high activity of SiO_2 caused by equilibration with metastable polymorphs of quartz (cristobalite, amorphous silica, etc) might lead to the formation of specific zeolites. Volcanic glass does not always lead to zeolite assemblage upon alteration [Otalora (149) reported that in parts of Puerto Rico glass was altered to chlorite but crystal tuffs were altered to zeolite], nor certainly are zeolites confined to glassy protolith; however, the large coincidence between zeolite minerals and glass does suggest some causal relationship.

Metastability is also a significant factor in exchange equilibrium, where equilibria among partial sets of chemical components are possible, a phenomenon much taken advantage of in experimental studies. Many transitions, both homogeneous and heterogeneous, involving the phyllosilicates may be partial equilibria, for instance the progressive reaction of mixed-layer clay minerals. The isotopic equilibrium of oxygen, previously mentioned, may be another example. Recognition of these various types and levels of chemical equilibrium, not necessarily coupled, should be useful in understanding some of the low grade metamorphic reactions.

CORRELATION OF ISOGRADS

Correlation of isograds for low grade rocks of different chemical compositions is difficult for several reasons. First, many specific chemical reactions that must have taken place in the rocks have not been fully identified yet. Second, experimental data bearing on phase equilibrium of low grade mineral assemblages are only beginning to become available. Third, the problem of partial or metastable equilibrium is probably acute for many low grade rocks. Fourth, few of the variables that must have operated during much of regional metamorphism may be safely assumed to be generally not affected by sharp local variations.

Many low grade reactions depend on the chemical activities of various volatile components. As different reactions will be affected to different extents by a given change of these values, isograds or facies boundaries defined by different reactions will move relative to one another. Thus there can be no simple scheme of correlating isograds for different rocks. In addition, some reactions, notably those which are zone markers only, may depend on kinetic factors [transformation of white micas, Frey (49); of chlorite, Hayes (67)].

Nevertheless, some broad correlations can be made provided we do not try to read too much into them. In a series of papers, Kisch (100–103) made valiant efforts to tie together metamorphic zones described by the phyllosilicates and those described by calcium zeolites by means of degree of coalification. Kisch (103, p. 409; also 1970, personal communication) realized that the degree of coalification is partly a rate process, but pointed out that temperature is the other main variable, whereas pressure and even the evolved volatile components are only of secondary importance. The data from Kisch's summary diagram (103) combined with those of Frey & Niggli (53) are shown in Table 1, together with additional interpretations.

Table 1 Approximate correlations of coal rank, phyllosilicate assemblage, zeolite assemblage, illite crystallinity, and graphite crystallinity. Modified from Kisch (102, 103), Frey & Niggli (53), and adopting the data from Puerto Rico (Otalora, 149; Jolly, 95), the Keuper beds (Frey, 49), New Zealand (Steiner, 168; Browne & Ellis, 17), and Rhode Island (Grew, 62).

Grade of coal (Kisch, 102; Frey & Niggli, 53)			Zeolite minerals (Frey & Niggli, 53, this report)	Illite crystallinity (Kubler, 104; Frey & Niggli, 53; Frey 49)	Clay minerals (Kisch, 102, 103)	Graphite (002) (Grew, 62)		Suggested further correlations
Name	Volatile %	Carbon %				d(002) (Ceylon graphite = 3.354)	1/2–ht width in arb. units (Ceylon graphite = 0.20)	
Flame coal	50–40		analcime heulandite					? --- 1 M mica ---?; illite ← mixed layer clay ← montmorillonite; celadonite-illite; mordenite
		80						
Gas flame coal	40–35			unmetamorphosed				
		85						
Gas coal	35–28				kaolinite + albite; kaolinite + Fe, Mg carbonate; kaolinite + quartz			
		88	laumontite	7.5				
Fat coal	28–19							
		90						
Ess coal	19–14			anchimetamorphosed	chlorite + calcite	Not known	Not known	---wairakite---?
		~91						
Lean coal	14–10		prehnite-pumpellyite					2 M mica ---?
		~92						
Anthracite	10– 4					3.59 ± 0.01	5.6 ± 0.2	
		~93		4.0				
Meta-anthracite	< 4		actinolite epidote albite chlorite	epimetamorphosed	paragonite ---?; biotite---?; pyrophyllite ---?	3.393 ± 0.005	1.0 ± 0.03	albite + adularia

Neruchev & Parparova (142) recently added to the depth dependence of coalification. Hosterman et al (87) also added to the correlation of coal petrography and the nature of the associated micaceous minerals. Because coal beds and zeolitic minerals are not uncommonly associated, and mica is almost ubiquitous, the use of coal grade seems a promising way to get internally calibrated correlation of metamorphic grades under reducing conditions and should be further tested. In particular, the relation of the coal grade to nature of mixed-layer clay mineral transition should be studied in detail. In the absence of coal beds, there might still exist some correlation between the silicate phase assemblages and the crystallinity of the carbonaceous matter in the rock; the works of French (48), Landis (107), and Grew (62) are promising starts.

If the zeolite and the prehnite-pumpellyite assemblages occur, they tend to appear before the biotite zone is reached in adjacent pelitic rocks. In this connection, defining zeolite "facies" by single zeolite minerals (Coombs, 29; Seki, 158; see also Miyashiro & Shido, 136) can be risky because, in general, different zeolites do not behave as polymorphs, and the appearance of different zeolites may merely mark zones. For the same reason, the transitions from zeolite to prehnite-pumpellyite assemblages, as commonly suggested, do not define "facies" (see also Coombs, in Flanigen & Sand, 46).

Consideration of actual isograds must employ realistic chemical compositions. For instance, for the transition of prehnite-pumpellyite facies to typical greenschist-facies assemblages, many reactions have been suggested (see Hashimoto, 65; Bishop, 9). Some of Hashimoto's reactions involve only Mg-components for the mafic minerals. Other suggestions in the literature (e.g. Coombs et al, 32) treat MgO and FeO as a single component. These reactions may be useful models but cannot fully describe natural facies boundaries.

To conclude this paper, we suggest the following mineralogic reactions as possibly useful zone markers; some may even define isograds and these are indicated by asterisks. In parentheses we indicate the important variables affecting the reactions. We suggest that these reactions could be a starting point of systematic comparison and correlation of metamorphic grades for rocks of different bulk compositions.

1. Disappearance of specific mixed-layer clay minerals (μ_W, T)
2. Structural state of authigenic feldspar (T)

*3. Kaolinite + quartz = pyrophyllite (T, P, μ_W)

*4. Hematite + chlorite = magnetite + chloritoid (T, P, μ_W, μ_{O_2})

*5. Hematite = magnetite (T, μ_{O_2})

6. Chlorite + paragonite = chloritoid + albite (T, P, μ_W)

*7. Calcite + ilmenite + quartz = sphene + iron oxide phase (T, P, μ_W, μ_{CO_2}, μ_{O_2})

*8. Analcime + quartz = albite (T, P, μ_W, μ_{SiO_2})

9. Dolomite + muscovite + quartz = calcite + phlogopite + chlorite (T, P, μ_W, μ_{CO_2})

*10. Dolomite + quartz = calcite + actinolite (T, P, μ_W, μ_{CO_2})

*11. Dolomite + kaolinite = calcite + Mg-chlorite (T, P, μ_W, μ_{CO_2})

12. Dolomite + chlorite = Al-actinolite + calcite (excess quartz; T, P, μ_W, μ_{CO_2})

*13. Rutile + hematite = ilmenite (T, P, μ_{O_2})
*14. Calcite + rutile + quartz = sphene (T, P, μ_{CO_2})
15. Disappearance of stilpnomelane (T, ???)
16. Appearance of biotite, including phlogopite ($T, P, \mu_W, \mu_{O_2}, \mu_{CO_2}$)
17. Appearance and disappearance of prehnite (T, P, μ_W, ???)
18. Appearance and disappearance of pumpellyite (T, P, μ_W, μ_{O_2}, ???)
19. Appearance of epidote and zoisite ($T, P, \mu_W, \mu_{O_2}, \mu_{CO_2}$?)
20. Various coal grades (T, ???)
*21. Various calcium zeolite reactions in the presence of quartz (T, P, μ_W, ???)
*22. Various specific reactions involving prehnite and pumpellyite ($T, P, \mu_{O_2}, \mu_W, \mu_{CO_2}$)

ACKNOWLEDGMENTS

This paper benefitted greatly from thoughtful reviews by N. D. Chatterjee, D. S. Coombs, Martin Frey, R. O. Fournier, B. A. Morgan, E. H. Roseboom, and D. R. Wones, and we are grateful to them. Zen wishes to thank H. J. Kisch and J. B. Thompson, in addition to those cited above, for many stimulating and instructive discussions on low grade metamorphism over a period of years. Thompson wishes to thank W. S. Fyfe and J. B. Thompson for stimulation and discussion on these and related problems.

Literature Cited

1. Albee, A. L. 1965. A petrogenetic grid for the Fe-Mg silicates of pelitic schists. *Am. J. Sci.* 263: 512–36
2. Albee, A. L. 1972. Metamorphism of pelitic schists: Reaction relations of chloritoid and staurolite. *Geol. Soc. Am. Bull.* 83: 3249–68
3. Althaus, E. 1966. Die Bildung von Pyrophyllit und Andalusit zwischen 2000 und 7000 Bar H_2O-Druck. *Naturwissenschaften* 53: 105–6
4. Banno, S. 1964. Petrologic study on Sanbagawa crystalline schists in the Bessi-Ino District, central Sikoko, Japan. *J. Fac. Sci. Univ. Tokyo, Sect. 2* 15: 203–319
5. Barrow, G. 1893. On an intrusion of muscovite-biotite gneiss in the south-eastern Highlands of Scotland. *Quart. J. Geol. Soc. London* 49: 330–58
6. Berryhill, H. L. Jr., Briggs, R. P., Glover, L. III 1960. Stratigraphy, sedimentation, and structure of Late Cretaceous rocks in eastern Puerto Rico —Preliminary report. *Am. Assoc. Petrol. Geol. Bull.* 44: 137–55
7. Birch, F., Roy, R. F., Decker, E. R. 1968. Heat flow and thermal history in New England and New York. In *Studies of Appalachian Geology, Northern and Maritime*, ed. E. Zen, W. S. White, J. B. Hadley, J. B. Thompson Jr., 437–51. New York: Interscience
8. Bird, G. W., Fawcett, J. J. 1973. Stability relations of Mg-chlorite-muscovite and quartz between 5 and 10 kb water pressure. *J. Petrology*. 14: 415–28
9. Bishop, D. G. 1972. Progressive metamorphism from prehnite-pumpellyite to greenschist facies in the Dansey Pass area, Otago, New Zealand. *Geol. Soc. Am. Bull.* 83: 3177–98
10. Black, P. M. 1969. Paragonite and phengitic muscovite from the New Caledonian blueschists. *Geol. Soc. Am. Abstr. Progr.* Pt. 7: 14–15
11. Boettcher, A. L. 1970. The system CaO-Al_2O_3-SiO_2-H_2O at high pressures and temperatures. *J. Petrology* 11: 337–79
12. Boettcher, A. L., Wyllie, P. J. 1969. Phase relationships in the system $NaAlSiO_4$–SiO_2–H_2O to 35 kilobars pressure. *Am. J. Sci.* 267: 875–909
13. Bowen, N. L. 1940. Progressive metamorphism of siliceous limestone and dolomite. *J. Geol.* 48: 225–74
14. Bradley, W. H., Eugster, H. P. 1969. Geochemistry and paleolimnology of the trona deposits and associated auth genic minerals of the Green River

Formation of Wyoming. *US Geol. Surv. Prof. Pap. 496-B.* 71 pp.
15. Brown, C. E. 1961. Prehnite-pumpellyite metagraywacke facies of Upper Triassic rocks, Aldrich Mountains, Oregon. *Geol. Surv. Res. 1961, US Geol. Surv. Prof. Pap. 424-C,* C146–47
16. Brown, E. H. 1967. The greenschist facies in part of western Otago, New Zealand. *Contrib. Mineral. Petrol.* 14: 259–92
17. Browne, P. R. L., Ellis, A. J. 1970. The Ohaki-Broadlands hydrothermal area, New Zealand: Mineralogy and related geochemistry. *Am. J. Sci.* 269:97–131
18. Burnham, C. W., Holloway, J. R., Davis, N. F. 1969. Thermodynamic properties of water to 1,000°C and 10,000 bars. *Geol. Soc. Am. Spec. Pap. 132.* 96 pp.
19. Burst, J. F. 1969. Diagenesis of Gulf coast clayey sediments and its possible relation to petroleum migration. *Am. Assoc. Petrol. Geol. Bull.* 53:73–93
20. Campbell, A. S., Fyfe, W. S. 1965. Analcime-albite equilibria. *Am. J. Sci.* 263:807–16
21. Cann, J. R. 1968. Geological processes at mid-ocean ridge crests. *Geophys. J. Roy. Astron. Soc.* 15:331–41
22. Carmichael, D. M. 1969. On the mechanism of prograde metamorphic reactions in quartz-bearing pelitic rocks. *Contrib. Mineral. Petrol.* 20:244–67
23. Chatterjee, N. D. 1970. Synthesis and upper stability of paragonite. *Contrib. Mineral. Petrol.* 27:244–57
24. Chatterjee, N. D. 1971. Preliminary results on the synthesis and upper stability limit of margarite. *Naturwissenschaften* 58:147
25. Chatterjee, N. D. 1971. Phase equilibria in the Alpine metamorphic rocks of the environs of the Dora-Maira massif, western Italian Alps. *Neues Jahrb. Mineral. Abh.* 114(2):181–210
26. Chatterjee, N. D. 1972. The upper stability limit of the assemblage paragonite + quartz and its natural occurrences. *Contrib. Mineral. Petrol.* 34: 288–303
27. Chernosky, J. V. 1973. The stability of clinochlore and the free energy of Mg-cordierite. (Abstr.) *EOS Trans. Am. Geophys. Union* 54:479
28. Coombs, D. S. 1954. The nature and alteration of some Triassic sediments from Southland, New Zealand. *Trans. Roy. Soc. N.Z.* 82(1):65–109
29. Coombs, D. S. 1961. Some recent work on the lower grades of metamorphism. *Aust. J. Sci.* 24:203–15
30. Coombs, D. S., Ellis, A. J., Fyfe, W. S., Taylor, A. M. 1959. The zeolite facies, with comments on the interpretation of hydrothermal syntheses. *Geochim. Cosmochim. Acta* 17:53–107
31. Coombs, D. S., Whetten, J. T. 1967. Composition of analcime from sedimentary and burial metamorphic rocks. *Geol. Soc. Am. Bull.* 78:269–82
32. Coombs, D. S., Horodyski, R. J., Naylor, R. S. 1970. Occurrence of prehnite-pumpellyite facies metamorphism in northern Maine. *Am. J. Sci.* 268:142–56
33. Crawford, W. A., Fyfe, W. S. 1965. Lawsonite equilibria. *Am. J. Sci.* 263: 262–70
34. Dungan, M. A., Vance, J. A. 1972. Metamorphism of ultramafic rocks in the upper Stillaguamish River area, north Cascades, Washington. *Geol. Soc. Am. Abstr. Progr.* 4:493
35. Dunoyer de Segonzac, G. 1969. Les minéraux argileux dans la diagenèse passage au métamorphisme. *Mém. Serv. Carte Géol. Alsace Lorraine 29.* 320 pp.
36. Ellis, A. J. 1967. The chemistry of some explored geothermal systems. In *Geochemistry of Hydrothermal Ore Deposits,* ed. H. L. Barnes, 465–514. New York: Holt, Rinehart & Winston
37. Ernst, W. G. 1963. Significance of phengitic micas from low-grade schists. *Am. Mineral.* 48:1357–73
38. Ernst, W. G., Seki, Y., Onuki, H., Gilbert, M. C. 1970. Comparative study of low-grade metamorphism in the California Coast Ranges and the outer metamorphic belt of Japan. *Geol. Soc. Am. Mem. 124.* 276 pp.
39. Eslinger, E. V., Savin, S. M. 1973. Mineralogy and oxygen isotope geochemistry of the hydrothermally altered rocks of the Ohaki-Broadlands, New Zealand geothermal areas. *Am. J. Sci.* 273:240–67
40. Eugster, H. P., Wones, D. R. 1962. Stability relations of the ferruginous biotite, annite. *J. Petrology* 3:82–125
41. Eugster, H. P., Skippen, G. B. 1967. Igneous and metamorphic reactions involving gas equilibria. In *Researches in Geochemistry, Vol. 2,* ed. P. H. Abelson, 492–520. New York: Wiley
42. Eugster, H. P. et al 1972. The two-phase region and excess mixing properties of paragonite-muscovite crystalline solutions. *J. Petrology* 13:147–79
43. Fawcett, J. J., Yoder, H. S. Jr. 1966. Phase relationships of chlorites in the system $MgO–Al_2O_3–SiO_2–H_2O$. *Am. Mineral.* 51:353–80

44. Fenner, C. N. 1936. Bore-hole investigation in Yellowstone Park. *J. Geol.* 44:225–315
45. Fisher, G. W. 1970. The application of ionic equilibria to metamorphic differentiation: An example. *Contrib. Mineral. Petrol.* 29:91–108
46. Flanigen, E. M., Sand, L. B., Eds. 1971. Molecular sieve zeolites. I. *Advan. Chem. Ser. 101.* 526 pp.
47. French, B. M. 1965. Some geological implications of equilibrium between graphite and a C-H-O gas phase at high temperatures and pressures. *NASA Goddard Space Flight Center Publ. X641–65–324, Greenbelt, Md.* 69 pp.
48. French, B. M. 1968. Progressive contact metamorphism of the Biwabik iron-formation, Mesabi Range, Minnesota. *Minn. Geol. Surv. Bull. 45.* 103 pp.
49. Frey, M. 1969. Die Metamorphose des Keupers vom Tafeljura bis zum Lukmanier-Gebiet. *Beitr. Geol. Karte Schweiz, Neue Folge 137, Lieferung.* 160 pp.
50. Frey, M. 1969. A mixed-layer paragonite/phengite of low grade metamorphic origin. *Contrib. Mineral. Petrol.* 24:63–65
51. Frey, M. 1970. The step from diagenesis to metamorphism in pelitic rocks during Alpine orogenesis. *Sedimentology* 15:261–79
52. Frey, M. 1972. Progressive low-grade metamorphism of a black-shale formation, central Swiss Alps. *Geol. Soc. Am. Abstr. Progr.* 4:512
53. Frey, M., Niggli, E. 1971. Illitkristallinität, Mineralfazien und Inkohlungsgrad. *Schweiz. Mineral. Petrogr. Mitt.* 51(1):229–34
54. Frey, M., Niggli, E. 1972. Margarite, an important rock-forming mineral in regionally metamorphosed low-grade rocks. *Naturwissenschaften* 59:214–15
55. Frey, M., Hunziker, J. C., Roggwiller, P., Schindler, C. 1973. Progressive niedriggradige Metamorphose glaukonitführender Horizonte in den helvetischen Alpen der Ostschweiz. *Contrib. Mineral. Petrol.* 39:185–218
56. Ganguly, J. 1969. Chloritoid stability and related parageneses: Theory, experiments, and applications. *Am. J. Sci.* 267:910–44
57. Goranson, R. W. 1942. Heat capacity; heat of fusion. In *Handbook of Physical Constants,* ed. F. Birch et al, 223–42. *Geol. Soc. Am. Spec. Pap. 36*
58. Gordon, T. M., Greenwood, H. J. 1970. The reaction: Dolomite + quartz + water = talc + calcite and carbon dioxide: *Am. J. Sci.* 268:225–42
59. Gordon, T. M., Greenwood, H. J. 1971. The stability of grossularite in H_2O-CO_2 mixtures. *Am. Mineral.* 56:1674–88
60. Greenwood, H. J. 1962. Metamorphic reactions involving two volatile components. *Carnegie Inst. Washington Yearb.* 61:82–85
61. Greenwood, H. J. 1967. Mineral equilibria in the system MgO-SiO_2-H_2O-CO_2. In *Researches in Geochemistry, Vol. 2,* ed. P. H. Abelson, 542–67. New York: Wiley
62. Grew, E. S. 1974. Carbonaceous material in some metamorphic rocks of New England and other areas. *J. Geol.* In press
63. Guidotti, C. V., Crawford, K. E. 1968. Determination of Na/Na + K in muscovite by x-ray diffraction and its use in the study of pelitic schists in northwest Maine (Abstr.). *Geol. Soc. Am. Spec. Pap.* 115:86
64. Hanshaw, B. B., Zen, E. 1965. Osmotic equilibrium and overthrust faulting. *Geol. Soc. Am. Bull.* 76:1379–86
65. Hashimoto, M. 1972. Reactions producing actinolite in basic metamorphic rocks. *Lithos* 5:19–31
66. Hay, R. L. 1966. Zeolites and zeolitic reactions in sedimentary rocks. *Geol. Soc. Am. Spec. Pap. 85.* 130 pp.
67. Hayes, J. B. 1970. Polytypism of chlorite in sedimentary rocks. *Clays Clay Miner.* 18:285–306
68. Helgeson, H. C. 1967. Solution chemistry and metamorphism. In *Researches in Geochemistry Vol. 2,* ed. P. H. Abelson, 362–404. New York: Wiley
69. Helgeson, H. C. 1968. Geologic and thermodynamic characteristics of the Salton Sea geothermal system. *Am. J. Sci.* 266:129–66
70. Helgeson, H. C. 1968. Evaluation of irreversible reactions in geochemical processes involving minerals and aqueous solutions. I. Thermodynamic relations. *Geochim. Cosmochim. Acta* 32:853–77
71. Helgeson, H. C. 1971. Kinetics of mass transfer among silicates and aqueous solutions. *Geochim. Cosmochim. Acta* 35:421–69
72. Helgeson, H. C., Garrels, R. M., Mackenzie, F. T. 1969. Evaluation of irreversible reactions in geochemical processes involving minerals and aqueous solutions. II. Applications. *Geochim. Cosmochim. Acta* 33:455–81

73. Helgeson, H. C., Brown, T. H., Nigrini, A., Jones, T. A. 1970. Calculation of mass transfer in geochemical processes involving aqueous solutions. *Geochim. Cosmochim. Acta* 34:569–92
74. Hellner, E., Hinrichsen, T., Seifert, F. 1965. The study of mixed crystals of minerals in metamorphic rocks. In *Controls of Metamorphism,* ed. W. S. Pitcher, G. W. Flinn, 155–68. Edinburgh: Oliver & Boyd
75. Hemley, J. J. 1959. Some mineralogical equilibria in the system K_2O-Al_2O_3-SiO_2-H_2O: *Am. J. Sci.* 257:241–70
76. Hemley, J. J. 1967. Stability relations of pyrophyllite, andalusite, and quartz at elevated pressures and temperatures. (Abstr.) *Trans. Am. Geophys. Union* 48: 224
77. Hemley, J. J., Jones, W. R. 1964. Chemical aspects of hydrothermal alteration with emphasis on hydrogen metasomatism. *Econ. Geol.* 59:538–67
78. Hemley, J. J., Hostetler, P. B., Gude, A. J., Mountjoy, W. T. 1969. Some stability relations of alunite. *Econ. Geol.* 64:599–612
79. Hemley, J. J., Montoya, J. W., Nigrini, A., Vincent, H. A. 1971. Some alteration reactions in the system CaO-Al_2O_3-SiO_2-H_2O. *Soc. Mining Geol. Jap., Spec. Issue* 2:58–63 (*Proc. IMA-IAGOD Meet. 1970, Joint Symp. Volume*)
80. Hewitt, D. A. 1973. Stability of the assemblage muscovite-calcite-quartz. *Am. Mineral.* 58:785–91
81. Hinrichsen, T., Schürmann, K. 1969. Untersuchungen zur Stabilität von Pumpellyit. *Neues Jahrb. Mineral. Monatsh.* 10:441–45
82. Hinrichsen, T., Schürmann, K. 1971. Synthese und Stabilität von Glimmern im System CaO-Na_2O-Al_2O_3-SiO_2-H_2O. (Abstr.) *Fortschr. Mineral.* 49(1): 21–22
83. Holdaway, M. J. 1971. Stability of andalusite and the aluminum silicate phase diagram. *Am. J. Sci.* 271:97–131
84. Holdaway, M. J. 1972. Thermal stability of Al-Fe epidote as a function of f_{O_2} and Fe content. *Contrib. Mineral. Petrol.* 37:307–40
85. Honda, S., Muffler, L. J. P. 1970. Hydrothermal alteration in core from research drill hole Y-1, Upper Geyser Basin, Yellowstone National Park, Wyoming. *Am. Mineral.* 55:1714–37
86. Hoschek, G. 1969. The stability of staurolite and chloritoid and their significance in metamorphism of pelitic rocks. *Contrib. Mineral. Petrol.* 22:208–32
87. Hosterman, J. W., Wood, G. H. Jr., Bergin, M. J. 1970. Mineralogy of underclays in the Pennsylvania anthracite region. *Geol. Surv. Res. 1970, US Geol. Surv. Prof. Pap. 700-C,* C89–97
88. Hsu, L. C. 1968. Selected phase relationships in the system Al-Mn-Fe-Si-O-H: A model for garnet equilibria. *J. Petrology* 9:40–83
89. Hubbert, M. K., Rubey, W. W. 1959. Mechanics of fluid-filled porous solids and its application to overthrust faulting. I. Role of fluid pressure in mechanics of overthrust faulting. *Geol. Soc. Am. Bull.* 70:115–66
90. Jäger, E. 1969. Geochronology of Phanerozoic orogenic belts, Guidebook to field trip, *Uebersichtsexkursion durch die Zentralalpen, 47de Jahrestagung, Deutsche Mineralogische Gesellschaft, Bern, Switzerland*
91. James, H. L., Clayton, R. N. 1962. Oxygen isotope fractionation in metamorphosed iron formations of the Lake Superior region and in other iron-rich rocks. In *Petrologic Studies—A Volume in Honor of A. F. Buddington,* ed. A. E. J. Engel, H. L. James, B. F. Leonard, 217–39. New York, N.Y.: Geol. Soc. Am.
92. Johannes, W. 1969. An experimental investigation of the system MgO-SiO_2-H_2O-CO_2. *Am. J. Sci.* 267:1083–1104
93. Johannes, W., Puhan, D. 1971. The calcite-aragonite transition, reinvestigated. *Contrib. Mineral. Petrol.* 31:28–38
94. Johannes, W., Orville, P. M. 1972. Zur Stabilität der Mineralparagenesen Muskovit + Calcit + Quarz, Zoisit + Muskovit + Quarz, Anorthit + K-Feldspat und Anorthit + Calcit. (Abstr.) *Fortschr. Mineral.* 50(1):46–47
95. Jolly, W. T. 1970. Zeolite and prehnite-pumpellyite facies in south central Puerto Rico. *Contrib. Mineral. Petrol.* 27:204–24
96. Jolly, W. T., Smith, R. E. 1972. Degradation and metamorphic differentiation of the Keweenawan tholeiitic lavas of northern Michigan, U.S.A. *J. Petrology* 13:273–309
97. Jones, B. F. 1965. The hydrology and mineralogy of Deep Springs Lake, Inyo County, California. *US Geol. Surv. Prof. Pap. 502-A.* 56 pp.
98. Keith, T. E. C., Muffler, L. J. P., Cremer, M. 1968. Hydrothermal epidote formed in the Salton Sea geothermal system, California. *Am. Mineral.* 53:1635–44
99. Kerrick, D. M. 1968. Experiments on the upper stability limit of pyrophyllite at 1.8 kilobars and 3.9 kilobars water

pressure. *Am. J. Sci.* 266:204–14
100. Kisch, H. J. 1966. Zeolite facies and regional rank of bituminous coals. *Geol. Mag.* 103:414–22
101. Kisch, H. J. 1966. Chlorite-illite tonstein in high-rank coals from Queensland, Australia: Notes on regional epigenetic grade and coal rank. *Am. J. Sci.* 264:386–97
102. Kisch, H. J. 1968. Coal rank and lowest-grade regional metamorphism in the southern Bowen Basin, Queensland, Australia. *Geol. Mijnbouw* 47(1): 28–36
103. Kisch, H. J. 1969. Coal-rank and burial-metamorphic mineral facies. *Advan. Org. Geochem., Proc. Int. Meet., 4th, 1968*, 407–25
104. Kubler, B. 1967. Anchimétamorphisme et schistosité. *Cent. Rech. Pau—SNPA, Bull.* 1:259–78
105. Kubler, B. 1968. Evaluation quantitative du métamorphisme par la cristallinité de l'illite. *Cent. Rech. Pau—SNPA, Bull.* 2:385–97
106. Laduron, D., Martin, H. 1969. Coexistence de paragonite, muscovite et phengite dans un micaschiste a grenat de la zone du Mont-Rose. *Ann. Soc. Géol. Belg.* 92:159–72
107. Landis, C. A. 1971. Graphitization of dispersed carbonaceous material in metamorphic rocks. *Contrib. Mineral. Petrol.* 30:34–45
108. Landis, C. A., Rogers, J. 1968. Some experimental data on the stability of pumpellyite. *Am. Mineral.* 53:1038–41
109. Lebedev, M. M., Bondarenko, V. N. 1962. K voprosu o vozraste i genezise metamorficheskikh porod tsentral'noy Kamchatki (On the problem of the age and genesis of metamorphic rocks of central Kamchatka). *Sov. Geol.* No. 11: 98–105
110. Lebedev, M. M., Tarazin, I. A., Lagovskaya, E. A. 1967. Metamorphic zones of Kamchatka as an example of the metamorphic assemblages of the inner part of the Pacific belt. *Tectonophysics* 4:445–61
111. Levi, B. 1969. Burial metamorphism of a Cretaceous volcanic sequence west from Santiago, Chile. *Contrib. Mineral. Petrol.* 24:30–49
112. Levi, B. 1970. Burial metamorphic episodes in the Andean geosyncline, central Chile. *Geol. Rundsch.* 59:994–1013
113. Liou, J. G. 1970. Synthesis and stability relations of wairakite, $CaAl_2Si_4O_{12}\cdot 2H_2O$. *Contrib. Mineral. Petrol.* 27: 259–82
114. Liou, J. G. 1971. Analcime equilibria. *Lithos* 4:389–402
115. Liou, J. G. 1971. P-T stabilities of laumontite, wairakite, lawsonite, and related minerals in the system $CaAl_2Si_2O_8$-SiO_2-H_2O. *J. Petrology* 12:379–411
116. Liou, J. G. 1971. Stilbite-laumontite equilibrium. *Contrib. Mineral. Petrol.* 31:171–77
117. Liou, J. G. 1971. Synthesis and stability relations of prehnite, $Ca_2Al_2Si_3O_{10}(OH)_2$. *Am. Mineral.* 56:507–31
118. Liou, J. G. 1973. Synthesis and stability relations of epidote, $Ca_2Al_2FeSi_3O_{12}(OH)$. *J. Petrology.* 14:381–413
119. Logvinenko, N. V., Osipova, Z. V. 1969. Tseolity v osadochnykh porodakh (zeolites in sedimentary rocks). *Litol. Polez. Iskop.* No. 3:134–40
120. Mahon, W. A., Finlayson, J. B. 1972. The chemistry of the Broadlands geothermal area, New Zealand. *Am. J. Sci.* 272:48–68
121. Marakushev, A. A. 1968. *Termodinamika metamorficheskoy gidratatsii mineralov* (*Thermodynamics of metamorphic hydration of minerals*). Moscow: Nauka. 200 pp.
122. Martini, J. 1968. Etude pétrographiques des Grès de Taveyanne entre Arve et Giffre (Haute-Savoie, France). *Schweiz. Mineral. Petrogr. Mitt.* 48(2):539–654
123. Martini, J. 1972. Le métamorphisme dans les chaînes alpines externes et ses implications dans l'orogenèse. *Schweiz. Mineral. Petrogr. Mitt.* 52:257–75
124. Martini, J., Vaugnat, M. 1965. Présence du faciès à zéolites dans la formation des "Grès" de Taveyanne (Alpes Franco-Suisses). *Schweiz. Mineral. Petrogr. Mitt.* 45(1):281–93
125. Martini, J., Vaugnat, M. 1970. Metamorphose niedrigst temperierten Grades in den Westalpen. *Fortschr. Mineral.* 47:52–64
126. Maxwell, D. T., Hower, J. 1967. High-grade diagenesis and low-grade metamorphism of illite in the Precambrian Belt Series. *Am. Mineral.* 52:843–57
127. Melson, W. G., Van Andel, T. H. 1966. Metamorphism in the Mid-Atlantic Ridge, 22°N latitude. *Mar. Geol.* 4:165–86
128. Metz, P., Winkler, H. G. F. 1964. Experimentelle Untersuchung der Diopsidbildung aus Tremolit, Calcit und Quarz. *Naturwissenschaften* 51: 1–3
129. Metz, P., Puhan, D., Winkler, H. G. F.

1968. Equilibrium reactions on the formation of talc and tremolite by metamorphism of siliceous dolomite. *Naturwissenschaften* 55:225–26

130. Metz, P., Trommsdorff, V. 1968. On phase equilibria in metamorphosed siliceous dolomites. *Contrib. Mineral. Petrol.* 18:305–9
131. Metz, P., Puhan, D. 1970. Experimentelle Untersuchung der Metamorphose von kieselig dolomitischen Sedimenten. *Contrib. Mineral. Petrol.* 26:302–14
132. Meyer, C., Hemley, J. J. 1967. Wall rock alteration. In *Geochemistry of Hydrothermal Ore Deposits,* ed. H. L. Barnes, 166–235. New York: Holt, Rinehart & Winston
133. Milton, C., Eugster, H. P. 1959. Mineral assemblages of the Green River Formation. In *Researches in Geochemistry, Vol. 1,* ed. P. H. Abelson, 118–50. New York: Wiley
134. Miyashiro, A. 1961. Evolution of metamorphic belts. *J. Petrology* 2: 277–311
135. Miyashiro, A. 1964. Oxidation and reduction in the earth's crust with special reference to the role of graphite. *Geochim. Cosmochim. Acta* 28:717–29
136. Miyashiro, A., Shido, F. 1970. Progressive metamorphism in zeolite assemblages. *Lithos* 3:251–60
137. Miyashiro, A., Shido, F., Ewing, M. 1970. Petrologic models for the Mid-Atlantic Ridge. *Deep Sea Res.* 17: 109–23
138. Miyashiro, A., Shido, F., Ewing, M. 1971. Metamorphism in the Mid-Atlantic Ridge near 24° and 30°N. *Phil. Trans. Roy. Soc. London Ser. A.* 268:589–603
139. Mossman, D. J., Bachinski, D. J. 1972. Zeolite facies metamorphism in the Silurian-Devonian fold belt of northeastern New Brunswick. *Can. J. Earth Sci.* 9:1703–09
140. Muffler, L. J. P., White, D. E. 1969. Active metamorphism of upper Cenozoic sediments in the Salton Sea geothermal field and the Salton Trough, southeastern California. *Geol. Soc. Am. Bull.* 80:157–81
141. National Academy of Sciences 1972. Understanding the Mid-Atlantic Ridge. *Nat. Acad. Sci. Ocean Affairs Bd. Rept.,* Washington, DC. 131 pp.
142. Neruchev, S. G., Parparova, G. M. 1972. Glubinnaya zonal'nost' metamorfizma ugley i organicheskogo veshchestva porod (Depth zonation of metamorphism of coals and organic matter in rocks). *Geol. Geofiz.* No. 9: 28–36
143. Newton, R. C. 1966. Some calc-silicate equilibrium relations. *Am. J. Sci.* 264: 204–22
144. Niggli, E. 1960. Mineral-Zonen der alpinen Metamorphose in den Schweizer Alpen. *Int. Geol. Congr. Rep. Sess. Norden, 21st, 1960, Pt. 13, Petrogr. Prov., Igneous Metamorphic Rocks,* 132–38
145. Nitsch, K.-H. 1968. Die Stabilität von Lawsonit. *Naturwissenschaften* 55:388
146. Nitsch, K.-H. 1971. Stabilitätsbeziehungen von Prehnit und Pumpellyithaltigen Paragenesen. *Contrib. Mineral. Petrol.* 30:240–60
147. Nitsch, K.-H. 1972. Das P-T-X_{CO_2}-Stabilitätsfeld von Lawsonit. *Contrib. Mineral. Petrol.* 34:116–34
148. Nitsch, K.-H., Storre, B. 1972. Zur Stabilität von Margarit in H_2O-CO_2-Gasgemischen. *Fortschr. Mineral.* 50(1):71–73
149. Otalora, G. 1964. Zeolites and related minerals in Cretaceous rocks of east-central Puerto Rico. *Am. J. Sci.* 262: 726–34
150. Reed, B. L., Hemley, J. J. 1966. Occurrence of pyrophyllite in the Kekiktuk Conglomerate, Brooks Range, northeastern Alaska. *Geol. Surv. Res. 1966, US Geol. Surv. Prof. Pap.* 550-C, C162–66
151. Reed, J. C. Jr., Morgan, B. A. 1971. Chemical alteration and spilitization of the Catoctin Greenstones, Shenandoah National Park, Virginia. *J. Geol.* 79: 526–48
152. Robie, R. A., Waldbaum, D. R. 1968. Thermodynamic properties of minerals and related substances at 298.15°K (25.0°C) and one atmosphere (1.013 bars) pressure and at higher temperatures. *US Geol. Surv. Bull. 1259.* 256 pp.
153. Roedder, E. 1972. *Data of Geochemistry, 6th Ed., Chapter JJ,* Composition of fluid inclusions. *US Geol. Surv. Prof. Pap. 440-JJ.* 164 pp.
154. Rook S. H., Williams, G. C. 1944. Imperial carbon dioxide gas field. Calif. Oil Fields, 28(2): 12–33
155. Saha, P. 1959. Geochemical and x-ray investigation of natural and synthetic analcites. *Am. Mineral.* 44:300–13
156. Sato, M. 1972. Intrinsic oxygen fugacities of iron-bearing oxide and silicate minerals under low total pressure. *Geol. Soc. Am. Mem.* 135: 289–307
157. Schwarcz, H. P., Clayton, R. N.,

Mayeda, T. 1970. Oxygen isotopic studies of calcareous and pelitic metamorphic rocks, New England. *Geol. Soc. Am. Bull.* 81:2299–2316
158. Seki, Y. 1969. Facies series in low-grade metamorphism. *Geol. Soc. Jap. J.* 75:255–66
159. Seki, Y. 1971. Some physical properties of analcime-wairakite solid solutions. *Geol. Soc. Jap. J.* 77:1–8
160. Seki, Y. et al 1969. Metamorphism in the Tanzawa Mountains, central Japan. *J. Jap. Assoc. Mineral. Petrologists Econ. Geol.* 61:1–75
161. Senderov, E. E. 1965. Features of the conditions of zeolite formation. *Geochem. Int.* 2:1143–55
162. Senderov, E. E. 1968. Experimental study of crystallization of sodium zeolites under hydrothermal conditions. *Geochem. Int.* 5:1–12
163. Sheppard, R. A., Gude, A. J. III 1968. Distribution and genesis of authigenic silicate minerals in tuffs of Pleistocene Lake Tecopa, Inyo County, California. *US Geol. Surv. Prof. Pap. 597.* 38 pp.
164. Skippen, G. B. 1971. Experimental data for reactions in siliceous marbles. *J. Geol.* 79:457–81
165. Smith, R. E. 1968. Redistribution of major elements in the alteration of some basic lavas during burial metamorphism. *J. Petrology* 9:191–219
166. Smith, R. E. 1969. Zones of progressive regional burial metamorphism in part of the Tasman geosyncline, eastern Australia. *J. Petrology* 10:144–63
167. Steiner, A. 1953. Hydrothermal rock alteration at Wairakei, New Zealand. *Econ. Geol.* 48:1–13
168. Steiner, A. 1968. Clay minerals in hydrothermally altered rocks at Wairakei, New Zealand. *Clays Clay Miner.* 16:193–213
169. Storre, B. 1970. Stabilitätsbedingungen Grossular-führender Paragenesen im System CaO-Al_2O_3-SiO_2-CO_2-H_2O. *Contrib. Mineral. Petrol.* 29:145–62
170. Storre, B., Nitsch, K.-H. 1972. Die Reaktion 2 Zoisit + 1 $CO_2 \leftrightarrows$ 3 Anorthit + 1 Calcit + $1H_2O$. *Contrib. Mineral. Petrol.* 35:1–10
171. Strens, R. G. J. 1968. Reconnaissance of the prehnite stability field. *Mineral. Mag.* 36:864–67
172. Surdam, R. C. 1973. Low-grade metamorphism of tuffaceous rocks in the Karmutsen Group, Vancouver Island, British Columbia. *Geol. Soc. Am. Bull.* 84:1911–22
173. Thomas, H. 1958. Geología de la Cordillera de la Costa entre el valle de La Ligua y la cuesta de Barriga: Santiago, Chile. *Inst. Invest. Geol. Chile, No. 2.* 86 pp.
174. Thompson, A. B. 1970. A note on the kaolinite-pyrophyllite equilibrium. *Am. J. Sci.* 268:454–58
175. Thompson, A. B. 1970. Laumontite equilibria and the zeolite facies. *Am. J. Sci.* 269:267–75
176. Thompson, A. B. 1971. P_{CO_2} in low-grade metamorphism; zeolite, carbonate, clay mineral, prehnite relations in the system CaO-Al_2O_3-SiO_2-CO_2-H_2O. *Contrib. Mineral. Petrol.* 33:145–61
177. Thompson, A. B. 1971. Analcite-albite equilibria at low temperatures. *Am. J. Sci.* 271:79–92
178. Thompson, A. B. 1972. Experiments on the lower stability of micas. *Progr. Exp. Petrology, London, Natur. Environ. Res. Counc. Publ. Ser. D* 2:55–56
179. Thompson, A. B. 1973. The instability of feldspar in metamorphism. In *The Feldspars,* ed. W. S. MacKenzie, J. Zussman, 654–72. Manchester, England: Manchester Univ. Press
180. Thompson, J. B. Jr. 1957. The graphical analysis of mineral assemblages in pelitic schists. *Am. Mineral.* 42:842–58
181. Thompson, J. B. Jr. 1972. Oxides and sulfides in regional metamorphism of pelitic schists. *Int. Geol. Congr. Rep. Sess. Montreal, 24th, Sec. 10, Geochem.,* 27–35
182. Thompson, J. B. Jr., Norton, S. A. 1968. Paleozoic regional metamorphism in New England and adjacent areas. In *Studies of Appalachian Geology, Northern and Maritime,* ed. E. Zen, W. S. White, J. B. Hadley, J. B. Thompson Jr., 319–27. New York: Interscience
183. Tilley, C. E. 1924. The facies classification of metamorphic rocks. *Geol. Mag.* 61:167–71
184. Tilley, C. E. 1925. A preliminary survey of metamorphic zones in the southern Highlands of Scotland. *Quart. J. Geol. Soc. London* 81:100–12
185. Tilley, C. E. 1948. Earlier stages in the metamorphism of siliceous dolomites. *Mineral. Mag.* 28:272–76
186. Tobschall, H. J. 1969. Eine Subfaziesfolge der Grünschieferfazies in den Mittleren Cévennen (Dep. Ardèche) mit Pyrophyllit aufweisenden Mineralparagenesen. *Contrib. Mineral. Petrol.* 24:76–91
187. Touray, J.-C., Jauzein, A. 1967. In-

clusions à méthane dans les quartz des "terres noires" de la Drôme. *C.R. Acad. Sci. Ser. D* 264:1957–60
188. Touray, J.-C., Sagon, J.-P. 1967. Inclusions à méthane dans les quartz des marnes de la région de Mauléon (Basses-Pyrénées). *C.R. Acad. Sci. Ser. D* 265:1269–72
189. Trommsdorff, V., Evans, B. W. 1972. Progressive metamorphism of antigorite schist in the Bergell tonalite aureole. *Am. J. Sci.* 272:423–37
190. Tu, K. 1956. Ts'ui yün mu ja shui tsung ho shi yen te ch'u pu chieh kuo (Preliminary results of hydrothermal synthesis of the brittle micas). *Ti Chih Hsüeh Pao* (*Acta Geol. Sinica*) 36:229–238. (In Chinese with Russian summary)
191. Turnock, A. C. 1960. The stability of iron chlorites. *Carnegie Inst. Washington Yearb.* 59:98–103
192. Vallance, T. G. 1967. Mafic rock alteration and isochemical development of some cordierite-anthophyllite rocks. *J. Petrology* 8:84–96
193. Vallance, T. G. 1969. Spilites again: Some consequences of the degradation of basalt. *Proc. Linn. Soc. N.S. Wales* 94:8–50
194. Velde, B. 1965. Experimental determination of muscovite polymorph stabilities. *Am. Mineral.* 50:436–49
195. Velde, B. 1965. Phengite micas: Synthesis, stability, and natural occurrence. *Am. J. Sci.* 263:886–913
196. Velde, B. 1971. The stability and natural occurrence of margarite. *Mineral. Mag.* 38:317–23
197. Velde, B. 1972. Celadonite mica: solid solution and stability. *Contrib. Mineral. Petrol.* 37:235–47
198. Walker, G. P. L. 1960. Zeolite zones and dike distribution in relation to the structure of the basalts of eastern Iceland. *J. Geol.* 68:515–28
199. Wise, W. S., Eugster, H. P. 1964. Celadonite: Synthesis, thermal stability and occurrence. *Am. Mineral.* 49:1031–83
200. Zen, E. 1959. Clay mineral—carbonate relations in sedimentary rocks. *Am. J. Sci.* 257:29–43
201. Zen, E. 1960. Metamorphism of lower Paleozoic rocks in the vicinity of the Taconic Range in west-central Vermont. *Am. Mineral.* 45:129–75
202. Zen, E. 1960. Petrology of lower Paleozoic rocks from the slate belt of western Vermont. *Int. Geol. Congr., Rep. Sess., Norden, 21st, Pt. 13*:362–71
203. Zen, E. 1961. The zeolite facies: An interpretation. *Am. J. Sci.* 259:401–9
204. Zen, E. 1963. Components, phases, and criteria of chemical equilibrium in rocks. *Am. J. Sci.* 261:929–42
205. Zen, E. 1967. Some topological relationships in multisystems of n+3 phases, Part 2, unary and binary metastable sequences. *Am. J. Sci.* 265:871–97
206. Zen, E. 1969. Stratigraphy, structure and metamorphism of the Taconic allochthon and surrounding autochthon in Bashbish Falls and Egremont quadrangles and adjacent areas. *N. Engl. Intercollegiate Geol. Conf., 61st Ann. Meet., Albany, NY, Guidebook.* 3/1–3/41
207. Zen, E. 1972. The Taconide zone and the Taconic orogeny in the western part of the northern Appalachian orogen. *Geol. Soc. Am. Spec. Pap.* 135. 72 pp.
208. Zen, E. 1972. Gibbs free energy, enthalpy, and entropy of ten rock-forming minerals: Calculations, discrepancies, implications. *Am. Mineral.* 57:524–53
209. Zen, E. 1974. Prehnite-pumpellyite bearing metamorphic rocks, west side of the Appalachian foldbelt, Pennsylvania to Newfoundland. *J. Petrology* 15(2): In press
210. Zen, E., Albee, A. L. 1964. Coexistent muscovite and paragonite in pelitic schists. *Am. Mineral.* 49:904–25

REGIONAL GEOPHYSICS OF THE BASIN AND RANGE PROVINCE

George A. Thompson
Department of Geophysics, Stanford University, Stanford, California 94305

Dennis B. Burke
U.S. Geological Survey, Menlo Park, California 94025

Of late years the most important contributions have come from the Physicists, and in their scales have been weighed the old theories of Geologists.

G. K. Gilbert (1874)

INTRODUCTION AND GEOLOGIC SETTING

Nearly one hundred years ago, Gilbert (23, 24) and other geologic pioneers introduced the idea that much of the seeming jumble of mountains and valleys in western North America was the result of far different processes than fold mountain systems such as the Appalachians or Alps. After a century of geologic and geophysical investigations in the region, it is now generally accepted that the physiography of the Basin and Range province (Figure 1) is one of sculptured and partially buried fault-bounded blocks that have been produced by the extension of the region during late Cenozoic time. Crustal blocks composed of complexly deformed, diverse pre-Cenozoic rocks and relatively undeformed, predominantly nonmarine volcanic rocks of early and middle Cenozoic age have been variously uplifted, tilted, and dropped along numerous normal faults throughout a broad region from Mexico to Canada—from as far west as California and Oregon to as far east as western Texas (e.g. Cook, 13; Gilluly, 27; Thompson, 76).

The distribution of late Cenozoic normal faults in the western United States is shown on Figure 2 (note that the regional extent of faulting is somewhat larger than the Basin and Range physiographic province of Figure 1). The recent seismicity (Figure 3) shows that small earthquakes are widespread in what Atwater (5) called a wide soft zone accommodating oblique divergence between the Pacific and North American plates. The net effect of fault movements within this region is a crustal extension oriented roughly WNW–ESE. The actual motion on individual faults is quite variable, however, and appears to be controlled by the orientation of faults with respect to this principal extension (Thompson & Burke, 78). In the northern portion of the region—across Nevada and western

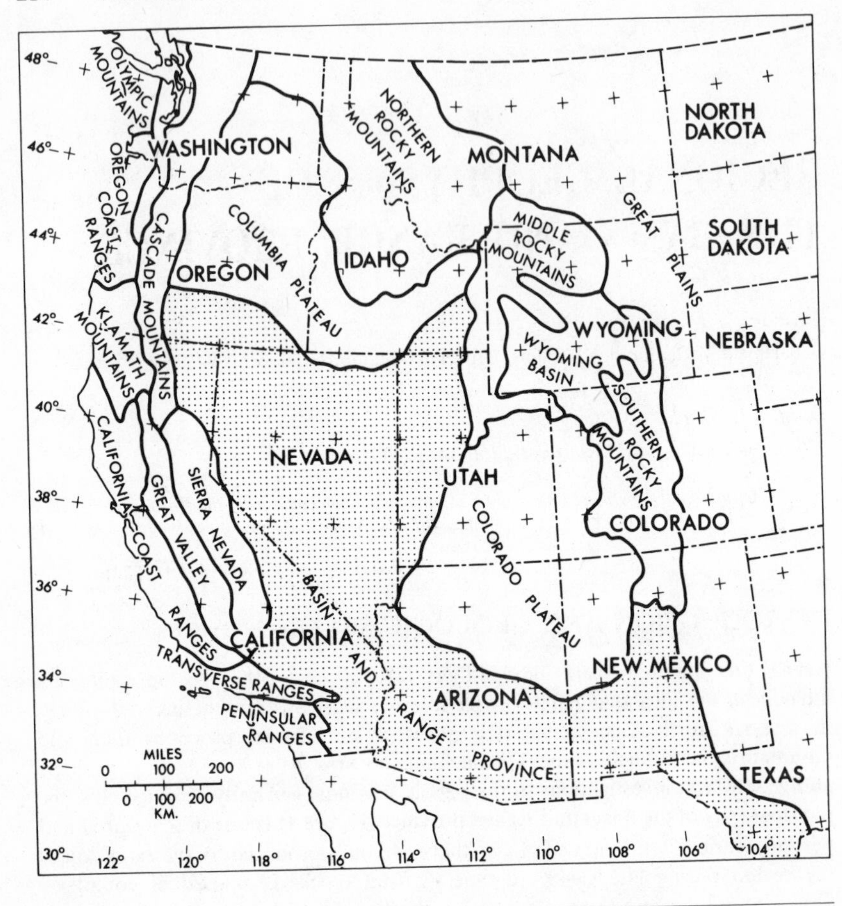

Figure 1 Physiographic provinces of the western United States (Fenneman, 21).

Utah—the domain of faulting is neatly confined between the Sierra Nevada of California and the Wasatch Mountains of north-central Utah. The relatively unfaulted Colorado Plateau separates the central portion from a zone of faulting in the Rio Grande trough in New Mexico and west Texas. Relative motion between the unextended and rather enigmatic mass of the plateau and the encircling faulted terrain is presumably accommodated by a component of right-lateral strike-slip along the southern plateau border. Faulted terrain extends southwards without interruption into Mexico and the Gulf of California. Faulting seems to die out to the north, and the manner in which relative motions are accommodated along the northern boundary remains a troublesome problem.

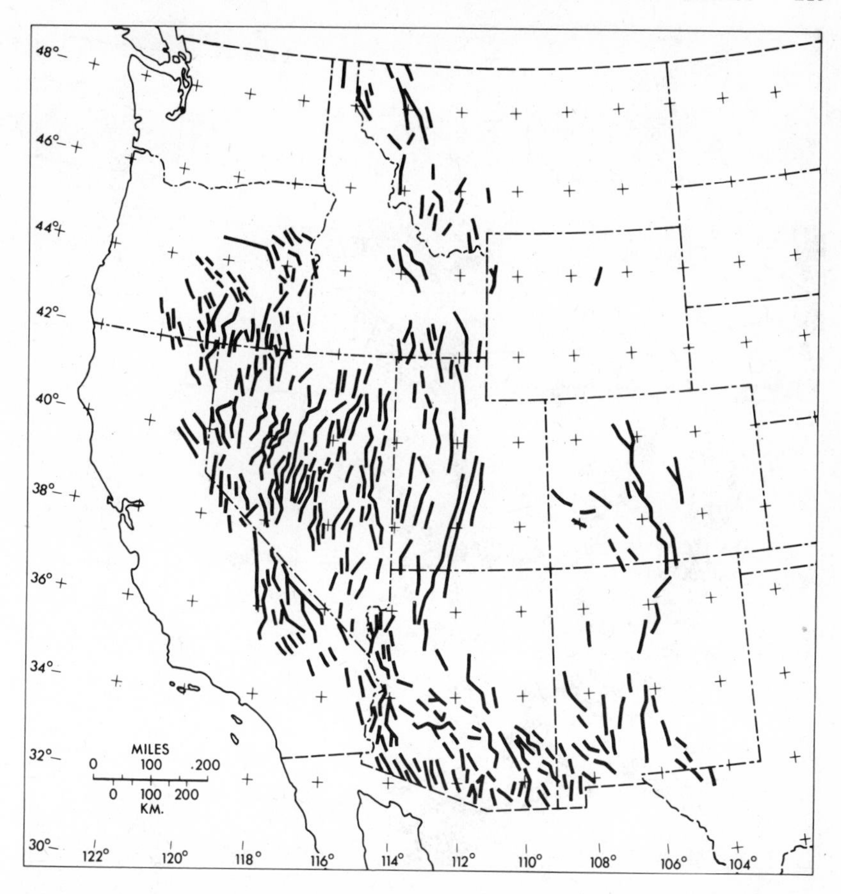

Figure 2 Predominantly normal (Basin and Range) faults of late Cenozoic age in the western United States (modified from Gilluly, 26).

Although the Basin and Range province is in many ways a unique physiographic and geologic entity, increasingly precise and reliable geophysical studies, together with advances in tectonic theory, highlight similarities between the province and other regions of past or present crustal extension. It has a high heat flow and widespread volcanism like other regions of active normal faulting, such as the Rift Valleys of Africa, the Lake Baikal depression of the USSR, the Rhine graben of Europe, the marginal basins of the western Pacific Ocean, and the worldwide system of oceanic ridges and rises. Along with the Sierra Nevada and Colorado Plateau, it forms a wide elevated region averaging 1–2 km above sea level and thus may resemble the elevated, thermally expanded oceanic ridges (Sclater &

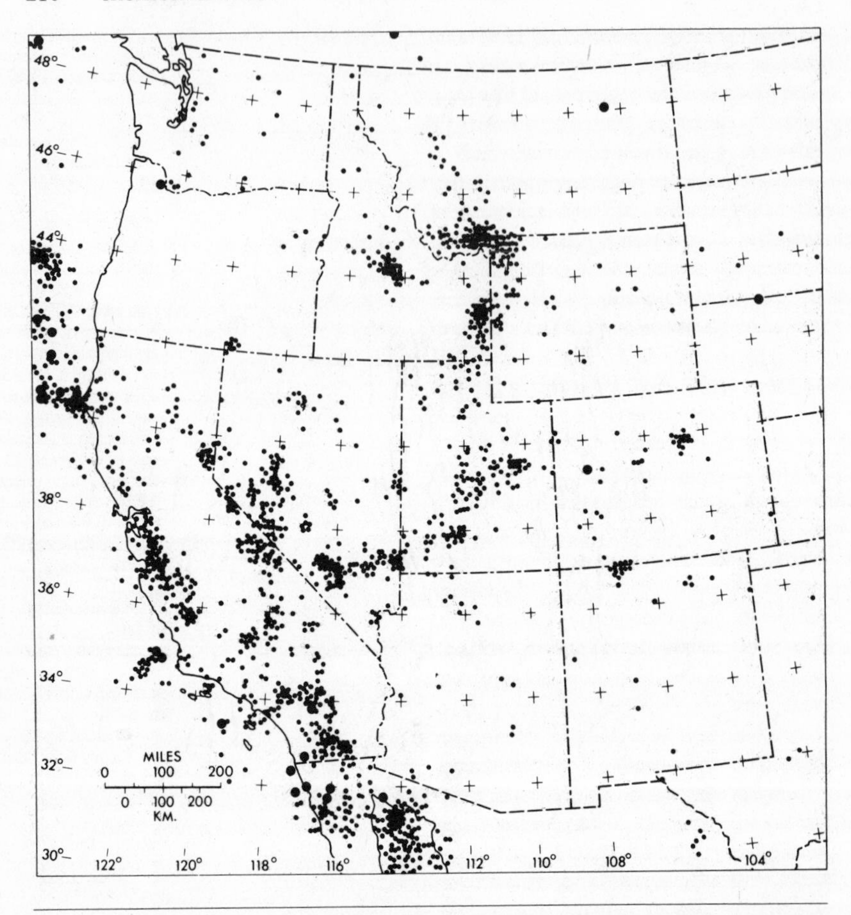

Figure 3 Earthquake epicenters in western North America for the period 1961–1970. Small dots represent earthquakes of magnitude about 3 to 5, large dots greater than 5. National Oceanographic and Atmospheric Administration epicenters replotted by J. C. Lahr and P. R. Stevenson of the US Geological Survey (personal communication, 1973).

Francheteau, 65). Also like some of these other regions, it has a thin crust and low mantle velocity.

Can regional geophysical data for the Basin and Range province, combined with interpretations of its geologic history, lead toward a better understanding of the tectonic processes that have controlled its development? To what extent have earlier geologic events in the region preordained the pattern of faulting that we now see in western North America? What constraints must be heeded in tectonic models of the region, and what aspects of the province allow these models to be compared with other portions of the global system of ever-changing lithosphere plates? We believe this last consideration to be of great importance, although it can only be touched on lightly here, because much understanding of the province derives from analogy with other regions of crustal extension. The currently most promising models relate Basin and Range structure to an earlier subducting plate at the western margin of North America, and they incorporate close physical comparisons with the marginal basins of the western Pacific.

REGIONAL CRUST AND MANTLE STRUCTURE

Crustal Thickness: Seismic Refraction

Seismic waves from explosions have provided the most reliable and detailed information on crustal thickness and indicate that the region of distinctive Basin and Range structures corresponds quite closely with a region of thin continental crust (Pakiser, 52; Prodehl, 55). Prior to the work of Tatel & Tuve (74) it was generally assumed that the crust would be thicker under this elevated region than in continental regions near sea level, a relationship that has been found in other mountain regions. It was thought that lateral variations of velocity and density in the mantle were unimportant, or at least inconvenient in seismic interpretation, and that isostatic compensation was accomplished mainly by variations in crustal thickness.

Tatel and Tuve found that the crust in northwestern Utah is an anomalously thin 29 km. Verification came from Berg et al (6), Diment et al (16), and Press (54), although these authors initially used a different definition of the crust. They found abnormally low *P*-wave velocities of 7.6 to 7.8 km/sec at shallow depth for what we have now come to identify as P_n, the wave traveling in the uppermost mantle below the *M* discontinuity.

Extensive explosion studies carried out by the US Geological Survey established the basic picture as we know it today. David H. Warren, of the USGS (personal communication, 1973) has compiled and interpreted these and other data into a contour map of crustal thickness (Figure 4). The contours are based on data of varying quality and on varying interpretation of velocity structures within the crust; nonetheless they represent a good first approximation. Almost the whole region from the Rocky Mountains westward has a thin but variable crust, roughly two thirds the thickness found in stable regions of comparable elevations. The eastern border of the Basin and Range province is marked by a fairly sharp gradient at the 35 km contour to a thicker crust under the Colorado Plateau. Southeast of the Colorado Plateau there is some indication of thinning beneath the Rio Grande trough of New Mexico and west Texas.

The crust is thicker beneath the Sierra Nevada to the west of the province [although this conclusion has been called into question by Carder (10)]. It is interesting to point out that in detail the thick crust of the Sierran region (Figure 5) extends into the Basin and Range province to the east of the Sierra Nevada. The eastward extent of thick crust does not correspond with the eastern border of the Mesozoic Sierra Nevada batholith (Figure 6), however; although a correlation of the low velocity zone with the border of the batholith is not ruled out.

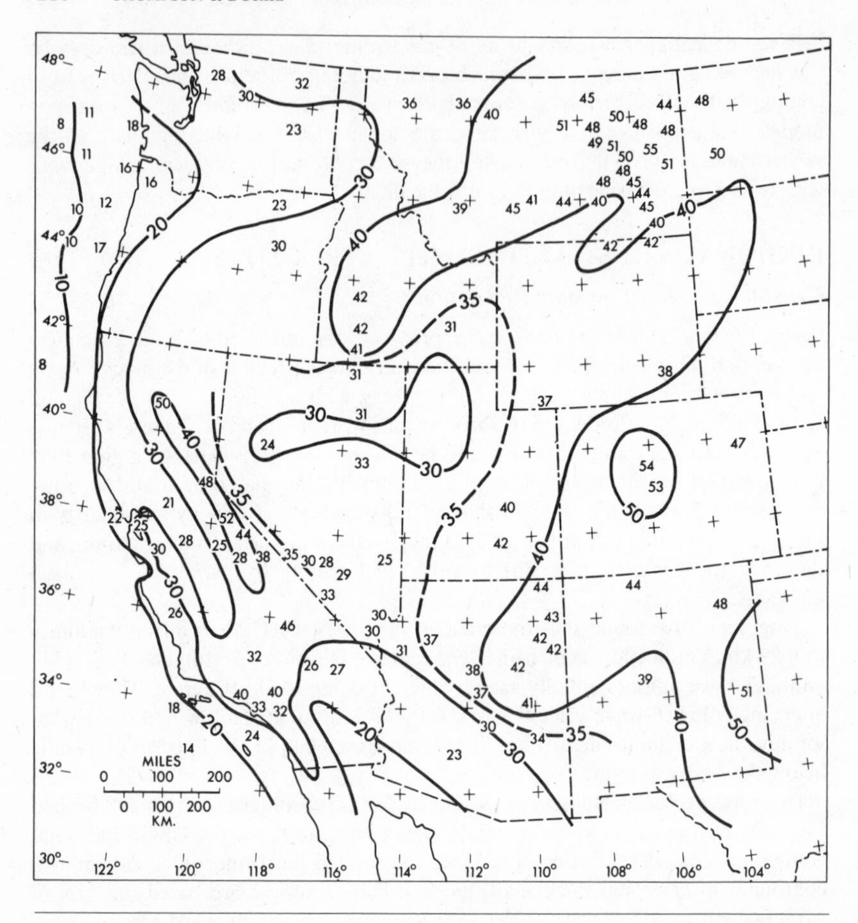

Figure 4 Contour map of crustal thickness (in kilometers) based on seismic refraction studies. Small numbers indicate individual thickness determinations. Compiled by David H. Warren from the following sources: 1, 4, 7, 11, 16, 19, 20, 22, 30, 34–39, 44, 55, 57–59, 62, 66, 67, 70, 72, 73, 82–85.

Upper Mantle Velocity and Implications From Gravity

When it was found that the crust is abnormally thin beneath the Basin and Range province and adjacent regions it was also discovered that P_n is anomalous. Its velocity of 7.7 to 7.9 km/sec is significantly less than the normal velocity of about 8.2 km/sec observed in stable regions (Pakiser, 52; Herrin & Taggart, 33; see Figure 7). Most of the Basin and Range province is characterized by the lowest P_n velocities, less than 7.8 km/sec.

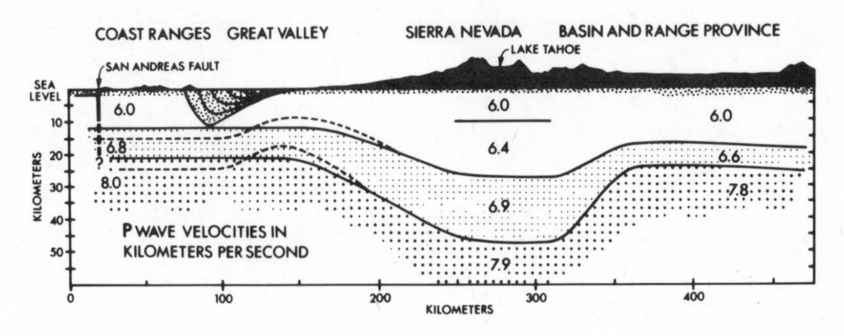

Figure 5 Crust and upper mantle structure in a section across central California and west-central Nevada as deduced from seismic-refraction studies. An alternative model beneath the Coast Ranges and Great Valley is shown by dashed lines; topography greatly vertically exaggerated (from Eaton, 20).

Gravity data supply a fundamental constraint on the amount of mass per unit area underlying any region. This information is particularly valuable because seismic refraction measurements do not by themselves allow interpretations of the thickness of the anomalous upper mantle of low P_n velocity. Gravity interpretation utilizes: 1. crustal thicknesses from seismic refraction, 2. crustal densities estimated from seismic velocities and geology, and 3. upper mantle densities estimated from P_n velocities. The gravity data then yield estimates of the thickness of anomalous mantle relative to stable regions (Thompson & Talwani, 79). The required thickness of low-density, low-velocity anomalous upper mantle is at least 20 km over much of the region.

In comparison with stable continental regions near sea level, most of the isostatic support for the high Basin-Range and adjacent regions is in the anomalous upper mantle. This material must surely be a key element in any tectonic model.

Isostatic gravity anomalies in the United States (Figure 8) show that most of the region from the west coast to the eastern limit of the Basin and Range province is deficient in mass, with an average anomaly of perhaps around -10 mgal. In this respect the region is similar to marginal basins of the western Pacific, which also tend to be isostatically negative.

The Lake Bonneville Experiment

A natural experiment in gravitational unloading of the Basin-Range crust occurred as pluvial Lake Bonneville, of late Pleistocene age, dried up, leaving the Great Salt Lake as its principal remnant. Prominent shorelines around the edge and on former islands mark the successively lower levels of Lake Bonneville. These shorelines are domed up toward the center as much as 64 m as a result of the unloading (Gilbert, 25; Crittenden, 14).

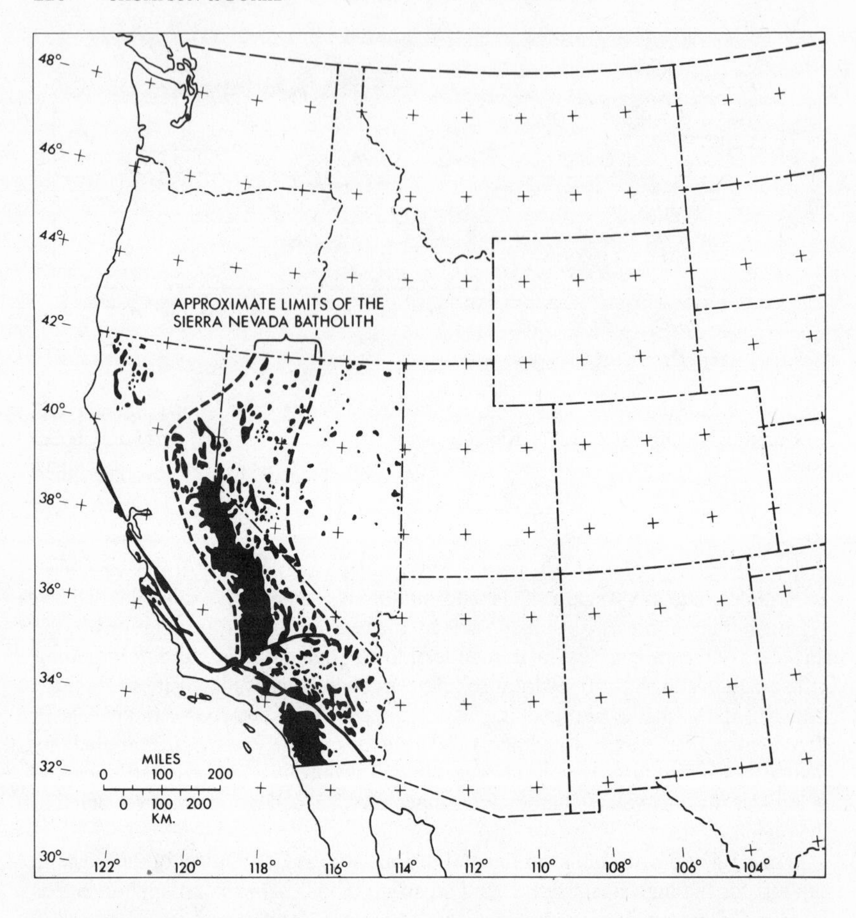

Figure 6 Distribution of granitic rocks in California and Nevada. Solid lines represent major active strike-slip faults (from Crowder et al, 15).

Using data from this natural experiment and a simple model of an elastic lithosphere floating on a fluid asthenosphere, Walcott (81) has computed the apparent flexural rigidity of the lithosphere and compared it with that of other regions subjected to various kinds of loading and unloading. Walcott's results, as shown in Table 1, illustrate that the flexural rigidity of the Basin and Range lithosphere is unusually low. He suggests that the anomaly may be explained by a "very thin lithosphere, only about 20 km thick, with hot, lower crustal material" acting as part of the asthenosphere. In contrast, the flexural rigidity of stable continental and oceanic regions suggests lithosphere thicknesses of 110 km and

Table 1 Apparent flexural rigidity of the lithosphere (from Walcott, 81)

Data	Region	Apparent flexural rigidity, Newton meters	Characteristic time, years
Lake Bonneville	Basin and Range province	5×10^{22}	10^4
Caribou Mountains	Stable continental platform	3×10^{23}	5×10^6
Interior Plains	Stable continental platform	4×10^{23}	5×10^6
Boothia uplift	Stable continental platform	7×10^{22}	5×10^8
Lake Algonquin	Stable continental platform	6×10^{24}	10^3
Lake Agassiz	Stable continental platform	9×10^{24}	10^3
Hawaiian archipelago	Oceanic lithosphere	2×10^{23}	10^7
Island arcs	Oceanic lithosphere	2×10^{23}	10^7

75 km or more, respectively. The low P_n velocity and high heat flow (discussed in a later section) are consistent with Walcott's interpretation.

Anomalous Mantle and the Low-Velocity Zone

Several studies have indicated that the Basin-Range region has an unusually well-developed upper mantle low-velocity zone (LVZ) for both *P*- and *S*-waves. The relationship is not always clear between the accentuated LVZ (as defined by waves refracted at deeper levels in the mantle) and the anomalous upper mantle (as defined by low P_n velocity). In a study applicable to the central part of the Basin and Range province in Nevada and western Utah, Archambeau and associates (2) derived a model (Figure 9) in which the *M* discontinuity is at a depth of 28 km and the P_n velocity just below it is 7.7 km/sec. This low velocity remains nearly constant to a depth of 130 km, where it undergoes a rapid transition to 8.3 km/sec. Thus the LVZ is about 100 km thick; it begins at the top of the mantle and coincides with the anomalous upper mantle.

In comparison, the same investigators derived three models applicable to regions northeast and east of the Basin and Range province, including the Colorado Plateau (Figure 9). These models have in common a "lid" of higher velocity material (P_n about 8.0 km/sec) above the LVZ, which is only about half as thick as in the Basin-Range model.

In the foregoing discussion a single model has been assumed to represent the Colorado Plateau, and this assumption seems reasonable because of the geological uniformity of the Plateau. However, within the limited resolution of the data, P_n velocities (Figure 7) appear to vary markedly over the Plateau and would not allow a single upper mantle model. This seeming conflict invites further research.

Helmberger (31) developed a new technique for studying regional variations of the LVZ. The method makes use of the nearly constant velocity of the *PL* wave in the crustal wave guide and the regional variation in the velocity of long-period *P*-waves. Results are mapped on Figure 10 (York & Helmberger, 87) as observed time differences minus the time difference predicted from a model LVZ roughly

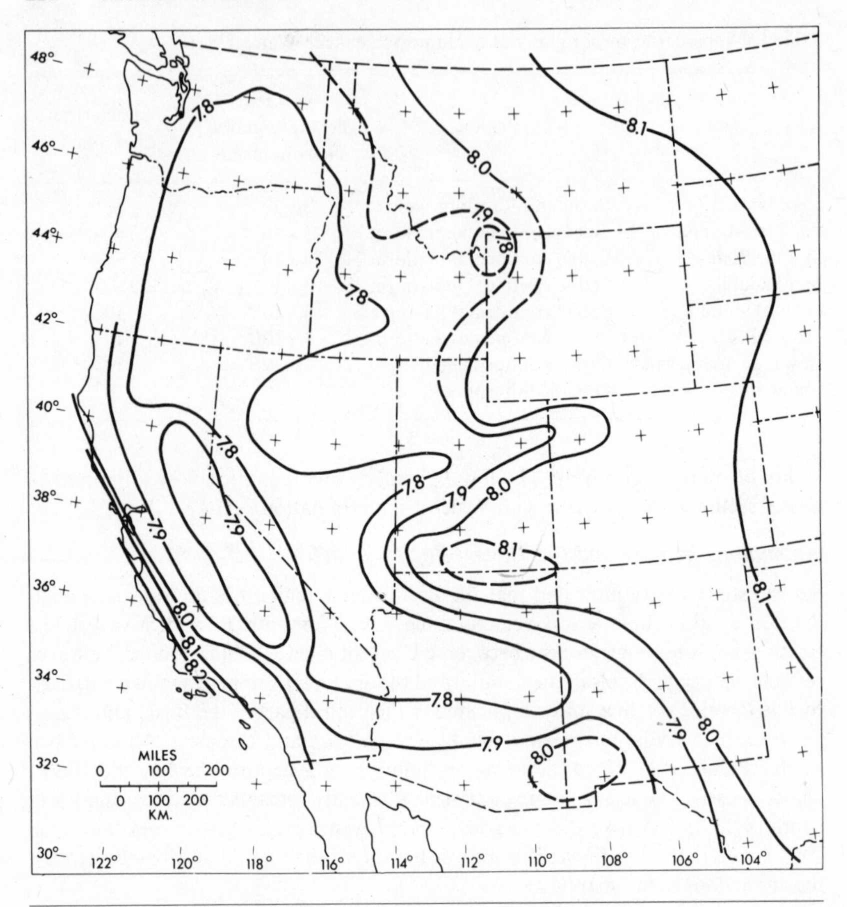

Figure 7 Contour map of P_n (upper mantle) velocities (in kilometers per second) (from Herrin, 32).

comparable to the Colorado Plateau model of Figure 9. Progressively more negative Δt values (delays of the long-period *P*-wave relative to the model) represent progressively thicker LVZ or lower upper mantle velocity. Positive values represent thinner or higher velocity LVZ relative to the model. Two main zones of thick LVZ within the −3 sec contour trend northward through eastern Nevada and western Utah and northeastward into the Rio Grande trough in New Mexico. These zones join to the southwest and continue across southern California and northern Mexico toward the continental borderland off southern California (generally considered to have Basin-Range structure) and the Gulf of California. The Colorado Plateau is strikingly outlined by the zero contour, which is expected because the reference model resembles the Colorado Plateau mantle.

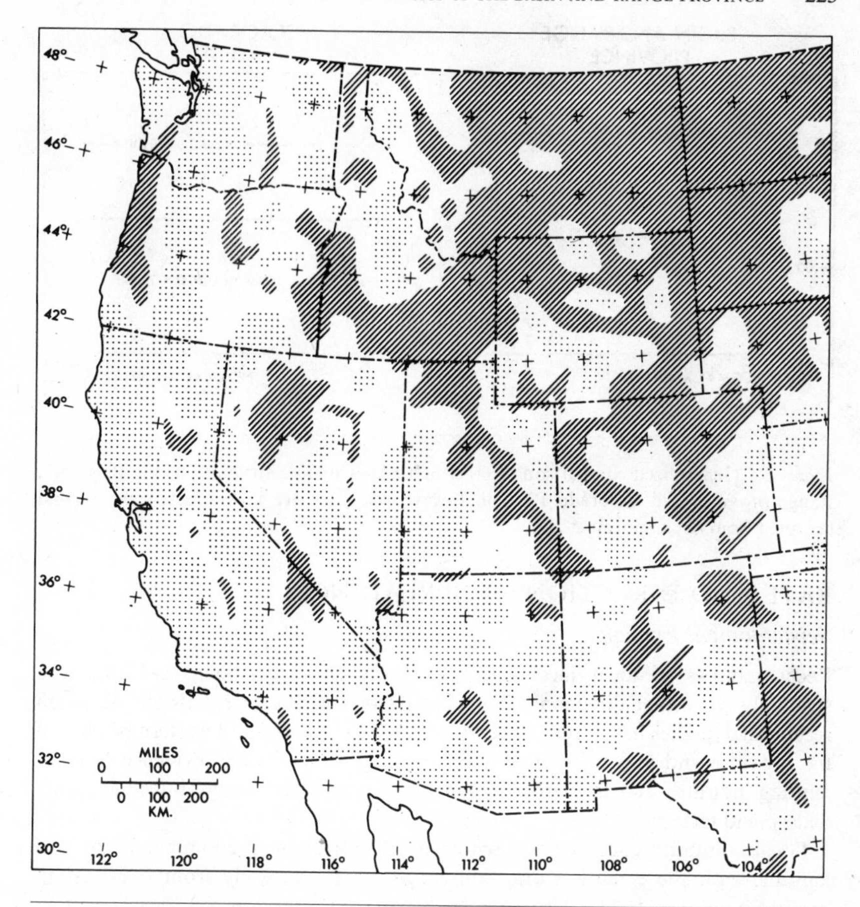

Figure 8 Regional isostatic gravity anomaly (based on Airy-Heiskanen concept with standard column 30 km). Line pattern, greater than +10 mgal; stippled pattern more negative than −10 mgal (from Woollard, 86).

In other important investigations Robinson & Kovach (56) studied upper mantle *S*-waves in the Basin and Range province, and Herrin (32) compared the Basin and Range upper mantle with that of a stable region, the Canadian Shield. Using direct measurements of the travel time gradient, Robinson and Kovach found a thin lid zone (9 km) of shear velocity 4.5 km/sec at the top of the mantle, overlying a low velocity zone with a minimum velocity at 100 km. Herrin's comparative model for the Canadian Shield contains no LVZ for *P*-waves and only a weak one for *S*-waves. The comparison is important because it emphasizes a degree of similarity between the Basin and Range and Colorado Plateau mantles relative to the stable region.

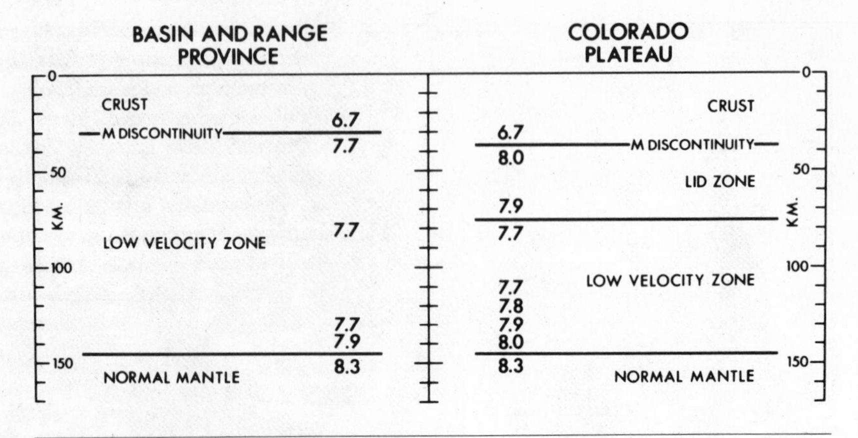

Figure 9 Generalized comparison of crust and upper mantle structure in the Basin and Range province and Colorado Plateau. *P*-wave velocities are in kilometers per second (adapted from Archambeau et al, 2).

RATE AND DIRECTION OF SPREADING

Seismological Evidence

Recent studies of focal mechanisms of many small earthquakes highlight a strikingly consistent direction of ongoing Basin and Range extension. Although recent earthquakes have been concentrated near the eastern and western borders of the province and in a belt across southern Utah and Nevada, evidence of older faulting indicates that they are a reasonable sample of this longer but much more widespread tectonic activity.

Focal solutions compiled by Scholz et al (64) show predominantly normal faulting, with the extension direction ranging approximately from east-west to northwest-southeast. The few examples of strike-slip motion are also consistent with this extension direction.

Only a few of the larger historical earthquakes were accompanied by surface ruptures large enough for the amount of offset to be directly observed, and these larger shocks (Figure 11) probably account for most of the total deformation. The main north-south zone of historical faulting in Nevada and adjacent California is nearly continuous. Horizontal extension across the faults ranges from a few centimeters to a few meters (Thompson, 75) and is greatest near the north and south ends of the zone. This wide range in extension, plus the existence of unfaulted gaps, shows that the 100-yr historical period is too short for measuring a meaningful rate of extension.

Dixie Valley, a Type Basin

Near the northern end of the zone of historical faulting, at the site of the 1954 faulting in Dixie Valley (Figure 12), two measures of long-term displacement have

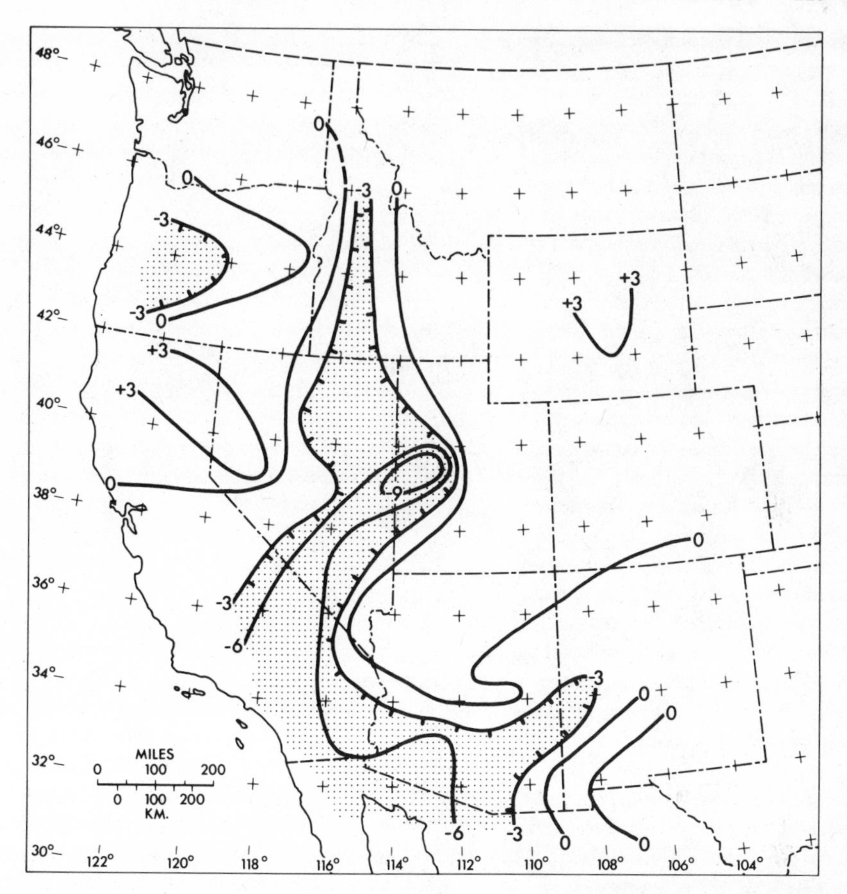

Figure 10 Relative development of upper mantle LVZ (low velocity zone), expressed as contours of time difference in seconds with respect to a model LVZ similar to that of Colorado Plateau. Stipple pattern accentuates region of pronounced (thicker or lower velocity) LVZ (from York & Helmberger, 87).

been investigated (Thompson & Burke, 78): 1. Displacements of the shoreline of a late Pleistocene lake supply a measure of extension during the last 12,000 years (Figure 13), and 2. fault displacements determined from geophysical exploration of the valley give the total amount of extension for late Cenozoic time, at least 5 km in 15 m.y. The average spreading rates are 1 mm/yr for the short interval and at least 0.4 mm/yr for the total displacement. The spreading direction we obtained from large slickenside grooves on fault planes is approximately N55°W–S55°E,

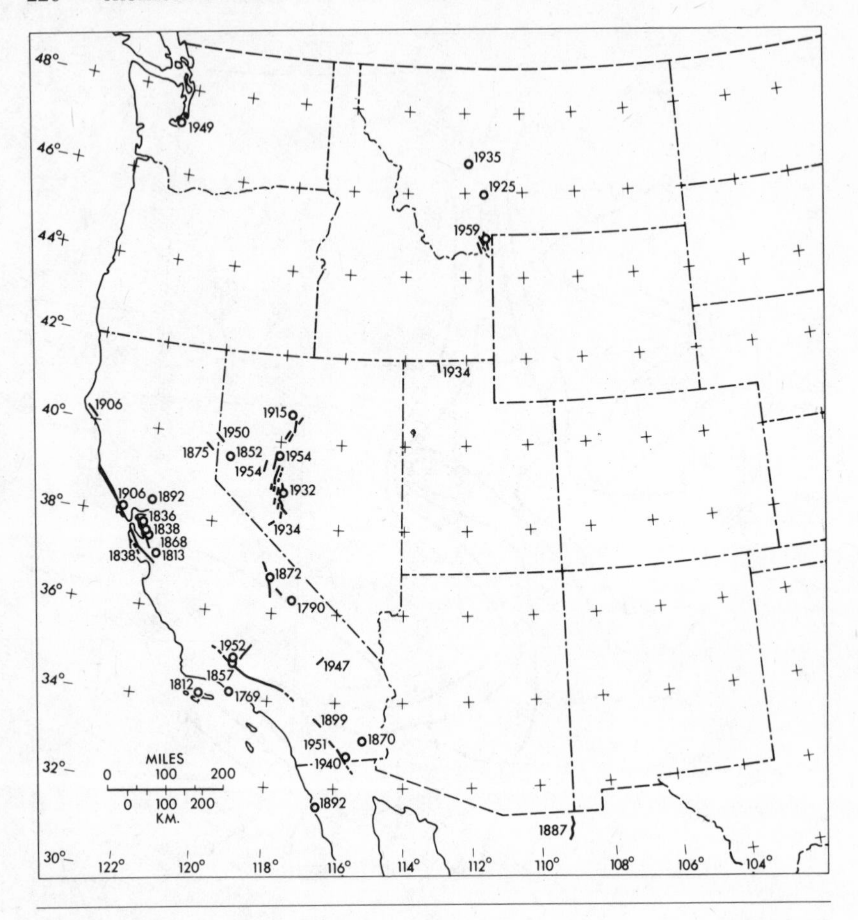

Figure 11 Historic surface offsets and epicenters for earthquakes of greater than about magnitude 7 in the western United States (from Ryall et al, 61).

which corresponds well with the range of directions obtained from earthquake focal mechanisms.

Dixie Valley is the only basin for which this much data is available. A simple extrapolation to 20 major basins across this part of the province suggests a total Basin-Range spreading of about 100 km (10% increase in crustal area) and a spreading rate of 8 mm/yr. On somewhat different assumptions, Gilluly (28) estimated that the areal expansion ranges from 4% to 12% over most of the province. Stewart (71) estimated 50 to 100 km (5% to 10%) of extension on the basis of a careful analysis of all available data. Hamilton & Myers (29) suggest that the extension may be as great as 300 km (30%). More subsurface data on many basins is needed to improve these estimates.

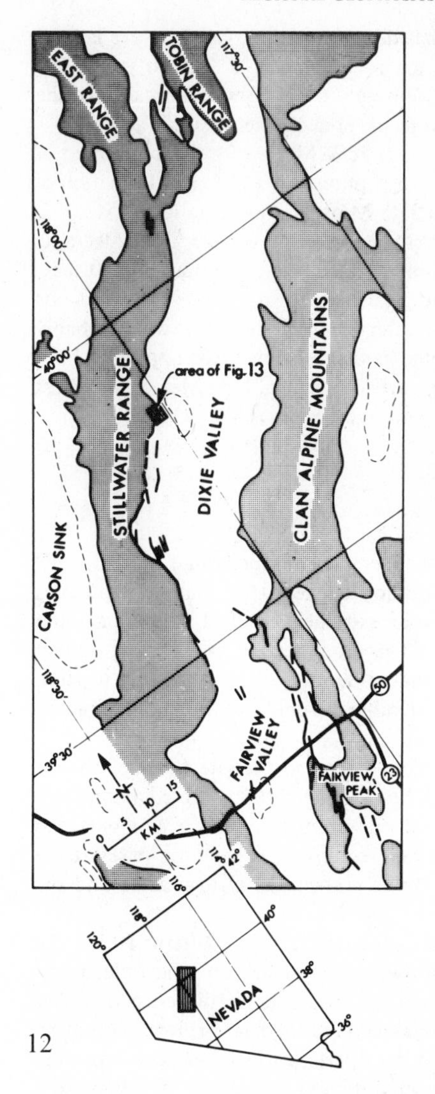

12

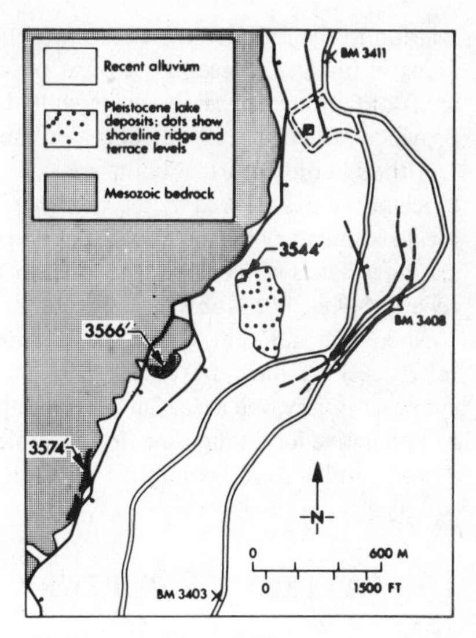

13

Figure 13 (*above*) Map of offset lake shorelines in west-central Dixie Valley. The relative vertical spacing of beach ridges around the valley demonstrates that the highest beach ridge preserved in this area (3544 ft) marks—like the tufa cemented terrace deposits on bedrock—the highest lake stand. The age of the high shoreline is 12,000 years (from Thompson & Burke, 78).

Figure 12 (*left*) Dixie Valley region. Fault scarps formed or reactivated in 1903, 1915, and 1954 are shown (from Thompson & Burke, 78).

Locus and Time of Basin-Range Faulting

The present seismicity (Figure 3) is a misleading guide to even the geologically youngest faulting. Fault scarps of Quaternary age are widespread and bear little relation to the seismicity. Slemmons (69) has documented this fact for Nevada with maps of faults in three age groups covering roughly the last 100,000 years. The locus of faulting appears to have shifted randomly over the whole breadth of the province rather than having been confined to the area of recent seismic activity.

Although older normal faults are known (Burke & McKee, 9), the main onset of block faulting is marked by the widespread disruption of drainage and formation of local sedimentary basins about middle or late Miocene time. The lower Miocene ash-flow sheets which cover broad areas were deposited on surfaces of low tectonic relief (McKee, 46; Noble, 50). The inception of Basin-Range faulting over at least Nevada and adjacent California is dated at 15 to 17 m.y. (see Noble, 50, for references). It must be emphasized that after faulting began it was probably sporadic in any one area. On physiographic evidence, some areas appear to have been inactive for a long time (for example, parts of Arizona, New Mexico, and west Texas), while activity continued to the present in other areas.

THE PATTERN OF RUPTURE

Basin-Range faults are often described in a general way as high-angle normal faults striking north to northeast, but the impression conveyed by that description is highly misleading. Individual faults tend to be extremely crooked in map plan and the fault pattern is more nearly rhomboid or even rectilinear. Some mountain ranges are bounded by en echelon faults that strike diagonal to the range (eastern front of Sierra Nevada for example). Considerable warping and tilting of the blocks accompany the faulting, particularly near the ends of elongate basins.

Nowhere is the fault pattern better exhibited than in the late Cenozoic basalt flows of south-central Oregon (Figure 14), but similar patterns are common from Nevada (Figures 12 and 13) to Texas. Moreover, the roughly rhomboid map pattern of faulting is characteristic of other regions of present or past crustal extension, such as the African Rifts, the Rhine graben, the Oslo graben, and the Triassic basins of eastern North America.

No well-founded explanation for the complex rupture pattern is known. Alternative hypotheses include changes in the stress system with time, influence of older structures, and anisotropy in mechanical properties of the crust. Another possibility is that the pattern is roughly analogous to the near-orthogonal pattern formed by oceanic ridges and transform faults, a pattern which has been explained as offering minimum resistance to plate separation (Lachenbruch & Thompson, 42). Oldenburg & Brune (51) dramatically reproduced the near-orthogonal oceanic pattern in a laboratory model with a thin crust of wax forming on molten wax, and Duffield (18) observed similar patterns forming on the solidified crust of a convecting lava lake.

The simple application of the minimum resistance theory to the Basin and Range province would suggest a series of northeast-trending grabens (normal to the spreading direction) and northwest-trending transform faults. The actual mechanics are more complex, and the faults commonly are hybrid, having components of

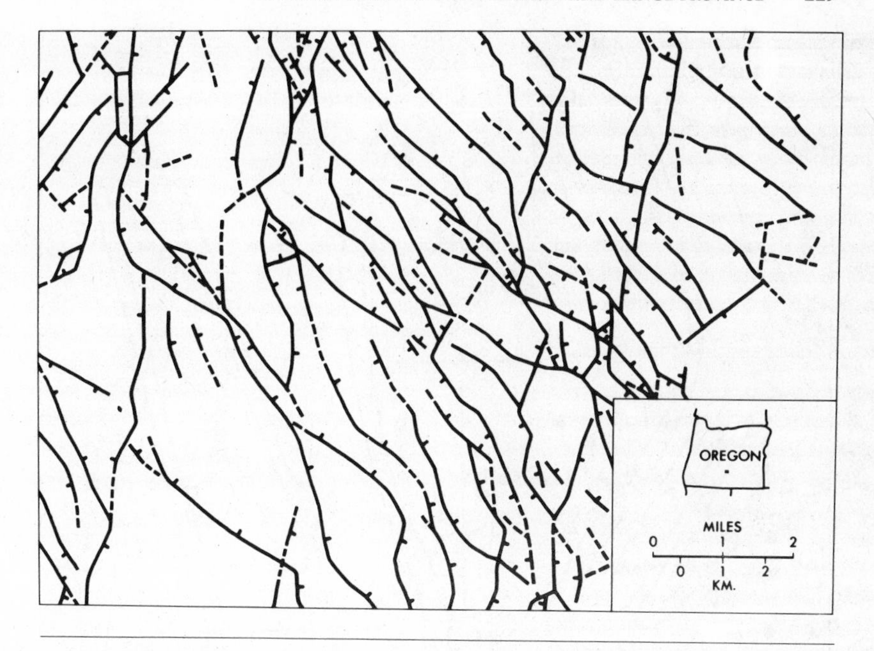

Figure 14 Rhomboid pattern of rupture expressed by late Cenozoic normal faults in south-central Oregon. Barbs on downthrown side of faults; faults dashed where inferred (from southeast portion of plate 3 of Donath, 17).

both dip- and strike-slip. The pattern is not simple and the question of the rupture pattern is far from resolved.

In addition to the problem of the pattern of faulting, the question of whether the normal faults systematically flatten with depth has been much debated, in part because such changes would imply greater regional extension. The seismic focal mechanisms lend no support to the notion of major decreases in dip, however, and serious geometric problems would ensue at the ends of basins if such decreases did occur. Therefore the low dipping to subhorizontal normal faults that have been observed in surface exposures and mine workings seem best ascribed to gravitational sliding and tilting in response to deeper primary faulting. The problem has been explored by Stewart (71) and Moore (48). Armstrong (3) interprets low-angle faults in eastern Nevada as gravitational sliding features of late Cenozoic age.

HEAT FLOW AND CRUSTAL TEMPERATURE

Regional Variation of Heat Flow

A region of anomalously high heat flow comprises the entire Basin and Range province and extends across the Columbia Plateau and part of the Rocky Mountain province (Figure 15). Heat flow values greater than 2 HFU [heat flow

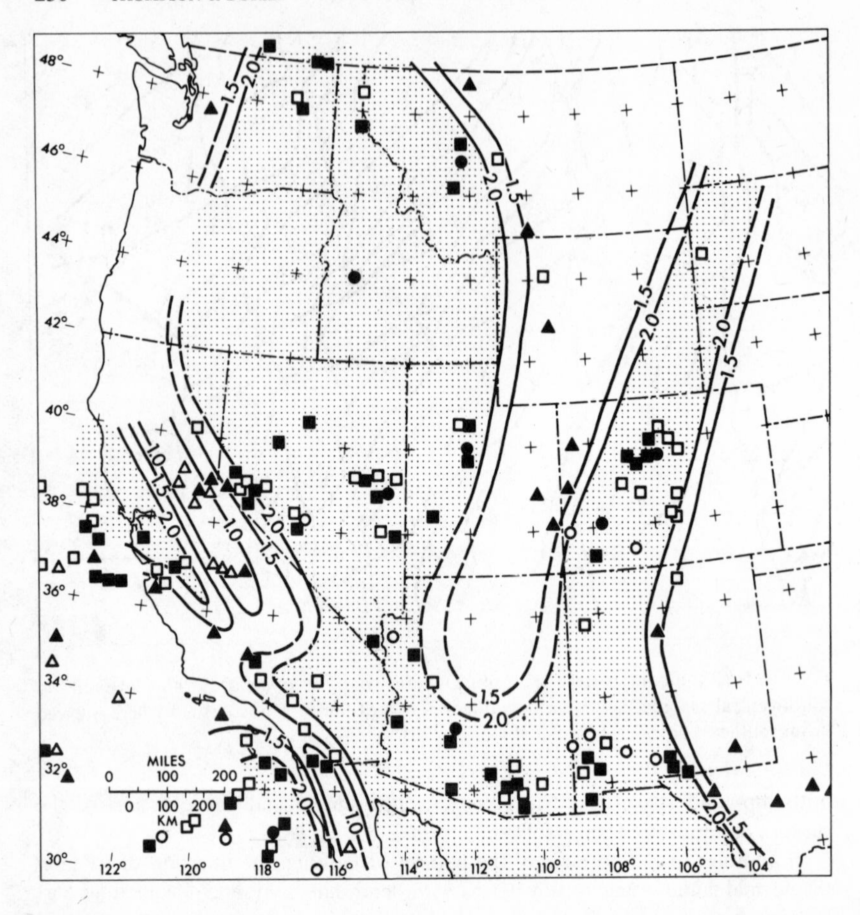

Figure 15 Contour map of heat flow. Contours in Heat Flow Units (μcal cm^{-2} sec^{-1}); dashed where extended on the basis of meager data. Data points shown as open triangles are measured heat flows in the range 0 to 0.99; solid triangles, 1.0 to 1.49; open squares, 1.5 to 1.99; solid squares, 2.0 to 2.49; open circles, 2.5 to 2.99; solid circles, 3.0 and larger (from Roy et al, 60).

units (HFU), μcal cm^{-2} sec^{-1}] characterize this broad region, in contrast to normal average values of about 1.5 HFU.

Although the Colorado Plateau is at least partly an area of normal heat flow, the distribution of measurements is inadequate to explore its boundaries with the Basin and Range province. The boundary with the Sierra Nevada appears to be surprisingly sharp.

Another compilation of the regional heat flow, by Sass and associates (63),

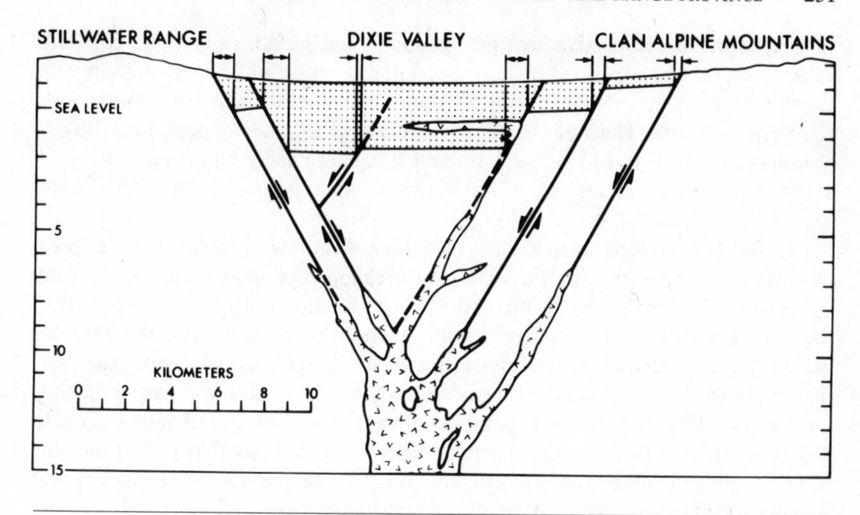

Figure 16 Cross section of Dixie Valley, Nevada. The subsurface structure to the depth of the sedimentary fill (stippled) is based on geophysical exploration. Dike at depth is hypothesized to accommodate surface extension, as shown by arrows (based on Burke, 8; Thompson, 76).

although more conservatively contoured, contains important additional details. One cluster of consistently high values (mostly above 3 HFU), the "Battle Mountain high" in northern Nevada, is interpreted as a transient effect of fairly recent crustal intrusion. To the south in Nevada, a cluster of values less than 1.5 HFU, the "Eureka low," is thought to be the result of unusual deep circulation of ground water. These examples emphasize the importance of nonconductive heat transfer. We point out that spreading of the grabens may be accompanied by intrusion of dikes at depths of a few kilometers (Thompson, 76), and these intrusions may be important in the heat transfer (Figure 16). The Battle Mountain high is on the projection of the active zone of spreading (historic fault breaks) at its north end (Figure 11).

The thermal transition to the Sierra Nevada may occur within a lateral distance of only 10 or 20 km (Sass et al, 63). If this proves to be the case, it will require shallow heat sources and will strengthen the hypothesis of intrusions beneath the grabens. Furthermore, present evidence suggests that the heat flow boundary with the Sierra Nevada follows in detail the irregular boundary of the normal faulting and not the generalized physiographic or topographic boundary.

Heat Production and the Linear Heat Flow Relation

A surprising and remarkably simple relationship has been found between heat flow and the heat production of surface rocks in plutonic areas; that is, within areas such as the Sierra Nevada and Basin and Range province, the heat flow varies linearly with the radioactive heat production at the surface (Roy et al, 60). This relationship is best explained by an exponential decrease of heat production with depth in the crust, combined with an additional flow of heat from the mantle (Lachenbruch, 41). The flow from the mantle—called the reduced heat flow—amounts to 1.4 ± 0.2 HFU in the Basin and Range province, compared to 0.8 ± 0.1 HFU in the United States east of the Rocky Mountains and only 0.4 HFU in the Sierra Nevada (Roy et al, 60).

Crustal temperature profiles for the three heat-flow provinces have been calculated by Lachenbruch (41), based on the exponential model. Temperatures at a depth of 30 km in the Basin and Range province range from 700–1000°C (depending on surface heat flow or heat production), as compared to 400–600°C in eastern United States. Temperatures in the Basin-Range crust may thus reach the melting range for granite, and temperatures in the upper mantle may reach melting for basalt. These high temperatures, combined with widespread late Cenozoic volcanism, form a basis for the generally accepted hypothesis that partial melting is responsible for the thin lithosphere and for the shallow, accentuated low velocity zone (asthenosphere) of the Basin and Range province.

The conductive model will need to be modified if much heat is carried into the crust by intrusions beneath spreading centers, as we have suggested (Figure 16).

Hot-spots and Mantle Plumes?

The Yellowstone volcanic region in northwestern Wyoming may represent a hot-spot above an upwelling convective plume in the mantle (Morgan, 49). According to Morgan's theory the North American lithosphere as a whole is moving west-southwest with respect to the mantle. The trail of the persistent Yellowstone hot-spot across its mantle plume would be marked by the older volcanics west-southwest of Yellowstone (in the Snake River part of the Columbia Plateau province). Other possible hot-spots have been suggested within the Basin and Range province.

A significant point about the theory should be kept in mind regarding the origin of the fault-block structures. If the Yellowstone plume is a driving mechanism for the structures and the lithosphere is moving westward across it, the locus of Basin-Range tectonic activity should be migrating eastward; and we know of no strong evidence for an eastward march of tectonic activity. Westward movement of the lithosphere at a rate on the order of 1 cm/yr would have produced a movement of 150 km in the 15 m.y. since the inception of Basin-Range faulting.

MAGNETIC AND ELECTRICAL ANOMALIES

Anomalies in the regional magnetic field and in electrical conductivity generally support other evidence of a hot upper mantle in the Basin and Range province, but the resolution of lateral variations has so far been very limited.

Zietz (88) showed that from the Sierra Nevada to the Rocky Mountains, magnetic anomalies are subdued in amplitude, and that long-wavelength anomalies are absent. This fact suggests that the lower crust and mantle may be above the Curie temperature (578° for magnetite). Surprisingly, the magnetic field over the Colorado Plateau does not appear to differ significantly from that over the Basin and Range province, in contrast to results from other kinds of studies.

Porath & Gough (53) explored variations in mantle electrical conductivity from the eastern and southern Basin and Range province to the Great Plains by measuring geomagnetic fluctuations. The anomalies are well represented by variations in depth to a half-space of conductivity 0.2 $(\text{ohm m})^{-1}$. The top of this conductor is inferred to correspond approximately with the 1500° isotherm. Depths to the surface of the conductor are 190 km under the Basin and Range province and 350 km under the Colorado Plateau, with a ridge of depth 120 km at the boundary. The depth under the Rio Grande trough is 120 km, that under the southern Rocky Mountains is 150 km, and that under the Great Plains is 350 km. Although such models are naturally not unique, they strengthen the interpretations of regional heat-flow variations and add another dimension to the unusual properties of the Basin and Range province.

PETROLOGIC RELATIONS

Three important relationships among the rocks deserve special emphasis:

1. Prior to Basin and Range faulting, lower and middle Cenozoic volcanoes erupted largely intermediate-composition rocks that become more alkalic toward the continental interior (Lipman et al, 43). This pattern is similar to volcanics now being erupted around the Pacific margin in association with convergent plate margins.

2. A major change to fundamentally basaltic volcanism (including bimodal mafic-silicic associations) took place during late Cenozoic time at about the inception of Basin-Range faulting (Christiansen & Lipman, 12). The transition to this new volcanism began in the southeastern part of the region and moved northwestward. The time of transition may be correlated with the initial intersection of the East Pacific Rise with the continental-margin trench system, an intersection which Atwater (5) also interprets as having progressed northwestward.

3. The composition of the crust and upper mantle as it existed beneath the Colorado Plateau prior to Basin-Range faulting has been ingeniously reconstructed from crystalline rock fragments in a breccia-filled diatreme, which is about 30 m.y. old (McGetchin & Silver, 45). The crust contained about 31% intermediate and acidic igneous rocks, 66% basic metaigneous rocks, and 3% eclogite. The upper mantle to a depth of about 100 km contained about 75% peridotite and pyroxenite and 25% eclogite. It is especially interesting that the mantle 30 m.y. ago contained this much eclogite, because eclogite is capable of converting into gabbro with a volume expansion of about 10% in response to a rise in temperature or decrease in pressure.

Eclogite may be a key to an understanding of late Cenozoic uplift of the broad region that includes the Sierra Nevada, Basin and Range province, and Colorado Plateau. The expansion of eclogite in only 60 km of mantle could produce an uplift of 1.5 km (60 × 25% × 10%). The former eclogite may now be represented by gabbro dispersed in the mantle low velocity zone, or by crustal additions of basic metaigneous rock, or by basaltic volcanics.

SYNTHESIS AND TECTONIC MODEL

The regional geophysical data put many useful constraints on speculations about the fundamental tectonic processes of the Basin and Range province. Among these data the heat flow is central; the volcanism, thin crust, low mantle velocity, accentuated low velocity zone, generally high elevation, subdued magnetic anomalies, high electrical conductivity, and great breadth of the seismically active zone can logically be associated with high temperatures and high heat flow.

The gravity data—coupled with the estimated extension—supply an interesting constraint that does not seem to have been widely recognized (Thompson, 77). If a 30 km crustal plate were simply attenuated by a horizontal extension of 10%, a negative isostatic anomaly of more than 300 mgal would be produced. If the attenuated plate were only 10 km thick the anomaly would still be 100 mgal. Because the regional isostatic anomalies average no more than about 10 mgal, the gravity emphatically indicates that the circuits of mass flow must be closed. Near-surface crustal spreading is almost perfectly matched by lateral backflow in the mantle.

If we imagine a vertical fence surrounding the Basin and Range province and extending through the crust and mantle, the integrated flux of mass through the fence must be zero, despite the outward flow by extension in the upper crust. We now need to find out how the deep lateral inflow takes place. Is the lateral flow in the low velocity zone? Is it a deeper mantle flow associated with narrow upwelling convective plumes, analogous to a thunderhead in the atmosphere? Is the flow related to former subduction of an oceanic lithospheric plate at the continental margin? At present these questions lead rather quickly into speculation.

The regional geophysical characteristics, geologic history, and petrology rather strongly suggest a link with plate-tectonic interactions at the western edge of the continent going back to early Cenozoic time. Analogies with spreading marginal basins of the western Pacific are especially promising (Karig, 40; Matsuda & Uyeda, 47; Scholz et al, 64; Sleep & Toksöz, 68; Thompson, 77; Uyeda & Miyashiro, 80).

The general idea is that in a broad belt on the continental side of an arc-trench system, a descending lithospheric plate either generates magma along its upper surface or creates a convecting subcell by viscous drag. The rising magma or convection current helps to move the arc away from the continent, creating a spreading marginal basin. The situation is somewhat different along the central coast of North America in that subduction ceased when a spreading ridge reached the trench in middle Cenozoic time. But because the descending young lithosphere would still be very hot when the ridge reached the trench, and because conductive heat transfer is very slow, it is easy to imagine that the thermal effects of a past subduction process are still being felt in the Basin and Range province.

ACKNOWLEDGMENTS

We thank David Warren for allowing us to use his unpublished map of crustal thickness[1]; John Lahr and Peter Stevenson for their map of earthquake epicenters; and Kimberly Bailey, James Baxter, Robert Daniel, Allan Lindh, Dohn K. Riley, Don C. Riley, and Donald Steeples for stimulating discussions in a seminar which ranged widely over many of the topics discussed here.

Literature Cited

1. Aldrich, L. T. et al 1960. The earth's crust. *Carnegie Inst. Washington Yearb.* 1959–1960: 202–8
2. Archambeau, C. B., Flinn, E. A., Lambert, D. G. 1969. Fine structure of the upper mantle. *J. Geophys. Res.* 74: 5825–65
3. Armstrong, R. L. 1972. Low-angle (denudation) faults, hinterland of the Sevier orogenic belt, eastern Nevada and western Utah. *Geol. Soc. Am. Bull.* 83: 1729–54
4. Asada, T. et al 1961. See Ref. 1, 1960–1961: 244–50
5. Atwater, T. 1970. Implications of plate tectonics for the Cenozoic tectonic evolution of western North America. *Geol. Soc. Am. Bull.* 81: 3513–36
6. Berg, J. W., Cook, K. L., Narans, H. D., Dolan, W. M. 1960. Seismic investigation of crustal structure in the eastern part of the Basin and Range province. *Seismol. Soc. Am. Bull.* 50: 511–36
7. Berg, J. W. et al 1966. Crustal refraction profile, Oregon Coast Range. *Seismol. Soc. Am. Bull.* 56: 1357–62
8. Burke, D. B. 1967. Aerial photograph survey of Dixie Valley, Nevada. In *Geophysical study of Basin-Range structure, Dixie Valley region, Nevada.* US Air Force Cambridge Res. Lab. Spec. Rep., 66–848
9. Burke, D. B., McKee, E. H. 1973. Mid-Cenozoic volcano-tectonic features in central Nevada. *Geol. Soc. Am. Abstr. with Programs* 5: 18
10. Carder, D. S. 1973. Trans-California seismic profile, Death Valley to Monterey Bay. *Seismol. Soc. Am. Bull.* 63: 571–86
11. Carder, D. S., Qamar, A., McEvilly, T. V. 1970. Trans-California seismic profile, Pahute Mesa to San Francisco Bay. *Seismol. Soc. Am. Bull.* 60: 1829–46
12. Christiansen, R. L., Lipman, P. W. 1972. Cenozoic volcanism and plate-tectonic evolution of the western United States. II Late Cenozoic. *Phil. Trans. Roy. Soc. London Ser. A* 271: 249–84
13. Cook, K. L. 1969. Active rift system in the Basin and Range province. *Tectonophysics* 8: 469–511
14. Crittenden, M. D. Jr. 1963. Effective viscosity of the earth derived from isostatic loading of Pleistocene Lake Bonneville. *J. Geophys. Res.* 68: 5517–30
15. Crowder, D. F., McKee, E. H., Ross, D. C., Krauskopf, K. B. 1973. Granitic rocks of the White Mountains area, California-Nevada: age and regional significance. *Geol. Soc. Am. Bull.* 84: 285–96
16. Diment, W. H., Stewart, S. W., Roller, J. C. 1961. Crustal structure from the Nevada Test Site to Kingman, Arizona, from seismic and gravity observations. *J. Geophys. Res.* 66: 201–14
17. Donath, F. A. 1962. Analysis of Basin-Range structure, south-central Oregon. *Geol. Soc. Am. Bull.* 73: 1–16
18. Duffield, W. A. 1972. A naturally occurring model of global plate tectonics. *J. Geophys. Res.* 77: 2543–55
19. Eaton, J. P. 1963. Crustal structure from San Francisco, California, to Eureka, Nevada, from seismic-refraction measurements. *J. Geophys. Res.* 68: 5789–5806
20. Eaton, J. P. 1966. Crustal structure in northern and central California from seismic evidence. In Geology of northern California, ed. E. H. Bailey. *Calif. Div. Mines Geol. Bull., 190,* 419–26
21. Fenneman, N. M. 1946. Physical divisions of the United States. *US Geol. Surv. Map* (1: 7,000,000 scale)
22. Gibbs, J. F., Roller, J. C. 1966. Crustal structure determined by seismic-refraction measurements between the Nevada Test Site and Ludlow, Cali-

[1] Warren, D. H., Healy, J. H. Structure of the crust in the conterminous United States. *Tectonophysics.* In press.

fornia. *US Geol. Surv. Prof. Pap. 550-D*, D125–31
23. Gilbert, G. K. 1874. *US Geogr. Geol. Surv. W. 100th Meridian Progr. Rep.*
24. Gilbert, G. K. 1875. *US Geogr. Geol. Surv. W. 100th Meridian*, Rep. 3
25. Gilbert, G. K. 1890. Lake Bonneville. *US Geol. Surv. Monogr. 1*, 438 pp.
26. Gilluly, J. 1963. The tectonic evolution of the western United States. *Quart. J. Geol. Soc. London* 119:133–74
27. Gilluly, J. 1970. Crustal deformation in the western United States. In *The Megatectonics of Continents and Oceans*, ed. H. Johnson, B. L. Smith, 47–73. New Jersey: Rutgers Univ. Press
28. Gilluly, J. 1972. Tectonics involved in the evolution of mountain ranges. In *The Nature of the Solid Earth*, ed. E. Robertson, 406–39. New York: McGraw. 677 pp.
29. Hamilton, W., Myers, W. B. 1966. Cenozoic tectonics of the western United States. *Rev. Geophys.* 4:509–49
30. Healy, J. H. 1963. Crustal structure along the coast of California from seismic-refraction measurements. *J. Geophys. Res.* 68:5777–87
31. Helmberger, D. V. 1972. Long period body wave propagation from 4° to 13°. *Seismol. Soc. Am. Bull.* 62:325–41
32. Herrin, E. 1972. A comparative study of upper mantle models: Canadian Shield and Basin and Range province. See Ref. 28, 216–34
33. Herrin, E., Taggart, J. 1962. Regional variations in *Pn* velocity and their effect on the location of epicenters. *Seismol. Soc. Am. Bull.* 52:1037–46
34. Hill, D. P. 1963. Gravity and crustal structure in the western Snake River Plain, Idaho. *J. Geophys. Res.* 68: 5807–19
35. Hill, D. P., Pakiser, L. C. 1966. Crustal structure between the Nevada Test Site and Boise, Idaho, from seismic-refraction measurements. In the earth beneath the continents, ed. J. S. Steinhart, T. J. Smith. *Geophys. Monogr. 10*, Am. Geophys. Union, Washington DC, 391–419
36. Jackson, W. H., Pakiser, L. C. 1965. Seismic study of crustal structure in the southern Rocky Mountains. In Geological Survey Research 1965. *US Geol. Surv. Prof. Pap. 525-D*, D85–92
37. Jackson, W. H., Stewart, S. W., Pakiser, L. C. 1963. Crustal structure in eastern Colorado from seismic-refraction measurements. *J. Geophys. Res.* 68:5767–76
38. Johnson, L. R. 1965. Crustal structure between Lake Mead, Nevada, and Mono Lake, California. *J. Geophys. Res.* 70:2863–72
39. Johnson, S. H., Couch, R. W. 1970. Crustal structure in the North Cascade Mountains of Washington and British Columbia from seismic refraction measurements. *Seismol. Soc. Am. Bull.* 60:1259–69
40. Karig, D. E. 1971. Origin and development of marginal basins in the western Pacific. *J. Geophys. Res.* 71:2542–61
41. Lachenbruch, A. H. 1970. Crustal temperature and heat production: implications of the linear heat-flow relation. *J. Geophys. Res.* 75:3291–3300
42. Lachenbruch, A. H., Thompson, G. A. 1972. Oceanic ridges and transform faults: their intersection angles and resistance to plate motion. *Earth Planet. Sci. Lett.* 15:116–22
43. Lipman, P. W., Prostka, H. J., Christiansen, R. L. 1972. Cenozoic volcanism and plate-tectonic evolution of the western United States. I. Early and middle Cenozoic. *Phil. Trans. Roy. Soc. London Ser. A* 271:217–48
44. McCamy, K., Meyer, R. P. 1964. A correlation method of apparent velocity measurement. *J. Geophys. Res.* 69:691–99
45. McGetchin, T. R., Silver, L. T. 1972. A crustal-upper mantle model for the Colorado Plateau based on observations of crystalline rock fragments in the Moses Rock dike. *J. Geophys. Res.* 77:7022–37
46. McKee, E. H. 1971. Tertiary igneous chronology of the Great Basin of western United States—implications for tectonic models. *Geol. Soc. Am. Bull.* 82:3497–3502
47. Matsuda, T., Uyeda, S. 1971. On the Pacific-type orogeny and its model—extension of the paired belts concept and possible origin of marginal seas. *Tectonophysics* 11:5–27
48. Moore, J. G. 1960. Curvature of normal faults in the Basin and Range province of the western United States. In Geological Survey Research, 1960. *US Geol. Surv. Prof. Pap. 400-B.* B409–11
49. Morgan, W. J. 1972. Deep mantle convection plumes and plate motions. *Am. Assoc. Petrol. Geol. Bull.* 56:203–13
50. Noble, D. C. 1972. Some observations on the Cenozoic volcano-tectonic evolution of the Great Basin, western United States. *Earth Planet. Sci. Lett.* 17:142–50
51. Oldenburg, D. W., Brune, J. N. 1972. Ridge transform fault spreading in freezing wax. *Science* 178:301–4
52. Pakiser, L. C. 1963. Structure of the crust

and upper mantle in the western United States. *J. Geophys. Res.* 68:5747–56
53. Porath, H., Gough, D. I. 1971. Mantle conductive structures in the western United States from magnetometer array studies. *Geophys. J. Roy. Astron. Soc.* 22:261–75
54. Press, F. 1960. Crustal structures in the California-Nevada region. *J. Geophys. Res.* 65:1809–14
55. Prodehl, C. 1970. Crustal structure of the western United States from seismic-refraction measurements in comparison with central European results. *Z. Geophys.* 36:477–500
56. Robinson, R., Kovach, R. L. 1972. Shear wave velocities in the earth's mantle. *Phys. Earth Planet. Interiors* 5:30–44
57. Roller, J. C. 1964. Crustal structure in the vicinity of Las Vegas, Nevada, from seismic and gravity observations. *US Geol. Surv. Prof. Pap. 475-D.* D108–11
58. Roller, J. C. 1965. Crustal structure in the eastern Colorado Plateaus province from seismic-refraction measurements. *Seismol. Soc. Am. Bull.* 55:107–19
59. Roller, J. C., Healy, J. H. 1963. Seismic-refraction measurements of crustal structure between Santa Monica Bay and Lake Mead. *J. Geophys. Res.* 68:5837–48
60. Roy, R. F., Blackwell, D. D., Decker, E. R. 1972. Continental heat flow. In *The Nature of the Solid Earth,* ed. E. Robertson, 506–43. New York: McGraw. 677 pp.
61. Ryall, A., Slemmons, D. B., Gedney, L. D. 1966. Seismicity, tectonism, and surface faulting in the western United States during historic time. *Seismol. Soc. Am. Bull.* 56:1105–36
62. Ryall, A., Stuart, D. J. 1963. Travel times and amplitudes from nuclear explosions, Nevada Test Site to Ordway, Colorado. *J. Geophys. Res.* 68:5821–35
63. Sass, J. H., Lachenbruch, A. H., Monroe, R. J., Greene, G. W., Moses, T. H. Jr. 1971. Heat flow in the western United States. *J. Geophys. Res.* 76:6376–6413
64. Scholz, C. H., Barazangi, M., Sbar, M. L. 1971. Late Cenozoic evolution of the Great Basin, western United States, as an ensialic interarc basin. *Geol. Soc. Am. Bull.* 82:2979–90
65. Sclater, J. G., Francheteau, T. 1970. The implications of terrestrial heat flow observations on current tectonic and geochemical models of the crust and upper mantle of the earth. *Geophys. J. Roy. Astron. Soc.* 20:509–42
66. Shor, G. G. Jr. 1955. Deep reflections from southern California blasts. *Trans. Am. Geophys. Union* 36:133–138
67. Shor, G. G., Raitt, R. W. 1958. Seismic studies in the southern California continental borderland. In *Tomo 2 Geof. Aplicada: Int. Geol. Congr. 20th., Mexico, D. F., 1956 (Trabajos), Sect. 9,* 243–59
68. Sleep, N. H., Toksöz, M. N. 1971. Evolution of marginal basins. *Nature* 233:548–50
69. Slemmons, D. B. 1967. Pliocene and Quarternary crustal movements of the Basin and Range province, USA. *J. Geosci. Osaka City Univ.* 10:91–103
70. Steinhart, J. S., Meyer, R. P. 1961. Explosion studies of continental structure. Univ. of Wisconsin, 1956–1959. *Carnegie Inst. Wash. Publ. 622.* 409 pp.
71. Stewart, J. H. 1971. Basin and Range structure: a system of horsts and grabens produced by deep-seated extension. *Geol. Soc. Am. Bull.* 82:1019–44
72. Stewart, S. W. 1968. Preliminary comparison of seismic travel times and inferred crustal structure adjacent to the San Andreas fault in the Diablo and Gabilan Ranges of central California. In *Proc. Conf. Geol. Prob. San Andreas Fault System,* ed. W. R. Dickinson, A. Grantz, 218–30. Stanford, California: Stanford Univ.
73. Stewart, S. W., Pakiser, L. C. 1962. Crustal structure in eastern New Mexico interpreted from the GNOME explosion. *Seismol. Soc. Am. Bull.* 52:1017–30
74. Tatel, H. E., Tuve, M. A. 1955. Seismic exploration of a continental crust. *Geol. Soc. Am. Spec. Pap.* 62:35–50
75. Thompson, G. A. 1959. Gravity measurements between Hazen and Austin, Nevada, a study of Basin-Range structure. *J. Geophys. Res.* 64:217–30
76. Thompson, G. A. 1966. The rift system of the western United States. In The world rift system, ed. T. N. Irvine. *Geol. Surv. Can. Dep. Mines Tech. Surv. Pap.* 66-14:280–90
77. Thompson, G. A. 1971. Cenozoic Basin Range tectonism in relation to deep structure. *Geol. Soc. Am. Abstr.* 3(2):209. 1972 *Proc. Int. Geol. Congr., 24th Sect.* 3:84–90
78. Thompson, G. A., Burke, D. B. 1973. Rate and direction of spreading in Dixie Valley, Basin and Range province, Nevada. *Geol. Soc. Am. Bull.* 84:627–32
79. Thompson, G. A., Talwani, M. 1964. Crustal structure from Pacific basin to central Nevada. *J. Geophys. Res.* 68:4813–37

80. Uyeda, S., Miyashiro, A. Plate tectonics and Japanese Islands. Unpublished
81. Walcott, R. I. 1970. Flexural rigidity, thickness and viscosity of the lithosphere. *J. Geophys. Res.* 75:3941–54
82. Warren, D. H. 1969. A seismic-refraction survey of crustal structure in central Arizona. *Geol. Soc. Am. Bull.* 80:257–82
83. Warren, D. H., Healy, J. H., Bohn, J., Marshall. P. A. 1973. Crustal structure under LASA from seismic refraction measurements. *J. Geophys. Res.* 78: 8721–34
84. Warren, D. H., Jackson, W. H. 1968. Surface seismic measurements of the project GASBUGGY explosion at intermediate distance ranges. *US Geol. Surv. Open File Rep.* 45 pp.
85. Willden, R. 1965. Seismic-refraction measurements of crustal structure between American Falls Reservoir, Idaho, and Flaming Gorge Reservoir, Utah, *US Geol. Surv. Prof. Pap. 525-C.* C44–50
86. Woollard, G. P. 1972. Regional variations in gravity. See Ref. 28, 463–505
87. York, J. E., Helmberger, D. V. 1973. Low-velocity zone variations in the southwestern United States. *J. Geophys. Res.* 78:1883–86
88. Zietz, I. 1969. Aeromagnetic investigations of the earth's crust in the United States. In The Earth's Crust and Upper Mantle, ed. P. J. Hart, 404–15, *Am. Geophys. Union Monogr. 13.* 735 pp.

CLAYS AS CATALYSTS FOR NATURAL PROCESSES

J. J. Fripiat and M. I. Cruz-Cumplido
Department of Geology, University of Illinois, Urbana, Illinois 61801

INTRODUCTION

Clay minerals are important constituents of the earth's crust not only because of their abundance but merely because of their chemical activity. Clays found in soils and in sediments are distributed among three main mineralogical groups: the kaolin, hydrated micas, and smectite groups. Allophanes derived from volcanic glasses and amorphous Al-Si or Fe-Si mixed oxide gels are also abundant in large areas. These clays and clay-like materials share two common characteristics: a high specific surface area and the presence of exposed cations on their surface. In the natural environment these cations are either Na^+, K^+, Ca^{2+}, or Mg^{2+}, and/or polynucleic Al or Fe basic cations such as $[Al(OH^-)_x(H_2O)_y]_n^{m+}$ or $[Fe(OH^-)_x(H_2O)_y]_n^{m+}$. They balance the excess negative charge of the lattice, generated by isomorphic substitutions (for instance substitution of Si^{4+} by Al^{3+} in tetrahedral position, or Al^{3+} by Mg^{2+} in octahedral position).

A clay microcrystal may thus be considered as the salt of a weak acid, whose anion (i.e. the lattice) has an infinite radius of curvature. In this situation the electrical charges of the cations generate a high electrostatic field because of the very incomplete screening of the positive charges by the anionic lattice charges. Since the lattice charges have fixed positions, the electrical neutrality is easily achieved by monovalent alkali cations, while for divalent cations the situation is much less favorable.

Oxygen atoms or hydroxyl groups are the main surface constituents. They may play an important role by forming hydrogen bonds with proton donor or acceptor molecules. In addition, van der Waals (or dispersion) forces are expected to be quite active because of the magnitude of the surface area. In the natural environment a variable number of layers of water molecules but also polar organic molecules do form an adsorbed phase in which many kinds of chemical transformations may occur. The surface constituents and the adsorbed water may show a high catalytic activity in these transformations. The aim of this contribution is to review the recent progress of our knowledge in this field.

After a paragraph in which the surface properties of clays and clay-like minerals are recalled, the role of clays in the origin of petroleum, in the origin of life, and in chemical reactions in soils is examined. To remain within the limits assigned to this review, the catalytic properties of clays as used in some industrial processes are not considered, in spite of the fact that deep similarities with mechanisms observed in natural processes are often apparent.

Our main goal is to show that it is possible to classify into a very restricted number of mechanisms, most of the observations reported in the three areas that we intend to cover and that these mechanisms may be predicted on the basis of the fundamental properties of clay surfaces.

In this review there are important omissions, the most important probably being that we do not mention the interactions between clays and microorganisms. The increasing interest attached to enzyme engineering, as outlined by Wingard (42), and the use of clays for enzyme immobilization will be likely to contribute to the increase of our knowledge in this controversial domain.

SURFACE PROPERTIES OF CLAYS

In many instances clay minerals behave as acid catalysts. This aspect of their reactivity originates from the strong electrical field of the charge balancing cations and of the action of this field on the water molecules in the cation hydration shell: The higher the electrical charge of the cation, the smaller its radius and the stronger the field. Divalent cations promote a stronger surface acidity than monovalent cations, as shown originally by Mortland et al (28). Trivalent cations, because of their facile hydrolysis, are still more active, as emphasized by Solomon and co-workers (38) and by Hawthorne & Solomon (22).

To understand this mechanism, consider the simplest case of a Wyoming Na-bentonite in the dehydrated state. According to Pezerat & Mering (31) the exchangeable cations are slightly embedded in the hexagonal holes in the 001 lattice oxygen planes and this position is maintained when a monolayer of water molecules is adsorbed between the sheets. In this situation the water molecules are tetrahedrally distributed as revealed by the electron density distribution along the normal to the 001 planes. There are at least two kinds of environment in such a distribution: 1. water molecules in contact with the Na^+ cations and sharing one of their lone pairs of electrons with an emptied orbital of these cations and 2. those not in direct contact with the cations. The latter are hydrogen-bonded to each other. Water molecules in this situation are represented by different symbols in the following reactions.

$$(H_2O)+[Na^+(H_2O)_n] \leftrightarrows [Na^+OH^-(H_2O)_{n-1}]+(H_3O^+) \qquad 1.$$
$$(H_3O^+)+(M) \leftrightarrows (MH^+)+(H_2O) \qquad 2.$$
$$(MH^+) \rightarrow M'+H^+ \qquad 3.$$
$$H^+ +(H_2O) \rightarrow (H_3O^+) \qquad 4.$$
$$(H_3O^+)+(H_2O) \rightarrow (H_2O)+(H_3O^+) \qquad 5.$$
$$(H_2O)+[Na^+(H_2O)_{n-1}M] \rightarrow [Na^+OH^-(H_2O)_{n-1}]+(MH^+) \qquad 6.$$

The parenthesis indicates the adsorbed species which are outside the coordination shell, while the brackets stand for this coordination shell. According to Calvet (5), the H_2O-cation binding energy would be of the order of 19 kcal $mole^{-1}$ while the water molecules outside the coordination sphere would be linked to each other with an energy of the order of 10 kcal $mole^{-1}$. The protons attached to molecules outside or inside the coordination shell exchange rapidly and it has been shown by Touillaux et al (40) that the frequency of these proton jumps at 248°K is approximately 10^{10} sec^{-1}. Such a high figure has to be accounted for by a H_3O^+/H_2O ratio of the order of 10^{-2}–10^{-3}, i.e. approximately six orders of magnitude higher than in bulk neutral water. Reaction scheme 1 suggests the probable origin of such an enhancement of the degree of dissociation. Under the influence of the cationic electrostatic field, a water molecule in the hydration sphere is polarized to such an extent that it dissociates, leaving an hydroxyl anion in close approach to the cation while an hydronium cation is formed outside this shell. This hydronium exchanges protons with surrounding water molecules and eventually with the OH^- in the hydration shell.

This picture can change deeply if another type of molecule is present. Let M be such a molecule and assume that M may compete with H_2O either to occupy a ligand position in the cation coordination shell or any position in the adsorbed phase. This may be the case for instance if M is an electron donor. M gets protonated either through reaction 2, i.e. by the action of H_3O^+, or through reaction 6, i.e. when in the hydration shell. Depending on the chemical nature of M, the protonated form may be stable or unstable. If MH^+ is stable, proton transfers 2 or 6 are reversible while if MH^+ is unstable, the transformation shown by reaction 3 occurs and the proton reenters the cycle shown by reactions 1, 2, or 6 through reaction 4. The net result is the catalytic transformation of M into another chemical species M′, the reaction intermediate being some form of carbonium cation, MH^+. Some transformations occurring probably through carbonium intermediates are shown by reactions 8 and 16.

The high acidity of adsorbed water and an eventual effect of the cation polarizing field on M are responsible therefore for this transformation. In opposition to the usual industrial use of acid catalysts, the mechanism depicted above is a low temperature catalytic process since the presence of water is essential and an increase in temperature, let us say, above 300°C would remove most of the adsorbed water. The situation may, however, be different under hydrothermal conditions in the presence of a high water vapor pressure.

Entirely different catalytic processes may be initiated on clay surfaces by charge transfers between the adsorbed species and either polyvalent exchangeable cations and/or oxidizing or reducing cations in lattice positions. Very interesting surface charge-transfer complexes have been described by Doner & Mortland (13) and Mortland & Pinnavia (29, 32). Benzene, toluene, and other methyl-substituted benzenes coordinate through π electrons to the exchangeable copper II cations on the surface of montmorillonite. Two types of such complexes have been observed. In type I, the aromatic ring remains planar and aromaticity is retained on coordination. In type II, the ring is distorted and aromaticity is lost. Benzene forms types I and II complexes while toluene and other methyl-substituted benzenes form type I complexes only.

The degree of hydration of Cu^{2+} in type I is higher than in type II but the two kinds of complexes (with benzene) can be interconverted by adding or removing controlled amounts of water. Heterocycles form similar complexes and as an example of this, it has been shown recently by Cloos & Van de Poel (9) that thiopene forms also a charge transfer complex with copper II on montmorillonite.

Aromatic molecules forming type II complexes might be expected to show a chemical reactivity different from that known in homogeneous medium. More generally speaking such a change in reactivity could be anticipated each time the electronic configuration of a molecule is modified by its introduction into the coordination shell of an exchangeable cation.

Charge transfers may also end up in some oxidation or reduction processes. This may result in an irreversible modification of the surface and such a process cannot therefore be considered as catalytic. Because in the natural environment the amount of catalyst is in general in large excess with respect to the amount of compounds to be transformed, a noncatalytic process of this kind may well be implied in the transformation of huge amounts of materials. Very little is known on the redox properties of cations on lattice sites. It has been shown by Miller et al (27) that ferric cation in octahedral position is responsible for the oxidizing properties of attapulgite clay, but more research would be needed in this area. Exchangeable cations may also confer redox properties to a surface as illustrated by the mechanism of the cobalt III hexammine decomposition, studied by Fripiat & Helsen (16). This complex, which is very stable in aqueous solution, is readily destroyed under the action of residual water at room temperature when adsorbed on montmorillonite. Cobalt III is reduced into cobalt II hydroxide; NH_4^+ replaces the original complex cation on the exchange positions; excess NH_3 escapes in the gas phase and a part of it is transformed into N_2, along with the reduction of Co III.

Finally it should be also mentioned that, on the surface of layer lattice silicates, the orientation of the adsorbed species may be in some instances perfectly defined. A high degree of preferential orientation may enhance or deplete the chemical reactivity. These secondary effects might be important, for instance, in surface polymerization reactions.

CLAYS AND THE ORIGIN OF PETROLEUM

The role played by clays in the reactions leading to the transformation of organic residues into petroleum has been studied and discussed for a long time. However, it is relatively recent that some simple reactions representing important steps in these transformations have been shown to be catalyzed by clay surfaces.

Shimoyama & Johns (23, 37) have recently investigated the chain of transformations of fatty acids since it is well known that decarboxylation of these acids yields alkanes. In addition, it has become clear that some catalytic action must be invoked to obtain significant chemical transformations within reasonable geological times at temperatures prevailing during burial diagenesis. Experimental models using montmorillonite to promote degradation of fatty acids in alkane have shown that two major reactions occur sequentially:

1. Catalytic decarboxylation of fatty acids, producing alkanes

$$CH_3\text{–}(CH_2)_n\text{–}C{\overset{\displaystyle O}{\underset{\displaystyle OH}{\lessgtr}}} \rightarrow CH_3\text{–}(CH_2)_{n-1}\text{–}CH_3 + CO_2 \qquad 7.$$

and 2. catalytic cracking of these alkanes, producing a spectrum of shorter chain alkanes and carbonaceous residues. For instance, it might be suggested that a reaction like the following occurs:

$$CH_3\text{–}(CH_2)_{n-1}\text{–}CH_3 \rightarrow CH_3\text{–}(CH_2)_x\text{–}CH_3 + CH_3\text{–}(CH_2)_y\text{–}CH_3 + C \qquad 8.$$

where $x + y = n - 4$, or also that some olefins are produced

$$CH_3\text{–}(CH_2)_{n-1}\text{–}CH_3 \rightarrow CH_3\text{–}CH = CH\text{–}(CH_2)_x\text{–}CH_3 + CH_3\text{–}(CH_2)_y\text{–}CH_3 \qquad 9.$$

where $x + y = n - 4$.

Johns and co-workers have studied the cracking of a C_{21} alkane and got an activation energy of 46.5 kcal $mole^{-1}$. This value is lower than the theoretical result expected for cracking a single C–C bond, namely 57–58 kcal $mole^{-1}$. No experimental data about the decarboxylation of the docosanoic (C_{22}) acid are available. As far as we are concerned, we do not believe that these data permit one to conclude unambiguously to a real catalytic activity of clay minerals. Johns and co-workers consider that decarboxylation should be catalyzed by Lewis sites at exposed Al^{3+} or Fe^{3+} sites at the edges of montmorillonite crystallites while cracking could occur through a carbonium type mechanism, catalyzed by the acidity of residual water. The first hypothesis does not seem to be correct because decarboxylation occurs in the presence of enough water to fill up the Al^{3+} or Fe^{3+} hydration shell. It may be more appropriate to suggest that the Al^{3+} or Fe^{3+} basic salts of the long-chain fatty acids decarboxylate more easily than the acids.

In spite of some controversial arguments, the sequence of events that may occur during the burial diagenesis of a montmorillonite sediment and its adsorbed fatty acids, as derived by Johns and co-workers from their theoretical model is quite enlightening. According to them two distinct zones can be defined. In the first one (0–1000 m) excess interlayer water is expelled through the compaction of the mud and decarboxylation occurs, producing alkanes (reaction 7). Temperature ranges between 30 and 65°C. In the second zone (1000–3000 m) cracking of alkanes occurs, by some overall transformation of the types depicted by reactions 8 or 9. Temperatures in this zone are between 65 and 160°C. Simultaneously, montmorillonite would be transformed into illite. The connection between alkane cracking and the illitization of montmorillonite is not clear to us. It tends eventually to show that cracking is not a true catalytic process since the catalyst is destroyed.

Beside fatty acids, the possible contribution of amines to the formation of petroleum has also been examined recently. This possibility is suggested by Chaussidon & Calvet (7) who showed that in the adsorbed state on montmorillonite surfaces, alkylammonium cations are partially transformed at temperatures considerably below the decomposition temperatures of amines and of their chlorhydrates, and that some hydrocarbons result from these transformations. Durand et al (14) have reexamined extensively these experiments in an inert atmosphere and also in the presence of oxygen with much longer time periods. Fripiat & Lambert-Helsen (18) still more recently have extended these studies to alkylammonium cations adsorbed by near-faujasite molecular sieves. According to the experimental data reported in the last two references, the main chemical transformations in an inert atmosphere may be described as transalkylation reactions. Undoubtedly surface acidity is in both cases at the origin of the catalytic action. In the clay systems no transformation is observed in absence of water, at least in the temperature range investigated here, namely below 220°C. In the zeolite systems the surface acidity is provided by the incorporation of the proton of the alkylammonium into the oxygen network, so water is not required. Otherwise, the reactions in these two systems are very similar and only those operating on the surface of the clay will be reported here (reactions 10–18).

A surface hydronium cation reacts with an alkyl chain, producing the active intermediate, i.e. an oxonium ion, as shown in reaction 10. Simultaneously an NH_4^+ cation, balancing the lattice charge liberated by the removal of the alkylammonium, appears on the exchange site. Similar mechanisms may be written for di-, tri-, and tetra-alkylammonium (reactions 11–13).

Monoalkylamines

$$M^{-+}NH_3R + H_3O^+ \rightleftarrows M^{-+}NH_4 + RH_2O^+ \qquad 10.$$

Dialkylamines

$$M^{-+}NH_2R_2 + H_3O^+ \rightleftarrows M^{-+}NH_3R + RH_2O^+ \qquad 11.$$

Trialkylamines

$$M^{-+}NHR_3 + H_3O^+ \rightleftarrows M^{-+}NH_2R_2 + RH_2O^+ \qquad 12.$$

Quaternary ammoniums

$$M^{-+}R_4 + H_3O^+ \rightleftarrows M^{-+}NHR_3 + RH_2O^+ \qquad 13.$$

Alcohols

$$ROH \pm H_3O^+ \rightleftarrows H_2O + RH_2O^+ \qquad 14.$$

Ethers

$$R_2O + H_3O^+ \rightleftarrows ROH + RH_2O^+ \qquad 15.$$

Unsaturated hydrocarbons

$$(R\text{–}H) + H_3O^+ \rightleftarrows RH_2O^+ \qquad 16.$$

Saturated hydrocarbons

$$RH_2O^+ + O_2 \rightarrow H_3O^+ + R'\text{–}C\begin{smallmatrix} \nearrow O \\ \searrow OH \end{smallmatrix} \qquad 17.$$

$$R'\text{–}C\begin{smallmatrix} \nearrow O \\ \searrow OH \end{smallmatrix} \rightarrow CO_2 + R'H \qquad 18.$$

In combining reactions 10–13, all the observed transalkylation processes may be accounted for. As suggested by reactions 14 and 15, alcohol and ether are easily generated while unsaturated hydrocarbons would result from alcohol dehydration (reaction 16). Saturated hydrocarbons may appear also as a result of the oxonium sensitivity to oxidation and from the presence of chemisorbed oxygen (reactions 17 and 18).

Table 1 Extent of the transalkylation reactions observed in montmorillonite systems, expressed in percentage of carbon in the EA, DEA, and TEA form (14)[a]

Starting material	Time temp.	160°C			200°C			220°C		
		EA	DEA	TEA	EA	DEA	TEA	EA	DEA	TEA
EA	24 hr	98.7	1.3		62.5	67.5		23.8	72.6	0.9
	77 hr	96	4		31	66.5	0.7	16.8	76.5	5.5
	240 hr	89	11		16.8	79.5	2	14.7	72.3	11.6
	720 hr	69.9	30.1	*	15.6	76.3	7.2	13.6	70.5	14.4
DEA	77 hr	2	95.5	2.5	6.4	74	19.6	8	60	32
	720 hr	5.7	80.6	13.7	6.7	58.6	34.7	4.7	42.6	52.6
TEA	77 hr		6.1	93.9	*	16.3	83.7	*	18.8	81.2
	720 hr	9.8	13.2	76	5.9	22.2	71.8	14.2	28.2	57.6
TTEA	77 hr		9.7	90.3	*	7.4	92.6	0.4	11.3	88.3
	720 hr		5.8	94.2	0.9	15.9	83.2	2.7	29.7	67.6

[a] Note: EA—Ethylammonium cation; DEA—Diethylammonium cation; TEA—Triethylammonium cation; TTEA—Tetraethylammonium cation; * = traces.

Table 1, taken from Durand et al (14), shows the distribution of the alkylammonium cations after various periods of time, for various temperatures, and the reaction being carried out under a 30 mbar water pressure in the absence of air. Table 2 (14) shows that the carbon loss, due to the formation of volatile compounds (reactions 15–18), is never very important, except in the case of the tetraethylammonium system. In general a longer heating period increases the amount of nondesorbable (not cationic) carbon residues.

The main hydrocarbons in the gas phase are in order of importance: ethylene > propylene > methane. After 720 hr reaction at 200°C, the amounts of ethylene are in the following order, with respect to the nature of the starting material: TTEA (23200) > TEA (8700) > DEA (900) > EA (300), the numerical figures having a relative meaning only. In an inert atmosphere, the formation of unsaturated hydrocarbons is favored. In the presence of air, the carbon loss is more pronounced, but this may be partially due to a complete oxidation into CO_2 of unsaturated, as well as saturated, hydrocarbons.

Alkylamines, in the natural environment could originate from the decarboxylation of amino acids. If this was the main mechanism, then the model proposed by Johns and co-workers to explain the transformation of fatty acids and, in particular, the existence of two zones could be as well applied to the transformation of alkylamines. Decarboxylation (reaction 7) would occur in zone I while transalkylations and the side reactions (reactions 16–18) leading to the formation of unsaturated and saturated hydrocarbons could take place in zone II.

Table 2 Measurements of the carbon content after heating, in weight percentage of carbon referring to clay heated at 1200°C (14)[a]

		Ethylammonium Mt					Diethylammonium Mt		
		Initial value	24 hr	77 hr	240 hr	720 hr	Initial value	77 hr	720 hr
160°C	1	3.04	3.21	2.89	2.96	2.99	5.68	5.42	5.64
	2	0.11	0.26	0.26	0.31	0.26	0.11	0.16	0.22
200°C	1	3.04	3.05	3.03	2.95	2.97	5.68	5.38	5.60
	2	0.11	0.26	0.39	0.37	0.31	0.11	0.25	0.49
220°C	1	3.04	2.91	2.95	2.96	2.94	5.68	5.36	5.64
	2	0.11	0.31	0.38	0.40	0.52	0.11	0.42	0.87
		Triethylammonium Mt					Tetraethylammonium Mt		
160°C	1	8.27	—	8.59	—	8.67	10.49	10.44	10.14
	2	0.21	—	0.95	—	1.93	2.75	2.91	2.94
200°C	1	8.27	—	8.55	—	8.76	10.49	9.69	8.66
	2	0.21	—	2.36	—	2.99	2.75	2.71	2.67
220°C	1	8.27	—	8.34	—	7.93	10.49	8.53	8.69
	2	0.21	—	2.67	—	3.42	2.75	2.17	3.73

[a] Note: The sodium montmorillonite used as starting material contained 0.06% carbon. Measurement No. 1 corresponds to the amount of organic carbon after heating as indicated. Measurement No. 2 is obtained after three successive desorptions by a 0.05 N cobaltihexammine chloride solution.

CLAYS AND THE ORIGIN OF LIFE

Although the idea that clays or similar materials have been catalysts for the abiotic synthesis of molecules of significant biological interest is rather old (Bernal, 2, 3; Cairns-Smith, 4), quantitative or at least semiquantitative data on the possible mechanisms are rather scarce. Recently, however, it seems that more attention has been devoted to experimental studies in this area. Two main approaches can be distinguished. The first deals with the products of the interaction of gases reputed to have been present in the primeval earth atmosphere (such as NH_3, CO, H_2, N_2, CH_4, CO_2, and H_2O) and the surface of oxides and clays. It is of course suggested that these surfaces may have played the role of catalysts. The second approach is concerned with the intervention of clays or clay-like materials in the evolution of amino acid and purine bases into complex macromolecules such as protenoids or proteins. As emphasized by Calvin (6), the motivation for both kinds of approach is the recognition that "in general, the existence of a phase boundary, that is a boundary between two different physical phases, leads to the formation of a region of unique chemical composition and molecular structure between the two phases."

Degens & Matheja (12) have contributed noticeably to the study of the catalyzed formation of peptides by using a large number of amino acids; the bases of the purines and pyrimidines, urea, sugars; and a great variety of minerals such as clays, oxides, phosphates, carbonates, and sulfates. Clays and especially kaolinite were observed to be very efficient polymerization catalysts when used in aqueous suspension at about 80°C. In the presence of this clay, the polymer obtained from a mixture of amino acids, had the following composition—

glutamic 100 : aspartic 100 : glycine 1 : alanine 20 : phenylalanine 1

In the presence of montmorillonite, the composition was obviously very different—

glutamic 100 : aspartic 4 : glycine 4 : alanine 2 : phenylalanine 1

The polymers have molecular weights between 500 and 5000. The technique used by Degens and co-workers consists in analyzing the free and combined amino acids. Surprisingly enough, as far as condensation is concerned, under aqueous conditions, serine, proline, valine, lysine, and histidine remain inert.

An alternative route followed by Degens and co-workers was to heat a dry mixture of clay and amino acids at 140°C for rather long periods of time (63 hr to 90 days). Afterwards the same type of analytical determination was carried out. In the presence of kaolinite no free amino acid was left after this treatment, while montmorillonite again seems to be a poorer condensation catalyst. Higher molecular weight compounds with maxima between 1000 and 2000 were obtained with the dry treatments. In these so-called dry systems the amount of water molecules still present is large enough to complete at least a monolayer. This means that acid conditions prevail on the clay surfaces (kaolinite as well as montmorillonite). It is most unfortunate that no attention seems to have been paid to the nature of the charge balancing cations.

Degens & Matheja (12) consider that the formation of the amino acid polymer most likely proceeds via carboxyl activation. This activation is depicted by reaction 19, shown hereafter. Fundamentally an ionized carboxyl reacts either with an exposed Al^{3+} or with a basic $>Al(OH)^+$ cation. It is supposed also that the cationic form of the amino acids could be formed through the transfer on the amine group of a silanolic proton (reaction 20). Since in the usual pH range (3–9) the surfaces of kaolinite and montmorillonite are negatively charged, the reaction depicted by reaction 19 would occur via exchangeable aluminum or polynucleic aluminum cations. On the other hand the surface protonation of the amine groups occurs more likely through the action of residual water. The carboxyl activation, as

$Al-OH^{\oplus}\cdots{}^{\ominus}O(O)C-CH(R)-NH_2$

$Al^{\oplus}\cdots O^{\ominus}-C(=O)-CH(R)-NH_2$ (19)

$Al^{\oplus}\cdots O^{\ominus}-C(=O)-CH(NH_2)-CH_2-CH_2-C(O)O^{\ominus}\cdots{}^{\oplus}HO-Al$

$Si-O^{\ominus}\cdots H^{\oplus}-N(\text{imidazole})-CH_2-CH(NH_2)-COOH$

$Si-O^{\ominus}\cdots H_3^{\oplus}N-CH_2-CH_2-CH_2-CH_2-CH(NH_2)-COOH$ (20)

$(Si)_2O\cdots H_3^{\oplus}N-CH(R)-COO^{\ominus}$

shown by reaction 19, has some similarity with the process proposed by Johns & Shimoyama (23) to explain the catalytic decarboxylation of fatty acids.

In an earlier study, Cloos et al (8, 15) had shown by infrared spectroscopy that at least two mechanisms should be invoked to account for the adsorption of amino acids by a Na, Ca, or H montmorillonite, namely (*a*) cation exchange process with the Na and/or Ca clay and (*b*) a proton transfer mechanism with the acid clay. The surface concentration in zwitterions seems to be always higher than it could have been anticipated from the concentration in the equilibrium solution, and upon dehydration of the clay an additional increase in the zwitterion content is observed in the adsorbed phase. This observation is also in line with recent data by Sung Do Jang & Condrate (39).

The zwitterions could well be the active species in the polymerization reaction because of their strong tendency to associate, and eventually to react, if the temperature is increased.

$$H_3N^+\text{–R–COO}^- + H_3N^+\text{–R–COO}^- \rightarrow H_3N^+\text{–R–}\overset{\overset{\displaystyle O}{\|}}{C}\text{–O–NH–R–COO}^- + H_2O \qquad 21.$$

To explain what could be the catalytic action of the surface in such a transformation, back and forth proton jumps from the acidic residual water onto the amino and carboxyl groups should be invoked:

$$\begin{aligned} &H_2N\text{–R–COOH} + H_3O^+ \rightleftarrows H_3N^+\text{–R–COOH} + H_2O \\ &H_3N^+\text{–R–COOH} + H_2O \rightleftarrows H_3N^+\text{–R–COO}^- + H_3O^+ \end{aligned} \qquad 22.$$

This could open the way for the dehydration represented by equation 21. The rapid turnover of protons within the layer of adsorbed water supports this hypothesis. Ordering of the zwitterions and a favorable orientation on the surface might facilitate the peptide link formation. As already outlined, the same kind of reactions would occur regardless of the nature of the source of H_3O^+. In particular, the hydrolysis of surface aluminum cations on kaolinite may be such a source. It was observed also by Fripiat et al (15) that the zwitterion concentration was higher on the surface of silica gel than on montmorillonite with, as a consequence, more active polymerization reactions on the former surface.

Still more interesting is a note by Degens, Matheja & Jackson (11) about the asymmetric polymerization of aspartic acids. L, D, and D–L aspartic acids were adsorbed on kaolinite and treated for varying lengths of time (3 days to 4 weeks) at 90°C. In the L or D aspartic acid solutions, little optical rotation was left in the supernatant after 4 weeks, indicating an almost complete racemization while the polymer formed on the clay surface contained 25% L form as compared with less than 3% for the D enantiomer. If this result could be confirmed, and especially if higher contents in one of the optical forms could be introduced into the surface polymer, an entirely new domain of investigation would be open. The reason why a polar silicate (kaolinite?) would generate some dissymmetry in the course of a chemical reaction is not clear to us and the explanation by Degens et al (11) about the difference in polarity between the tetrahedral and the octahedral layer as the origin of the asymmetry is not convincing. At the crystal edges, the polarity of the Al–OH and of the Si–OH bonds directed outside the solid should balance each other while the two extreme 001 surfaces are located too far apart from each other.

In the different domain of the synthesis *ab initio* of molecules of biological interest, Harvey, Degens & Mopper (21) have shown that kaolinite treated with CO_2, NH_3, and H_2O at 80°C for 2 weeks catalyzes the formation of cytosine, uracil, cyanuric acid, and an amino derivative of cyanuric acid. Urea was a major component. As the CO_2 pressure is not specified, it is impossible to appreciate the extent of the catalytic activity. Harvey et al (21) suggest that an activated, not further specified, aluminum carbonate complex is formed on the clay surface and that this complex reacts with NH_3 to generate cyanuric acid and urea.

More recently Fripiat et al (17) have shown that the catalytic properties of zeolitic materials may play a role in the *ab initio* synthesis of simple biological molecules from gases commonly found in the extraterrestrial atmosphere.

A synthetic near-faujasite molecular sieve, which can be considered as a good model for allophane-like material has been cation exchanged by Ca^{2+} or by Fe^{3+} cations and treated for a few days at temperatures of the order of 300°C or less in the presence of 300 torr ($CO+NH_3$) and of traces of water. Gaseous HCN is formed while in the beginning of the reaction a polymeric substance which absorbed strongly in the infrared domain of 1400 to 1800 cm^{-1} and above 2000 cm^{-1} is formed. Later on, discrete absorption bands show up progressively in regions where carbonyl, carboxyl, amide II, and C=N stretching bands are expected. At the final stage, the infrared spectrum is very similar to that observed for an Na-montmorillonite, in which Na has been exchanged by the glycil-glycine cation. In addition to large amounts of urea, the 0.1 *N* acid extract of the solid contains adenine and up to 10 different amino acids. The nature of the cation on the surface seems to influence the spectrum of the identified amino acids. As a continuation of this work, Fripiat, Poncelet & Van Assche (19) have tried to obtain a more quantitative picture by using $C^{14}O$. This has allowed them to show that the amino acids hydrolyzed by the acid treatment of the solid, after reaction, were in concentration of the order of a micromole per gram of catalyst. In addition it was clearly established that the observed amino acids did not result from an accidental contamination.

The reaction scheme proposed to explain these observations may be summarized as follows:

1. CO and NH_3 form charge transfer complexes with the surface cations (Ca^{2+} or Fe^{3+}). In the coordination sphere, activated CO reacts with NH_3 and forms formamide, which in turn decomposes into HCN and H_2O.

$$\left[CO + NH_3\right] \rightarrow \left[H\text{-}C\begin{matrix}{\scriptstyle \nearrow O} \\ {\scriptstyle \searrow NH_2}\end{matrix}\right] \rightarrow HCN + H_2O \qquad 23.$$

2. HCN polymerizes and tautomerizes in aminocyanocarbene, which polymerizes further

$$2\,HCN \rightarrow \overset{\delta-}{N}{\equiv}C{=}\overset{\delta+}{C}{=}NH_2 \rightarrow \left(\begin{array}{c}-C=C=N-\\ |\\ NH_2\end{array}\right)_x \qquad 24.$$

The step represented by reaction 24 would be essential. It would open the way to the chain of reactions proposed in homogeneous medium by Matthews & Moser (26). The polyaminocyanocarbene could be responsible for the infrared spectrum observed at the early stage of the reaction.

3. Because of the acid character of the residual water (in small amount, i.e. smaller than the monolayer content), the polymer would progressively hydrolyze, the nitrile functions giving rise to carboxyl groups, the $>C{=}NH$ function to the carbonyl groups, etc. Immonium intermediates ($>\overset{+}{N}{=}C<$) could be active in this step. The evolution of the infrared spectra would reflect this slow hydrolysis process.

4. Eventual decarboxylation could produce CO_2 that would react with NH_3 to yield urea; CO_2 could also be observed from a water shift reaction. Adenine, according to Oró's view (30), could result from the straight polymerization of HCN.

In the framework of this hypothesis, the formation of the surface peptide would not occur via the prerequisite synthesis of individual amino acids and their further condensation. Instead the nature of the amino acids and their sequence would be determined by the position of the active groups in the initial surface polyamino-cyanocarbene which are affected by the hydrolysis. Perhaps in this way the asymmetry of some surface properties (for instance the asymmetric cationic environment in polyionic clays) would generate some molecular dissymmetry in the surface polymer.

Although several interesting new facts have been presented in the last ten years about the intervention of clays in prebiotic synthesis, the reader is already probably convinced that there is still a long way to go before it could be claimed without ambiguity that clays have been catalysts in the first steps of the formation of living things. For the time being, the only reasonable conclusion is that clays have probably added various ingredients to the primordial soup.

CLAY MINERALS AND CHEMICAL TRANSFORMATION IN SOILS

Under this heading the fate of organic matter and also of some pesticides buried in the soils will be examined from the catalytic viewpoint. Factors influencing the adsorption, desorption, and movement of pesticides in soils have been excellently reviewed by Bailey & White (1).

Because pesticides are much less complex than the natural soil organic matter, some of the facts about them that have been well documented recently are reviewed first, even though it may be argued that their transformations are not, strictly speaking, natural processes. The discussion is limited to transformation of *s*-triazines. Table 3 summarizes the chemical structures and some properties of the main

Table 3 Chemical structure and some properties of the main triazines (24)

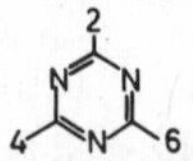

Common name	Substitution on triazine ring at positions 2	4	6	pK value	Solubility in water (ppm at 20–25°C).
—	SCH_3	$NHiC_3H_7$	$N(C_2H_5)_2$	4.43	—
Chlorazine	Cl	$N(C_2H_5)_2$	$N(C_2H_5)_2$	1.74	9
Simazine	Cl	NHC_2H_5	NHC_2H_5	1.65	5
Trietazine	Cl	NHC_2H_5	$N(C_2H_5)_2$	1.88	20
Norazine	Cl	$NHCH_3$	$NHiC_3H_7$	—	260
Atrazine	Cl	NHC_2H_5	$NHiC_3H_7$	1.68	33
Propazine	Cl	$NHiC_3H_7$	$NHiC_3H_7$	1.85	8.6
Ipazine	Cl	$NHiC_3H_7$	$N(C_2H_5)_2$	1.85	40
Simetone	OCH_3	NHC_2H_5	NHC_2H_5	4.17	3200
—	OCH_3	$N(C_2H_5)_2$	$N(C_2H_5)_2$	4.76	—
—	OCH_3	NHC_2H_5	$N(C_2H_5)_2$	4.51	40
Prometone	OCH_3	$NHiC_3H_7$	$NHiC_3H_7$	4.28	750
Ipatone	OCH_3	$NHiC_3H_7$	$N(C_2H_5)_2$	4.54	100
Noratone	OCH_3	$NHCH_3$	$NHiC_3H_7$	4.15	3500
Atratone	OCH_3	NHC_2H_5	$NHiC_3H_7$	4.20	1654
Simetryne	SCH_3	NHC_2H_5	NHC_2H_5	—	450
Prometryne	SCH_3	$NHiC_3H_7$	$NHiC_3H_7$	4.05	48
Ametryne	SCH_3	NHC_2H_5	$NHiC_3H_7$	—	193
Desmetryne	SCH_3	$NHCH_3$	$NHiC_3H_7$	—	580
Methoprotryne	SCH_3	$NHiC_3H_7$	$NH(CH_2)_3OCH_3$	—	320

members of this important family. The basic heterocycle of *s*-triazines is that of melamine which can be protonated as shown hereafter:

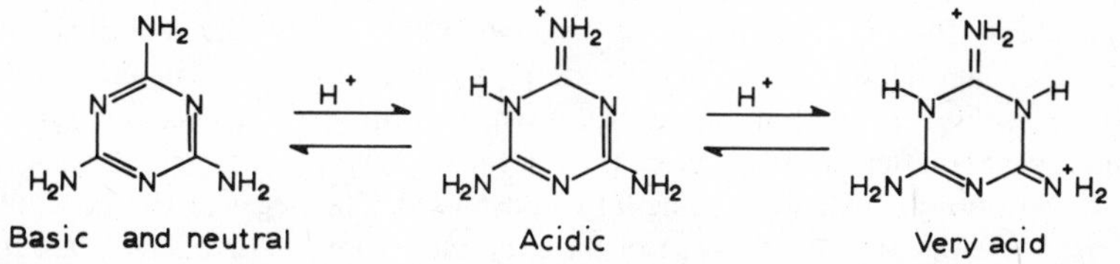

H^+ ⇌ H^+ ⇌

Basic and neutral Acidic Very acid

These protonation reactions are very important for the further transformations.

Weber (41) had shown that the maximum adsorption of molecules listed in Table 3 by an Na-montmorillonite suspension is obtained when the pH of that suspension is near the pK of the first acidic function. This can be expected if the adsorption proceeds mainly through a cation exchange process since, according to the usual chemical definition

$$\log(MH^+/M) = pK - pH$$

and that $MH^+ \approx M$ for $pK \approx pH$. However, the association of a basic or neutral form with surface cations through complexation is also possible.

Infrared studies by Cruz, White, and Russel (10, 33) have shown that protonated propazine, prometone, or prometryne adsorbed on montmorillonite are hydrolyzed easily in the presence of a sufficient amount of adsorbed water. The substituent in position 2 (Table 3) is affected by this process. For instance, the Cl atom or the $-OCH_3$ group is replaced by an hydroxyl group, leaving one of these keto forms shown hereafter:

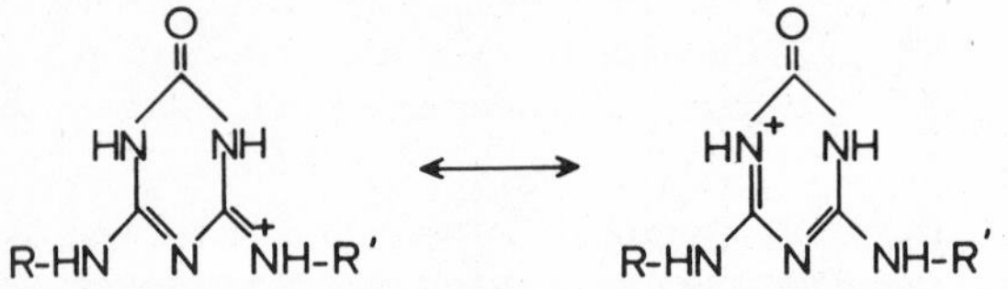

O HN NH R-HN N NH-R' ⟷ O HN$^+$ NH R-HN N NH-R'

NMR study by Russel et al (34) has suggested that the proton is probably on the nitrogen atom in the heterocycle. The keto forms have no biological activity and therefore the hydrolysis process occurring in the adsorbed state has the final consequence to provoke the degradation of a biologically active into an inactive molecule. Since it is known from the reactivity in solution that substitution by an OH in position 2 of the ring may be obtained at pH lower than 3, it is obviously the surface acidity that is responsible for the degradation process.

The practical consequence of this type of reaction is considerable because it provides an alternative pathway to microbial decomposition which has been considered for a long time as the prevailing route. The same kind of consideration applied probably to some extent to the transformation of natural soil organic matter. Under the denomination of "humic" substances are found a broad spectrum of components resulting from the primary alteration of plant, animal, and micro-organism remains. These components are intimately associated with the mineral components of the soils and, because of their adsorptive properties, they are in intimate contact with clay mineral surfaces. Humic substances are composed of humic acids (that can be extracted through alkali or organic solvents and precipitated at $pH \approx 1$, fulvic acids (not precipitated at $pH = 1$), and humin (fraction insoluble in alkali or organic solvents). According to Greenland (20) most of the organic matter in soils is present as negatively charged polymeric material, whose constitution is far from being fully understood. The negative charge would arise from the dissociation of some of the exposed carboxylic or phenolic groups of either the humic or fulvic acids. It is also likely that the interaction with the surface cations takes place through these ionized groups. Essentially the process involves the replacement of a water ligand by an oxygen of the anionic group:

$$[M^{2+}(H_2O)_n] + R\text{–}COOH \rightarrow H_3O^+ + [M^{2+}(H_2O)_{n-1}\,R\text{–}COO^-] \quad 26.$$

in a manner which has been invoked already several times in this paper. The anion could also be possibly associated more indirectly through coordination with water molecule as shown by reaction 27 (water bridge):

$$[M^{2+}(H_2O)_n] + R\text{–}COOH \rightarrow H_3O^+ + [M^{2+}(H_2O)_{n-1}(OH_2\text{–}OOC\text{–}R)] \qquad 27.$$

As in the usual pH range in soils, iron or aluminum cations form readily polynucleic complexes; it is probably through such complexes that reactions 26 and 27 do occur. In addition, when the clay-organic matter system gets less hydrated for one reason or another, the organic segments in R may approach the clay surface to a distance small enough for the van der Waals forces to produce very energetic effects. It is a matter of fact that when clays and organic matter have been dried together, the desorption of the organic fraction becomes increasingly difficult.

The question relevant to this review is the eventual chemical transformations that humic or fulvic acid adsorbed as described here above would undergo either spontaneously or under the influence of change in the hydration state or in temperature. It is obvious that to gain information on these processes, desorption products should be compared to the initial compounds. Because of the very complex and heterogeneous nature of these compounds such an analysis represents a difficult task.

Schnitzer & Kodama (36) have reported reactions of fulvic acid and Cu^{2+}-montmorillonite. Earlier Kodama and Schnitzer (25, 35) had found that fulvic acids could be adsorbed in the interlamellar space under favorable pH conditions, the nature of the interlayer cation being of great importance. At pH = 2.5, Cu^{2+}-montmorillonite swells from an initial d001 = 10 Å spacing to d001 = 15.1 Å. At pH > 3.5, no significant swelling is observed although the amount of adsorbed fulvic acid is still 50% that found at pH = 2.5. This means of course that acid conditions are required for interlayer adsorption. The infrared spectra of the complexes prepared at pH > 2.5 shows two bands at 1850 and 1520 cm^{-1} which were interpreted as experimental evidence for the interaction of β-diketone groups with Cu^{2+} forming chelates similar to acetylacetonates as schematically shown below:

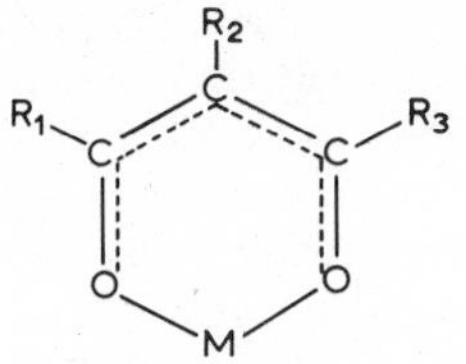

The resistance of such an adsorbed complex to oxidation seems to be very strong since Schnitzer & Kodama (36) have observed a weight loss smaller than 4% when heating up to 550°C. They conclude that this exceptional stability might explain the observed longevity of humic substances in soils and waters. The mechanism proposed for the fulvic acid copper (II) interaction could possibly apply to reactions between fulvic acid and montmorillonites saturated with other di- and trivalent metal ions. To obtain the structure shown above, a proton should be transferred to the surface in a manner that could be similar, *mutatis mutandis,* to that shown by reaction 22.

It would be very interesting to obtain more information on this kind of process since, in agreement with Jacks (quoted by Greenland, 20), "the union of mineral and organic matter to form the organo-mineral complex is a synthesis as vital to the continuance of life as, and less understood than, photosynthesis."

CONCLUSIONS

In spite of the complexity of the chemical transformations described in this review, the catalytic functions are paradoxically rather simple and very limited in number. Actually they may be classified into two groups, as summarized in the following:

(*a*) The surface acidity due to the high degree of dissociation of residual adsorbed water has been invoked to explain: the cracking of *n*-alkane; the transalkylation of alkylammonium cations; the formation of unsaturated and saturated hydrocarbons, as side products of this transalkylation; the surface protonation of amines and amino acids; the hydrolysis of a hypothetical polyaminocyanocarbene, resulting from the HCN surface polymerization; and the protonation and subsequent hydrolysis of *s*-triazines.

(*b*) The formation of surface cationic complexes and charge transfers in the coordination spheres of these complexes would be at the origin of: the decarboxylation reaction of fatty acids; the carboxyl activation, prerequisite to the amino acid polymerization; the activation of CO_2 in the *ab initio* synthesis of cytosine, cyanuric acid, urea, etc; the activation of CO and NH_3 for the synthesis of HCN, prerequisite to the formation of polyaminocyano carbene; the absorption of fulvic and humic acids and their eventual stabilization. Since the dissociation of residual water is provoked by the electrical field of the exposed surface cations and since the formation of surface cationic complexes results from the competition between water and other ligands in the coordination sphere, it may be concluded that the nature of the surface cations and the extent of surface hydration are the two factors ruling the catalytic activity of clays at low or moderate temperature.

With these conclusions in mind, suggestions for further work may be proposed: (*a*) The catalytic properties of polycationic clays should be thoroughly investigated. For instance clays partially exchanged by Ca^{2+} and Cu^{2+} or by Ca^{2+} and Fe^{3+} should be studied. (*b*) Clays with different charge distributions in octahedral or tetrahedral positions should be compared. (*c*) The exact nature of the mechanism proposed for the carboxyl activation through interaction with surface polynucleic Al^{3+} or Fe^{3+} should be established. (*d*) The redox properties of clay surface should receive more attention. Finally, it should be strongly recommended to those using clays as catalysts to be extremely cautious in defining the surface cations and their degree of hydration.

It is the hope of the authors of this review that we have stimulated rather than discouraged more people to contribute to this domain of geochemistry. The fundamental and practical implications of research in this area are very important for they concern the origin and the future of human beings.

Literature Cited

1. Bailey, G. W., White, J. L. 1970. Factors influencing the adsorption, desorption and movement of pesticides in soils. *Residue Rev.* 32:29–92
2. Bernal, J. D. 1951. *The Physical Basis of Life.* London: Routledge & Kegan Paul
3. Bernal, J. D. 1967. *The Origin of Life.* New York: World
4. Cairns-Smith, A. G. 1966. The origin of life and the nature of the primitive gene. *J. Theoret. Biol.* 10:53–88
5. Calvet, R. 1972. *Hydratation de la montmorillonite et diffusion des cations compensateurs.* Thèse de doctorat, CNRA Versailles, France
6. Calvin, M. 1969. *Chemical Evolution.* Oxford: Oxford Univ. Press
7. Chaussidon, J., Calvet, R. 1965. Evolution of amine cations adsorbed on montmorillonite with dehydration of the mineral. *J. Phys. Chem.* 69:2265–68
8. Cloos, P., Calicis, B., Fripiat, J. J., Makay, K. 1966. Adsorption of amino-acids and peptides by montmorillonite. I Chemical and x-ray diffraction studies. *Int. Clay Conf., Jerusalem* I:223–32. Israel Program for Scientific Translations
9. Cloos, P., Van de Poel, D. 1973. A copper II thiopene complex on the interlamellar surfaces of montmorillonite. *Nature* 243:54–55
10. Cruz, M. I., White, J. L., Russel, J. D. 1968. Montmorillonite-*s*-triazine interactions. *Isr. J. Chem.* 6:315
11. Degens, E. T., Matheja, J., Jackson, T. A. 1970. Template catalysis: Asymmetric polymerization of amino-acids on clay minerals. *Nature* 227:492–93
12. Degens, E. T., Matheja, J. 1971. Formation of organic polymers on inorganic templates. *Prebiotic and Biochemical Evolution,* ed. A. P. Kimball, J. Oró, 39. Amsterdam: North-Holland
13. Doner, H. E., Mortland, M. M. 1969. Benzene complexes with copper (II) montmorillonite. *Science* 166:1406–7
14. Durand, R., Pelet, R., Fripiat, J. J. 1972. Alkylammonium decomposition on montmorillonite surfaces in an inert atmosphere. *Clays Clay Miner.* 20:21–35
15. Fripiat, J. J., Cloos, P., Calicis, B., Makay, K. 1966. Adsorption of amino-acids and montmorillonite. II Identification of adsorbed species and decay products by infrared spectroscopy. *Int. Clay Conf., Jerusalem* I:233–46. Israel Program for Scientific Translations
16. Fripiat, J. J., Helsen, J. 1966. Kinetics of decomposition of cobalt coordination complexes on montmorillonite surfaces. *Clays and Clay Minerals,* ed. S. W. Bailey, 163–79. 14th Nat. Conf. New York: Pergamon
17. Fripiat, J. J., Poncelet, G., Van Assche, A. T., Mayaudon, J. 1972. Zeolite as catalysts for the synthesis of amino acids and purines. *Clays Clay Miner.* 20:331–40
18. Fripiat, J. J., Lambert-Helsen, M. M. 1973. Transalkylation of alkylammonium cations in Y zeolite. *Advan. Chem. Ser.* 121:518–28
19. Fripiat, J. J., Poncelet, G., Van Assche, A. T. Synthesis of biological molecules on molecular sieves using C^{14}. *Int. Conf. Origin of Life, 4th, Barcelona.* In press
20. Greenland, D. J. 1971. Interactions between humic and fulvic acids and clays. *Soil Sci.* 111:34–41
21. Harvey, G. R., Degens, E. T., Mopper, K. 1971. Synthesis of Nitrogen heterocycles on Kaolinite from CO_2 and NH_3. *Naturwissenschaften* 12:624–25
22. Hawthorne, D. G., Solomon, D. H. 1972. Catalytic activity of sodium kaolinites. *Clays Clay Miner.* 20:75–78
23. Johns, W. D., Shimoyama, A. 1972. Clay minerals and petroleum forming reactions during burial diagenesis. *Int. Clay Conf., Madrid.* Preprints, II:233–41
24. Jordan, L. S. 1970. Foreword to the International symposium held at the University of California, Riverside, on triazine herbicide-soil interactions. *Residue Rev.* 32: Ibidem VII, XIII
25. Kodama, H., Schnitzer, M. 1969. Thermal analysis of a fulvic acid-montmorillonite complex. *Proc. Int. Clay Conf., Tokyo* 1:765–74
26. Matthews, C. N., Moser, R. E. 1967. Peptide synthesis from hydrogen cyanide and water. *Nature* 215:1230–34
27. Miller, J. G., Haden, W. L. Jr., Oulton,

T. D. 1964. The oxidizing power of the surface of attapulgite clay. *Clays and Clay Minerals,* ed. W. F. Bradley, 382–96. 12th Nat. Conf. New York: Pergamon
28. Mortland, M. M., Fripiat, J. J., Chaussidon, J., Uytterhoeven, J. 1963. Interaction between ammonia and the expanding lattices of montmorillonite and vermiculite. *J. Phys. Chem.* 67:248–58
29. Mortland, M. M., Pinnavia, T. J. 1971. Formation of copper II arene complexes on the interlamellar surfaces of montmorillonite. *Nature* 229:75–77
30. Oró, J., Kamat, S. S. 1962. Synthesis of purines under possible primitive Earth condition. Purine intermediates from HCN. *Arch. Biochem. Biophys.* 96:293–313
31. Pezerat, H., Mering, J. 1967. Recherches sur la position des cations échangeables et de l'eau dans les montmorillonites. *C. R. Acad. Sci.* 265(D):529–32
32. Pinnavia, T. J., Mortland, M. M. 1971. Interlamellar metal complexes on layer silicates. I copper (II)—Arene complexes on montmorillonite. *J. Phys. Chem.* 75:3957–62
33. Russel, J. D., Cruz, M. I., White, J. L. 1968. The adsorption of 3-aminotriazole by montmorillonite. *J. Agr. Food Chem.* 16:21
34. Russel, J. D. 1968. Mode of chemical degradation of *s*-triazines by montmorillonite. *Science* 160:1340
35. Schnitzer, M., Kodama, H. 1967. Reactions between a Podzol fulvic acid and Na-montmorillonite. *Soil Sci. Soc. Am. Proc.* 31:632–35
36. Schnitzer, M., Kodama, H. 1972. Reactions between fulvic acid and Cu^{2+}-montmorillonite. *Clays Clay Miner.* 20:359–67
37. Shimoyama, A., Johns, W. D. 1971. Catalytic conversion of fatty acids to petroleum-like paraffins and their maturation. *Nature* 232:140–44
38. Solomon, D. H., Swift, J. D., Murphy, A. J. 1971. Acidity of clay minerals in polymerization and related reactions: *J. Macromol. Sci. Chem.* 5:587–601
39. Sung Do Jang, Condrate, R. A. 1972. The I. R. spectra of lysine adsorbed on several cation-substituted montmorillonites. *Clays Clay Miner.* 20:79–82
40. Touillaux, R., Salvador, P., Vandermeersche, C., Fripiat, J. J. 1968. Study of water layers adsorbed on Na and Ca montmorillonite by the pulsed nuclear magnetic resonance technique. *Isr. J. Chem.* 6:337–49
41. Weber, J. B. 1970. Mechanisms of adsorption of *s* triazines by clay colloids and factors affecting plant availability. *Residue Rev.* 32:93–130
42. Wingard, L. B. Jr. 1972. Enzyme engineering. *Advances in Biochemical Engineering,* ed. T. K. Ghose et al, 1–48. Berlin: Springer-Verlag

MARINE DIAGENESIS OF SHALLOW WATER CALCIUM CARBONATE SEDIMENTS

×10023

R. G. C. Bathurst
Department of Geology, University of Liverpool, Liverpool, U.K.

INTRODUCTION

In the last decade there has been a dramatic expansion in our understanding of marine diagenesis. The breadth and depth of this advance are revealed clearly in the remarkable Volume 12 of *Sedimentology,* in which the then new Editor, H. Füchtbauer, brought together not only the fundamental new discoveries regarding Recent marine carbonate diagenesis, but also a number of studies revealing clear evidence of marine diagenesis in limestones as old as the Devonian. It is to the more recent evolution of these researches, through the last three or four years, that the remainder of this article is devoted.

RECENT MARINE DIAGENESIS

Intraparticle Cementation

The most widespread occurrence of carbonate cement appears to be in voids within carbonate grains. The voids may be vacated living spaces of animals or plants, such as the tiny chambers in foraminiferids or the large body chamber of a mollusk. Other voids are bore holes of sponges such as *Cliona,* of bivalves like *Lithophaga,* and of noncalcareous algae. Much of this cement can be seen with the light microscope (as in Glover & Pray, 30; Pingitore, 57; Shinn, 74, 75) and even more might be seen if petrographers were to develop the delicate skills of thin section preparation of Mutvei (52). Nevertheless, the important revelations have come from the use of the scanning electron microscope (SEM), with its invaluable depth of focus such that, for example, Loreau (45) could show distinct needles of aragonite less than 2 μ long.

The breakthrough came with the work of Alexandersson (2) who examined with SEM the ultrastructures of skeletal sand grains from seven localities in the Caribbean and the Mediterranean. The sediments were collected from places where the water was sufficiently turbulent to induce bottom traction and where there were no oolites nor any evidence of intergranular cementation. Two void-filling minerals could clearly be recognized, aragonite needles (Figure 1) and calcite rhombs (Figure 2). X-ray diffraction confirmed the mineralogy and, with the additional aid of a microelectron probe, the calcite was shown to have 15–17 mole% $MgCO_3$. This is a typical high-magnesian calcite of diagenetic origin as noted earlier by Purdy (58), Land & Goreau (43), and others. The work of Glover & Pray (30) with a probe confirms these mineralogical conclusions.

The distribution of these two cements, encrusting the surfaces of voids, shows a preference for closed (or nearly closed) voids: pits on open surfaces or the body cavities of ostracods or bivalves are less favored sites. In many instances it is apparent that the host substrate is encrusted by a cement of the same mineralogy, but this is not always so. Glover & Pray (30) noted that the magnesium content in the host calcite is commonly reflected in the calcite cement. Crystal fabrics are described on p. 267. There seems no reason to doubt that cementation is occurring now: cements were seen in the chambers of living organisms, a situation confirmed by Glover & Pray (30). Attempts to detect organic material which might influence

Figure 1 Needles of aragonite in an algal borehole in a Recent molluscan grain from the Bahamas. Fracture surface SEM photograph courtesy of T. Alexandersson.

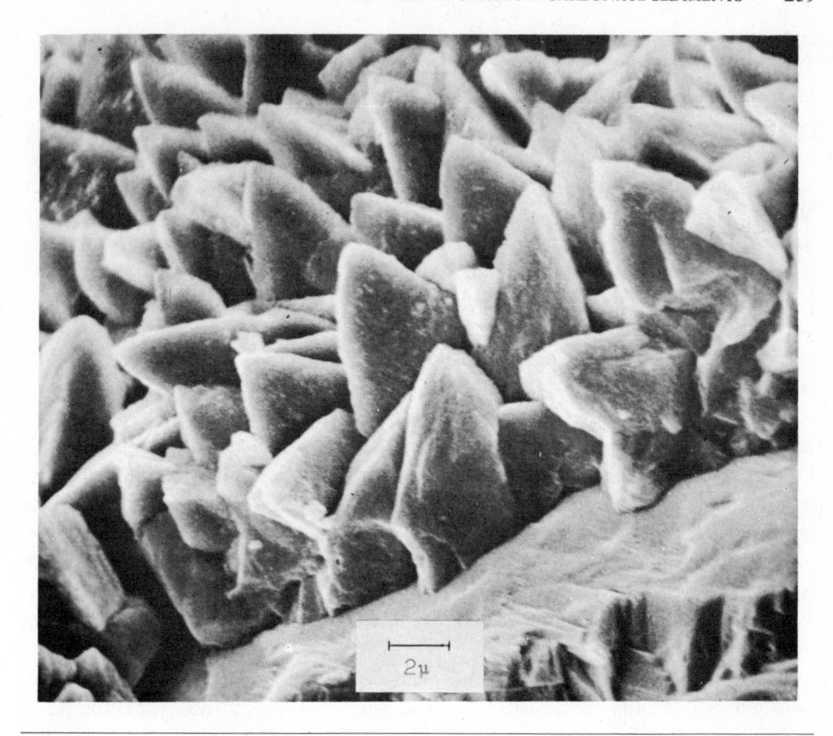

Figure 2 High-magnesium calcite cement rhombs in a Recent beach rock from Cyprus: grain surface below. Fracture surface SEM photograph courtesy of T. Alexandersson.

the nucleation of cement were unsuccessful, as were Andersson's attempts to detect tissue residue in these microenvironments.

A point of special significance for the study of ancient limestones is the crystal size of the high-magnesian calcite cement, mostly 2–4 μ. This falls within the category of micrite. It is distinguishable from detrital micrite, in thin section with the light microscope, only by its tendency to line voids. In voids completely filled with micrite the distinction is thus impossible in thin section. Indeed, conversations with Alexandersson have revealed that the situation is even more complex. In some cavities micritic skeletal debris has accumulated during micritic calcite cementation and the resultant filling is a mixture of sediment and cement, each having similar crystal sizes. The task of interpreting the recrystallized products of such diverse accumulations in, say, a Jurassic limestone is formidable.

Micritization

Recent carbonate grains on the Bahamas-Florida shallow-water platform undergo a more or less extensive centripetal replacement by micrite, as grains are bored by endolithic microscopic algae and the vacated bores are filled with micrite (Bathurst, 6). This interpretation of the replacement process has been confirmed by the work of Kendall & Skipwith (39) and Taylor & Illing (83). The bores, which are commonly 6–8 μ wide, are filled in many cases with aragonite (Bathurst, 6; Glover in Bathurst, 8), but Winland (88) and Alexandersson (3) have also recorded calcite. The tendency for the mineralogy of the cement filling to follow that of the host substrate was noted by Purdy (58) and Glover & Pray (30), though there are numerous exceptions. Application of SEM showed that the fillings are typical intraparticle cements. Loreau (45), Lloyd (44), and Margolis & Rex (46) found cement fillings of fibrous aragonite in vacated bores, and Alexandersson (3) made a particularly detailed and well illustrated study and found aragonite and high-magnesian calcite with the characteristic habits and fabrics of intraparticle subtidal cements (Figure 1).

Boring organisms other than algae certainly invade carbonate materials, but no assessment has yet been made of the extent of their involvement in micritization. Fungal hyphae are widely distributed and, being independent of light, are not restricted to the photic zone in the sea (Kohlmeyer, 40; Golubić et al, 31; Perkins & Halsey, 56; Rooney & Perkins, 63). An endolithic sponge is active in producing a dendritic arrangement of perforations in carbonate grains in the Great Barrier Reef complex (Rooney & Perkins, 63). The fungal bores have tube diameters about 1–4 μ; the sponge bores described by Rooney & Perkins have a maximum diameter of about 20 μ.

The possibility of a relationship between depth of water and the vigor of algal boring has naturally interested geologists who view the presence of micritized grains in ancient limestones as a possible depth criterion. In the Great Barrier Reef complex there is an extensive bathymetric variation and Swinchatt (82) found that endolithic microborers are not significantly active in sediments below 40 m and are only vigorously active in less than about 15–18 m of water. On the other hand, Rooney & Perkins (63) detected heavy infestation by a septate green alga (8–10 μ) in the Barrier Reef area down to 30 m. Off the coast of the Carolinas Perkins & Halsey (56) found algae active to 25 m; they have summarized the observations of other workers and have concluded that 50 m depth is probably a world maximum for boring photosynthesizing algae.

The assumption that micrite-filled bores are all endproducts of perforation by photosynthesizing algae has been questioned by Friedman et al (22) who have found evidence of boring below the photic zone. Heterotrophic algae were noted by Frémy (21).

Interparticle Cementation (Grapestone)

Grains held in contact in algal mats (Bathurst, 8) may be cemented to form lithified grain clusters (grapestone). Unpublished work by F. Fabricius of the Technische Universität, München, using SEM, has shown that boring algae (above) play a critical role, secreting rod-shaped bodies in their mucous sheaths which are similar in size and shape to those in the ooid cortex. Pores formed by the clustering of grains in the Bahamas contain radial-fibrous aragonite cement (R. G. C. Bathurst, unpublished).

Beach Rock Cementation

Details of occurrence, morphology, and petrography of beach rocks are given by Bathurst (8). The most recent researches have confirmed that there are two cements, aragonite and a high-magnesian calcite, with the variety of crystal habits known in other marine cements.

Early thinking on the cementation process tended to favor precipitation as a result of enhanced supersaturation caused by evaporation of seawater from the surface of the beach. Increasingly, however, it has been felt that the exposed beach rock is not growing but is, rather, being destroyed by a mixture of organic boring and wave action. As fabric and chemical evidence have accumulated, it has become obvious that the cementation takes place under a cover of unconsolidated sediment, maybe some 10–30 cm thick. Subsurface cementation has been examined most recently by Taylor & Illing (83), Schmalz (66), and Davies & Kinsey (15).

There is no doubt whatever that cementation is taking place now. In the Pacific the beach rocks contain cemented relics of World War II (skulls, fragments of military equipment) and both there and elsewhere, in the Bahamas and in the Great Barrier Reef, cemented beer bottles and Coca-Cola cans display additional fossil evidence of the recent spread of civilization. However, we have yet to learn how much of the precipitation is phreatic marine, how much is phreatic freshwater, and how much is, indeed, vadose. Certainly vadose cement has been recognized in Recent beach rocks, having a characteristic stalactitic distribution, with thicker fringes underneath grains (Taylor & Illing, 83; Loreau & Purser, 60). Moreover, not all the solid pore filling is cement. Both Taylor & Illing (83) and Davies & Kinsey (15) have noted that carbonate mud adheres to the pore walls, trapped originally by organic films probably of algal origin.

The mechanism of cementation in beach rock is not yet clear and this is not altogether surprising. The intertidal environment has long been known to biologists as perhaps the most variable in the world. Conditions in the sediment vary from saturation with water to total dryness. There is a hydrodynamic change from the turbulence of the surf zone, to immobility above water thence to wind transport. Temperature varies diurnally, as do photosynthesis and respiration of micro-organisms. The locations within the beach of phreatic seawater and phreatic land-derived freshwater change laterally and vertically with the tides. There must be a zone where these two waters mix and the existence of this zone may be critical for the cementation process.

Three recent studies shed valuable light on the cementation process while exposing also the complexity of the chemical environment. Davies & Kinsey (15) have made a thorough study of the occurrence of beach rock on Heron Island, in the latitude of the tropic of Capricorn, Great Barrier Reef. In this beach rock the quantity of the cement (aragonite) increases downward such that the porosity is reduced by as much as 20–50% in a vertical range of a meter. Despite the presence of an overlying algal mat, the authors could detect no relationship between the activity of the algae and cementation. Their chemical work included the monitoring of water chemistry in a pool on the beach rock and in an artificial polythene pool used as a control. Water was replenished or replaced in each pool as the tide rose and water flowed over the pool rim. The natural pool was floored by an algal mat; tidal water did not seep in or out through the floor. One cannot do justice here to the authors' long and interesting discussion of their chemical data: suffice to say that the cementation does not appear to be a temperature-aided or photosynthetic cycle. It may even be a dominantly nocturnal process, since there is evidence that at night an exchange occurs between water in the natural pool and the pore water in the underlying beach rock. Certainly, calcium increases in the natural pool water throughout daylight but decreases at night, a variation interpreted as implying dissolution of beach rock by day and precipitation of calcium by night.

Schmalz (66) investigated the chemistry of the zone of mixing of the landward freshwater and the seawater in beach rock on Eniwetok Atoll, Marshall Islands, in the Pacific. Water was sampled from specially prepared wells over eleven days during which there were rainy and dry periods. The meteoric (brackish) water was characterized by a low salinity of 24–29‰, an oxygen content much below that of seawater, and a carbon dioxide pressure from 5 to 15 times that in the seawater. The seawater had a salinity of 30–36.5‰. The pressure of carbon dioxide in all well waters increased with time unless diluted with rain, suggesting biological decay. Rain flow naturally caused a movement of the meteoric water seaward, whereas the incoming tide reversed the process. Rain also reduced salinity. Mixing of the two water masses occurred at mid-beach. Schmalz argued that the meteoric water is initially supersaturated with aragonite by about 300%, the seawater by about 250%. Where the two solutions mixed, the level of supersaturation was observed to fall to only 30%, indicating substantial precipitation of calcium carbonate. Existing data suggest that mixing of this kind could lead to a local short-lived supersaturation of aragonite of about 600%. [Other works on precipitation from mixed saturated waters may be found in Bathurst (8).]

Moore (51) has also followed the mixing of meteoric and seawaters, using wells in beach rock on Grand Caymen Island, West Indies. The mixed waters had an anomalously high ratio Sr/Ca and low ratio Mg/Ca compared to those of seawater, because of the seaward lateral inflow of meteoric water from the local Pleistocene limestones. This meteoric water is high in strontium, conceivably a result of dissolution of aragonite in the Pleistocene limestones and the release of strontium. The low magnesium content follows from the dilution of seawater. Moore's analyses of waters and cements (aragonite and high-magnesian calcite by electron probe) indicate a clear relation between water chemistry and cement chemistry. In the cements the Sr/Ca and Mg/Ca ratios follow the water ratios, respectively increasing and decreasing landward. The $\delta^{13}C$ indicates equilibrium with seawater. The $\delta^{18}O$ is low compared to that of Jamaican marine cement, possibly a result of inflow of meteoric water with more negative $\delta^{18}O$. The even distribution of cement over the grains indicates phreatic rather than vadose precipitation.

It would seem from these three researches that precipitation of beach rock cement may be a nocturnal process, unrelated to the temperature or photosynthetic cycles, but showing a distinct relation between the mixing of meteoric with seawaters and the precipitation of cement. It should, however, not be forgotten that precipitation of aragonite and high-magnesian calcite cements is a normal subtidal process in which no water mixing is involved. On the other hand, the intertidal mixing processes may well be responsible for the extremely rapid rate of beach rock cementation compared to that in its slower subtidal equivalent.

Reef Cementation

The cements in reef rock are the same as those in the other shallow marine carbonate environments where the water is supersaturated with calcium carbonate. Internal circulation of water in the reef body is of course essential, with liberal exchange with the open sea. This is normally made possible by the high primary porosity of the coral-algal framework and the pumping action of waves and tidal flow. The porosity is greatly augmented by the boring activities of mollusks (e.g. *Lithophaga*), clionid sponges, and endolithic algae. Despite the vigorous hydraulic environment of the exposed surface of the reef, the internal cavity system forms a quiet environment with sufficiently reduced flow and turbulence to allow accumulation of silt- and clay-grade internal sediments on the floors of cavities (Figure 3). Papers by Ginsburg et al (26, 28) and Land & Goreau (43) on the Bermudan and Jamaican reefs show that cementation must be proceeding now.

Figure 3 Section through cup reef rock, Bermuda. Undulose white laminae of coralline algae and laminae of *Millepora*. The three thin arrows point to small encrusting forams, *Homotrema rubrum*; the two larger arrows to gastropods, *Dendropoma irregulare*. The ellipses about a centimeter in diameter are large lined *Lithophaga* bores with geopetal sediment. Photograph courtesy of J. H. Schroeder.

Reef debris is cemented, as are internal sediments; primary and secondary voids are filled with aragonite and calcite; cemented internal sediments are themselves bored and layers of internal sediment alternate with layers of cement. Jamaican and Bermudan high-magnesian calcite cements gave $\delta^{13}C$ and $\delta^{18}O$ values indicating equilibrium with seawater (Land & Goreau, 43; Ginsburg et al, 28). The variety of habits of the aragonite and high-magnesian calcite cements, the time sequences of their precipitation, and the influences of the various substrates on cement mineralogy and fabric have been described by Ginsburg et al (27), Schroeder (68), and Friedman, Amiel & Schneidermann (23). Naturally, SEM photography has contributed substantially to understanding in this area.

Subsurface Nodules and Sheets

Contemporary subsurface cementation of subtidal marine sediments into nodules (concretions) and subsequently, by coalescence, into sheets (hardgrounds) was observed off the Trucial Coast of the Persian Gulf, by Taylor & Illing (83) and Shinn (74). Precipitation of the cements from seawater, supersaturated for calcium carbonate, seems to depend again on wave and tide induced circulation, but only among undisturbed grains buried below the sediment surface (De Groot, 16).

Alexandersson (1) has described examples from the Mediterranean of rocky sea bottoms in the photic zone, hardened by encrustations of coralline algae and other calcareous organisms. This hardening yields a crust which has a mixed structure of organic frame and loose sediment. Below the present surface, the crust has been altered by boring and filling with a micritic cement of high-magnesian calcite. Relics of the original frame remain quite unaltered and sharply separated from the micritic matrix, as one would expect from alteration by boring as distinct from recrystallization. The micrite cement itself is bored and the new bores filled with new micrite. Fine sediment is trapped and mixed with the micritic cement.

Ooid Precipitation

The subject of ooid growth has already been discussed in detail (Bathurst, 8). An ooid consists typically of a detrital nucleus with successive coats (lamellae) of aragonite. The aragonite has the form of needles with preferred orientations of three kinds: tangential (this is dominant), radial, or haphazard. Lamellae may be replaced by masses of secondary micrite formed by the algal micritization process already discussed. In addition to calcium carbonate, ooids contain organic matter, several percent by weight (much more by volume). Some of this is the product of endolithic algal infestation, fungi, or bacteria. Some protein is, apparently, intimately mixed within the aragonite needles (Mitterer, 50). The provision of supersaturation and nuclei presents no problem in warm shallow shelf seas. The degree of agitation required is uncertain since oolitic lamellae are known to grow on grains in an algal mat in Bimini Lagoon, Bahamas (Bathurst, 7). The local pattern of wave and tidal movement allied to bed form is sufficient to keep ooids in the growth region for long periods.

The actual growth process is not easy to determine. Theory and the experiments of Weyl (87) suggest that the growth occurs while ooids are free to move at the sediment-water interface, although they spend most of their maritime existences buried inside megaripples. Associated with this problem of growth is the extraordinarily strong preferred tangential orientation of the common lamellae. It is not clear that this fabric fits any of the three current theories of ooid growth, namely that of snowball accumulation (Sorby, 7), that of mechanical modification of an earlier radial-fibrous fabric by grain collisions (Rusnak, 64; Usdowski, 85), or that which suggests guidance of the needles into preferred orientations by the distribution of electric charges on substrate and needle (Bathurst, 8; Lippmann in Bathurst, 8).

In the last few years the most significant progress in solving these questions has come from the application of SEM, particularly by Fabricius & Klingele (20), Loreau (45), and Loreau & Purser (60). These workers have shown that the surface of an ooid with tangential aragonite is an accumulation of aragonite needles with an unusual appearance (Figure 4). They are rod-like, elongate, about 1–2 μ, and parallel sided. The ends are flattened perpendicular to the length, giving an appearance of sliced match sticks. The crystals seem to be slightly rounded: no

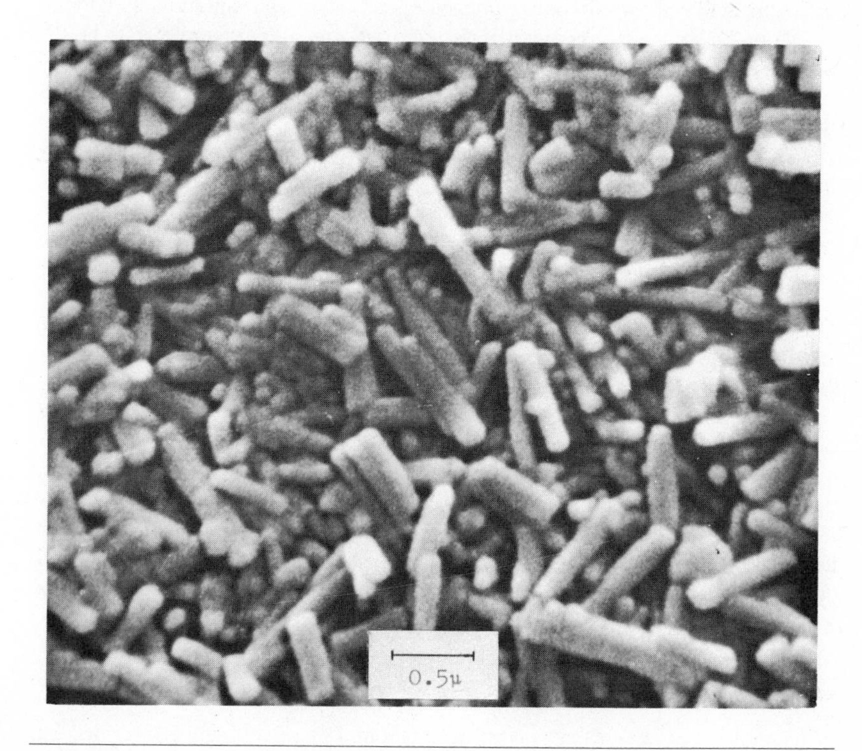

Figure 4 Rodlike aragonite needles in the surface of a Recent ooid from the Persian Gulf. SEM photograph courtesy of J. -P. Loreau.

distinct faces are visible. The crystals are thus unlike the typical marine aragonite cement crystals (Figure 1) with well defined faces and resemble them only in crude dimensions. Furthermore, the texture of the crystal accumulation is utterly at variance with the well-known organized, divergent bundles or pallisade structures of acicular cements. I must admit that the aragonite fabric of the ooid suggests to me a more or less compacted accumulation of loose rods. This conception certainly is related to that of Loreau & Purser (60) on the ooids of the Abu Dhabi region, Persian Gulf. Using SEM, they found that the aragonitic rods on the crests or bars or delta levées have a preferred tangential orientation with tight packing. In the quietest, least agitated lagoonal waters, the rods have a well developed radial-fibrous orientation and rather loose packing and the ooids tend to be larger and irregular in shape. Regions of intermediate hydraulic energy give mixed fabrics commonly with a more or less haphazard orientation. The loose radial-fibrous fabric may be changed to a compact tangential fabric when an ooid is moved onto a bar or levée. This corresponds to some extent with Donahue's (18) findings that, for spherulitic radial-fibrous cave ooids, there is a positive correlation between perfection of crystal orientation, compactness of fabric, and agitation. Finally, in these Abu Dhabi ooids, organic debris is mixed with the aragonite rods in the cortex, as in Jenkyns' (36) Jurassic ooids.

We are faced, therefore, with an apparent paradox. The radial-fibrous growth fabric suggests more or less normal intergranular cement, while the tangential growth fabric suggests highly selective mechanical accumulation accompanied by some form of adhesion. Or is this begging the question? Is the tangential fabric really a growth form developed during precipitation, or a modification of such a growth form? If it is, for example, a modification caused by grain-to-grain collision and pushing of broken radial needles into a tangential position, then how is this accomplished while maintaining adhesion? Or should we hark back to the ideas of Dangeard (14) and Nesteroff (53–55) and enlist the help of cyanophyte algae in the processes of precipitation and adhesion? Recent unpublished work by Fabricius with SEM has shown an organized packing of the rods which resembles organic secretion. Fabricius favors a phytogenetic origin. On the other hand a less direct biological control of precipitation is indicated by the work of Mitterer (47–50) who showed that the proteins in ooids are similar to those which direct much skeletal calcification. Indeed, Suess & Fütterer (81) were able to grow radial-fibrous ooids in the laboratory by precipitating aragonite in the presence of dissolved humic acid in supersaturated artificial seawater. These ooids had alternating concentric lamellae of aragonite (0.5–5.0 μ thick) and humate membranes (<0.2 μ thick). One is reminded here of the organic lamellae detected in Jurassic ooids by Shearman et al (73).

Dissolution in Temperate Waters

In the cold sea waters of the Swedish Kattegat and Skagerrak, undersaturated for calcium carbonate, grains of skeletal carbonate undergo selective dissolution (Alexandersson, 3). Rates of dissolution vary with the type and orientation of microfabric (e.g. layers of crossed-lamellar structure in mollusks). The result is a highly porous grain, brittle, friable, chalky white in reflected light, dark in transmitted light. The destruction of grains is aided by microboring of algae and clionid sponges, but these bores, unlike their warm water counterparts, remain empty.

Controls of Cement Precipitation

Ideas regarding the factors which control the site and type of precipitation of aragonite and calcite cements are still largely embryonic. In seawater supersaturated with calcium carbonate, why is precipitation of cement not universal on all surfaces? Instead, cements are found only on certain preferred surfaces, normally in small cavities, in grapestone, or on ooid surfaces.

Mineralogy and habit are variable, differing commonly among adjacent cavities in the same grain or reef-rock. Aragonite is acicular with long axes from about 0.1 μ to about 150 μ (Figure 1). The aragonite rods in ooids have their distinctive rounded form with lengths about 1–2 μ. The high-magnesian calcite (commonly 15–18 mole% $MgCO_3$) is rhombic (0.5–8.0 μ), equant (Figure 2) or radial-fibrous parallel to *c*, or bladed 30–40 μ. Calcite crystal faces may be curved or plain.

The order in which the various cements are deposited is variable. In some regions high-magnesian calcite always precedes aragonite, in others aragonite precedes calcite. The two polymorphs are occasionally seen to be mixed. Generally a more organized uniform fringe of needle bundles, or a pallisade structure of aragonite, or of micritic rhombic calcite, is succeeded by a less orderly arrangement of larger aragonite needles or irregular masses of micritic calcite. Commonly the calcite rhombs grow as globular rosettes with a pellet-like appearance. Whereas the aragonite pallisade or bundle cements show a drusy increase of crystal size from older to younger, there is normally no sign of such a vectorial change in the rhombic calcites. (Are these randomly distributed crystallographically?) Other distribution patterns are the meniscus cement of Dunham (19) and the stalactitic and stalagmitic forms of Taylor & Illing (83) and Loreau & Purser (60).

The influence of the substrate mineralogy on that of the encrusting cement commonly results in matching between host and cement with, for aragonite, in some instances a clear lattice continuity. Deposition of mud commonly accompanies cementation and may interrupt the mineralogical growth sequence. Organic films, perhaps monolayers invisible in SEM photographs, may either interrupt the sequence or positively induce a new mineralogy or habit. Replacement of aragonite cements by high-magnesian calcite has been reported by Taylor & Illing (83) and Shinn (74).

The microenvironments are obviously difficult to monitor despite use of SEM and the microelectron probe. Mass studies of pore water offer a complementary approach. Supersaturation alone will not assure precipitation and some kind of substrate, mineralogic or organic, is essential to permit heterogeneous nucleation (discussion in Bathurst, 8; Wollast, 89). As from 3,000 to 100,000 unit volumes of seawater are needed to produce one unit volume of cement (discussion in Bathurst, 8; Schroeder, 68), circulation of water is essential. Thus cementation is most rapid adjacent to turbulent water, as on the west side of the Bahama Bank

(Cloud, 13), on reef flats and fore-reef escarpments (down to 110 m), and on groove and spur (Ginsburg et al, 29; James et al, 35). Against the need for circulation must be balanced the need for stability. Cementation is slow enough to need a surface free from collision-abrasion and, for intergranular cementation, absence of bed load traction. Thus intragranular pores, lightly buried sediments, grains embedded in algal mats, and reef voids are favored sites. In time circulation can be blocked by the very cementation it encourages. Growth can also be inhibited, it would seem, by organic films. Nevertheless, the almost total absence of cement on so many external grain surfaces is mysterious (ooids may be regarded as an exception). Alexandersson (2) has tentatively attributed this to the organism-dependent diurnal variations in carbon dioxide tension and in carbonate saturation level described by Revelle & Emery (62), Schmalz (65), Schmalz & Swanson (67), and Weber & Woodhead (86). As dissolution is more rapid than precipitation, frequent occurrences of undersaturation could cause destruction of the supersoluble embryonic crystals. It is necessary, however, to beware of over generalizing, as sites of cementation vary enormously between beach rock, subtidal floor, and reef interiors.

Factors controlling the mineralogy of the cement must be subtle, as aragonite and high-magnesian calcite can be mixed in one place. Wollast (90) calculated that the Sr/Mg ratio in the solution is influential. De Groot (17) has suggested that the well-known inhibitory effect of magnesium ions on calcite precipitation (discussion in Bathurst, 8) is perhaps reduced to insignificance at very low supersaturations of aragonite and high-magnesian calcite. Friedman, Amiel & Schneidermann (23) have noted the role of high pH.

The influences of organic matter and organic processes are being increasingly studied. Whereas Ginsburg et al (28) found that $\delta^{18}O$ and $\delta^{13}C$ values for calcite cement indicated inorganic precipitation in equilibrium with seawater, Lloyd (44) found values of $\delta^{13}C$ sufficiently positive in ooids and cryptocrystalline grains to indicate isotopic fractionation and thus perhaps organic control. The possibility of an organic effect on ooid growth has already been mentioned. The roles of organic materials are diverse. Schroeder (69) has recently shown how the filaments of the endolithic alga *Ostreobium,* when living free in a cavity, become calcified. Chave (11) and Chave & Suess (12) have shown that organic films, naturally occurring on many carbonate grains, can isolate these grains and inhibit surface reactions with the enclosing seawater. For example, a hydrophobic adsorbed layer can be made with a fatty acid, the carboxyl end interacting with the carbonate surface but the water-repellent hydrocarbon end projecting outward (Suess, 80). Carboxyl groups can substitute for carbonate ions in the crystal lattice, perhaps accounting for the binding of aspartic acid-rich protein to carbonate surfaces. Jackson & Bischoff (34) found that amino acids with free hydroxyl groups are selectively adsorbed on carbonate compounds. Acidic amino acids inhibit the wet transformation of aragonite to calcite at 66°C, by such overgrowths as aspartic and glutamic acids. Basic and neutral amino acids, on the other hand, tend to accelerate the transformation catalytically. Organic layers, protein-polysaccharide complexes, clearly act as substrate-templates that control the mineralogy of an overlying precipitate (Towe & Cifelli, 84; Mitterer, 48). The high charge density on a large protein molecule could not only balance numerous grain surface charges, but with its own surplus ionizable groups could complex dissolved calcium ions. A sufficient quantity of adjacent adsorbed protein chains might raise the local concentration of calcium enough to start nucleation of a mineral phase (Mitterer, 50). In this general sphere research such as that of Hubbard (33) must be of increasing importance. She has demonstrated the complex histories and relationships of primary cavities in modern scleractinians and secondary cavities formed by boring. She has shown the varied influences that living tissues, skeletal precipitation, sediment filling, organic coating versus dead cavity surfaces, and the degree of free circulation have on the chemical microenvironments.

APPLICATION TO THE ANCIENT

Application of the advances discussed in the previous section to the development of a deeper understanding of diagenetic processes in the past has shown marked progress in the last few years, though much thinking remains tentative and little has yet been published. A brief attempt to summarize the current trends of thought will be made here, based largely on informal discussion of work still in progress.

There is a widespread realization that a micrite is just as likely to be an altered marine cement as an altered lime mud. The old fable about the value of the presence or absence of micrite as a criterion for "low" or "high" energy deposition is exposed as the hydrodynamic nonsense it nearly always was. The sheer complexity of carbonate sediments and the need for a broadminded approach to interpretation has been vividly demonstrated by George (25). We must look for micrite fringes on grains and irregular clusters of pellety micrite or patches of geopetal micrite in the pores of calcarenites. Even an apparently simple micrite matrix in a biomicrite may be an altered cement. A pale brown color may be inherited from an original cement. Recently, while I was teaching a class, a thin section of a Jurassic biomicrite reminded me of a section I had seen of a Recent hardground. So I looked more closely: I saw an irregular brown line of iron oxide across the slide which clearly truncated fossil debris on one side only. On this side also there were characteristic bores filled with geopetal sediment—a miniature hardground. So my "low" energy biomicrite became a marine (?) cemented "high" energy lime sand. Stalactitic clusters of micritic pellets may indicate a marine vadose environment; so may cement fringes preferentially thicker on the undersides of grains (e.g. Purser, 59). A great many micrites have a pellety structure with individual clots perhaps 10–30 μ in diameter. Could these be one-time rosettes of high-magnesian calcite cement? Any micrite occupying a microcavity, such as the chamber of a foraminiferid, is at once suspect as a cement fill. Probably most geopetal lime muds were cemented soon after deposition.

Fibrous fabrics or fabrics even vaguely reminiscent of fibers need to be examined particularly closely. If traces of the lamellar structure of a mollusk shell can survive meteoric in situ calcitization, then why not the fabric of an originally fibrous cement? Fibrous calcite in ancient limestones may be a relic fabric, for example in the elongate crystals with undulose extinction found in radiaxial fibrous mosaic (Bathurst, 5, 8, 9; Kendall & Tucker, 37). Ziz-zag lines of inclusions may mark old crystal growth fronts. A major advance in the interpretation of such fabrics as these has recently been made in a very carefully reasoned paper by Kendall & Tucker (38). They examined, in thin section, elongate calcite cement crystals in longitudinal section and cross section, noting patterns of optic axes and inclusions. They argued convincingly that originally fibrous cements were replaced in situ by larger elongate sparry crystals of low magnesian calcite, which have internal microfabrics (such as undoluse extinction and inclusions) that reflect in fine detail the original acicular bundles.

Hitherto unidentifiable blobs of micrite can now often be matched with Recent micritized grains; irregular clusters of grains, especially micritic peloids, can be related to the intensely micritized Recent grapestones.

Great advances have been made in the study of ancient reef-rock, using this term broadly. Students of modern reefs[1] have emphasized that much reef-rock is the product of a dynamic relationship between organic frame building accompanied by organic encrustation and alteration by boring, internal sedimentation, and cementation (Zankl, 91; Garrett et al, 24; Zankl & Schroeder, 93; Scoffin, 72). In the ancient we must seek, not frameworks, but their complex alteration products (Figure 3). Excellent illustrations of these varied and complicated relationships are given by Zankl (92), Krebs (41, 42), Scoffin (70, 71), and Schroeder (68), who bring their experience of the Recent to bear upon ancient reefoid or biohermal structures. Valuable evidence for early marine cementation in a Waulsortian-type bioherm is given by Stone (78).

Nodular structures, or concretions, which may coalesce locally to give continuous sheets, have aroused increased interest since the discovery of presumed modern equivalents in the Persian Gulf. Hallam (32) gave clear evidence for early exhumation, encrustation, and boring of Jurassic concretions, a story amply confirmed by Bromley (10) on Chalk concretions coated with glauconite. R. G. Bromley[2] and C. V. Jeans[3] have found much evidence of an early, precompactive nodular lithification of the Chalk, associated with shallowing of the sea and substantially reduced sedimentation rates. Jeans' studies of the ferroan pigment in red Chalk point to a strongly negative Eh for the early pore water and the action of anaerobic bacteria. Well-known nodular limestones such as the Ammonitico Rosso are being carefully reexamined in this context. The work of Raiswell (61) on the growth of concretions must have considerable influence.

A word must be said about the important developments in our knowledge of tepee structures which are, at least in part, a result of complex marine diagenetic

[1] T. P. Scoffin (University of Edinburgh) and J. H. Schroeder (Technische Universität, Berlin) in a recent unpublished seminar in Britain; also P. Garrett (University College of Aberystwyth), R. N. Ginsburg (University of Miami), L. Land (University of Texas), and H. Zankl (Philipps Universität, Marburg).

[2] University of Copenhagen.

[3] University of Cambridge.

processes. Assereto (4) and Smith (76), in particular, have described nontectonic, polygonal anticlinal structures from the Permian of the Guadalupe Mountains and many other formations from Silurian to Recent. In plan these vary from ~1 m to ~20 m across. The amplitudes of the anticlines vary from a few centimeters to about 7 m. The internal structures of the anticlines are a complex mixture of multiple generations of cement, brecciated and overthrust sedimentary laminae, and internal sediments of marine origin. Sturani (79) has found in possibly similar structures in the Venetian Alps, Bajocian internal sediments filling cavities in brecciated Liassic lime mudstones. Assereto and Smith have noted that the laminae, which are fine grained and normally fenestrate, may locally attain dips of 50° or even be vertical or overturned. Tepees typically develop on top of earlier tepees giving a vertical chevron structure. Their possible relation to somewhat similar Recent polygons described by Shinn (74) off the Trucial Coast in the Persian Gulf has not escaped attention. They are commonly dolomitic. Current thinking indicates a shallow subtidal to low supratidal origin, associated with crystal growth pressure of cements and desiccation.

EPILOGUE

Finally, it is necessary to stress that the researches here described are but tentative early advances in a largely new field—so new that the greatest obstacle to progress is simply a lack of data. Faced with new problems we must go back to the beginning and patiently start again the task of describing and measuring before we can hope to make secure generalizations.

Acknowledgments

I warmly acknowledge the invaluable help of my wife, Dr. J. H. Schroeder, and Dr. T. P. Scoffin in criticizing an earlier draft of the review; also the generous donations of photographs by Dr. T. Alexandersson, Dr. J.-P. Loreau, and Dr. J. H. Schroeder.

Literature Cited

1. Alexandersson, T. 1969. Recent littoral and sublittoral high-Mg calcite lithification in the Mediterranean. *J. Sediment. Petrology* 12:47–61
2. Alexandersson, T. 1972. Intragranular growth of marine *aragonite* and Mg-*calcite*: evidence of precipitation from supersaturated seawater. *J. Sediment. Petrology* 42:441–60
3. Alexandersson, T. 1972. Micritization of carbonate particles: processes of precipitation and dissolution in modern shallow marine sediments. *Bull. Geol. Inst. Univ. Uppsala* N.S. 3:201–36
4. Assereto, R. L., Kendall, C. G. St. C. 1971. Megapolygons in Landinian limestones of Triassic of southern Alps: evidence of deformation by penecontemporaneous desiccation and cementation. *J. Sediment. Petrology* 41:715–23
5. Bathurst, R. G. C. 1959. The cavernous structure of some Mississippian *Stromatactis* reefs in Lancashire, England. *J. Geol.* 67:506–21
6. Bathurst, R. G. C. 1966. Boring algae, micrite envelopes and lithification of molluscan biosparites. *Geol. J.* 5:15–32
7. Bathurst, R. G. C. 1967. Oölitic films on low energy carbonate sand grains, Bimini Lagoon, Bahamas. *Mar. Geol.* 5:89–109
8. Bathurst, R. G. C. 1971. *Carbonate*

Sediments and their Diagenesis. Amsterdam: Elsevier. 620 pp.

9. Bathurst, R. G. C. 1971. Radiaxial fibrous mosaic. *Carbonate Cements,* ed. O. P. Bricker, 292–93. Baltimore: Johns Hopkins. 360 pp.
10. Bromley, R. G. 1965. *Studies in the lithology and conditions of sedimentation of the Chalk Rock and comparable horizons.* PhD thesis. Univ. London. 355 pp. Unpublished
11. Chave, K. E. 1965. Carbonates: association with organic matter in surface seawater. *Science* 148:1723–24
12. Chave, K. E., Suess, E. 1967. Suspended minerals in seawater. *Trans. NY Acad. Sci. Ser. II* 29:991–1000
13. Cloud, P. E. Jr. 1962. Environment of calcium carbonate deposition west of Andros Island, Bahamas. *US Geol. Surv. Prof. Pap.* 350:1–138
14. Dangeard, L. 1936. Étude des calcaires par coloration et décalcification. Application à l'étude des calcaires oolithiques. *Bull. Soc. Géol. Fr.* (5)6:237–45
15. Davies, P. J., Kinsey, D. W. 1973. Organic and inorganic factors in recent beach rock formation, Heron Island, Great Barrier Reef. *J. Sediment. Petrology* 43:59–81
16. De Groot, K. 1965. Inorganic precipitation of calcium carbonate from seawater. *Nature* 207:404–5
17. De Groot, K. 1969. The chemistry of submarine cement formation at Dohat Hussain in the Persian Gulf. *Sedimentology* 12:63–68
18. Donahue, J. D. 1969. Genesis of oölite and pisolite grains: an energy index. *J. Sediment. Petrology* 39:1399–1411
19. Dunham, R. J. 1971. Meniscus cement. *Carbonate Cements,* ed. O. P. Bricker, 297–300. Baltimore: Johns Hopkins. 360 pp.
20. Fabricius, F. H., Klingele, H. 1970. Ultra strukturen von Ooiden und Oolithen: zur Genese und Diagenese quartärer Flachwasserkarbonate des Mittelmeeres. *Verh. Geol. Bundesanst. B.-A.* 4:594–617
21. Frémy, P. 1945. Contribution à la physiologie des Thallophytes marins perforant et cariant les roches calcaires et les coquilles. *Ann. Inst. Océanogr. (Paris)* 22:107–43
22. Friedman, G. M., Gebelein, C., Sanders, J. E. 1971. Micritic carbonate grains are not exclusively of photosynthetic algal origin. *Sedimentology* 16:89–96
23. Friedman, G. M., Amiel, A. J., Schneidermann, N. Submarine cementation in reefs: example from the Red Sea. *J. Sediment. Petrology* In press
24. Garrett, P., Smith, D. L., Wilson, A. O., Patriquin, D. 1971. Physiography, ecology, and sediments of two Bermuda patch reefs. *J. Geol.* 79:647–68
25. George, T. N. 1972. The classification of Avonian limestones. *J. Geol. Soc. London* 128:221–56
26. Ginsburg, R. N., Shinn, E. A., Schroeder, J. H. 1968. Submarine cementation and internal sedimentation within Bermuda reefs (Abstr.). *Geol. Soc. Am. Spec. Pap.* 115:78–79
27. Ginsburg, R. N., Marszalek, D. S., Schneidermann, N. 1971. Ultrastructure of carbonate cements in a Holocene algal reef in Bermuda. *J. Sediment. Petrology* 41:472–82
28. Ginsburg, R. N., Schroeder, J. H., Shinn, E. A. 1971. Recent synsedimentary cementation in subtidal Bermuda reefs. *Carbonate Cements,* ed. O. P. Bricker, 54–56. Baltimore: Johns Hopkins. 360 pp.
29. Ginsburg, R. N., James, N. P., Marszalek, D. S., Land, L. S., Lang, J. 1973. Sedimentation and diagenesis in the deep forereef, British Honduras barrier and atoll reefs (Abstr.). *Am. Assoc. Petrol. Geol. Bull.* 57:781
30. Glover, E. D., Pray, L. C. 1971. High-magnesian calcite and aragonite cementation within modern subtidal carbonate sediment grains. *Carbonate Cements,* ed. O. P. Bricker, 80–87. Baltimore: Johns Hopkins. 360 pp.
31. Golubić, S., Brent, G., Lecampion, T. 1970. Scanning electron microscopy of endolithic algae and fungi using a multipurpose casting embedding technique. *Lethaia* 3:203–9
32. Hallam, A. 1969. A pyritized limestone hardground in the Lower Jurassic of Dorset (England). *Sedimentology* 12:231–40
33. Hubbard, J. A. E. B. 1972. Cavity formation in living scleractinian reef corals and fossil analogues. *Geol. Rundsch.* 61:551–64
34. Jackson, T. A., Bischoff, J. L. 1971. The influence of amino acids on the kinetics of the recrystallization of aragonite to calcite. *J. Geol.* 79:493–97
35. James, N. P., Ginsburg, R. N., Marszalek, D. S., Choquette, P. W. 1973. Subsea cementation of shallow British Honduras reefs (Abstr.). *Am. Assoc. Petrol. Geol. Bull.* 57:786
36. Jenkyns, H. C. 1972. Pelagic "oolites" from the Tethyan Jurassic. *J. Geol.* 80:21–33
37. Kendall, A. C., Tucker, M. E. 1971.

Radiaxial fibrous calcite as a replacement after syn-sedimentary cement. *Nature Phys. Sci.* 232:62–63

38. Kendall, A. C., Tucker, M. E. 1973. Radiaxial fibrous calcite: a replacement after acicular carbonate. *Sedimentology* 20:365–89
39. Kendall, C. G. St. C., Skipwith, P. A. d'E. 1969. Holocene shallow-water carbonate and evaporite sediments of Khor al Bazam, Abu Dhabi, southwest Persian Gulf. *Am. Assoc. Petrol. Geol. Bull.* 53:841–69
40. Kohlmeyer, J. 1969. The role of marine fungi in the penetration of calcareous substances. *Am. Zool.* 9:741–46
41. Krebs, W. 1969. Early void-filling cementation in Devonian fore-reef limestones (Germany). *Sedimentology* 12: 279–99
42. Krebs, W. 1972. Facies and development of the Meggen reef (Devonian, West Germany). *Geol. Rundsch.* 61:647–71
43. Land, L. S., Goreau, T. F. 1970. Submarine lithification of Jamaican reefs. *J. Sediment. Petrology* 40:457–62
44. Lloyd, R. M. 1971. Some observations on Recent sediment alteration ("micritization") and the possible role of algae in submarine cementation. *Carbonate Cements,* ed. O. P. Bricker, 72–79. Baltimore: Johns Hopkins. 360 pp.
45. Loreau, J.-P. 1970. Ultrastructure de la phase carbonatée des oolithes marines actuelles. *C. R. Acad. Sci.* 271:816–19

45a. Loreau, J.-P., Purser, B. H. 1971. Distribution and ultrastructure of Holocene ooids in the Persian Gulf. *The Persian Gulf—Holocene Carbonate Sedimentation and Diagenesis in a Shallow Epicontinental Sea,* ed. B. H. Purser, 279–328. Heidelberg: Springer. 471 pp.

46. Margolis, S., Rex, R. W. 1971. Endolithic algae and micrite envelope formation in Bahamian oölites as revealed by scanning electron microscopy. *Geol. Soc. Am. Bull.* 82:843–52
47. Mitterer, R. M. 1968. Amino acid composition of organic matrix in calcareous oolites. *Science* 162:1498–99
48. Mitterer, R. M. 1971. Influence of natural organic matter on $CaCO_3$ precipitation. *Carbonate Cements,* ed. O. P. Bricker, 252–58. Baltimore: Johns Hopkins. 360 pp.
49. Mitterer, R. M. 1972. Calcified proteins in the sedimentary environment. *Advances in Organic Geochemistry 1971,* ed. H. R. V. Gaertner, H. Wehner, 441–51. Oxford: Pergamon. 736 pp.
50. Mitterer, R. M. 1972. Biogeochemistry of aragonite mud and oolites. *Geochim. Cosmochim. Acta* 36:1407–22
51. Moore, C. H. Jr. Intertidal carbonate cementation Grand Cayman, West Indies. *J. Sediment. Petrology.* In press
52. Mutvei, H. 1964. On the shells of *Nautilus* and *Spirula* with notes on the shell secretion in non-cephalopod molluscs. *Ark. Zool.* 16:221–78
53. Nesteroff, W. D. 1955. De l'origine des dépôts calcaires. *C. R. Acad. Sci.* 240: 220–22
54. Nesteroff, W. D. 1956. De l'origine des oolithes. *C. R. Acad. Sci.* 242:1047–49
55. Nesteroff, W. D. 1956. La substratum organique dans les dépôts calcaires, sa signification. *Bull. Soc. Géol. Fr.* 6(6): 381–90
56. Perkins, R. D., Halsey, S. D. 1971. Geologic significance of microboring fungi and algae in Carolina shelf sediments. *J. Sediment. Petrology* 41:843–53
57. Pingitore, N. E. Jr. 1971. Submarine precipitation of void-filling needles in Pleistocene coral. *Carbonate Cements,* ed. O. P. Bricker, 68–71. Baltimore: Johns Hopkins. 360 pp.
58. Purdy, E. G. 1968. Carbonate diagenesis: an environmental survey. *Geol. Rom.* 7:183–228
59. Purser, B. H. 1969. Syn-sedimentary marine lithification of Middle Jurassic limestones in the Paris Basin. *Sedimentology* 12:205–30
60. Purser, B. H., Ed. See Ref. 45a
61. Raiswell, R. 1971. The growth of Cambrian and Liassic concretions. *Sedimentology* 17:147–71
62. Revelle, R., Emery, K. O. 1957. Chemical erosion of beach rock and exposed reef rock. *US Geol. Surv. Prof. Pap.* 260-T: 699–709
63. Rooney, W. S., Perkins, R. D. 1972. Distribution and geologic significance of microboring organisms within sediments of the Arlington Reef Complex, Australia. *Geol. Soc. Am. Bull.* 83:1139–50
64. Rusnak, G. A. 1960. Some observations of recent oolites. *J. Sediment. Petrology* 30:471–80
65. Schmalz, R. F. 1967. Kinetics and diagenesis of carbonate sediments. *J. Sediment. Petrology* 37:60–67
66. Schmalz, R. F. 1971. Formation of beach rock at Eniwetok Atoll. *Carbonate Cements,* ed. O. P. Bricker, 17–24. Baltimore: Johns Hopkins. 360 pp.
67. Schmalz, R. F., Swanson, F. J. 1969. Diurnal variations in the carbonate saturation of sea water. *J. Sediment. Petrology* 39:255–67

68. Schroeder, J. H. 1972. Fabrics and sequences of submarine carbonate cements in Holocene Bermuda cup reefs. *Geol. Rundsch.* 61:708–30
69. Schroeder, J. H. 1972. Calcified filaments of an endolithic alga in Recent Bermuda reefs. *Neues Jahrb. Geol. Palaeontol. Monatsch.* H.1:16–33
70. Scoffin, T. P. 1971. The conditions of growth of the Wenlock reefs of Shropshire (England). *Sedimentology* 17:173–219
71. Scoffin, T. P. 1972. Cavities in the reefs of the Wenlock Limestone (Mid-Silurian) of Shropshire, England. *Geol. Rundsch.* 61:565–78
72. Scoffin, T. P. 1972. Fossilization of Bermuda patch reefs. *Science* 178:1280–83
73. Shearman, D. J., Twyman, J., Karimi, M. Z. 1970. The genesis and diagenesis of oolites. *Proc. Geol. Assoc.* 81:561–75
74. Shinn, E. A. 1969. Submarine lithification of Holocene carbonate sediments in the Persian Gulf. *Sedimentology* 12:109–44
75. Shinn, E. A. 1971. Holocene submarine cementation in the Persian Gulf. *Carbonate Cements,* ed. O. P. Bricker, 63–65. Baltimore: Johns Hopkins. 360 pp.
76. Smith, D. B. 1974. The origin of tepees in Upper Permian shelf carbonates of the Guadalupe Mountains. *Am. Assoc. Petrol. Geol. Bull.* In press
77. Sorby, H. C. 1879. The structure and origin of limestones. *Proc. Geol. Soc. London* 35:56–95
78. Stone, R. A. 1972. Waulsortian-type bioherms (reefs) of Mississippian age, central Bridger Range, Montana. *Ann. Field Conf. Montana Geol. Soc. 21st,* 37–55
79. Sturani, C. 1971. Ammonites and stratigraphy of the "*Posidonia Alpina*" beds of the Venetian Alps. *Mem. Ist. Geol. Mineral. Padova* 28:1–190
80. Suess, E. 1970. Interaction of organic compounds with calcium carbonate. I. Association phenomena and geochemical implications. *Geochim. Cosmochim. Acta* 34:157–68
81. Suess, E., Fütterer, D. 1971. Aragonitic ooids: experimental precipitation from seawater in the presence of humic acid. *Sedimentology* 19:129–39
82. Swichatt, J. P. 1969. Algal boring: a possible depth indicator in carbonate rocks and sediments. *Geol. Soc. Am. Bull.* 80:1391–96
83. Taylor, J. M. C., Illing, L. V. 1969. Holocene intertidal calcium carbonate cementation, Qatar, Persian Gulf. *Sedimentology* 12:69–107
84. Towe, K. M., Cifelli, R. 1967. Wall ultrastructure in the calcareous foraminifera: crystallographic aspects and a model for calcification. *J. Paleontol.* 41:742–62
85. Usdowski, H.-E. 1963. Der Rogenstein des norddeutschen Unteren Buntsandsteins, ein Kalkoölith des marinen Faziesbereichs. *Fortschr. Geol. Rheinl. Westfalen* 10:337–42
86. Weber, J. N., Woodhead, P. M. J. 1971. Diurnal variations in the isotopic composition of dissolved inorganic carbon in sea water from coral reef environments. *Geochim. Cosmochim. Acta* 35:891–902
87. Weyl, P. K. 1967. The solution behaviour of carbonate materials in sea water. *Stud. Trop. Oceanogr., Univ. Miami* 5:178–228
88. Winland, H. D. 1968. The role of high Mg calcite in the preservation of micrite envelopes and textural features of aragonite sediments. *J. Sediment. Petrology* 38:1320–25
89. Wollast, R. 1971. Kinetic aspects of the nucleation and growth of calcite from aqueous solutions. *Carbonate Cements,* ed. O. P. Bricker, 264–73. Baltimore: Johns Hopkins. 360 pp.
90. Wollast, R., Debouverie, D., Duvigneaud, P. H. 1971. Influence of Sr and Mg on the stability of calcite and aragonite. *Carbonate Cements,* ed. O. P. Bricker, 274–77. Baltimore: Johns Hopkins. 360 pp.
91. Zankl, H. 1968. Sedimentological and biological characteristics of a Dachsteinkalk reef complex in the Upper Triassic of the Northern Calcareous Alps. In *Recent Developments in Carbonate Sedimentology in Central Europe,* ed. G. Müller, G. M. Friedman, 215–18. Berlin: Springer. 255 pp.
92. Zankl, H. 1969. Structural and textural evidence of early lithification in fine-grained carbonate rocks. *Sedimentology* 12:241–56
93. Zankl, H., Schroeder, J. H. 1972. Interaction of genetic processes in Holocene reefs off North Eleuthera, Bahamas. *Geol. Rundsch.* 61:520–41

EARTHQUAKE MECHANISMS AND MODELING[1]

James H. Dieterich
National Center for Earthquake Research, U.S. Geological Survey,
Menlo Park, California 94025

INTRODUCTION

In a general discussion of earthquake source processes, Kasahara (43) states that "The function of an earthquake is often compared to that of a machine which accumulates energy from a deeper source to convert it instantaneously into kinetic energy or seismic disturbance." Our understanding of this machine draws extensively from the findings of a number of disciplines within the general fields of tectonophysics and seismology. The past ten or fifteen years have seen especially vigorous activity in these fields, and advances in the understanding of earthquakes owe much to the dramatic growth and development of experimental rock deformation, seismic source theory, and the concepts of plate tectonics and sea-floor spreading during this period. Following Kasahara's analogy, we may explain that the intent of this paper is to explore the current theories for the underlying principles of operation of the earthquake machine. Emphasis is given, specifically, to instability mechanisms that are the basis for the functioning of the machine and to some details of mechanical interactions in the source region. Our interest here is with possible earthquake mechanisms deduced from theoretical and experimental studies, including numerical models that incorporate the instability mechanisms and source interactions.

Earthquake modeling is a diverse subject that could be taken to include laboratory analogue models of earthquake processes, numerical and dislocation analyses of static displacements near shallow earthquake sources, and dynamic analyses that simulate the generation and propagation of seismic waves from the source region. The emphasis in this paper is restricted to a class of deterministic numerical models that incorporate physical mechanisms for earthquakes to help explain observed earthquake phenomena.

ELASTIC REBOUND THEORY

The beginnings of modern theory for the origins of earthquakes must be ascribed to Reid (65, 66) with the formulation of the elastic rebound theory for the 1906

[1] Publication authorized by the Director, U.S. Geological Survey.

California earthquake. Reid proposed that earthquakes are caused by sudden slip along a fault that releases stored elastic strain energy in the form of seismic radiation. Prior to Reid's work, a causal relationship between earthquakes and fault movements was occasionally noted, but it was generally held that fault movements resulted from earthquakes in much the manner that landslides or other damage resulted from earthquakes [see Richter (67) for historical summary].

Reid based his theory on observations of slip along the San Andreas fault and on accurate surveys in the vicinity of the fault made in the years 1851–1865, 1874–1892, and 1906–1907. These observations allowed Reid to deduce the nature of ground displacements prior to and after the earthquake. On examination of these data Reid found that, in general, ground movements were parallel to the fault. From his analysis, Reid concluded that the process consisted first of a slow accumulation of strain energy, perhaps over a period of 100 years prior to the earthquake, as the rocks some distance on either side of the fault were displaced parallel to the fault and in opposite directions. Strain energy continued to accumulate until a critical stress for fracturing along the fault was reached. With the onset of fracturing an elastic rebound occurred, accompanied by rapid slip and the release of the accumulated strain energy in the form of seismic waves.

INSTABILITY MECHANISMS

Brittle fracture is a familiar instability process and is attractive as an earthquake mechanism. However, a difficulty with the elastic rebound theory has been the apparent contradiction of employing a mechanism that requires fracturing of already faulted rock. Reid discussed the problem to a certain extent in terms of a slow healing of the fault following an earthquake, perhaps by cementation or some other process.

Early experimental support for Reid's mechanism came from experiments by Bridgman (11) to pressures of 50 kbar in a shear anvil testing apparatus. With this apparatus, thin wafer-shaped specimens are compressed between opposing pistons that operate on an anvil. To induce shear, the anvil is rotated with respect to the pistons. In experiments on brittle rock-forming minerals, Bridgman sometimes observed snapping accompanied by sudden stress drops as the shearing progressed. Because the snapping occurred repeatedly, it may be concluded that the materials were in a highly degraded state and that the process was not simple loss of cohesion by fracturing of intact materials. Somehow, healing restored the cohesion in the material between snaps. Although the phenomena observed by Bridgman may not be related to fracture (33), the process, if it is applicable to naturally deforming rocks at high confining pressures, provides a mechanism for episodic slip and earthquake generation along the general lines proposed by Reid.

At confining pressures in the range 2 to 10 kbar, the mechanisms of earthquakes on shallow faults were clarified substantially by Brace & Byerlee (7) who show that slip on previously faulted surfaces of rock does not always proceed in a smooth fashion, but rather frequently occurs intermittently with a slow episodic buildup of stress, followed by rapid slip and a near instantaneous drop of stress (Figure 1). This

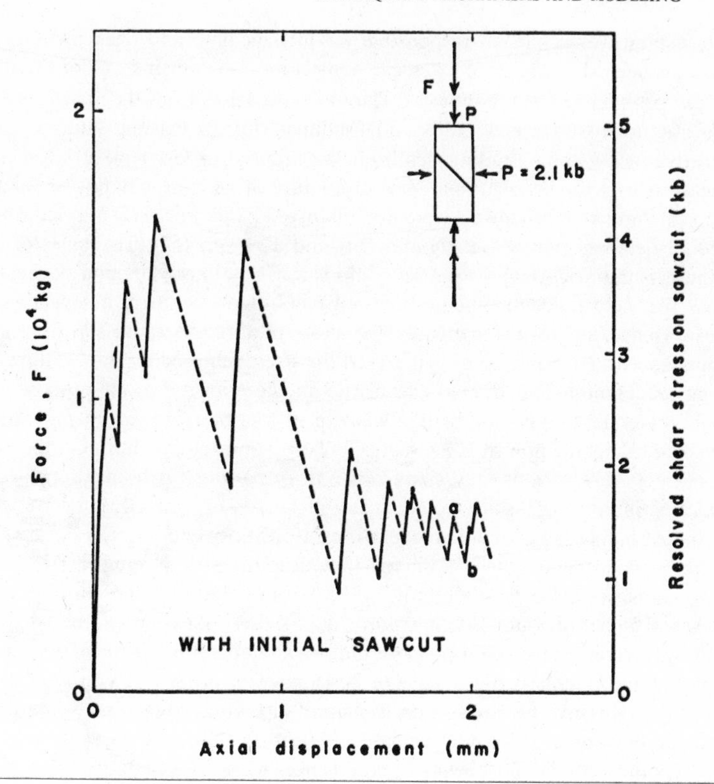

Figure 1 Force-displacement curve illustrating stick-slip on sawcut of Westerly granite at 2.1 kbar continuing pressure (6).

behavior is termed stick-slip, a designation long used for similar frictional instability on metallic surfaces. The mechanism of stick-slip, while not well understood in itself, overcomes the vagueness of Reid's original concept of slow healing of a fault following an earthquake and demonstrates that fracture-like instabilities are possible on existing faults. Additionally, Brace and Byerlee note that stick-slip, as compared to fracture at equivalent confining pressures, takes place at somewhat lower shear stress and with lower stress drops. Stress drop is an important consideration because the stress drops computed for earthquakes usually fall in the range 1 to 100 bars, which is orders of magnitude below the stress drops that can occur during fracture. The problem has not been entirely resolved, however; in the laboratory, stick-slip often displays stress drops of one or two orders of magnitude greater than that generally observed for earthquakes. In general, stick-slip or fracture is presently recognized as the principal instability mechanism for shallow earthquakes, and recent emphasis in laboratory studies has been on the examination of factors that control or affect these processes.

The instability mechanism responsible for intermediate and deep focus earthquakes has been the subject of extensive conjecture and remains one of the more persistent problems in tectonophysics. The difficulty arises from the argument that simple fracture to form new faults and frictional slip on existing faults become inoperative below some limiting depth on the order of a few tens of kilometers. The reasons why the conventional view of fracture or friction may not be suitable as deep earthquake mechanisms were first discussed by Jeffreys (42) and have been treated in detail by Griggs & Handin (34) and Orowan (57). The basis of this hypothesis is that frictional resistance to slip is a nearly linear function of pressure whereas the yield point for ductile flow is by comparison nearly independent of pressure. Hence, as pressure increases, the stress required to cause slip on a fault also increases and at some point will exceed the stress required for ductile flow.

Griggs & Handin (34) discuss the brittle-ductile transition and propose that geologic materials that exhibit brittle behavior at low pressures and temperatures will deform by ductile flow at some higher pressure and temperature. On the basis of laboratory observations they characterize deformations in the brittle range as culminating in faulting that is accompanied by sudden drop in stress from loss of cohesion. At higher pressures or temperatures, transitional behavior is observed in laboratory experiments whereby ductile flow accompanies faulting. In this case, faulting is distributed over some finite width with no loss of cohesion and stress drop. At still higher pressure or temperature, ductile flow takes place to the exclusion of faulting. Griggs and Handin propose that rocks undergoing deformation in the transitional and ductile regimes will not be capable of producing earthquakes by mechanisms that involve fracture or frictional slip, concluding that some other mechanism must operate for intermediate and deep earthquakes.

The hypothesis that ductile behavior commences when the resistance to frictional slip exceeds the yield strength has been examined by Maurer (50), Mogi (52), and Byerlee (15). For the most part these studies confirm this hypothesis and show that laboratory friction measurements yield results that conform with observations on the brittle ductile-transition (Figure 2). However, some ambiguity exists for the application of the data to the earthquake problem. To begin with, a limiting pressure has not been shown for brittle behavior of all materials within the experimental range of these studies. Additionally, the relationship of stick-slip to the brittle-ductile transition has not yet been clarified. For example, Byerlee (14, 15) assumes the brittle-ductile transition to be the pressure at which the stress required to form a new fault equals the stress required to cause slip on an existing fault. For Westerly granite this occurs when the normal stress is about 17 kbar. However, at this stress Byerlee (14) observes stick-slip behavior on previously faulted surfaces. Finally, the experiments of Bridgman (11) to 50 kbar exhibit phenomena that superficially, at least, resemble frictional stick-slip. The pressures of these experiments are much greater than that normally assumed for the brittle-ductile transition.

The effect of temperature on stick-slip and its transition to stable sliding has been examined by Brace & Byerlee (7). Their experiments indicate that stick-slip gives way to stable sliding with increasing temperature. They propose that the depth limit of 15 km for California earthquakes may be determined by the thermal gradients which place the rocks below that depth into the pressure-temperature range favoring stable sliding.

Theories for deep earthquakes fall into two broad groupings. The approach followed in the first group has been to abandon the concept of elastic rebound in favor of collapse mechanisms that depend upon rapid solid-solid phase changes in the lower crust and upper mantle. The other group has adhered to the general outlines of the elastic rebound theory and has studied several possible instability

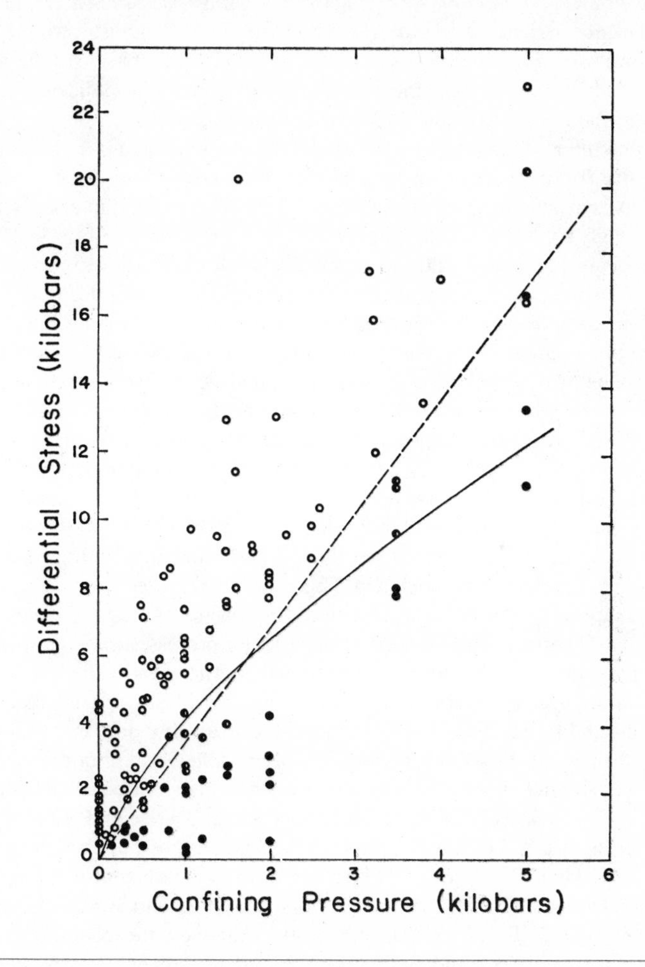

Figure 2 Compressive strength versus confining pressure for silicate rocks (5). Open symbols indicate brittle failure; half-closed symbols, transitional; closed symbols, ductile. The dashed and solid curves are the boundaries between brittle and ductile behavior based on friction data as proposed by Mogi (52) and Byerlee (15), respectively.

mechanisms for slip on a surface or over a narrow zone that may operate at conditions appropriate to intermediate and deep earthquakes.

The mechanism of rapid collapse or implosion caused by sudden changes to denser phases is a recurring theory for deep earthquakes that was first suggested by Bridgman (12). More recently, a number of workers have referred to and discussed this mechanism, including Benioff (3), Evison (29), Rieker & Rooney (68), Ringwood (69), and Dennis & Walker (22). Within the framework of observations for plate tectonics and sea-floor spreading this mechanism is attractive because deep earthquakes occur beneath island arcs where material is almost certainly being carried downward. In these areas solid-solid phase changes must be taking place. However, aside from this point there is little direct evidence to indicate that solid-solid phase changes are responsible for earthquakes.

As pointed out and discussed by Sykes (75) the most damaging evidence against the implosive source mechanism and in favor of a shear dislocation mechanism comes from analysis of the seismic radiation field of deep earthquakes. The data overwhelmingly favor the double couple mechanism required of a shear dislocation, and no evidence has been found for an implosive source. Studies by Knopoff and Randall (47, 64) indicate that in detail the radiation field from some deep earthquakes may contain components of motion described by the compensated linear-vector dipole superimposed on the double couple radiation. For the few events they studied, small components of volume change were indicated. As the volume changes were both positive and negative, however, these components of radiation, if real, cannot be ascribed to decrease in volume by phase changes to denser minerals.

A number of different processes have been proposed that might contribute to the mechanism of shear instability for deep earthquakes. Some of the processes are not necessarily mutually exclusive and more than one may contribute to the instability.

Byerlee & Brace (19) have proposed that all earthquakes including deep focus earthquakes may originate by stick-slip. This proposal represents an extreme point of view that appears to be at variance with the arguments of Griggs & Handin (34) and Orowan (57) on the unlikelihood of dry friction processes existing at the high confining pressures appropriate to deep earthquakes. The question here may be one of nomenclature because Byerlee and Brace do not specifically restrict stick-slip to frictional instability but discuss it in terms of the stress-displacement response. A proposal of frictional stick-slip is nonetheless permissible because of the ambiguity posed by the absence of direct laboratory evidence that frictional stick-slip may terminate for all rocks beyond some limiting pressure. Byerlee and Brace note the resemblance to stick-slip of the snapping phenomena observed by Bridgman (11) to pressures of 50 kbar. Pressures of 50 kbar are outside the range usually presumed for brittle behavior and might therefore support Byerlee and Brace's hypothesis. However, Griggs & Baker (33) have recently examined the conditions of the Bridgman experiments, concluding that the snapping phenomena may be caused by creep-induced shear melting instability.

The role of fluid pressure in reducing the total effective pressure may be critical in extending the depth to which brittle behavior may operate. The concept of effective pressure in the context of faulting was first discussed by Hubbert & Rubey (39). Simply stated, the principle of effective pressure holds that, for the purpose of determining the yield properties of a permeable rock, the influence of external confining pressure is reduced by an amount equal to the pore fluid pressure. The effective stress law has been verified in the laboratory by Handin et al (36), Brace (5), Brace & Martin (8), and Byerlee (14). Experimental confirmation of fluid pressure acting to extend the range of brittle behavior to higher pressures is given by the results of Raleigh & Paterson (63) and Heard & Rubey (37). In these experiments, fluid pressures are obtained by dehydration reactions at elevated temperatures. In the experiments by Raleigh and Paterson the dehydration of serpentinite was examined. Ductile behavior was observed at temperatures below the dehydration point, but with the onset of dehydration at higher temperatures, brittle behavior commences, accompanied by rapid stress drops (Figure 3). Similar findings were obtained by Heard & Rubey (37) for gypsum dehydration. Raleigh & Paterson

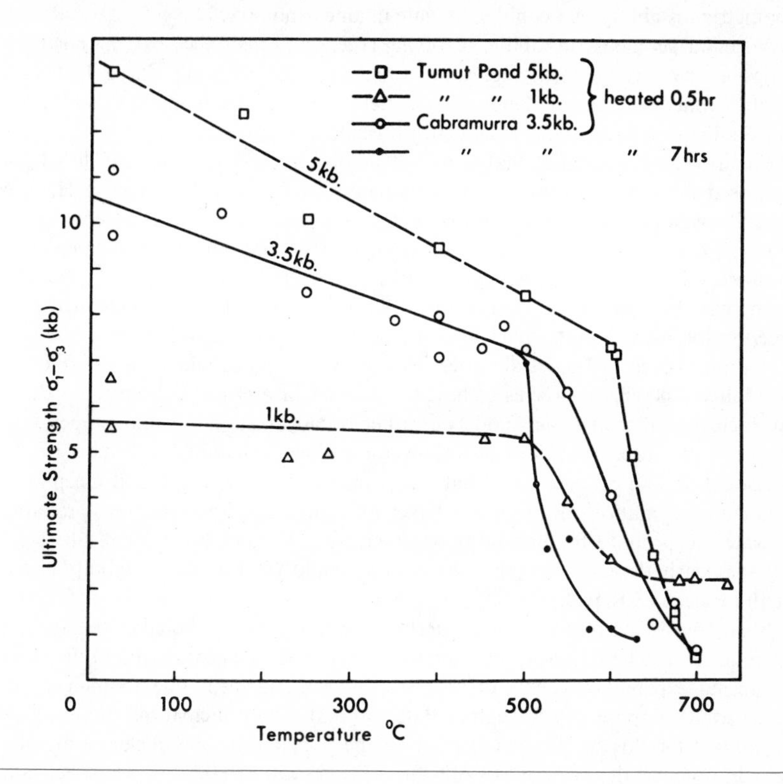

Figure 3 Ultimate strength of antigorite-chrysotile serpentinites at different temperatures and pressures showing the drop in strength arising from dehydration (63).

(63) and Raleigh (62) point out that in the earth a rise in temperature bringing about dehydration reactions would occur in zones where convection brings shallow rocks to greater depths, such as in island arcs where deep earthquakes occur. Additionally, water could be released from the dehydration of other minerals including the micas and amphiboles.

Raleigh (61) and Mogi (53) speculate that partial melt distributed along grain boundaries might also serve as the fluid phase under pressure to reduce the effective pressure and enhance embrittlement at great depths.

Partial melt has also been suggested to assist shear instability in other ways. Based upon Orowan's (57, 58) discussion of creep-induced porosity, Raleigh (61, and personal communication) suggests that partial melt would tend to accumulate in the pores, if present, as creep in the lower crust or upper mantle progresses. Because partial melt is weaker than surrounding mineral grains, a concentration of partial melt would tend to accelerate creep in that area, resulting in smearing out and an increase in rate of enlargement of the pores. This process has an inherent geometric instability that could culminate in an earthquake.

A similar geometric instability involving creep- and melt-filled cavities has been proposed by Savage (71). According to Savage, the melt fills large lenticular cavities that form along planes of weakness inherited from shallow faulting in downgoing plates of oceanic crust beneath island arcs.

Finally, creep instability that does not involve geometric instability has been proposed as the mechanism of deep earthquakes by Griggs (32), Griggs & Handin (34), Orowan (57, 58), and Griggs & Baker (33). The basis of this mechanism is that creep can induce local physical changes that promote further creep. For example, creep is thermally activated, and the temperature of creeping rocks tends to increase because of the dissipation of strain energy. Thus it is suggested that creep in the mantle would tend to concentrate into progressively narrower zones with progressively greater strain rates. This process could culminate in an explosive instability. The characteristics of heat conduction in areas of concentrated shear are such that in the simpler models a rather broad maximum in the temperature profile is produced that does not allow for sudden instability (34). This led to investigation (34) of local flaws that might initiate more concentrated creep. The detailed computations of Griggs & Baker (33) show that shear melting probably operated to produce the instability observed by Bridgman, but their results have not yet clearly demonstrated that this process could yield similar instability in the earth for deep earthquakes.

In summary, the mechanism of intermediate and deep earthquakes remains an open question. The seismic evidence favors a rapid shear dislocation and the most promising explanation seems to be given by embrittlement due to high water pressures as proposed by Raleigh & Paterson (63). The principal questions yet to be studied for this mechanism pertain to the availability of sufficient water by dehydration and the rates of loss of water as rocks are carried downward beneath island arcs. The problems raised by these questions may be eased if fluid pressures of partial melt along grain boundaries can also operate to reduce the effective pressure and enhance brittle behavior.

ROCK MECHANICS AND SHALLOW EARTHQUAKES

There can be little doubt that the instability mechanism for shallow earthquakes obtains occasionally from fracture of initially intact rock, but usually by a process of stick-slip that may be viewed as refracturing along already existing faults. In addition to stick-slip and fracture, other deformation processes of possible general interest to shallow earthquakes include dilatant changes and the effects of pore fluid flow on the stress field. Uncertainty arising from the absence of experimental data precludes further discussion of intermediate and deep focus earthquakes. However, if embrittlement at the depth of intermediate and deep earthquakes by the fluid pressure mechanisms proposed by Raleigh & Paterson (63) is indeed the correct explanation for these earthquakes, then the findings at conditions presently accessible in the laboratory may be generally applicable to all earthquakes.

The gross frictional characteristics of rocks usually conform with the more extensive findings for friction in metals and are surprisingly uniform. The principal quantity measured in a friction experiment is the coefficient of friction, μ, defined by

$$\tau = \mu\sigma \qquad 1.$$

where τ is the shear stress required for slip across the sliding surface and σ is the component of stress acting normal to the surface.

Generally, as a first approximation μ is assumed to be independent of σ. Detailed data on μ suggest a weak dependence on σ (see, for example, 15 and 50), although this may in part be an artifact of the definition given in equation 1 (27a). An interesting property of rock friction is that μ is apparently independent of rock type for crystalline silicate rocks and nearly always falls within the range of values from 0.5 to 1.0 (for example, 14, 24, 41, and 50). However, some recent work (27a) indicates that for other rock types μ can be closely related to mineral composition and that the effects of composition may exceed those of either displacement rate or temperature. The variation of μ seems to be more strongly affected by experimental procedure and surface characteristics than by rock type or hardness. Time dependent effects, the presence of small amounts of gouge, and the amount of displacement all have some effect on the frictional response within the normal range of values for μ (14, 24, 27a, 38, 73).

The role of pore fluid pressures in reducing the effective pressure, described above, has been verified in the laboratory by Byerlee (14, 17). In the presence of pore fluid pressure, P, equation 1 takes the form

$$\tau = \mu(\sigma - P) \qquad 2.$$

In detail, laboratory observations on rock friction display a notorious lack of consistency and exceptions to nearly any detailed observation of rock friction can be found. It is clear that many factors may subtly influence rock friction and much of the current laboratory effort is being directed toward delineating possible sources of variability.

A principal area of variability pertains to the transition between the contrasting behavior of stick-slip and stable sliding. It is important to understand the underlying causes of stable sliding and its natural counterpart, fault creep, in order to understand the conditions that lead to stick-slip.

In the course of laboratory experiments stable sliding has been well documented in three rather different situations. At low to very low confining pressures numerous experimenters including Byerlee (17), Byerlee & Brace (18, 19), and Scholz, Molnar & Johnson (73) reported that stable slip may occur to the exclusion of any stick-slip instability. This stable sliding, however, gives way to stick-slip with increasing confining pressure. Because the transition pressure seems to be different for different experiments, it would appear that this effect must depend in part upon the characteristics of the experimental apparatus. Additional controlling factors probably include surface finish (24, 38) and mineral composition (18). The low pressure transition provides one possible explanation for fault creep on the San Andreas fault of California. By this explanation fault creep might be a surficial phenomenon restricted to low pressures at shallow depths and only indirectly related to earthquake activity.

A different explanation for fault creep on portions of the San Andreas fault has been proposed by Scholz, Wyss & Smith (74) and Scholz, Molnar & Johnson (73) based on the second experimental situation in which stable sliding has been observed. In this case stable sliding sometimes occurs premonitory to stick-slip instability (Figure 4). Here again there is considerable variability, the observations of Byerlee (14) and Byerlee & Brace (19) often show relatively large slip prior to stick-slip whereas Scholz et al (73) found only very small displacements on the order of 5–10 μm and Wu, Thompson & Kuenzler (80) reported small displacements a few tens to hundreds of micro-seconds prior to stick-slip. It has been proposed (73, 74) that on a segment of the San Andreas fault the observed aseismic fault creep may represent premonitory stable sliding prior to an impending major earthquake. A difficulty with this explanation concerns the uncertainty of scaling the laboratory stable sliding displacements prior to stick-slip to the observed fault creep displacements which are several orders of magnitude larger.

Finally, stable sliding sometimes occurs in conjunction with the time dependent friction effects reported by Dieterich (23, 24), Scholz, Molnar & Johnson (73), and Brace & Stesky (10). In the experiments by Dieterich the static coefficient of friction for gouge-coated surfaces was found to increase with the logarithm of the duration of stationary contact. Dieterich observed that in some cases for short contact intervals (1–2 sec) slip always proceeded by stable sliding to the complete exclusion of stick-slip. For longer stationary contact intervals, stick-slip always occurs because the static friction has increased with time. In these experiments stable sliding, once initiated, tends to persist and takes place at stresses below that required for stick-slip. Fault creep by this type of stable sliding would imply that an earthquake on a creeping segment of a fault could occur only if the fault locally hangs up and ceases to slip for a period of time sufficient to allow the static friction and the stress to increase.

Of interest for numerical modeling studies is the ratio of the coefficient of static friction μ_s, to the coefficient of sliding friction, μ_k, for stick-slip. Hoskins, Jaeger

& Rosengren (38) infer μ_k from the stress drops for stick-slip by following the assumption that energy dissipation is accounted for entirely by work done in sliding. On this basis their data give values for μ_s/μ_k from 1.025 to 1.26 at normal stresses to 50 bars. Similarly, Byerlee obtains for Westerly granite an average value of 1.30 for μ_s/μ_k at normal stress from 2 to 12 kbar. Scholz, Molnar & Johnson (73) point out that the stress drop in stick-slip is controlled by seismic efficiency, η, which is the fraction of the total energy dissipation in the form of seismic radiation. It follows from their discussion that

$$\frac{\mu_s}{\mu_k} = \frac{\tau}{(\tau - \Delta\tau/2)(1-\eta)} \qquad 3.$$

where τ is the shear stress at initiation of slip and $\Delta\tau$ is the stress drop. Assuming $\eta = 0$, as did Hoskins, Jaeger & Rosengren (38) and Byerlee (16), the data from the Scholz et al experiments gives values from 1.015 to 1.15 for μ_s/μ_k. Employing the relatively large value for η of 0.10, equation 3 yields values for μ_s/μ_k of 1.14–1.40, 1.44, and 1.13–1.28 for the data of Hoskins, Jaeger & Rosengren (38), Byerlee (16), and Scholz, Molnar & Johnson (73), respectively.

An independent estimate of μ_s/μ_k can be made using the time-dependent friction

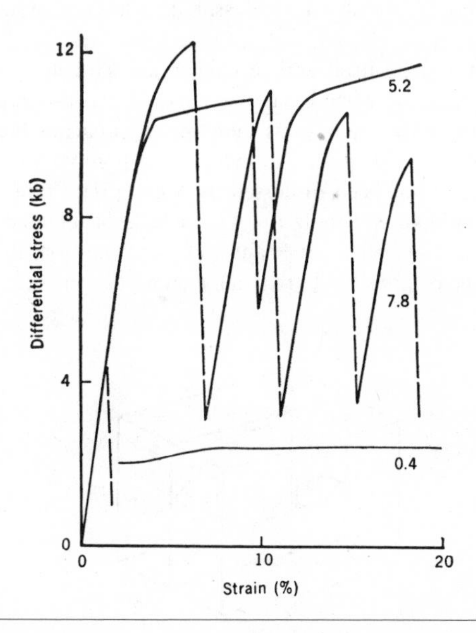

Figure 4 Stress-strain curves for serpentinized gabbro at different pressures from Byerlee & Brace (19). The dotted portion of the curves represent sudden stress drops during the slip phase of stick-slip. The linear portion of the solid curves show the accumulation of elastic strain and the nonlinear portion of the solid curves at higher stresses probably represent premonitory stable slip.

data of Dieterich (24). Because μ_k in these experiments was measured directly at low slip velocity no assumptions are made for radiation efficiency. However, this approach does assume an independence of μ_k with slip velocity. The logarithmic increase of static friction with time between stick-slip events observed in these experiments implies that μ_s/μ_k also increases logarithmically with time because μ_k is taken as a constant. These data give average values for μ_s/μ_k of 1.03 to 1.15 for stick intervals of 10 to 10^5 sec, respectively. Extrapolation to intervals of 10^8 sec (~ 3 yr) yields $\mu_s/\mu_k = 1.25$.

In addition to reducing effective pressure, interstitial fluids may significantly influence the mechanical response of stressed rocks near an earthquake source by other processes. The interstitial volume of a rock associated with pores or open cracks is a function of the interstitial fluid pressures and the state of stress. This leads to a coupling of the equations for the stress-strain fields with the equations for the interstitial fluid flow. Nur & Booker (55) examine the possible changes in pore pressures near a fault that might arise from earthquake-caused stress changes. Booker (4) extends this analysis and computes that the fluid flow that might arise in response to the changes in fluid pressure following an earthquake could generate time-dependent changes in the stress field analogous to a viscoelastic stress recovery. Noorishad, Witherspoon & Maini (54) use the finite element method to examine the coupling of flow and fluid pressure in a fracture system with the stress field.

A process of interest by itself and in connection with interstitial fluids is the phenomenon of dilatancy, whereby the volume of a rock or granular material increases relative to elastic changes as the stress approaches the failure strength. The increase in volume is attributed to increases in pore volume. Dilatancy in the presence of pore fluids has been discussed by Mead (51), Frank (30, 31), Orowan (59), and Brace, Paulding & Scholz (9). Two effects of dilatancy in the presence of pore fluids are distinguished by Frank (30, 31). The first, dilatancy pumping, involves the transfer of pore fluids that may take place in response to dilatancy

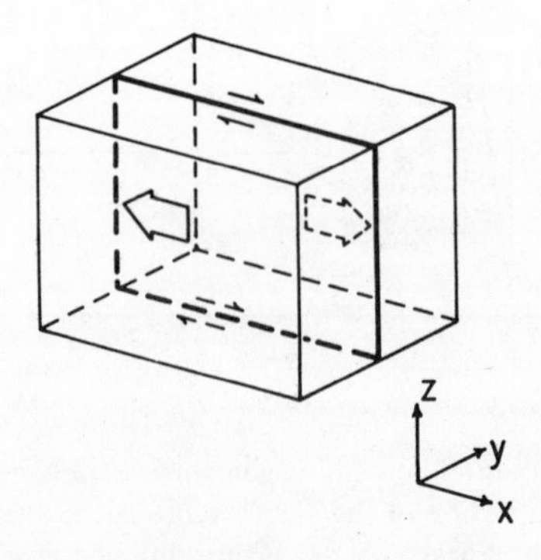

Figure 5 Diagram of a general model of a seismically active fault (26, 27).

induced fluid pressure changes. The second, dilatancy hardening, can occur if the amount of interstitial fluid is held constant. In this situation, the tendency for dilatation results in a reduction of fluid pressure and hence, strengthening of the rock as the effective pressure increases. Fluid pressures and effective pressures, therefore, depend upon rates of deformation, strain gradients, and rates of fluid flow within and near the zone undergoing dilatancy.

These effects in the presence of interstitial fluids can potentially lead to a number of significant processes in the earthquake source area. Perhaps the most interesting aspect of dilatancy and related phenomena concerns the possibility of providing a predictive capability for impending earthquakes. At present, possible precursory phenomena either directly or indirectly related to dilatancy are receiving attention at a number of laboratories. The near future should see interesting developments in this field.

NUMERICAL MODELING

Earthquake-related observations do not directly yield information on the mechanics of the earthquake source, and laboratory observations do not provide a direct indication of the relations between the deformational parameters of rocks and earthquake source characteristics. This gap may be bridged by numerical models that incorporate mechanical variables to deterministically compute the static and dynamic deformations arising from earthquake sequences on a fault.

Figure 5 illustrates a general model of a seismically active fault. Friction, which controls motion at the fault, is assumed to vary with position so that some portions of the fault are stronger than others. Sliding friction is everywhere less than the static friction. Boundary displacements indicated schematically, increase with time, induce elastic distortions, and ultimately cause fault slip. The slip, and hence an earthquake, occurs when the shear stress at some point on the fault exceeds the local static friction. Once initiated, a rupture tends to propagate until the stress at the tip of the rupture no longer exceeds the static friction. Stress concentrations arise because the friction along the fault is irregular and slip for any single event is not uniform and affects only a portion of the fault. If the boundary displacements continue to increase with time, successive rupture events that simulate earthquake sequences will be generated. Because of the lateral variations of friction and stress, the seismic events occur at different locations and have different magnitudes.

All computational models for friction-controlled seismic faulting have incorporated simplifications of the general model. The model developed and studied by Burridge & Knopoff (13) and employed in modified form by Dieterich (25) consists of a one-dimensional mechanical analogue (Figure 6). This model represents the area near a fault as an assemblage of massive frictional elements (blocks in Figure 6) interconnected by elastic elements represented (in Figure 6) as coil springs. The friction elements are attached to a driving block by additional elastic elements shown (in Figure 6) as leaf springs. Variable properties may be specified for the elastic and friction elements. Operation of this model follows along the lines described for the generalized model. The driving block is displaced at a constant rate and distorts the

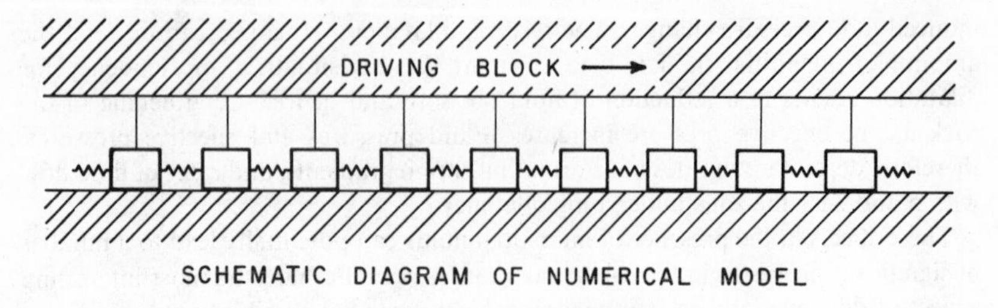

Figure 6 One-dimensional analogue of a seismic-fault based on the model of Burridge & Knopoff (13).

leaf springs to stimulate the slip events. Successive earthquakes simulated with these models have been found to vary in size and location. Burridge & Knopoff (13) and Knopoff, King & Burridge (46) show that the statistical properties of the simulated earthquake sequences possess many of the features observed for natural earthquakes including similar frequency-magnitude relationships. Although the one-dimensional models are deterministic the applications of these models are limited because strain fields and elastic wave propagation cannot be represented.

Two- and three-dimensional numerical models of seismic faulting have been studied by Dieterich, Raleigh & Bredehoeft (27) and Dieterich (26) that do not have the major disadvantages found in the one-dimensional representation. The two- and three-dimensional models treat the continuum surrounding the fault and, therefore, quantitative results may be obtained for strain fields, source motions, elastic wave propagation, and energy changes related to earthquake sequences in the models.

The configuration of the two-dimensional model is illustrated in Figure 7. The computations treat the plane-strain deformations for a slice perpendicular to the fault and parallel to the direction of slip. Because the fault is assumed to be initially planar with uniform elastic properties, it is necessary to treat the deformations on only one side of the fault. Periodicity in the direction parallel to the fault has been introduced to avoid undesired conditions at the ends (*ac* and *bd* in Figure 7) of the model. The periodicity serves to preserve continuity at the ends

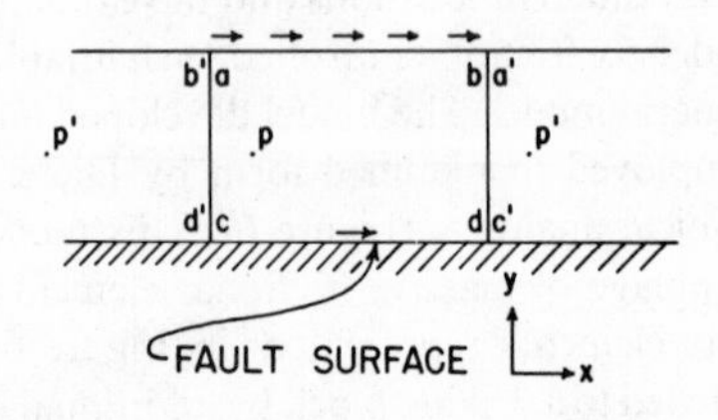

Figure 7 Diagram of two-dimensional plane strain fault model of a seismic-fault with periodicity in the x-direction (26, 27).

of the fault. Thus in Figure 7 the side *ac* is equivalent to *bd* and any point *P* is repeated in the direction parallel to the fault. The displacement boundary conditions that distort the model and drive motion on the fault are applied on *ab* parallel to the fault. As with the one-dimensional model, static and sliding friction parameters are specified at various points on the fault. These parameters may vary by location. Additionally, multi-event simulations are performed that give events of different size at different locations on the fault.

Finally, Dieterich (26) has studied a three-dimensional model (Figure 8), for analysis of strike-slip faulting that is similar to the plane-strain model. The surface *aa′c′c* represents the fault surface; the surface *aa′b′b* is a free surface that represents the surface of the earth. Periodicity in the horizontal direction parallel to the fault is again assumed to avoid undesired end conditions on the fault. The boundary conditions for loading the fault are applied on the surface *bb′d′d*. Numerical analysis of earthquake of the three-dimensional model requires considerable amounts of computer time; hence, only single event earthquake simulations have been examined.

For the two- and three-dimensional models, analysis of the quasi-static displacement fields prior to and after earthquake events and the acceleration fields at repeated intervals during an earthquake employ the finite element method. The discretization process implicit in the finite element method is analogous to the discretization of the one-dimensional model. Hence, the equations employed for numerical analysis of all three models have the same general form. For the quasi-static deformations, displacements at nodal points for the two- and three-dimensional finite element mesh are obtained from systems of simultaneous equations of the form

$$\{F\} = [K]\{\delta\} \qquad 4.$$

where $\{F\}$ is a vector array that lists n components of force at nodal points, $\{\delta\}$ is the corresponding list of n nodal displacements, and $[K]$ is the $n \times n$ stiffness matrix that is derived from the deformational and geometric characteristics of the elements that comprise the finite element mesh. For solution of a specific problem,

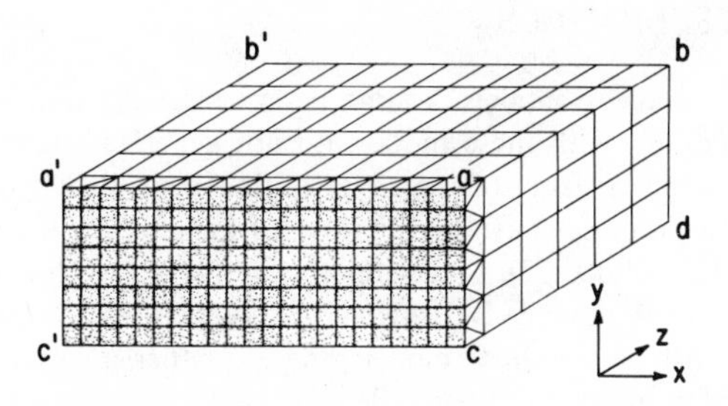

Figure 8 Three-dimensional model of a seismic-fault. The shaded surface represents the fault surface (26, 27).

displacement and force boundary conditions are incorporated directly into $\{\delta\}$ and $\{F\}$, respectively. For static analysis of the fault models, nodes along the fault are not permitted to slip, but warping of the fault is allowed by holding the component of force perpendicular to the fault constant on the fault surface. A seismic event begins when the component of force acting parallel to the fault at a fault node equals the static strength of the node. For the one-dimensional model, all displacements are specified and equation 4 is used only to determine $\{F\}$ for the friction elements.

When a fault node slips during a seismic event, the force acting on the node is opposed by the sliding friction. Therefore, force boundary conditions are operative on nodes along the fault that are moving. Applying the principle of d'Alembert, the dynamic problem is reduced to the static case by introducing the forces due to the accelerations of the nodes. Adding the dynamic force to equation 4 yields

$$\{F\} = [K]\{\delta\} + [M]\{\ddot{\delta}\} \tag{5.}$$

where $\{\ddot{\delta}\}$ represents the nodal accelerations and $[M]$ is the mass matrix.

A number of numerical techniques are available for solution of equation 5. With the approach employed by Dieterich (26) the variation with time of the accelerations is determined from equation 5 by performing a stepwise forward integration. This yields simultaneous equations for the acceleration fields at successive time steps, which are solved until the termination of all slip on a fault. After each time step the velocities and displacements are computed from the accelerations. The force acting on each stationary fault node is determined and if the force exceeds the static friction, the node is allowed to begin movement during the next time step. Similarly, if the velocity of a moving fault node has gone to zero during the preceding time step, the node is then held stationary until the force again exceeds the static friction. Repeated solutions for the acceleration fields are obtained until all fault nodes cease to move. A static solution is then obtained to determine the static equilibrium displacements at the end of the event.

Specific results from studies with the above models are discussed in the remaining sections of this paper.

AFTERSHOCK MODELS

In simple form, the above models assume perfect elasticity and simple frictional properties. Simulations with these models do not generate aftershock sequences. Aftershocks imply the existence of some type of time-dependent process. A number of time-dependent processes have been proposed for aftershock generation in the work of Benioff (2), Scholz (72), Lieber & Braslau (48), Burridge & Knopoff (13), Knopoff (45), Knopoff, King & Burridge (46), Dieterich (25), Nur & Booker (55), and Booker (4). All of the these models employ either a mechanism for time-controlled partial recovery of stress following an earthquake or time-controlled weakening of the fault leading to failure at reduced stresses.

Although many of the aftershock models have a mathematical foundation, only the model of Burridge & Knopoff (13) and Knopoff, King & Burridge (46) and the model of Dieterich (25) have been tested by carrying out detailed deterministic computations for earthquake sequences. These models are based on the one-dimensional fault analogue described above.

The aftershock model of Burridge & Knopoff (13) and Knopoff, King & Burridge (46) incorporates time dependency by assigning linearly viscous properties to some of the fault elements. These elements produce stable sliding (fault creep) in response to an applied load. The velocity of slip depends linearly upon the magnitude of the applied load. If a large earthquake-related slip event displaces one of the friction elements adjacent to the viscous element, a load will be transferred to the viscous element and it will begin to creep at the time of the earthquake. Displacement of the creeping block transfers load back to the adjacent block on the slipped segment of the fault and to the other adjacent element outside the earthquake zone. The time-dependent increase of load on the adjacent friction elements may then induce an aftershock. The aftershock in turn increases the load on the viscous element and the process may be repeated. Burridge & Knopoff (13) find that aftershocks simulated with this model have many of the statistical characteristics of real aftershock sequences. Although this model serves to demonstrate how aftershocks might originate by a viscous process, the viscous elements employed in the model cannot be identified in nature and the model requires the interaction of fault creep outside the original source area. This interaction has not been demonstrated for natural aftershock sequences.

The aftershock model proposed by Dieterich (25) is based on partial stress recovery following an earthquake by a mechanism of viscoelastic recovery of the rocks near the source. Stress recovery leading to aftershocks by this general mechanism was first proposed by Benioff (2). Experimental evidence shows that rocks can, under a variety of conditions, display the necessary viscoelastic recovery (e.g. Evans, 28; Lomnitz, 49; and Rummel, 70). Alternatively, viscoelastic recovery could originate by the mechanism of post-earthquake fluid pressure adjustments discussed by Booker (4). As employed by Dieterich, viscoelastic elements replace the elastic springs in the linear analogue. Because viscoelastic recovery can only partially restore the stress drop that occurs during an earthquake, it is necessary to employ a mechanism whereby the fault will be weaker following an earthquake than it was prior to the earthquake. A laboratory basis for this weakening is provided by the time-dependent friction observations of Dieterich (24), Scholz, Molnar & Johnson (73), and Brace & Stesky (10). These observations suggest that a segment of a fault that has remained stationary for some period of time will be weaker immediately after a slip event than it was prior to the slip event. Hence, aftershocks may be triggered at lower stress levels than that required for the main shock.

It was found from computations with this model that many of the characteristics of natural aftershock sequences could be duplicated: The aftershocks are confined to the portion of the fault that slipped during the main shock. The duration of the aftershock sequences increases with the size of the main shock. The aftershock frequency decreased with time. The details of the decay of aftershocks, however, were definitely not in agreement with the usual observations on aftershock

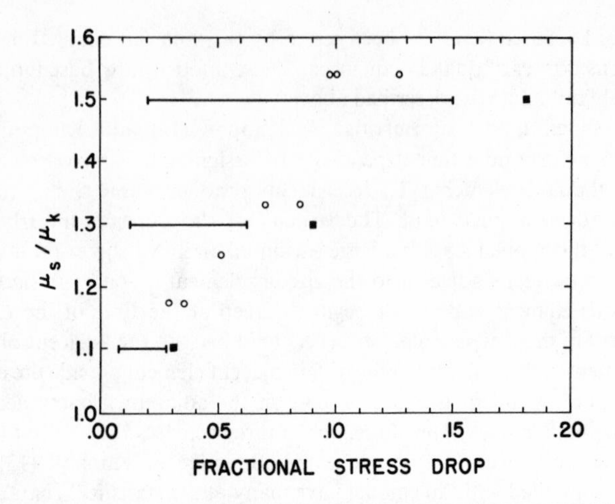

Figure 9 Fractional stress drop versus μ_s/μ_k from Dieterich et al (27). The open circles give results from simulations with the three-dimensional finite element model; the closed squares, single event simulations with the two-dimensional model; horizontal bars, the range of values for multi-event simulations with the two-dimensional model.

frequency (e.g. Omori, 56; Utsu, 79). This discrepancy may be attributed to the rough nature of the approximation of viscoelastic recovery employed for the model.

SOURCE PARAMETERS

The numerical models described above employ fault friction as the principal parameter that governs the details of fault slip and hence the source characteristics of an earthquake. In a general way the models can be employed to examine qualitatively the relations that exist between friction parameters and earthquake source parameters. However, many of the more interesting potential applications of the numerical models would be enhanced by quantitatively reliable input data. The availability of these data largely rests on the applicability of laboratory friction experiments on small samples to large and complicated fault zones. It might be argued that the insensitivity of gross frictional properties to rock type and experimental conditions would allow for application of the experimental data without serious consequence. On the other hand, the disparity between the stress drops computed for earthquakes and the stress drops observed for stick-slip has led to some uncertainty concerning the validity of laboratory stick-slip for earthquake mechanisms. Earthquake data (Chinnery, 20; Wyss & Brune, 81; King & Knopoff, 44; Thatcher & Hanks, 76) show that stress drops are usually less than 100 bars while, by comparison, the stress drops for stick-slip are commonly many kilobars (Figures 1 and 4).

The stress drop problem has been examined with the two- and three-dimensional numerical models. For the numerical models, stress drop, $\Delta\tau$, is taken as the difference between the average initial shear stress, τ, and the average final shear stress on the rupture surface. Fractional stress drop is defined as $\Delta\tau/\tau$. An implication of scaling equations 4 and 5 for the numerical analysis (Dieterich, 26, 27) is that fractional stress drop is independent of the magnitude of τ but depends on μ_s/μ_k. A plot (Figure 9) from Dieterich (27), gives values for $\Delta\tau/\tau$ against μ_s/μ_k for different simulations with two- and three-dimensional models. The range of values for μ_s/μ_k from 1.1 to 1.5 is in accord with experimental values determined from the stick-slip data described earlier. In Figure 9 the closed squares give the data for single bilateral rupture events with the plane strain model. For these events the friction and the initial stress on the rupture surface were uniform. The open circles give the fractional stress drops for similar events in the three-dimensional models. The data for the three-dimensional models scatter because different length to depth ratios for the rupture surface were employed. The horizontal lines give the limits for fractional stress drop determined for multi-event simulations with the two-dimensional model. For the multi-event simulations, strength varied randomly with location on the fault and, following the first rupture event, the shear stress was not uniform.

A striking feature of the data given in Figure 9 is the low values for fractional stress drop. For example, at μ_s/μ_k of 1.3 or less, the models give fractional stress drops in the range 0.01 to 0.10. Taking 100 to 1000 bars as a possible range of values for the shear strength of a fault zone, the models would give stress drops that vary from 1 to 100 bars. This range of stress drops contains the values of stress drop that have been determined for most earthquakes. It would appear therefore that the observed difference between stress drops for laboratory stick-slip and earthquakes obtains because of differences in the conditions under which slip takes place and not because of differences in the inherent frictional properties.

In the models, and presumably for real earthquakes, two processes contribute to produce low stress drops. The first arises because slip in an earthquake takes place on a limited segment of a fault. In a dynamic system involving the acceleration of mass, stress drop is proportional in part to the stress available to drive the accelerations. For experimental stick-slip, where the conditions are quite uniform on the slip surface, the shear stress at the instant slip commences is taken to be equal to the static friction and therefore the driving stress is simply the difference between the static and sliding friction. For slip on a fault, however, the conditions cannot be uniform because strong areas or areas with low initial shear stress are required to limit slip to a finite area. As a result, the average initial shear stress on the segment of a fault that ultimately breaks is less than the average static friction. Therefore the driving stress for earthquake faulting is properly taken as the difference between the average initial shear stress and the average sliding friction. This driving stress is significantly less than the driving stress that applies to laboratory stick-slip. The second process that contributes to the low stress drops operates because displacements on the slip surface during an earthquake do not take place as a simultaneous rigid body translation. For the simplest earthquake simulations the ruptures start at a point and propagate outward as a growing dislocation. Similarly, the mechanism of arresting slip on the fault surface begins at some location (e.g. the ends of the rupture) and propagates inward. For more complicated earthquake events with variable shear stress and variable friction the initiation and termination of slip propagate in a very erratic fashion. Often a single point on the fault may start and stop several times during the course of an earthquake. The result of the nonsimultaneity of slip during an earthquake is that a point on the fault, once it stops sliding, will experience an increase in the applied stress because other nearby points continue to move. This process applies to all points along the dislocation surface except for the last point to slip.

In addition to stress drop, other parameters that describe the earthquake source include earthquake magnitude, source dimensions, displacements, and seismic moment. A number of studies including Tocher (77), Iida (40), Press (60), Wyss & Brune (81), Chinnery (21), and King & Knopoff (44) have proposed a variety of possible empirical relationships among these parameters. For example, Figure 10 illustrates some of the relationships that have been proposed for the possible dependence of magnitude on rupture length.

The relationships among source parameters for the numerically simulated earthquakes have been examined by Dieterich (26). The analysis is based on scaling of the equations employed for the finite-element computations. This scaling allows the computations to be performed with arbitrary values for the grid dimensions, stresses (or friction constants), elastic constants, and density, and the results may then be scaled at a later time for any specifically desired set of values for these model parameters. The analysis leads to scaling of the source parameters in terms of the other source parameters and physical model parameters. Scaling laws for the

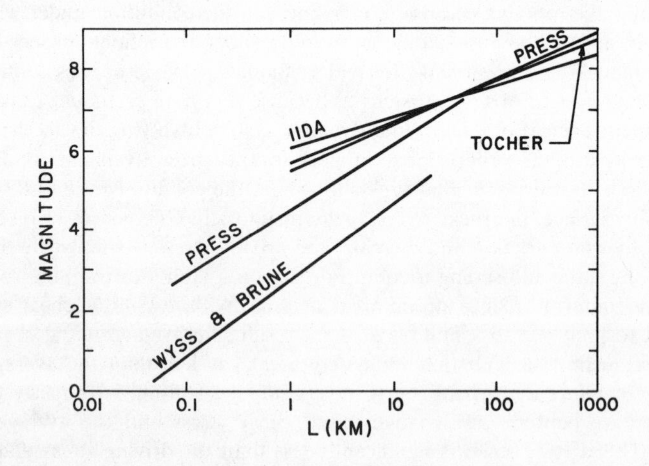

Figure 10 Magnitude-rupture length correlations proposed by Tocher (77), Iida (40), Press (60), and Wyss & Brune (81).

source parameters derived by this approach are general and may be obtained independent of the computational procedure using the method of dimensional analysis. Implicit in the scaling are assumptions of geometric similarity of the source, independence of the friction constants on slip velocity or displacement, and a constant Poisson's ratio.

Scaling for average displacement, D, on the rupture surface in terms of rupture length, L, stress drop, $\Delta\tau$, and shear modulus, G, is given by

$$D = K_1 \Delta\tau L/G \quad 6.$$

where K_1 is a constant that is evaluated from each specific computation. Equation 6 is of the same general form as that obtained from dislocation analysis (e.g. Chinnery, 20).

Total change of strain energy, E_T, for a simulated event is obtained from the static displacement field immediately before and after the earthquake, taking into account the two sides of the fault and, for the two-dimensional models, an assumed depth of rupture. Work expended by sliding on the fault surface is similarly obtained from the displacements and the sliding friction. Seismic energy is then obtained from

$$E_S = E_T - W \quad 7.$$

Energy terms are found to scale by

$$E_S = K_2 L^3 \Delta\tau^2/G \quad 8.$$

where K_2 is again a constant that must be evaluated numerically. Combining equation 6 with 8 yields:

$$E_S = K_3 L D^2 G \quad 9.$$

Equating energy with earthquake magnitude, equation 9 has the same form as the relation suggested by King & Knopoff (44). Seismic moment, M_O, is defined by Aki (1) as

$$M_O = ADG \quad 10.$$

where A is the area of the rupture. Because A scales by L^2:

$$M_O = K_4 G L^2 D \quad 11.$$

where K_4 is a simple geometric constant that relates L to A. Combining equation 9 with 11 we obtain:

$$E_S = K_5 M_O^2/GL^3 \quad 12.$$

The scaling laws in equations 8, 9, and 12 are presented graphically in Figure 11 which plots E_S against L for different values of $\Delta\tau$, D, and M_L. The values for K_2, K_3, and K_5 employed for locating the curves in Figure 11 are average values from the two-dimensional multi-event simulations. The results from the three-dimensional simulation are similar. Seismic energy in Figure 11 has been equated to surface wave magnitude M_S and local magnitude M_L following the energy-magnitude relations of Gutenberg & Richter (35) and Thatcher & Hanks (76), respectively.

The data plotted in Figure 11 include earthquakes with surface breaks, from the compilations of King & Knopoff (44) and Chinnery (20); source parameters determined by Thatcher & Hanks (76) from analysis of seismic spectra; and the data for the San Fernando earthquake (78) which is not included in those compilations. These data are plotted by magnitude and rupture length. Agreement of the displacement data with displacements predicted by the scaling laws are indicated by the different symbols in Figure 11.

Figure 11 illustrates the source of difficulty that has been encountered in studies that have attempted to find a simple linear relationship between L, D, or M_o and earthquake magnitude. Any such characterization will carry the assumption that stress drop is the same for all earthquakes or that stress drop varies regularly

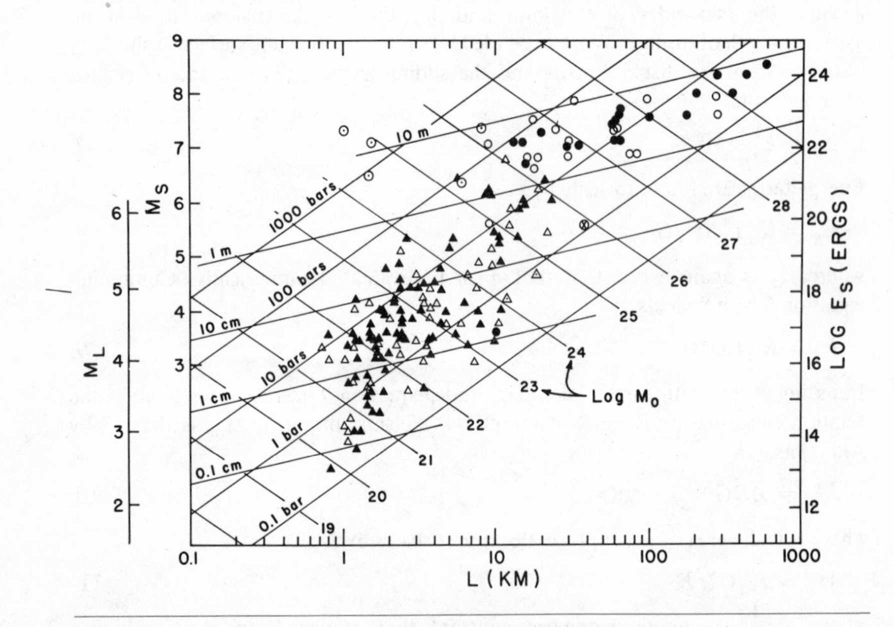

Figure 11 Graphic representation of scaling laws for seismic energy, E_S; rupture length, L; seismic moment, M_o; average displacement; and average stress drop (in bars) from Dieterich (26). The circles give data for earthquakes with observed surface breaks (20, 44, 78) and the triangles give data derived from seismic spectra (76). The data are plotted by rupture length and earthquake magnitude. Agreement of the displacement data with displacements predicted by the scaling laws (contours sloping gently to the left) is indicated by the type of symbol employed. Agreement by less than a factor of two, between a factor of two and five, and more than a factor of five is indicated by the closed, open, and open with dot symbols, respectively. The Parkfield, California earthquake of June, 1966 which had estimated displacements from 0.05 to 0.5 m is indicated by the circle with an x.

with magnitude. Stress drop data (20, 44, 76, 81) show considerable scatter and provide scant indication of regular variation with magnitude.

An interesting implication of the scaling laws is that only two independent source parameters are needed to infer the remaining source parameters.

EARTHQUAKE TRIGGERING BY FLUID INJECTION

The agreement of stress drops for the numerical models with earthquake stress drops and the correlation of source parameters for the simulated earthquakes with the earthquake data, provide support for the numerical model and attest to the applicability of laboratory stick-slip experiments to shallow earthquake mechanics. Additional confirmation of these conclusions has been provided by the results of studies of earthquakes at the Rangely, Colorado oil field. At Rangely the local earthquake activity is apparently linked to water injection operations that are being conducted for the purposes of secondary recovery of oil. This study is of interest because it provided a hitherto unique opportunity to study earthquakes in a controlled situation. Independent measurements were available for stresses, fluid pressures, and frictional properties of rocks in the source region.

The earthquakes at Rangely occur along a portion of a pre-existing fault where water injection has increased the fluid pressures to levels that are above the pressures that existed prior to the beginning of oil field operations. The earthquakes occur at and below the depth of the Weber sandstone which is the producing formation of the Rangely anticline (Raleigh, Healy & Bredehoeft, 62). Simulations of these earthquakes have been performed with the two-dimensional plane strain model described above (Dieterich, Raleigh & Bredehoft, 27). The simulations are based on the hypothesis that slip on the fault at Rangely obeys the effective pressure law of equation 2. Hence, earthquakes are believed to occur when the strength is reduced to the ambient stress by the increasing fluid pressures. The analysis was performed by first applying the displacement boundary conditions to load the fault to a specified stress level. Following this the frictional strength along the fault is varied with time as a function of changes in fluid pressure by equation 2.

For the specific simulation of the Rangely earthquakes, the variation with time of the fluid pressures is obtained from independent computations with a numerical reservoir simulator. These computations for fluid pressure are based on detailed modeling of the reservoir, and employed several years of data for observed fluid pressure changes in the oil field. Friction values for the simulations were obtained from results of experiments by Byerlee (17) on the Weber sandstone. The applied loads used the data from the stress measurements summarized by Raleigh, Healy & Bredehoeft (62).

Results of the computations may be summarized as follows: First, earthquakes were simulated by the model using the input parameters described above; second, the rate of the simulated earthquake activity agreed in a rough way with the observed rates of activity; and third, the simulated earthquakes were restricted to the same portions of the fault as that observed at Rangely.

SUMMARY

It appears that the process of stick-slip observed for laboratory experiments is the relevant instability mechanism for earthquakes on shallow faults. The instability mechanism for intermediate and deep focus earthquakes remains a point of conjecture with some evidence favoring mechanisms of embrittlement by high fluid pressures and creep induced instability. It has been shown that the apparent discrepancy between stress drops for stick-slip and stress drops for earthquakes is not necessarily caused by fundamental differences in the instability processes but can be explained by differences in the conditions under which slip takes place in these two situations. When friction parameters for laboratory stick-slip are employed as the basis for numerical earthquake faulting simulations, results for earthquake stress drop and other earthquake source parameters conform very well with the earthquake data.

The technique of numerical modeling provides insight into the complicated mechanical interactions that occur in the vicinity of the earthquake source. The models described above are especially useful in illuminating the relationships between parameters that describe the mechanical characteristics of the different components of the system and observations of earthquake parameters. As such the models provide a useful quantitative bridge between laboratory observations, earthquake source theory, and field observations of earthquake phenomena. This is illustrated by the simulations of earthquakes at Rangely, Colorado. The results of these computations indicate that measurements for the input parameters, including the laboratory friction measurements, accurately reflect the conditions in the earthquake source area and that earthquakes may be triggered by injection of fluids by the process of reducing the effective pressure in the source area.

Literature Cited

1. Aki, K. 1966. Generation and propagation of G waves from the Niigata earthquake of June 6, 1964. Part 2. Estimation of earthquake moment, released energy and stress-strain drop from G-wave spectrum. *Bull. Earthquake Res. Inst. Tokyo Univ.* 44:73–88
2. Benioff, H. 1951. Earthquakes and rock creep. *Bull. Seismol. Soc. Am.* 41:31–62
3. Benioff, H. 1964. Earthquakes source mechanisms. *Science* 143:1399–1406
4. Booker, J. R. Time-dependent strain following fracture of a porous medium. *J. Geophys. Res.* In press
5. Brace, W. F. 1969. The mechanical effects of pore pressure on fracture of rocks. *Geol. Surv. Can. Pap.* 68–52:113–23
6. Brace, W. F., Byerlee, J. D. 1966. Stick-slip as a mechanism of earthquakes. *Science* 153:990–92
7. Brace, W. F., Byerlee, J. D. 1970. California earthquakes: why only shallow focus? *Science* 168:1573–75
8. Brace, W. F., Martin, R. J. 1968. A test of the law of effective stress for crystalline rocks of low porosity. *Int. J. Rock Mech. Mineral. Sci.* 5:415–426
9. Brace, W. F., Paulding, B. W., Scholz, C. 1966. Dilatancy in the fracture of crystalline rocks. *J. Geophys. Res.* 71:3939–53
10. Brace, W. F., Stesky, R. M. 1973. Time-dependence of frictional sliding in gabbro at high temperature and pressure. *Trans. Am. Geophys. Union* 54:466
11. Bridgman, P. W. 1936. Shearing phenomena at high pressure of possible importance to geology. *J. Geol.* 44:653–69
12. Bridgman, P. W. 1945. Polymorphic

transitions and geological phenomena. *Am. J. Sci.* 243:90–97
13. Burridge, R., Knopoff, L. 1967. Model and theoretical seismicity. *Bull. Seismol. Soc. Am.* 57:341–71
14. Byerlee, J. D. 1967. Frictional characteristics of granite under high confining pressure. *J. Geophys. Res.* 72:3639–48
15. Byerlee, J. D. 1968. Brittle-ductile transition in rocks. *J. Geophys. Res.* 73:4741–50
16. Byerlee, J. D. 1970. Static and kinetic friction of granite at high normal stress. *Int. J. Rock Mech. Mineral. Sci.* 7:577–82
17. Byerlee, J. D. 1971. The mechanical behavior of Weber sandstone. *Trans. Am. Geophys. Union* 52:343
18. Byerlee, J. D., Brace, W. F. 1968. Stick-slip stable sliding, and earthquakes—effect of rock type, pressure, strain rate, and stiffness. *J. Geophys. Res.* 73:6031–37
19. Byerlee, J. D., Brace, W. F. 1969. High-pressure mechanical instability in rocks. *Science* 164:713–15
20. Chinnery, M. A. 1967. Theoretical fault models. *Symp. Processes Focal Region.* 37:211–23. Publ. Dominion Observatory, Ottawa
21. Chinnery, M. A. 1969. Earthquake magnitude and source parameter. *Bull. Seismol. Soc. Am.* 59:1969–82
22. Dennis, J. G., Walker, C. T. 1965. Earthquakes resulting from metastable phase transition. *Tectonophysics* 2:401–7
23. Dieterich, J. H. 1970. Time dependence in stick-slip sliding. *Trans. Am. Geophys. Union* 51:423
24. Dieterich, J. H. 1972. Time-dependent friction in rocks. *J. Geophys. Res.* 77:3690–97
25. Dieterich, J. H. 1972. Time-dependent friction as a possible mechanism for aftershocks. *J. Geophys. Res.* 77:3771–81
26. Dieterich, J. H. 1973. A deterministic near field source model. *Proc. World Conf. Earthquake Eng., 5th.* Pap. no. 301, 13 pp.
27. Dieterich, J. H., Raleigh, C. B., Bredehoeft, J. D. 1972. Earthquake triggering by fluid injection at Rangely, Colorado. *Proc. Int. Soc. Rock Mech. and Int. Assoc. Eng. Geol. Symp.: Percolation through Fissured Rock.* Pap. T2-B, 12 pp.
27a. Donath, F. A., Fruth, L. S. Jr., Olsson, W. A. 1973. Experimental study of frictional properties of faults. *Proceedings of the Fourteenth Symposium on Rock Mechanics,* 189–222. New York: Am. Soc. Civil Eng.
28. Evans, R. H. 1936. The elasticity and plasticity of rocks and artificial stone. *Proc. Leeds Phil. Lit. Soc.* 3:145–58
29. Evison, F. F. 1963. Earthquakes and faults. *Bull. Seismol. Soc. Am.* 53:873–91
30. Frank, F. C. 1965. On dilatancy in relation to seismic sources. *Rev. Geophys.* 3:485–503
31. Frank, F. C. 1966. A further note on dilatancy in relation to seismic sources. *Rev. Geophys.* 4:405–8
32. Griggs, D. T. 1954. High-pressure phenomena with applications to geophysics. In *Modern Physics for the Engineer,* ed. L. N. Ridinour, 272–305. New York: McGraw-Hill
33. Griggs, D. T., Baker, D. W. 1969. The origin of deep-focus earthquakes. In *Properties of Matter under Unusual Conditions,* 23–42. New York: Interscience
34. Griggs, D., Handin, J. 1960. Observations on fracture and a hypothesis of earthquakes. *Geol. Soc. Am. Mem.* 79:347–64
35. Gutenberg, B., Richter, C. F. 1956. Magnitude and energy of earthquakes. *Ann. Geofis.* 9:1–15
36. Handin, J., Hager, R. V., Friedman, M., Feather, J. W. 1963. Experimental deformation of sedimentary rocks under confining pressure; pore pressure tests. *Am. Assoc. Petrol. Geol. Bull.* 47:717–55
37. Heard, H. C., Rubey, W. W. 1966. Tectonic implication of gypsum dehydration. *Geol. Soc. Am. Bull.* 77:741–60
38. Hoskins, E. R., Jaeger, J. C., Rosengren, K. J. 1968. A medium-scale direct friction experiment. *Int. J. Rock Mech. Mineral. Sci.* 5:143–54
39. Hubbert, M. K., Rubey, W. W. 1959. Mechanics of fluid-filled porous solids in application to overthrust faulting. *Geol. Soc. Am. Bull.* 70:115–66
40. Iida, K. 1965. Earthquake magnitude, earthquake fault and source dimensions. *J. Earth Sci. Nagoya Univ.* 13:115–32
41. Jaeger, J. C., Rosengren, K. J. 1969. Friction and sliding of joints. *Aust. Inst. Mining Met.* 229:93–104
42. Jeffreys, H. 1936. Note on fracture. *Proc. Roy. Soc. Edinburgh Sect. A* 56:158–63
43. Kasahara, K. 1967. Focal processes and various approaches to their mechanism. *Symp. Processes Focal Region.* 37:187–89. Publ. Dominion Observatory, Ottawa

44. King, C. Y., Knopoff, L. 1968. Stress drop in earthquakes. *Bull. Seismol. Soc. Am.* 58:249–57
45. Knopoff, L. 1972. Model for aftershock occurrence. In *Flow and Fracture of Rocks. Geophys. Monogr. 16,* 259–63. Am. Geophys. Union
46. Knopoff, L., King, C. Y., Burridge, R. 1967. A physical basis for earthquake statistics. *Symp. Processes Focal Region.* 37:224–34. Publ. Dominion Observatory, Ottawa
47. Knopoff, L., Randall, M. J. 1970. The compensated linear-vector dipole: a possible mechanism for deep earthquakes. *J. Geophys. Res.* 75:4957–63
48. Lieber, P., Braslau, D. 1965. On an earthquake and aftershock mechanism relating to a model of the crust and mantle. *Rep. Am-65-8, Off. Res. Serv. Univ. Calif., Berkeley,* 141 pp.
49. Lomnitz, C. 1956. Creep measurements in igneous rocks. *J. Geol.* 64:473–79
50. Maurer, W. C. 1965. Shear failure of rock under compression. *J. Soc. Petrol. Eng. J.* 5:167–75
51. Mead, W. J. 1925. The geologic role of dilatancy. *J. Geol.* 33:685–98
52. Mogi, K. 1966. Pressure dependence of rock strength and transition from brittle fracture to ductile flow. *Bull. Earthquake Res. Inst. Tokyo Univ.* 44:215–32
53. Mogi, K. 1967. Earthquakes and fractures. *Tectonophysics* 5:35–55
54. Noorishad, J., Witherspoon, P. A., Maini, Y. W. T. 1972. The influence of fluid injection on the state of stress in the earth's crust. *Proc. Int. Soc. Rock Mech. Int. Assoc. Eng. Geol. Symp.: Percolation through Fissured Rock,* Pap. T2-H, 11 pp.
55. Nur, A., Booker, J. R. 1972. Aftershocks caused by pore fluid flow? *Science* 175: 885–87
56. Omori, F. 1894. On the aftershocks of earthquakes. *J. Coll. Sci. Imp. Univ. Tokyo* 7:111–200
57. Orowan, E. 1960. Mechanism of seismic faulting. *Geol. Soc. Am. Mem.* 79:323–45
58. Orowan, E. 1964. Continental drift and origins of mountains. *Science* 146:1003–10
59. Orowan, E. 1966. Dilatancy and the seismic focal mechanism. *Rev. Geophys.* 4:395–404
60. Press, F. 1967. Dimensions of the source region for small shallow earthquakes. *Proc. VESIAC Conf. on Shallow Source Mechanisms, VESIAC Rep.* 7885-1-X 155–63
61. Raleigh, C. B. 1967. Tectonic implications of serpentinite weakening. *Geophys. J.* 14:113–18
62. Raleigh, C. B., Healy, J. H., Bredehoeft, J. D. 1972. Faulting and crustal stress at Rangely, Colorado. In *Flow and Fracture of Rocks, Geophys. Monogr. 16,* 275–84. Am. Geophys. Union
63. Raleigh, C. B., Paterson, M. S. 1965. Experimental deformation of serpentinite and its tectonic implications. *J. Geophys. Res.* 70:3965–85
64. Randall, M. J., Knopoff, L. 1970. The mechanism at the focus of deep earthquakes. *J. Geophys. Res.* 75:4965–76
65. Reid, H. F. 1910. *The mechanics of the earthquake, the California earthquake of April 18, 1906, Report of the State investigations commission,* 2:16–28. Washington DC: Carnegie Inst.
66. Reid, H. F. 1911. The elastic-rebound theory of earthquakes. *Univ. Calif. Publ. Geol. Sci.* 6:413–44
67. Richter, C. F. 1958. *Elementary Seismology.* San Francisco: Freeman. 768 pp.
68. Rieker, R. E., Rooney, T. P. 1966. Shear strength, polymorphism, the mechanical behavior of olivine, enstatite, diopside, labradorite, and pyrope garnet: Tests to 920°C and 60 kb. *AFCRL Environ. Res. Pap. W-216*
69. Ringwood, A. E. 1967. The pyroxene-garnet transformation in the earth's mantle. *Earth Planet. Sci. Lett.* 2:255–63
70. Rummel, F. 1968. Studies of time-dependent deformation of some granite and eclogite rock samples under uniaxial constant stress and temperatures to 400°C. *Z. Geophys.* 35:17–42
71. Savage, J. C. 1969. The mechanics of deep-focus faulting. *Tectonophysics* 8: 115–27
72. Scholz, C. H. 1968. Microfractures, aftershocks, and seismicity. *Bull. Seismol. Soc. Am.* 58:1117–30
73. Scholz, C. H., Molnar, P., Johnson, T. 1972. Detailed studies of frictional sliding of granite and implications for earthquake mechanism. *J. Geophys. Res.* 77:6392–406
74. Scholz, C. H., Wyss, M., Smith, S. W. 1969. Seismic and aseismic slip on the San Andreas fault. *J. Geophys. Res.* 74: 2049–69
75. Sykes, L. R. 1968. Deep earthquakes and rapidly-running phase changes, a reply to Dennis and Walker. *J. Geophys. Res.* 73:1508–10
76. Thatcher, W., Hanks, T. C. *Bull. Seismol. Soc. Am.* In press

77. Tocher, D. 1958. Earthquake energy and ground breakage. *Bull. Seismol. Soc. Am.* 48:147–52
78. US Geological Survey and National Oceanic and Atmospheric Administration 1971. The San Fernando earthquake of February 9, 1971. *Geol. Surv. Prof. Pap.* 733. 254 pp.
79. Utsu, T. 1961. A statistical study on the occurrence of aftershocks. *Geophys. Mag.* 30:521–605
80. Wu, F., Thompson, K. C., Kuenzler, H. 1972. Stick-slip propagation velocity and seismic source mechanism. *Bull. Seismol. Soc. Am.* 62:1621–28
81. Wyss, M., Brune, J. 1968. Seismic moment, stress and source dimensions for earthquakes in the California-Nevada region. *J. Geophys. Res.* 73:4681–98

SOLAR SYSTEM SOURCES OF METEORITES AND LARGE METEOROIDS

George W. Wetherill
Department of Planetary and Space Science, Department of Geology, and Institute of Geophysics and Planetary Physics, University of California, Los Angeles, California 90024

INTRODUCTION

About 10^{10} g/yr of extraterrestrial matter impacts the earth annually. Much of this is in the form of fine dust (micrometeoroids) and the larger (~ 1 g) objects which produce typical visual meteors. However, most of this mass is received in the form of larger objects: as meteorites, typically in the 100–10^6 g range; as fireballs of similar mass which fail to survive passage through the earth's atmosphere; and as larger crater-forming bodies, ranging in mass up to $\sim 10^{18}$ g. This review is concerned with our knowledge of the sources of these objects in the range of 10^2 to 10^{12} g.

The term meteoroid will be used as a general term for bodies impacting the earth (and other planets). Meteorites are meteoroids which survive passage through the atmosphere. The terms meteor and fireball refer to the atmospheric phenomena associated with the entry of a meteoroid, and, when the context is clear, will also refer to the meteoroid which caused these phenomena. More detailed estimates of the flux of impacting material in this mass range as a function of mass have been reported (29, 35, 44, 59). For unknown reasons, the extraterrestrial flux reported by lunar seismometry (44) as well as time-averaged lunar cratering is less than estimates based on terrestrial data (59) by a factor of ~ 100. This cannot be explained by differences in the mass or heliocentric velocity of the earth and moon. The actual flux on these bodies is, almost certainly, nearly identical. One or both of the measured flux values must be in error.

It is known that there is considerable variety in the physical and chemical nature of the impacting objects. The meteorites which fall to earth and are recovered include masses of nearly pure nickel-iron, chemically differentiated rocks similar in many ways to terrestrial as well as lunar igneous rocks (basaltic achondrites), and a variety of types of objects more similar to solar composition with regard to the nonvolatile elements (chondrites). Detailed descriptions of these and other types of meteorites may be found in standard sources (54). In addition, about a hundred times as many objects of similar mass enter the atmosphere but fail to reach its surface because of atmospheric ablation. Either they are too fragile or have too high an entry velocity, or both (57). These objects are also of different kinds, as shown by differences in their ablation characteristics (61) and spectra (13, 58). Among the larger objects ($\sim 10^{12}$ g) some produce ~ 1 km craters such as Meteor Crater, Arizona, and are associated with iron meteorites identical to smaller objects of this class. Others of similar mass, such as the great Tunguska (Siberia) meteor of 1908 (43) release their energy ($\sim 10^{24}$ erg) entirely in the atmosphere, and fail to produce a crater or major quantities of recoverable fragments.

Historically, there have been attempts to associate many or all of these objects to a common source or parent body which experienced a catastrophic disruption at some time in the past. For reasons which will emerge in subsequent sections, this is no longer an attractive hypothesis. The starting point of the present discussion is to assume that the only thing these various classes of objects have in common is that their orbits in space intersected that of the earth. The problem of identifying their sources is that of discovering sources which yield a sufficient quantity of matter, and satisfactory mechanisms for placing these fragments in appropriate orbits which intersect that of the earth. Of course, the earth is not alone in receiving a flux of these interplanetary objects. The cratered surfaces of the moon and Mars show that these bodies experience a similar bombardment.

An understanding of the sources of the meteoroids and the mechanisms by which they become earth intersecting is of great importance for a number of reasons. Among these are:

1. Such an understanding will permit direct laboratory studies of samples of identifiable extraterrestrial bodies. In this regard they are similar to returned lunar samples. Most meteoroids are probably samples of bodies even more primitive than the moon.

2. Knowledge of the orbital history of these objects will facilitate their use as natural space probes, containing within them the record of the galactic and solar particle flux of those regions of the solar system which they have traversed.

3. Understanding of the interplanetary meteoroid flux is necessary for the establishment of cratering time scales for planetary surfaces for which direct age measurements are difficult or impossible.

EARTH-CROSSING ORBITS

As mentioned above, the orbits of these meteoroids must have intersected that of the earth, otherwise they would not have collided with the earth and been observed. The question then arises: perhaps no source is necessary, why could there not have been objects of their present size in earth-crossing orbits for the entire history of the solar system? Although until about twenty years ago there was little basis for rejection of this possibility, consideration of cosmic ray exposure ages, dynamical lifetime in earth-crossing orbit, and collisional lifetime renders this no longer a possible solution to the problem. Discussion of these reasons will also serve to introduce several concepts needed in subsequent sections.

Cosmic Ray Exposure Ages

While in interplanetary space the meteoroid will not be shielded by the earth's atmosphere and will be bombarded by energetic particles (principally protons) of galactic and solar origin. The range of the more energetic particles will be a few meters, and these particles will therefore penetrate bodies of meteoritic size. Interaction of these cosmic rays with the atomic nuclei of the meteoroid will cause nuclear reactions, and the production of both stable and radioactive nuclear species. Among the stable isotopes are those of the inert gases. Because the meteoritic abundance of these highly volatile elements is extremely low, the meteoritic contents of some of these isotopes is almost entirely produced by cosmic ray bombardment (cosmogenic). Knowledge of the present cosmic ray flux and simulation of the production rate of these isotopes using particle accelerators, as well as measurement of the concentrations of radionuclides in the meteorite, permit calculation of the rate at which the cosmogenic inert gases are produced. Measurement of the meteoritic concentration of the cosmogenic inert gases and their production rates then permits calculation of the length of time the object has been bombarded. More complete reviews of cosmic ray exposure ages have been given by Anders (2), Wänke (82), and Zähringer (102).

The results of these measurements show that the stone meteorites have been exposed to cosmic ray bombardment for only a few tens of millions of years. Typical iron meteorites have been bombarded for hundreds of millions of years. In either case, the bombardment time is short compared to the 4.6×10^9 age of the last chemical equilibration of the meteoritic matter (84, 91). This result implies that relatively recently in solar system history the present meteoritic fragment broke off or otherwise became dissociated from a larger body of at least sufficient size (~ 10 m) to shield the meteorite from cosmic rays. This may have occurred in various ways, e.g. chipping off of the fragment by collision in the asteroid belt or by sublimation of icy shielding material covering a cometary nucleus. In any case, the cosmic ray exposure age clearly shows that a recent event occurred, reducing the meteorite to an object of approximately its present size. The physical formation of these objects was not confined to early solar system history, but is an active process occurring in the present solar system.

Dynamical Lifetime in Earth-Crossing Orbit

As stated above, collision of the meteoroid with the earth can occur only when the orbit of the meteoroid and that of the earth intersect. Even then, the two objects will not usually collide; however, given that the orbits intersect, the meteoroid will sooner or later come within one earth radius of the earth's center, and collision will result. Actually, a slightly more distant approach will suffice to cause a collision. The gravitational field of the earth will cause it to have a gravitational radius R_g:

$$R_g = R\sqrt{1+U_g^2/U^2} \qquad 1.$$

where R is the earth's physical radius, U_g is the earth escape velocity (11.2 km/sec) and U is the velocity of the body with respect to the earth, not including the increase in this velocity resulting from its falling into the gravitational field of the earth. Collision will occur if any part of the meteoroid comes within one gravitational radius of the center of the earth.

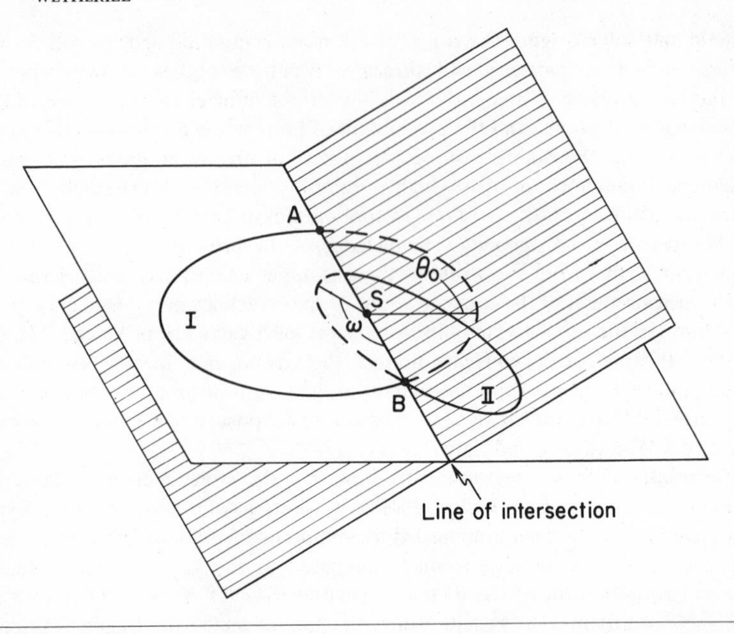

Figure 1 Crossed orbits: Elliptical orbit I is in the unshaded plane; orbit II is in the shaded plane. The two planes interesect along the line ASB at an angle i. Intersection of the two orbits can occur at points A or B at four values of ω.

In order for the orbit of the meteoroid to intersect the orbit of the earth, it is necessary that its perihelion lie within the orbit of the earth, and its aphelion beyond that of the earth. However, this condition in itself is not sufficient to produce intersecting orbits; the orbits may simply be linked, as in a chain. Such orbits are termed earth-crossing orbits. As explained more fully by Öpik (67), planetary perturbations, primarily by Jupiter, will cause the argument of perihelion of the meteoroid to precess, corresponding to rotation of the angle (ω) in Figure 1. As an earth-crossing orbit precesses, it will inevitably intersect that of the earth. Normally this can occur at four positions as shown in Figure 1. Under special circumstances, explained elsewhere (85; 88), intersection can occur for as many as eight values of ω, or as few as two. The important thing is that earth-crossing orbits will inevitably evolve into earth-intersecting orbits. Because the time scale for the rotation of ω through 360° is $\sim 10^4$ yr, the intersecting condition will occur every few thousand years.

Therefore objects in earth-crossing orbits can potentially collide with the earth. Öpik (67) derived an expression, on rather simple geometrical and dynamical grounds, for the probability (P) per year of a collision:

$$P = \frac{R^2 U}{2\pi^2 \sin i \, a_o a^2 (1-e^2)^{\frac{1}{2}} |\cot \alpha|} \tag{2.}$$

where R and U are defined in equation 1; a, e, and i are the semi-major axis, eccentricity, and inclination of the meteoroid; a_0 the semi-major axis of the earth; and α the angle between the radius vector from the sun and the direction of motion of the meteoroid at the point in its orbit which intersects the orbit of the earth. In this expression the earth is assumed to be in a circular orbit. A more general expression for the collision probability of two bodies, both in elliptical orbits, has been given (85) but for practical purposes the expression of Öpik is usually sufficient.

Substitution of numerical values into equation 2 gives the result that the annual probability of an earth-crossing object colliding with the earth is $\sim 10^{-8}$ yr^{-1}. If this were the only fate that could befall such an object, its lifetime in earth-crossing orbit would be $\sim 10^8$ yr. However, rather than actually strike the earth, it is much more probable that the meteoroid will make a close encounter with the earth, e.g. at a distance of 10 earth radii. Such a close approach will cause a strong perturbation of the object's orbit. Accumulation of a series of these perturbations will cause major orbital changes. These will cause most earth-crossing meteoroids to become Venus-crossing as well, on a time scale of $\sim 10^6$ yr; thereafter, close approaches and perturbations by that planet will also occur. Further accumulation of perturbations will frequently result in the object becoming Jupiter-crossing, following which it will usually be ejected from the solar system in a hyperbolic orbit within an additional $\sim 10^5$ yr. A statistical theory of the evolution of the orbit of a planet-crossing object as a consequence of planetary encounters was developed by Öpik (68). This was further refined by Arnold (5, 6) into a Monte Carlo method in which the evolution of an ensemble of initially very similar orbits was followed by iteration of the equations of Öpik. Further elaboration of this technique and the results found thereby will be given in subsequent sections. At this point it is sufficient to say that earth-crossing objects with aphelia ~ 3 AU will typically have a lifetime of $\sim 10^7$ yr. About 10% of such objects will strike the earth, another 10% will strike Venus. Almost all the rest will be ejected from the solar system following Jupiter-crossing. A small number will strike Mars, Mercury, the moon, and the outer planets, or make a sufficiently close approach to the sun to vaporize the meteoroid.

Collisional Lifetime

The measured cosmic ray exposure ages and the theoretical lifetime with respect to gravitational perturbations, discussed above, both show that the lifetime of a meteoroid in earth-crossing orbit is limited to $\sim 10^7$ yr. A less certain, but nevertheless possibly important, limitation on the lifetime is imposed by the length of time the meteoroid can survive bombardment by other meteoroids, primarily by those smaller than itself. This problem has been discussed recently by a number of authors (19, 27, 36, 85). This work shows that space erosion, i.e. reduction of the meteoroid size by continual microcratering is of negligible importance. Experimental confirmation of this theoretical conclusion is provided by the discovery of low energy solar cosmic-ray tracks in the surface layer (~ 1 cm) of meteorites with $\sim 10^7$ yr cosmic ray exposure ages (11, 72). The density of these tracks leads to a space erosion rate of $\lesssim 10^{-7}$ cm yr^{-1}. Even lower erosion rates are found for lunar samples (24). The most effective mechanism of collisional destruction is complete fragmentation in a single collision with smaller bodies. Laboratory studies by Gault & Wedekind (28) show that complete fragmentation requires about 10^6 erg of projectile energy per gram of fractured target. Use of this value leads to the result that in a typical collision of asteroidal material at 5 km sec^{-1}, a 1 g meteoroid will completely fragment a 10^5 g meteoroid. Collision of a higher velocity cometary meteoroid will destroy an even larger object. Although the quantity of laboratory data of this kind is limited, the result is probably applicable to rocks having the strength of ordinary chondritic and achondritic stone meteorites. The Type I and Type II carbonaceous chondrites will probably be more readily destroyed.

The principal problem in estimating the collisional lifetime of earth-crossing objects is the absence of adequate data on interplanetary matter in the size range (1–10 g) of the projectiles causing the fragmentation. The lifetime calculation has been made in several ways. The NASA model of the asteroid belt (40) is based on a calculation of the spatial density of visible asteroids, which is then extrapolated down to much smaller objects by using Dohnanyi's (19) steady-state population index $\alpha = 0.84$:

$$S = 10^{-15.79} m^{-0.84} \tag{3.}$$

where S is the number of asteroids of mass m (in grams) or greater per cubic meter. Correction factors are introduced to allow for the variation in density as a function of heliocentric distance, longitude, and latitude. The target asteroid is then considered as a body moving through a cloud of matter at an appropriate relative velocity. Another approach (85) taken was to use Öpik's collision formula (equation 2), modified for the case of crossing elliptical orbits. The target asteroid is assumed to be intersecting a field of projectiles in the same orbits as those of the visible asteroids, the total number of these asteroids greater than mass m (in grams) given by

$$N = Cm^{-.80}; \quad C = 3.62 \times 10^{18} \tag{4.}$$

The constants in equation 4 were fixed to match the distribution of observed asteroids at absolute magnitude 9.0, and for the total area of the asteroid belt to match the intensity of the gegenschein (30), with the distribution cut off by the Poynting-Robertson effect at a radius of 10^{-3} cm. Use of either of these approaches gives collisional lifetimes of the order of 10^7 yr for meter-size bodies colliding with asteroidal matter. The population of small higher velocity projectiles can be estimated using the measured flux of meteors (96) entering the earth's atmosphere, together with a theory relating their observed magnitude to their mass. However, all these calculations are seriously model dependent. Dohnanyi's steady-state population index (-0.84, or the similar value -0.80) may obtain only for smaller asteroids, and determination of C by matching to excessively large values of m will overestimate the number of 1 g objects. The gegenschein may be only partly asteroidal in origin. The albedo assumed for asteroids may be incorrect, and the mass-magnitude relationship for meteors is open to question. Considering all this, it is quite possible that the collision lifetime of one meter meteoroids is $\gtrsim 10^8$ yr. Direct measurement of the meteoroid population in the size range of 10^{-3} g to 1 g by spacecraft in the cruise mode of outer planet missions would be of great value in providing an experimental point to which these theoretical expressions could be fitted at the low mass end, but still above the size range in which nongravitational forces such as the Poynting-Robertson effect and radiation pressure become dominant.

In summary, collisional destruction may supplement or even supplant planetary perturbations as the limitation to the earth-crossing lifetime of even relatively large (1 m) meteoroids. On the other hand, these lifetimes could be primarily dynamically controlled, and collisions may only be important in removing stragglers which are stranded in especially stable orbits.

PLANETARY SOURCES

The most massive bodies in the vicinity of the earth and the moon are the other planetary bodies: Mars, Venus, and Mercury. The cratered surfaces of the moon and Mars, and probably Mercury as well, show that they are impacted by large (> 100 m) interplanetary meteoroids. The secondary ejecta from such cratering has been proposed as a source of some meteorites in the 100 g to 10^6 g range (78, 79, 83). In the work cited, Urey recognized the importance of Öpik's (67) calculations to the problem of the source of meteorites and correctly pointed out that a lunar source is in agreement with the observed exposure ages. Venus and the earth itself have not been proposed, presumably because of the problem of the ejecta penetrating the denser atmospheres of these planets, as well as their greater escape velocities. Ejecta from any of these bodies will have a high probability (15–50%) of impacting the earth.

The principal and probably overwhelming reason why these bodies are not a major source of meteoroids is the high escape velocities to which the ejected fragments must be accelerated. Laboratory studies and examination of the distribution of boulders in the vicinity of terrestrial and lunar craters (26) suggest that ~100 kg masses are ejected from ~1 km diameter craters at velocities of about 200 m/sec, far below the 2.4 km/sec escape velocity of even the moon. Fine pulverized spray and glass spherules are ejected at much higher velocities, up to 10 km/sec. Some of these materials must escape the moon and be captured by the earth. However, these very small particles will have very low luminosity, will enter the earth's atmosphere without being observed, and become so well mixed with terrestrial material that their identification will be extremely difficult, if not impossible.

Lin (45) has proposed a theory for the acceleration of massive objects to velocities of ~10 km/sec by the impact of comets with the earth. It could be that the gases released by the comet might permit acceleration with less severe shock effects than those usually accompanying such acceleration. If this reasoning is correct, such events could eject meteorites from the moon. The problems of scaling laboratory experiments to kilometer-size craters are so formidable that it is not possible to rule out such events on experimental or theoretical grounds. The principal reason for rejecting the moon as a significant source of meteorites is the fact that the lunar samples collected from eight lunar sites (25), orbital geochemical mapping of the entire moon (9, 63), and spectrophotometry (standardized to actual lunar samples) (56), show gross chemical dissimilarity between any abundant lunar rock type and most classes of meteorites. The only exceptions are the basaltic achondrites, similar in many ways to mare basalts. However, there are sufficient differences between the lunar rocks and the basaltic achondrites, such as their oxygen isotope ratios (76) and lead (75) isotope systematics, to rule out a lunar origin for these rocks as well. If the earth is not receiving a significant flux of lunar rocks, derivation of meteorites from other planets, e.g. Mars, is even more unlikely.

COMETARY SOURCES

Orbital History of Comets

Comets are small (radius $\sim$1–10 km) bodies containing volatile compounds (such as H_2O, CO_2, NH_3, CH_4) as major constituents (18). Almost all observed comets are in highly eccentric heliocentric orbits with perihelia inside the orbit of Jupiter, many inside that of Mercury. Most have aphelia beyond the orbit of Neptune and Pluto, with semi-major axes as large as 10^5 AU (1 light year). If it were not for their volatile constituents, the observation of comets would be very difficult. However, when they approach within about 2 AU of the sun, the volatile compounds are emitted as gases, forming a coma of radius $\sim 10^4$ km and usually a tail of ions and dust, sometimes as much as 10^8 km in length. This matter, lost from the comet, causes these otherwise inconspicuous bodies to become relatively bright, occasionally bright enough to be seen in daylight.

Because the observability of comets depends on release of their volatile components, usually by solar heating, there may be many comets with perihelia at greater distance from the sun which are never observed, and only those which chance to approach the sun are discovered. According to the theory of Oort (66), there are $\sim 10^{11}$ of these unobserved comets forming a cloud in orbits extending out to about 10^5 AU, and with perihelia beyond 100 AU. These comets were formed in the solar nebula during the formation of the solar system 4.6×10^9 yr ago and were perturbed into larger orbits: first, by planetary perturbations, and subsequently, by perturbations by the $\sim 10^4$ stars which probably have passed through the cometary cloud at distances of $\sim 10^4$ AU since the formation of the solar system. The low temperature of interstellar space results in the volatile compounds of the comet being in the solid state during the entire history of the solar system.

According to these ideas, stellar perturbations will occasionally cause the comet's perihelion to come sufficiently close to the sun for the comet to be visible. These are the so-called new comets which are discovered at the rate of about three per year. The period of these comets is $\sim 10^6$ yr. They will lose some of their volatile constituents at every perihelion passage. However, it is unlikely that they will become entirely depleted in these constituents before planetary perturbations, primarily by Jupiter, cause them to be ejected from the solar system in hyperbolic heliocentric orbits (48).

Sometimes planetary perturbations will have the opposite effect, and reduce the aphelion of the comet to within the orbit of Neptune, causing it to become a periodic comet, such as comet Halley, which reappears on a time scale which is short compared to human history, and is, therefore, observed on successive perihelion passages. An especially interesting subclass of these periodic comets are those of short period with aphelia in the vicinity of Jupiter, and perihelia usually within 3 AU (64 cases), within the orbit of Mars (45 known cases), or within the orbit of the earth (9 cases) (53). All the details of the capture of very long-period comets into orbits of this kind are not understood, but considerable progress has been made on this problem (22, 23, 42, 47). It is likely that these short-period comets are selected from the greater number of very long-period comets with random inclinations and perihelion distances, by virtue of their being in direct orbits of low inclination ($\lesssim 30°$) and with perihelia near Jupiter. Orbits of this kind have a high probability of being converted into those with aphelia near Jupiter on a time scale of $\sim 10^4$ yr. A corollary of this is that there should exist a family of comets complementary to the observed short-period comets, but having their perihelion near Jupiter, rather than their aphelion. Generally, these bodies will be unobservably faint. Comets in orbits of low eccentricity in the vicinity of Jupiter, such as Oterma, and Schwassman–Wachman II may be comets transitional between these two classes, which are visible at the distance of Jupiter because of their unusually great size.

The short-period comets have periods of about six years. Calculations of the typical time required for Jupiter to eject these comets in hyperbolic heliocentric orbits (69, 86) is 10^5–10^6 yr. Consequently these comets can make many more perihelion passages during their dynamic lifetime than new comets and those of longer period. For this reason it is probable that they will lose all of their volatile material (on a time scale of $\sim 10^3$–10^4 yr) while still in short-period orbits. If they retain significant quantities of nonvolatile compounds ($\sim 10\%$ of their initial mass, assuming solar composition of carbon, nitrogen, oxygen and less volatile elements), they will become dead or extinct comets, continuing to move in short-period heliocentric orbits, but much more difficult to observe. It is known that they do not retain all of their nonvolatile material, as the presence of dust tails, as well as elements such as Fe, Mg, and Na in spectra of meteors associated with observed comets, shows that some nonvolatile material is lost during their active phase. However, there is also evidence that some comets (Arend-Rigaux and Neujmin 1) are quite inactive at present (51) and that a residual core of nonvolatile compounds remains at least in a significant fraction of cases. Öpik (68) has presented reasons for believing that earth-crossing Apollo asteroids are extinct cometary cores. Asteroids such as 1580 Betulia, 944 Hidalgo, 719 Albert, and 887 Alinda, as well as the small body PL 6344 of the Palomar Leiden survey (80) are also likely candidates for identification as former comets.

In addition to the gravitational perturbations, principally by Jupiter, which modify the orbits of short-period comets, these comets are also subject to nongravitational perturbations (21, 50). Extrapolation of these effects into the future shows that nongravitational forces may reduce the aphelion of many short-period comets to within the orbit of Jupiter. The eight observed short-period comets with aphelion within Jupiter's perihelion, constituting 11% of those with well-determined orbits, may have evolved in this way, although gravitational perturbations may also perturb comets into orbits of this kind. The extreme example of this type of orbit is comet Encke, with aphelion at 4.10 AU, 0.8 AU within Jupiter's perihelion. If comets in such orbits become dead bodies with a nonvolatile core, their greater mean approach distance to Jupiter will lengthen their mean lifetime to 10^6–10^7 yr. If these assumptions are correct, the one live comet Encke may be accompanied by $\sim 10^3$–10^4 dead comets with radii ~ 1 km and moving in orbits with similar aphelia.

A physical theory for these nongravitational perturbations has been proposed by Whipple (94) as a jet effect associated with a tangential component of momentum of the material escaping from a rotating comet. This theory has been extended by Sekanina (52, 74) in terms of a modification of Whipple's icy-core comet model: a layered cometary model with a nonvolatile rich core mantled by predominantly volatile matter. The historical evolution of the orbit of comet Encke has been interpreted using this model, and it is concluded that this body has nearly reached its inactive phase, and that within ~ 100 yr only a nonvolatile core will remain. It will thereby join the group of Apollo asteroids even though its previous cometary origin will be documented. It will still be observable as indicated by the observation of its highly depleted nucleus near aphelion by Roemer (73).

Comets as Meteoroid Sources

Comets are the only established sources of meteoroids in the size range (> 100 g) under consideration in this article. The Prairie Network (57) has observed fireballs definitely associated with the Taurid stream, known to be derived from comet Encke. The ~ 100 kg preatmospheric mass of these objects definitely shows that there is massive nonvolatile matter in cometary nuclei. In addition, other Prairie Network fireballs are associated with the Leonids, Draconids, and Perseids, all associated with known periodic comets, showing that the Taurids are not unique in this regard. Material with similar ablation characteristics comprises the majority of the fireballs, and it is plausible that they are also of cometary origin, although in most cases their parent comet is no longer in the inner solar system, or is present only as an extinct comet. The high velocity of the Taurids upon entering the atmosphere (30 km/sec) precludes their survival and recovery as meteorites. However, similar fireballs of lower geocentric velocity, if sufficiently large, may reach the ground. The carbonaceous chondrites, at least those of type I and II (55) may be associated with these fireballs.

The only Prairie Network fireball actually recovered on the ground (Lost City) is an ordinary chondrite (65), H5 in the Van Schmus-Wood classification (81). Its ablation characteristics were different from those of the Taurids, facilitating its survival in the atmosphere. Relative to the Taurid type, half as many fireballs are of the "Lost City type" (61), although the distinction between the ablation characteristics may be gradational. Meteor theory, indicating that these fireballs had low density ($\rho < 1$ g/cm^3) has been revised (60) to fit the density of Lost City (3.7 g/cm^3). In addition, there is a class of fireballs, fewer in number than the Taurid and Lost City types, which are destroyed almost immediately by the atmosphere. The distribution of the orbital elements of these three types overlaps considerably, but on a statistical basis there are differences between the orbits of the three classes.

As mentioned above, some of the Taurid type fireballs are certainly of cometary origin and it is plausible that many, most, or all of them are. The similarity of the orbits of the Taurid and Lost City type fireballs has the consequence that it is difficult to distinguish between their sources on dynamical grounds. Therefore, from the dynamical point of view, the Lost City type in general, and the Lost City ordinary chondrite meteorite in particular, may also be of cometary origin. The remaining fireballs, more easily ablated, tend to be of higher velocity and more uniquely dynamically derivable from typical cometary sources. Therefore this line of reasoning suggests that all Prairie Network fireballs, including all chondritic meteorites, are of cometary origin. Öpik (70) has gone further and proposed that all meteorites, including those which have undergone major chemical differentiation, such as the achondrites and iron meteorites, are also derived from comets. Anders (4) has summarized chemical and mineralogical arguments against such an origin by presenting reasons for believing that chemically fractionated objects are more likely to be derived from the asteroid belt. Ceplecha (13) has shown that some Na-free, apparently pure nickel-iron meteoroids are moving in orbits similar to those of the Prairie Network fireballs. If one accepts Anders' chemical arguments this suggests that there must exist some mechanism for placing asteroidal fragments into orbits very similar to those known to be associated with short-period cometary material. On the other hand, one cannot exclude the possibility that comets contain chemically fractionated material. Harvey (34) has observed Fe-free spectra, consequently representing chemically fractionated nonvolatile material, from meteors which he infers are moving in orbits similar to those of long-period comets. More often, meteor spectra have solar ratios of these nonvolatile elements, and are therefore relatively unfractionated (64).

The chemical arguments against a cometary origin for meteorites are based ultimately on the assumption that complex chemical, mineralogical, and physical processes, for which evidence exists in meteorites, could not have occurred in the icy core of a comet. This is plausible; however, it is possible that these processes occurred prior to the formation of the present comet, since it is known that essentially all these processes took place during the first ~ 100 m.y. of solar system history (32, 39, 71, 93). A possible exception are the hypersthene chondrites; the parent body of at least some of these appears to have experienced a shock impact about 500 m.y. ago (37, 77). However, it is possible this event is not well dated and may have occurred much more recently (31). The origin of comets (and asteroids as well) is so poorly understood, and our certain knowledge of the detailed processes which took place during the formation of the solar system so meager, that it seems premature to conclude a priori that any particular type of object could not be found in the interior of a comet.

One approach to the problem of the origin of meteoroids with known orbital elements is to assume they are ejected from a hypothetical source moving in a particular orbit with a small spread in initial velocities, so that the orbit of the ejected fragments is essentially that of the source. The evolution of the orbits of the fragments is calculated forward in time until the fragments collide with the earth or are eliminated in some other way. Comparison of the calculated orbits with the observed orbits of the meteoroids at the time of earth impact may then be used to evaluate the plausibility of the proposed source.

The difficulty with a straightforward approach to this problem is that the usual methods of numerical integration are not applicable on the time scales (10^6–10^9 yr) indicated by the cosmic ray exposure ages. The orbital evolutions of particular comets for hundreds of years have been calculated by Marsden (49). After some thousands of years, uncertainties in the initial orbits, rounding off errors in computation, and propagation of errors resulting from the details of one close approach to a planet being strongly sensitive to the details of the previous close encounter, will preclude accurately following the exact history of a particular fragment. However, repetition of such calculations would still yield the statistical distribution of earth-impacting orbits, if computational errors are kept comparable to those resulting from variations in orbital evolution caused by the initial small velocity spread. The limitation to this approach is that calculation of a minimal number of initial orbits (e.g. 100) for 10^7 yr would cost millions of dollars of computer time, which would certainly be considered excessive by funding agencies.

There are alternative methods which forego, at the outset, the possibility of predicting an exact orbit such as would be used for an ephemeris, but should hopefully suffice to predict the statistical evolution of an ensemble of slightly different initial orbits. One such method is that used by Everhart (22) for the particular problem of the capture by Jupiter of very long-period comets into short-period orbits. The Monte Carlo method of Arnold (5, 6) is more directly applicable to the calculation of the very long-term orbital evolution of fragments initially moving in the orbit of a short-period comet. This method proceeds as follows:

1. An initial orbit is chosen and the planets, whose orbits are crossed by that orbit, are identified.

2. The probability of a close encounter with each of the planets at a given number n (e.g. $n = 50$) of gravitational radii is calculated by the expression of Öpik (equation 2) with some modification to include the effects of the planets' eccentricities and inclinations. The result is insensitive to the choice of n.

3. These probabilities are combined to calculate a mean time (τ) to the next encounter; the next planet to be encountered is chosen at random using probabilities weighted according to the individual planet's encounter probability calculated in equation 2.

4. The time to the next encounter is chosen at random assuming possible encounter times are distributed as $e^{-t_e/\tau}$, where t_e is the actual encounter time and τ is the mean encounter time calculated in step three.

5. The fragment is assumed to move on a trajectory passing through a random point on a target circle of radius $= n$ gravitational radii centered on the chosen planet.

6. The change in velocity of the fragment resulting from the close encounter is calculated, and from the new velocity components new orbital elements are calculated.

7. These new orbital elements are used as the new initial orbits in step one.

8. This procedure is continued until something terminates the history of the fragment, such as striking a planet (passing within one gravitational radius of the planet), having hyperbolic heliocentric orbital elements following a close encounter, or by coming within a preset distance (e.g. 0.03 AU) of the sun.

This entire procedure is repeated several hundred times for a given initial orbit, thereby generating a distribution of final orbits resulting from slight differences in the initial orbits. A distribution of this kind can be obtained for a given starting orbit at relatively little cost (i.e. $\sim$ \$100).

Of course there is a price to be paid for this saving of a factor of $\sim 10^5$ in computer time. The conditions under which these calculations should be valid have been discussed (5, 6). The method essentially assumes that the long-term evolution of a planet-crossing body is primarily determined by the relatively rare close encounters to the planets crossed, rather than by the continuing long range perturbations. Furthermore, it is assumed that the position on the target circle at a close encounter is independent of the position on the previous close encounter. Other similar assumptions of randomness are also involved. Except for perturbations caused by Jupiter (and possibly Saturn) the ordinary periodic and secular terms caused by the planets are probably small compared to the close encounters. A possible exception occurs when a small body has a commensurability relation between its period and that of earth or Venus. However, even this is not necessarily a major factor. Among the earth-crossing Apollo asteroids (possibly comets) only one, 1685 Toro, has been found to be in a commensurability resonance (8:5 with earth and 13:5 with Venus) (17). It has been shown (101) that the lifetime of this commensurability is not too long, being destroyed on a $\sim 10^6$ yr time scale by close approaches to Mars.

The problem of long-range Jupiter perturbations may be more serious. The secular terms can cause large changes in e and i (41, 98), significant close encounters occur even when the fragment is not Jupiter-crossing, and the time between successive close encounters to Jupiter is sufficiently short for the details of successive encounters to be possibly correlated. As will be discussed later, resonance with Jupiter's motion is important for some types of orbits. Efforts to reduce the effect of these problems have been made in subsequent applications of Arnold's method (89). Perturbations caused by encounters with Jupiter within its perihelion, but beyond 4.50 AU have been included in an approximate way. The free oscillation terms of the secular perturbations (those independent of Jupiter's eccentricity) have been introduced by parameterization of Kozai's and Williams' results. The effects of planetary eccentricity are included in a more exact way. For most orbits crossing Jupiter's perihelion, or with aphelia within $\sim$4 AU, these changes do not affect the results very much. It should not be concluded that either Arnold's or these revised procedures are entirely reliable representations of

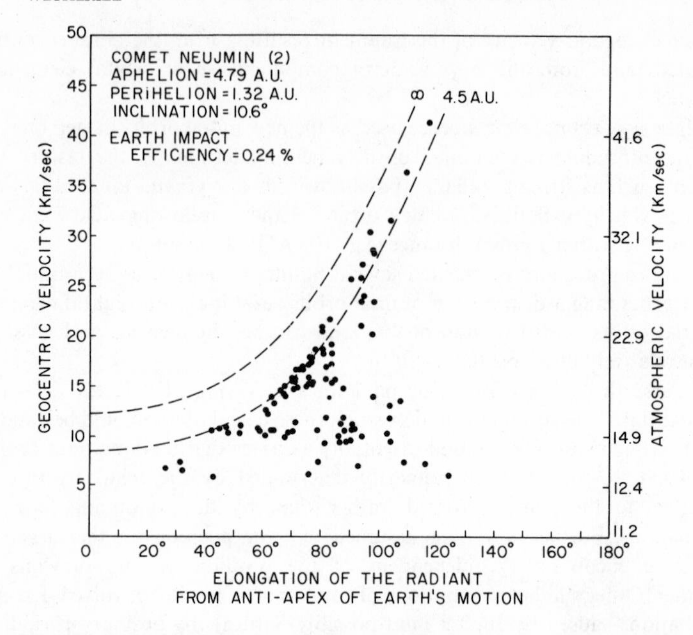

Figure 2 Calculated distribution of geocentric velocities and elongation of the radiants for earth impacts resulting from a starting orbit equivalent to the observed short-period comet Neujmin 2. The curve marked ∞ is the boundary between elliptic and hyperbolic orbits; the other curve is the locus of relatively low-inclination orbits with aphelia at 4.5 AU.

Jupiter perturbations; but, pending improvements in these methods (103), results obtained by these methods are plausible and suggest approximations to those which may in the future be obtained by a more exact treatment.

Monte Carlo calculations of the earth-impacting orbits of fragments initially in orbits similar to those of short-period comets have been published (81, 86, 89, 90). These are conveniently represented in a diagram (Figure 2) in which the geocentric velocity is plotted on the ordinate and the elongation of the radiant from the anti-apex of earth's motion (Figure 3) on the abscissa. The third variable, the azimuth of the radiant with respect to the direction of earth's motion changes rapidly once the object becomes earth-crossing, and is not diagnostic of the initial orbit. Therefore in representing the three parameters (a, e, i) defining the orbit in a two-dimensional diagram, this is the variable chosen for suppression.

As explained more fully elsewhere (90), on this diagram a scatter of points along the line corresponding to an aphelion of 4.5 AU, with little intrusion into the region of low velocity and large values of λ, is uniquely characteristic of objects which first became earth-crossing with their aphelion still near the orbit of Jupiter, e.g. 4.0–4.7 AU. The earth impacts resulting from short-period comets are of this

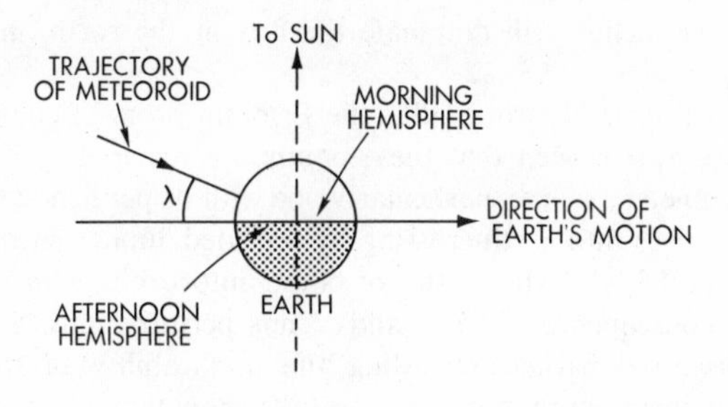

Figure 3 Definition of the elongation of the radiant λ.

kind, as shown in Figure 2. As the aphelion of the first earth-crossing orbit is extended to Jupiter's perihelion, and to Jupiter crossing, the aphelia of the fragments reaching the earth becomes somewhat larger (Figure 4). More important, the yield of earth impacts drops rapidly from 28% for $Q = 4.0$ AU, through 0.24% for $Q = 4.79$ AU to 0.0023% for $Q = 5.10$ AU. Therefore, those few comets with aphelia

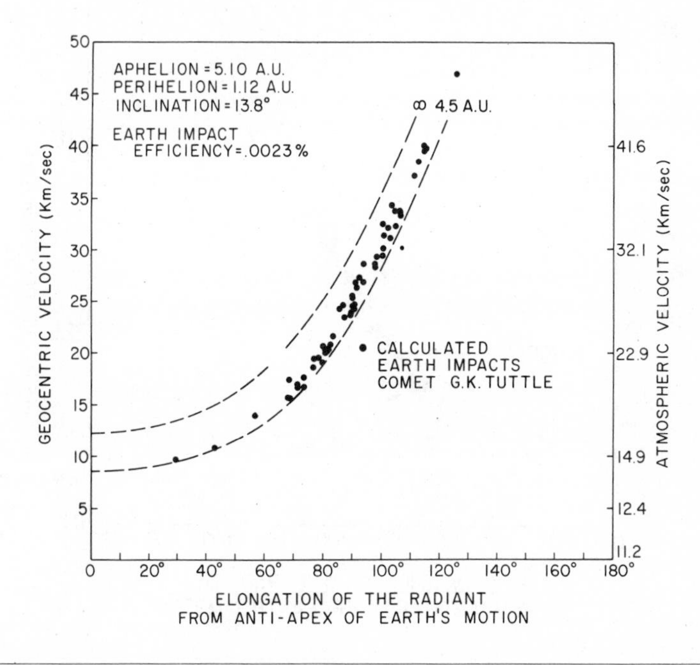

Figure 4 Calculated distribution of velocities and radiants for earth impacts resulting from a starting orbit equivalent to the observed short-period comet Giacobini-Kresak-Tuttle.

within Jupiter's perihelion will dominate the flux at the earth, in spite of their smaller number.

Data from the Prairie Network [plus the Pribram bronzite chondrite (12)] are plotted in Figure 5. It is seen that these points are remarkably similar to those predicted for a cometary source, particularly one with its perihelion slightly beyond 1.0 AU which evolves into earth-crossing by repeated Jupiter encounters near its aphelion at about 4.5 AU. The scatter of points into orbits with smaller aphelia is the expected consequence of earth and Venus perturbations. Subject to some reservations, discussed earlier, regarding the applicability of this method of calculation to Jupiter perturbations, these results show that a cometary source for

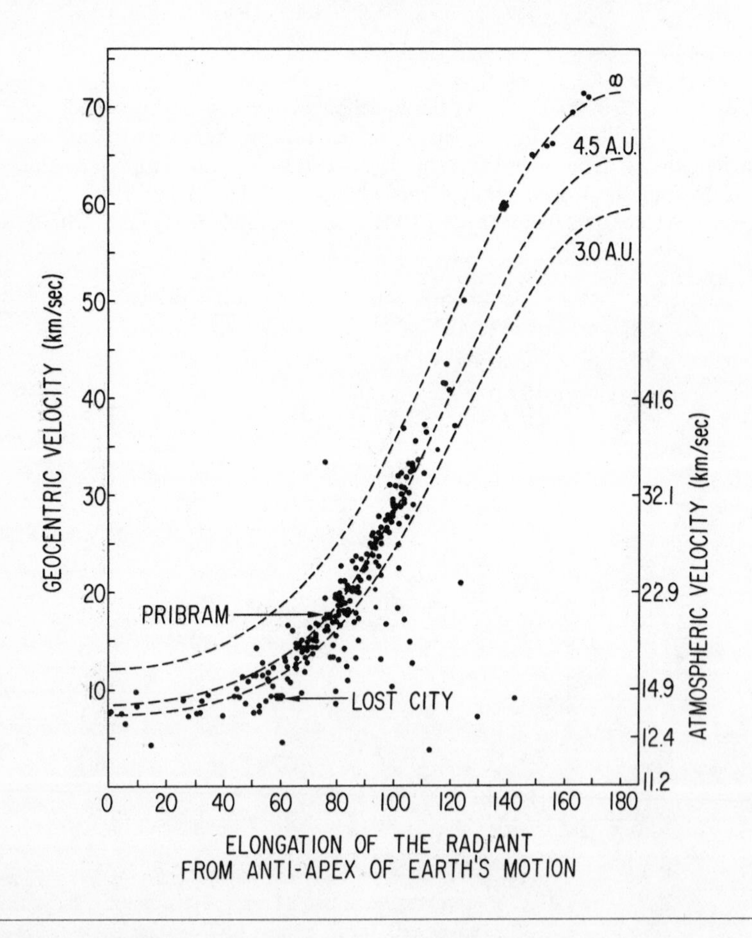

Figure 5 Observed distribution of geocentric velocities and radiants for Prairie Network fireballs and meteorites.

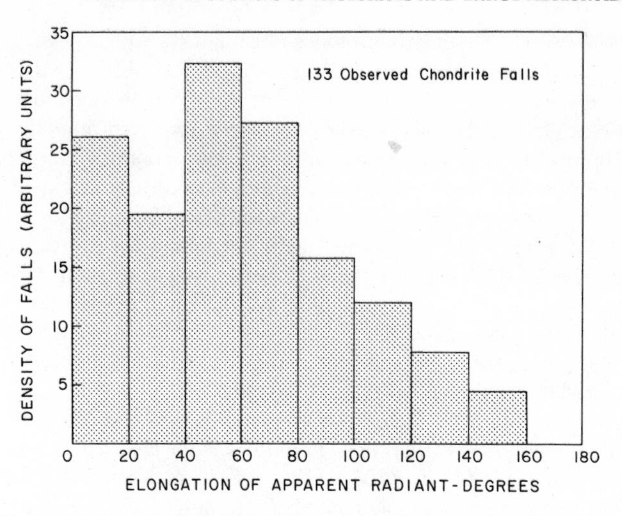

Figure 6 Observed distribution of 133 chondrite radiants.

the Prairie Network fireballs is entirely compatible with the orbital data. In the following section some mechanisms which can place asteroidal material into similar initial earth-crossing orbits are discussed. However, it is doubtful if the yield from these mechanisms can reach the values reported by McCrosky. When one considers the similarity of the ablation characteristics of the Prairie Network Taurids to those of other fireballs, the mass flux, and the orbital distribution, it appears that comets are the most probable source of most of these fireballs.

There does not exist a similar body of orbital data for meteorites. The two shown in Figure 5, Lost City (60) and Pribram (12), are the only meteorite orbits determined by astrometric methods. Their orbits are those of typical Prairie Network fireballs, but their number is too small to be very diagnostic of their initial orbit.

Although there are hundreds of visual observations of meteorite falls (primarily chondrites), these observations cannot be used to calculate reliable orbits since the velocity cannot be measured with sufficient accuracy. On the other hand, visual radiant determinations are sufficiently accurate to be very useful, and, for chondrites, exist in sufficient number to be statistically significant (7, 95). As seen in Figure 6, the elongations of these radiants are primarily $<90°$. This distribution of radiants is also of the type associated with the 2:1 preponderance of afternoon falls of chondrites (87).

The distribution of radiants differs from those found by the Prairie Network and predicted from cometary sources. However, when one includes the fact that it is unlikely that a meteoroid will survive penetration through the atmosphere if its initial velocity is much over 20 km/sec, these results are compatible with those of the Prairie Network. The relative absence of chondrites with $\lambda < 90°$ corresponds to the small number of low-velocity fireballs with $\lambda > 90°$. Monte Carlo calculations of cosmic ray exposure ages for a meteorite source with aphelion ~4.50 AU (Figure 7) are very similar to observed exposure ages, except for a long-lived tail which will probably be eliminated by collisional destruction.

From the point of view of orbital dynamics, a cometary origin of chondrites seems entirely plausible. There are insufficient data for most other classes of meteorites. For iron meteorites it seems likely that a cometary source would predict a larger number of short exposure ages. However, this conclusion depends on the relative number of dead comets with initial perihelia well beyond the earth, which could require a long time to become earth-crossing. In any case, there is no positive evidence associating the highly differentiated irons and achondrites with the initial orbits required by the fireball and chondritic data.

The principal arguments against a cometary source for chondrites are the chemical ones mentioned earlier. Related to these is the fact that at least at present (and presumably for the past ~10^7 yr) there are a small number (7) of classes of chondrites. These suggest a small number of sources. While suggestive of an

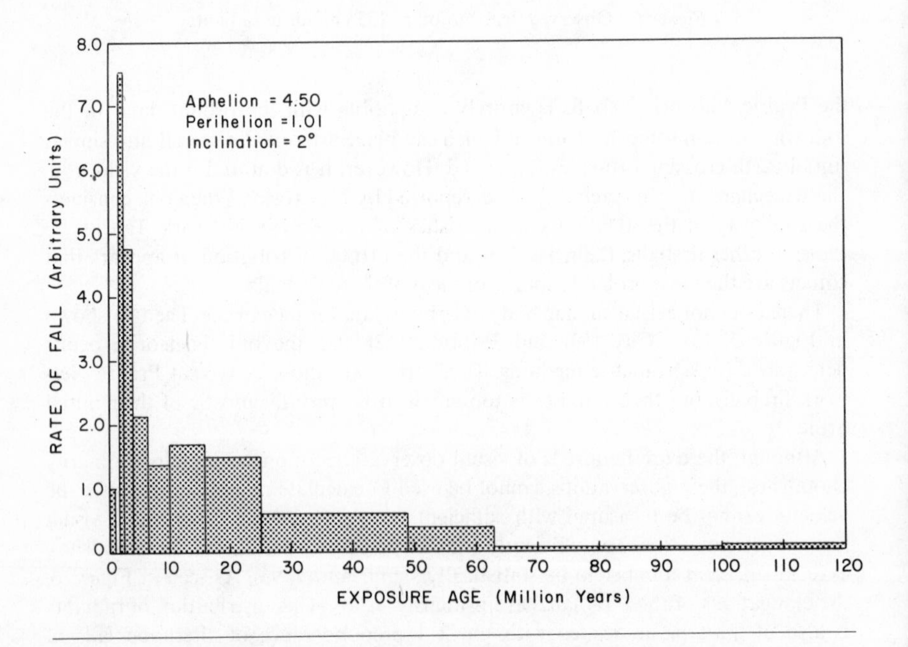

Figure 7 Calculated distribution of exposure ages for a starting orbit resulting in earth impacts similar to the low-velocity component of the Prairie Network flux: Aphelion = 4.50 AU, perihelion = 1.01 AU, inclination = 2°. This dynamically determined distribution is very similar to that observed for chondrites, except for those few with exposure ages greater than 50 m.y. Those will probably be removed by collisional destruction.

asteroidal origin, these are not compelling arguments against a few dominant dead cometary sources, associated with their achieving relatively small aphelia, while their perihelia were still beyond earth. It is also possible that discrete classes of cometary cores were established at the time of their origin.

ASTEROIDAL SOURCES

As discussed earlier, bodies moving in the asteroid belt will collide with one another, producing collision fragments, many in the meteorite size range. The exact quantity of these fragments is difficult to estimate, for the same reasons which limit our certain knowledge regarding the collisional lifetime of meteorites. Nevertheless reasonable estimates can be made; it is likely that the annual asteroidal production of fragments in the 10^2–10^6 g range is $\sim 10^{14}$ g. Only 1 in 10^5 of these fragments need impact the earth in order to produce the terrestrial flux of meteorites, 10–100 times as many are required to explain the terrestrial flux of meteoroids. Spectrophotometric studies (14) show that the surface composition of asteroids varies, thereby providing possible sources for a number of meteorite classes.

In the past few decades it has become almost conventional to regard asteroids as the source of meteorites, and meteor astronomers commonly refer to denser, more refractory meteors (including fireballs) as asteroidal as compared to the more fragile cometary meteors. This is an unfortunate usage as it necessitates such expressions as: "assuming that cometary meteors are of cometary origin." There was at least one serious problem which had to be overcome before the asteroid belt could be regarded as a major source of meteorites and meteoroids. This is the problem of removing the fragments from the asteroid belt at a sufficiently rapid rate to be compatible with the observed flux and $\sim 10^7$ yr cosmic ray exposure ages. Whatever mechanism is proposed, it should be much more effective in removing the earth-impacting fragments from the asteroid belt than in removing typical asteroids themselves. Otherwise the entire asteroid belt would disappear on a 10^7 yr time scale.

Meteorite fragments can be ejected only at velocities $\lesssim 1$ km/sec in order to avoid shock effects much greater than those usually observed. Direct collisional transfer of a fragment from the asteroid belt into earth-crossing orbit requires a velocity change of ~ 6 km/sec, far above this value. If meteorites could survive accelerations of this magnitude, the moon, with its lower (2.4 km/sec) escape velocity would also be a major meteorite source. This is now known not to be the case.

With the exception of those few asteroids already in earth-crossing orbits, or nearly so (to be discussed later), direct collisional transfer into earth-crossing orbits does not appear possible. Therefore more subtle mechanisms, probably involving more gentle gravitational effects must be invoked.

One such mechanism which has been extensively discussed is that of using Mars as an intermediary, whereby gravitational perturbations by Mars accumulate so as to eventually effect earth-crossing. The problems with this approach are twofold. First, if as suggested by Arnold (6), the original source of the fragment is an asteroid not crossing the orbit of Mars but with perihelion near Mars, only a small fraction ($\sim 1\%$) of the asteroids are sufficiently near to Mars aphelion to inject a fragment into Mars-crossing at probable (~ 200 m/sec) ejection velocities. Second, statistical and Monte Carlo calculations (5, 6, 68, 89) show that $\sim 10^9$–10^{10} yr must elapse before the fragment becomes earth-crossing. This is not compatible with the cosmic ray exposure ages, probably even for iron meteorites. Also, the collisional lifetime of stone meteorites is almost certainly too short to permit them to survive the slow accumulation of small perturbations by Mars required to effect earth-crossing.

A modification of this mechanism which avoids some of these difficulties is that proposed by Anders (3). This is that the meteorite sources are those asteroids which are already Mars-crossing. Their low velocity ejecta will also be Mars-crossing. If the orbits of these asteroids were distributed at random, then $\sim 10^{-3}$ of these fragments would become earth-crossing before suffering the collisional destruction necessary in order that this theory not conflict with the short observed exposure ages of stone meteorites. In fact, the orbits of observed Mars-crossing asteroids are not randomly distributed, but have their perihelia concentrated near the orbit of Mars. This introduces a finite delay of $\sim 10^9$ yr before the orbit of the fragment evolves into earth-crossing; for these bodies the yield of earth-impacting fragments with exposure ages of $\lesssim 10^7$ yr would be negligible. An additional difficulty is that many of the larger asteroids with perihelia inside the orbit of Mars cannot make close approaches to Mars (87, 92). At present the arguments of perihelion of these bodies are in the vicinity of 90° or 270°, causing their nodes to be well beyond the orbit of Mars. As the argument of perihelion precesses, so that the nodes move toward perihelion and aphelion, planetary perturbations by Jupiter will cause the eccentricity to decrease. By the time $\omega \sim 0°$ or 180°, i.e. the node is at perihelion, the eccentricity will decrease sufficiently for the perihelion to be well beyond the orbit of Mars. This phenomenon is a consequence of the free oscillation terms in the secular perturbation theory of Kozai (41) and Williams (98). The effect of this in limiting the number of true Mars-crossing asteroids has been discussed in more detail elsewhere (92). It is quite possible that the Mars-crossing asteroids are suitable sources of iron meteorites, although a detailed investigation of this possibility has not been made. The long exposure ages of most iron meteorites are at least compatible with the delay required for the fragment to become earth-crossing, and a penalty in yield is less likely to be imposed by collisional destruction of these stronger objects.

The remaining historically (i.e. pre-1971) proposed asteroidal sources are the earth-crossing Apollo asteroids, to which should be added Mars-crossers like 1580 Betulia. The radius of the node of this body will become earth (and Venus) intersecting as a result of free oscillations (88). There are now about 16 Apollo asteroids known, 4 having been discovered in the last 2 years. The long lost body Apollo (1932 HA) has also been recovered by McCrosky & Shao (62). The small radii of these asteroids (~ 1 km) reduces the observability enormously. The total number >0.5 km in radius has been estimated by Whipple (97) to be >50, quite likely there are ~ 1000 or more. Their low velocity ejecta are already in earth-crossing orbits and the probability of these fragments colliding with the earth is high ($\sim 10\%$). Also, it has been reported that the reflection spectra of the two bodies of this class studied so far, 1566 Icarus (20) and 1685 Toro (16), are much more similar to those of ordinary chondrites than any other asteroid. Some fragments of these bodies must strike the earth and appear as meteoroids. Some of these bodies will have sufficiently low geocentric velocities to permit fragments to survive penetration of the atmosphere if they are reasonably strong. At least some meteorites must be fragments of objects of this kind.

The unanswered question is whether or not these objects are the *dominant* source of meteorites of some unidentified class. As discussed elsewhere (88), if the observed orbits of these bodies are typical of the entire class, the meteorite yield will probably be too small ($\lesssim 10^7$ g/yr) and the cosmic ray exposure ages too long (up to 2×10^8 yr). Reducing the exposure ages by collisional destruction will lower the yield, and cause the predicted distribution of radiants and fall times, already presenting problems, to differ even more seriously than otherwise with observations.

Like the meteorites themselves, the dynamical lifetime of these bodies in earth-crossing orbits is limited to $\sim 10^7$–10^8 yr and they must have a source. It has been suggested that Mars-crossing asteroids may be the source, and earth-crossing is achieved by Mars perturbations. Öpik (68) and others (88) have objected to this on the basis that the observed number of Mars-crossers is insufficient to supply the observed number of earth-crossers. Anders (43) and Whipple (96) have criticized Öpik's argument on the basis that it is very dependent on uncertain assumptions, such as the index 0.84 in equation 3. Öpik's argument is strengthened by the fact that many of the Mars-crossers should not be counted as such, because they cannot make close approaches to Mars. A definitive answer to his question will require more complete searches for earth- and Mars-crossing asteroids, calculation of their proper elements, and better estimation of the completeness of discovery. It does seem that the number of Mars-crossers is too small to explain the number of earth-crossing bodies. The high velocity orbits of some of these bodies (1566 Icarus and Adonis) are also hard to produce by this mechanism of deflection by Mars.

One alternative is that at least some of the earth-crossing asteroids are cometary cores, as discussed in the previous section. In this case, the Apollo asteroid source becomes a special case of a cometary source. If meteorites are derived from cometary cores, there should also be a number of cometary cores which are not earth-crossing, from which meteorites could be derived, provided their aphelion remains sufficiently close to Jupiter for their fragments to be perturbed into earth-crossing. If the Apollo asteroids are cometary cores their observed distribution is probably completely unrepresentative of their actual distribution (88). Observational factors will preferentially select those bodies with small aphelia, as they will spend more of their time close to the sun and to the earth, whereas those with aphelia near 4.5 AU will usually be too distant to be seen.

The attempt to find an asteroidal source for meteoroids and meteorites which agrees with observation appears to have led us to the conclusion that an asteroidal source is satisfactory only if the asteroids under consideration are comets after all. Is this a necessary conclusion? The answer is probably no. However, those investigations discussed do lead to the conclusion that an acceptable asteroidal source for the most abundant meteoroids and chondrites necessarily involves a mechanism by which asteroidal fragments are placed in orbits which are usually thought of as cometary. It is necessary that the majority of these fragments first become earth-crossing in orbits subject to strong Jupiter perturbations near their aphelion, i.e. when their perihelion becomes less than 1 AU, their aphelion is greater than 4 AU. If ways can be found for placing a sufficient mass of asteroidal fragments into orbits which evolve in this way, then the asteroidal mechanism will be satisfactory. Recently two mechanisms have been proposed for achieving this.

The first, proposed by Zimmerman & Wetherill (103), invokes resonant perturbation by Jupiter in the vicinity of the 2:1 Kirkwood Gap. This occurs for objects with semi-major axis near 3.27 AU corresponding to their period being one half that of Jupiter. Although there is a marked deficiency of asteroids near the exact resonance period, there are a large number near it, primarily with semi-major axes slightly less than 3.27 AU. There are several ways by which material in these adjacent orbits can become subject to perturbations resulting in their acquiring orbits similar to those required.

Asteroidal collisions can cause moderate size (e.g. 10 m–1 km) fragments of ~ 10 km asteroids to be ejected at velocities of 50–200 m/sec resulting in their semi-major axis becoming near the resonance value. Then they will be subjected to strong perturbations by Jupiter which will increase their eccentricity to 0.3 or 0.4, increasing their aphelion distance and decreasing their perihelion. These larger aphelia would ordinarily result in close approaches to Jupiter. Close encounters would then lead to rapid orbital evolution very similar to that previously discussed for dead comets with aphelia ~ 4.5 AU. However, this will not necessarily occur. Near the exact resonance at 3.27 AU there is a libration region within which bodies will move in such a way as to avoid close encounters with Jupiter, even though they have large aphelia. When the body is near aphelion, Jupiter will tend to be on the other side of the sun. Alignment of the sun, the body, and Jupiter with both of the latter bodies on the same side of the sun will tend to occur only when the fragment is nearer perihelion, and hence far from Jupiter. This libration could permit fragments to remain in these highly eccentric orbits until they are destroyed by further collisions. Some additional contribution could be made by fragments of asteroids outside the libration region which acquire large aphelia simply as a consequence of their ejection velocity. Large fragments of this kind, if they exist, could evolve into Apollo asteroids. The evolution of these objects has not yet been studied in detail.

However, considerable attention has been given to the fragments which are injected into the libration region, or represent a residual population in this region remaining from the formation of the solar system. This has been done by development of techniques of calculation which replace the Monte Carlo methods discussed earlier, and still permit long-term evolution calculations to be made at a reasonable cost.

These librating bodies of < 1 km radius will be destroyed by further collisions during the history of the solar system, and a significant fraction of their ~ 1 m fragments will have even greater aphelia and smaller perihelia. The new calculations discussed above show that about 7% of these fragments will achieve earth-crossing orbits with aphelia beyond 4 AU, and thereby be similar to those cometary sources most effective in producing terrestrial meteoroids. Calculation of the magnitude of the asteroidal flux expected from the source shows it to be 10^8–10^9 g/yr. This is in agreement with the *meteorite* flux of Hawkins (35) and the lunar *meteoroid* flux reported by Latham et al (44), but below the value of $\sim 5 \times 10^{10}$ g/yr reported by McCrosky (59) for meteoroids in this mass range. More precise values of the predicted mass yield can be obtained by more detailed study of the orbital dynamics associated with this mechanism. It is possible that it will prove to be competitive with the cometary component, known to be significant, or could be responsible for a more minor component of the stone meteoroid flux, e.g. all or some achondrites. This mechanism has the advantage of identifying a few large asteroids (e.g. 530 Turandot, 175 Andromache, 805 Hormuthia, 927 Ratisbona, 511 Davida) in the vicinity of the Kirkwood Gap as candidate meteorite sources. Some of these may be fragments of the others. The limited number of sources may be related to the discrete number of meteorite classes. Comparison of their reflection spectra in the visible and near infrared with those of meteorites (15) may permit their identification with known meteorite classes.

An alternative mechanism for placing asteroid fragments into similar orbits has been proposed by Williams (100). This also involves resonant acceleration. However Williams' mechanism does not involve commensurability in orbital periods, but instead invokes secular resonances with the motion of Jupiter's perihelion. Williams' (98) detailed study of asteroidal perturbations by Jupiter first involves averaging over the short-term perturbations (periodic terms) which depend on the particular positions of the asteroid and Jupiter in their orbits. The remaining secular perturbations, independent of the starting positions, are divided into two components, the free oscillations and the forced oscillations.

The free oscillations are those secular perturbations which would occur if Jupiter's orbit were circular, rather than slightly elliptical. These perturbations preserve the z component of angular momentum of the asteroid resulting in the quantity

$$\theta = \sqrt{a(1-e^2)}\cos i = \text{constant} \qquad 5.$$

By Poisson's theorem (33) the secular change in a is small. However, it is found that e and i can undergo oscillations of large amplitude, but of opposite phase, a necessary condition for θ to remain constant. These free oscillations have already been discussed in connection with their effects on Mars-crossing asteroids. The free oscillations are of little direct importance in placing objects into earth-crossing orbits, although they must be included in any correct calculation of orbital evolution, as was done in the case of the Kirkwood Gap ejecta.

The forced oscillations represent the additional perturbations caused by the fact that Jupiter has a small but finite eccentricity. In the many-body problem of the real solar system they also arise in part from the fact that the planets are not exactly coplanar. Although the angular momentum of the entire system is still conserved, elimination of the symmetry of Jupiter's orbit breaks the symmetry which caused θ to be constant. The forced oscillations therefore do not require an inverse relationship between the maxima and minima of e and i, and the relative phases of these oscillations can shift. They also do not involve the correlation between these maxima and minima and the argument of perihelion, as was the case for the free oscillations. This lack of symmetry results in the forced oscillations being potentially more important in causing permanent changes in the orbit of an asteroid. However, because Jupiter's eccentricity is low, their amplitude is usually small ($e \pm \sim 0.01$).

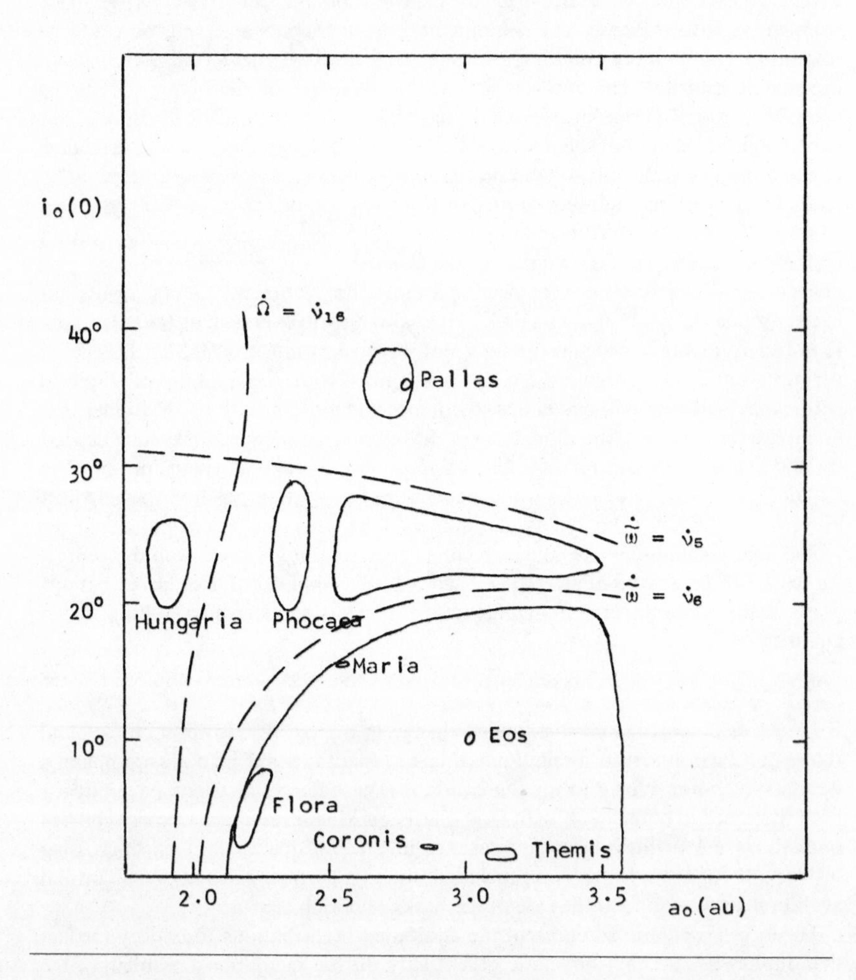

Figure 8 Resonances surface for proper eccentricity of 0.1 together with schematic representation of asteroid regions and families; from Williams (98).

There are exceptions to this. When the rate of precession of the longitude of perihelion ($\dot{\tilde{\omega}} = \dot{\omega} + \dot{\Omega}$), where Ω is the longitude of the node, is the same as that of one of the components of the velocity of Jupiter's longitude of perihelion, a resonant condition can exist. Similar resonances occur between the motion of the asteroidal nodes and Jupiter's nodes. Under these conditions, the forced oscillations can reach large amplitudes. A consequence of this is that there exist resonant surfaces in *a, e, i* space in the asteroid belt in the vicinity of which there are no observed asteroids (99). At a particular eccentricity these will define boundary lines on a diagram of *a* versus *i*. Figure 8 is a plot of this kind (98) together with some of the recognized families of asteroids. The resonant surfaces are unpopulated because any asteroidal fragments lying on these sufaces will be accelerated into an unstable orbit. Williams (100) has shown that collision fragments of asteroids near a particular one of these surfaces ($\dot{\tilde{\omega}} = \dot{v}_6$) will be accelerated into earth-crossing orbits with aphelion near Jupiter on a time scale of $\sim 10^6$ yr. Candidate asteroids are 6 Hebe, 89 Julia, 130 Elektra, and 8 Flora.

It is also possible that the proximity of some asteroids [e.g. the Flora group (10, 38)], to both Mars' aphelion and a resonant surface may have a synergistic effect, in which Mars' perturbation may move a fragment near the resonant surface, and the resonant effects may stimulate Mars-crossing. Although this mechanism needs detailed study in order to be acceptable, it is possible that it may provide a way of overcoming the $\sim 10^9$ yr lag required for an object barely crossing Mars' aphelion to become earth-crossing. Detailed calculations of the predicted final earth-impacting orbits have not been reported for Williams' mechanism. Although their orbital elements are not sufficiently precise to be certain of this, it is of interest that the two meteorites with determined orbits both appear to lie on one of the resonant surfaces: Pribram on $\dot{\tilde{\omega}} = \dot{v}_6$ [1] and Lost City on $\dot{\Omega} = v_{16}$ (46). Nodal resonances of the latter type are probably ineffective in establishing the initial earth-crossing orbit of the fragment but are likely to be important in its later evolution.

CONCLUDING REMARKS

During the last decade a massive quantity of new experimental and observational data has been obtained: on meteorites and meteoroids from the Prairie Network; from laboratory studies of the physical, chemical and isotopic properties of meteorites; and from spectrophotometric data on asteroids. Significant progress in the difficult problem of interpreting meteor spectra (64) is likely as a result of the laboratory work of Boitnott & Savage (8, 9). Hopefully, proper collection and analysis of residual dust from large meteoroids collected in the upper atmosphere will supplement laboratory studies of meteorites. Understanding of the acoustic effects of meteoroids entering the atmosphere may permit global sampling of the meteoroid flux. New theoretical techniques have permitted orbital evolution

[1] Williams (private communication, 1971) has calculated the proper elements of these meteorite orbits and showed that they lie on these resonant surfaces.

calculations to be extended beyond the time limitations imposed by numerical integrations.

The time may be near when tentative assignments of meteorite sources can be made. A speculation, no less likely to be true than others, would be that the Taurid-type fireballs and the Type I and Type II carbonaceous chondrites are derived from comets, the achondrites from asteroids by a resonant mechanism, and the iron meteorites from Mars-crossers. The remaining chondrites may be either cometary or asteroidal in origin.

Whether or not these assignments are correct, recent developments have brought us much nearer to an understanding of the sources of these objects in the solar system. Further progress on this important problem may be anticipated, possibly including data from unmanned missions to comets and asteroids. Such missions are well within the technological and fiscal scope of the present NASA and USSR planetary programs.

Literature Cited

1. Adler, I. et al. 1972. The Apollo 15 X-ray fluorescence experiment. *Proc. of the Third Lunar Sci. Conf.*, 3:2157–78. Cambridge: MIT Press. 3263 pp.
2. Anders, E. 1963. Meteorite ages. In *The Solar System* IV, ed. B. M. Middlehurst, G. P. Kuiper, 402–95. Chicago: Univ. Chicago. 810 pp.
3. Anders, E. 1964. Origin, age, and composition of meteorites. *Space Sci. Rev.* 3:583–714
4. Anders, E. 1971. Meteorites and the early solar system. *Ann. Rev. Astron. Astrophys.* 9:1–34
5. Arnold, J. R. 1964. The origin of meteorites as small bodies. *Isotopic and Cosmic Chemistry*, 347–64. Amsterdam: North-Holland. 553 pp.
6. Arnold, J. R. 1965. The origin of meteorites as small bodies, 2, the model; 3, general considerations. *Astrophys. J.* 141:1536–47
7. Astapovich, I. S. 1939. Some results of the study of 66 orbits of meteorites. *Astron. Zh.* 16:15–45
8. Boitnott, C. A., Savage, H. F. 1970. Light-emission of sodium at simulated meteor conditions. *Astrophys. J.* 161: 351–63
9. Boitnott, C. A., Savage, H. F. 1971. Light emission measurements of calcium and magnesium at simulated meteor conditions. *Astrophys. J.* 167:349–55
10. Brouwer, D. 1951. Secular variations of the orbital elements of minor planets. *Astron. J.* 56:9–32
11. Cantelaube, Y., Maurette, M., Pellas, P. 1967. Traces d'ions lourds dans les mineraux de la chondrite de Saint Severin. *Radioactive Dating and Methods of Low-Level Counting*, 215–29. Vienna: IAEA
12. Ceplecha, Z. 1961. Multiple fall of Pribram meteorite photographed. *Bull. Astron. Inst. Czech.* 12:21–47
13. Ceplecha, Z. 1966. Complete data on iron meteoroid (meteor 36221). *Bull. Astron. Inst. Czech.* 17:195–206
14. Chapman, C. R., McCord, T. B., Johnson, T. V. 1973. Asteroid spectral reflectivities. *Astron. J.* 78:126–40
15. Chapman, C. R., Salisbury, J. W. 1973. Comparisons of meteorite and asteroid spectral reflectivities. *Icarus* 19:507–22
16. Chapman, C. R., McCord, T. B., Pieters, C. 1973. Minor planets and related objects X. Spectrophotometric Study of the composition of (1685) Toro. *Astron. J.* 78:502–5
17. Danielsson, L., Ip, W. H. 1971. Capture resonance of the asteroid 1685 Toro by the earth. *Science* 176:906–7
18. Delsemme, A. H., Miller, D. C. 1971. Physico-chemical phenomena in comets-III, The continuum of Comet Burnham (1960 II). *Planet. Space Sci.* 19:1229–57
19. Dohnanyi, J. W. 1969. Collisional model of asteroids and their debris. *J. Geophys. Res.* 75:3468–93
20. Egan, W. G., Veverka, J., Noland, M., Hilgeman, T. 1973. Photometric and polarimetric properties of the Bruderheim chondritic meteorite. *Icarus* 19: 358–71
21. Encke, J. F. 1823. Fortgesetzte Nachricht über den Pons'chen Kometen. *Berliner Astron. Jahrbuch für 1826*. 51: 124–40

22. Everhart, E. 1973. Horseshoe and Trojan orbits associated with Jupiter and Saturn. *Astron. J.* 78:316–28
23. Everhart, E. 1973. Examination of several ideas of comet origins. *Astron. J.* 78:329–37
24. Fleischer, R. L., Hart, H. R. Jr., Comstock, G. M. Evwaraye, A. O. 1971. The particle track record of the Ocean of Storms. *Proc. of the Second Lunar Sci. Conf.,* 3:2559–68. Cambridge: MIT Press. 2818 pp.
25. Gast, P. W. 1972. The chemical composition and structure of the moon. *The Moon* 5:121–48
26. Gault, D. E., Shoemaker, E. M., Moore, H. J. 1963. Spray ejected from the lunar surface by meteoroid impact. *NASA TN D-1767.* 39 pp.
27. Gault, D. E. 1969. On cosmic ray exposure ages of stone meteorites. *Meteoritics* 4:177
28. Gault, D. E., Wedekind, J. A. 1969. The destruction of tektites by micrometeoroid impacts. *J. Geophys. Res.* 74:6780–94
29. Gault, D. 1970. Saturation and equilibrium conditions for impact cratering on the lunar surface: criteria and implications. *Radio Sci.* 5:273–91
30. Gindilis, L. M. 1962. The gegenschein as an effect produced by the scattering of light from particles of interplanetary dust. *Sov. Astron. AJ* 6:540–48
31. Gopalan, K., Wetherill, G. W. 1971. Rubidium-strontium studies on black hypersthene chondrites: Effects of shock and reheating. *J. Geophys. Res.* 76: 8482–92
32. Gray, C. M., Papanastassiou, D. A., Wasserburg, G. J. 1973. Primitive $^{87}Sr/^{86}Sr$ in the Allende carbonaceous chondrite. *EOS* 54:346
33. Hagihara, Y. 1972. *Celestial Mechanics,* II(1):164–74. Cambridge: MIT Press. 504 pp.
34. Harvey, G. 1973. Strongly differentiated material in long-period orbits. *Bull. Am. Astron. Soc.* 5:343
35. Hawkins, G. 1960. Asteroidal fragments. *Astron. J.* 65:318–22
36. Hellyer, B. 1971. The fragmentation of the asteroids II. *MNRAS* 154:279–91
37. Heymann, D. 1967. On the origin of hypersthene chondrites: ages and shock effects of black chondrites. *Icarus* 6: 189–221
38. Hirayama, K. 1928. Families of asteroids. *Jap. J. Astron. Geophys.* 5: 137–62
39. Hohenberg, C. M., Podosek, F. A., Reynolds, J. H. 1967. Xenon-iodine dating: Sharp isochronism in chondrites. *Science* 156:202–6
40. Kessler, D. J. 1970. Meteoroid environment model-1970 (Interplanetary and Planetary) *NASA SP-8038.* 66 pp.
41. Kozai, Y. 1962. Secular perturbations of asteroids with high inclination and eccentricity. *Astron. J.* 67:591–98
42. Kresak, L. The cometary and asteroidal origin of meteors. *IAU Colloquium,* No. 13, ed. C. L. Hemenway, A. F. Cook, P. M. Millman. *NASA SP-319.* In press
43. Krinov, E. L. 1966. *Giant Meteorites,* transl. J. S. Romankiewicz. Oxford: Pergamon. 397 pp.
44. Latham, G. V. et al 1972. Passive Seismic Experiment. *Apollo 16 Prelim. Sci. Rep. NASA SP-315.* Chap. 9
45. Lin, S. C. 1966. Cometary impact and the origin of tektites. *J. Geophys. Res.* 71:2427–37
46. Lowrey, B. E. 1971. Orbital evolution of Lost City Meteorite. *J. Geophys. Res.* 76:4084–89
47. Lowrey, B. E. 1973. The effect of multiple encounters on short period comet orbits. *Astron. J.* 78:428–37
48. Lyttleton, R. A., Hammersley, J. M. 1964. The loss of long-period comets from the solar system. *MNRAS* 127: 257–72
49. Marsden, B. G. 1967. One hundred periodic comets. *Science* 155:1–7
50. Marsden, B. G. 1970. Comets and nongravitational forces III. *Astron. J.* 75: 75–84
51. Marsden, B. G. 1971. Evolution of comets into asteroids? *Physical Studies of Minor Planets,* ed. T. Gehrels, 413–21. *NASA SP-267.* 687 pp.
52. Marsden, B. G., Sekanina, Z. 1971. Comets and non-gravitational forces IV. *Astron. J.* 76:1135–51
53. Marsden, B. G. 1972. *Catalogue of Cometary Orbits.* Cambridge: Smithson. Astrophys. Observ. 70 pp.
54. Mason, B. 1962. *Meteorites.* New York: Wiley. 274 pp.
55. Mason, B. 1971. The carbonaceous chondrites—a selective review. *Meteoritics* 6:59–70
56. McCord, T. B. et al 1972. Spectrophotometry (0.3 to 1.1 μ) of visited and proposed Apollo lunar landing sites. *The Moon* 5:52–89
57. McCrosky, R. E. 1967. Orbits of photographic meteors. *Smithson. Astrophys. Observ. Spec. Rep. 252.* 20 pp.
58. McCrosky, R. E. 1968. Meteors without sodium. *Smithson. Astrophys. Observ. Spec. Rep. 270.* 13 pp.

59. McCrosky, R. E. 1968. The distribution of large meteoritic bodies. *Smithson. Astrophys. Observ. Spec. Rep. 280.* 13 pp.
60. McCrosky, R. E., Posen, A., Schwartz, G., Shao, C.-Y. 1971. Lost City meteorite—its recovery and a comparison with other fireballs. *J. Geophys. Res.* 71:4090–4108
61. McCrosky, R. E. 1972. Meteoroid structure and composition. *EOS* 53:724
62. McCrosky, R. E., Shao, C.-Y. 1973. *1932 HA (Apollo). Centr. Bur. Astron. Telegrams, IAU,* ed. B. Marsden. Circular 2516
63. Metzger, A. E. et al 1973. Lunar surface radioactivity: Preliminary results of Apollo 15 and Apollo 16 gamma-ray spectrometer experiment. *Science* 179: 800–3
64. Millman, P. M. 1972. Giacobinid meteor spectra. *J. Roy. Astron. Soc. Can.* 66: 201–11
65. Nava, D. F., Walter, L. S., Doan, A. S. Jr. 1971. Chemistry and mineralogy of the Lost City meteorite. *J. Geophys. Res.* 76:4067–71
66. Oort, J. H. 1950. The structure of the cloud of comets surrounding the solar system, and a hypothesis concerning its origin. *Bull. Astron. Inst. Neth.* 11: 91–110
67. Öpik, E. J. 1951. Collision probabilities with the planets and the distribution of interplanetary matter. *Proc. Roy. Irish Acad.* 54A: 165–99
68. Öpik, E. J. 1963. Survival of comet nuclei and the asteroids. *Advan. Astron. Astrophys.* 2:219–62
69. Öpik, E. J. 1965. The dynamic aspects of the origin of comets. *Armagh Observ. Contr.* 53l
70. Öpik, E. J. 1966. The stray bodies in the solar system 2. The cometary origin of meteorites. *Advan. Astron. Astrophys.* 4:302–36
71. Papanastassiou, D. A., Wasserburg, G. J. 1969. Initial strontium isotopic abundances and the resolution of small time differences in the formation of planetary objects. *Earth Planet. Sci. Lett.* 5:316–76
72. Price, P. B., Rajan, R. S., Tamhane, A. S. 1967. On the preatmospheric size and maximum space erosion rate of the Patwar stony-iron meteorite. *J. Geophys. Res.* 72:1377–88
73. Roemer, E. 1972. Periodic comet Encke (1970 1). *Centr. Bur. Astron. Telegrams, IAU,* ed. B. Marsden. Circular 2435
74. Sekanina, Z. 1969. Total gas concentration in atmosphere of the short-period comets and impulsive forces upon their nuclei. *Astron. J.* 74:944–50
75. Tatsumoto, M., Knight, R. J., Doe, B. R. 1971. U-Th-Pb systematics of Apollo 12 lunar samples. *Proc. of the Second Lunar Sci. Conf.,* 2:1521–46. Cambridge: MIT Press. 2818 pp.
76. Taylor, H. P., Epstein, S. 1970. O^{18}/O^{16} ratios of Apollo 11 lunar rocks and minerals. *Proceedings of the Apollo 11 Lunar Science Conference,* 1613–26. New York: Pergamon. 2492 pp.
77. Turner, G. 1969. Thermal histories of meteorites by the ^{39}Ar–^{40}Ar method. In *Meteorite Research,* ed. P. Millman, 407–17. Dordrecht: Reidel. 941 pp.
78. Urey, H. C. 1959. Primary and secondary objects. *J. Geophys. Res.* 64: 1721–37
79. Urey, H. C. 1965. Meteorites and the moon. *Science* 147: 1262–65
80. Van Houten, C. J., Van Houten-Groeneveld, I., Herget, P., Gehrels, T. 1970. The Palomar-Leiden survey of faint minor planets. *Astron. Astrophys.* Suppl. 2:339–48
81. Van Schmus, W. R., Wood, J. A. 1967. A chemical petrologic classification for the chondritic meteorites. *Geochim. Cosmochim. Acta* 31:747–65
82. Wänke, H. 1966. Meteoritenalter und verwandte Probleme der Kosmochemie. *Fort. Chem. Forsch.* 7:322–408
83. Wänke, H. 1968. Radiogenic and cosmic-ray exposure ages of meteorites, their orbits and parent bodies. In *Origin and Distribution of the Elements,* ed. L. H. Ahrens, 411–21. Oxford: Pergamon. 1178 pp.
84. Wasserburg, G. J., Burnett, D. S. 1969. The status of isotopic age determinations on iron and stone meteorites. In *Meteorite Research,* ed. P. M. Millman, 467–79. Dordrecht: Reidel. 941 pp.
85. Wetherill, G. W. 1967. Collisions in the asteroid belt. *J. Geophys. Res.* 72: 2429–44
86. Wetherill, G. W. 1968. Dynamical studies of asteroidal and cometary orbits and their relation to the origin of meteorites. In *Origin and Distribution of the Elements,* ed. L. H. Ahrens, 425–43. Oxford: Pergamon. 1178 pp.
87. Wetherill, G. W. 1968. Time of fall and origin of stone meteorites. *Science* 159: 79–82
88. Wetherill, G. W., Williams, J. G. 1968. Evaluation of the Apollo asteroids as sources of stone meteorites. *J. Geophys. Res.* 73:635–48
89. Wetherill, G. W. 1969. Relationships between orbits and sources of chondritic meteorites. In *Meteorite Research,* ed. P. M. Millman, 573–89. Dordrecht:

Reidel. 941 pp.
90. Wetherill, G. W. 1971. Cometary versus asteroidal origin of chondritic meteorites. In *Physical Studies of the Minor Planets,* ed. T. Gehrels, 447–60. NASA SP-267. 687 pp.
91. Wetherill, G. W. 1972. Origin and age of chondritic meteorites. In *Contributions to Recent Geochemistry and Analytical Chemistry,* ed. A. I. Tugarinov, 22–34. Moscow: Nauka. (In Russian) 642 pp.
92. Wetherill, G. W. Problems associated with estimating the relative impact rates on Mars and the moon. *The Moon.* In press
93. Wetherill, G. W., Mark, R., Lee-Hu, C. 1973. Chondrites: Initial Sr^{87}/Sr^{86} ratios. *Science* 182:281–83
94. Whipple, F. L. 1950. A comet model I, the acceleration of comet Encke. *Astrophys. J.* 11:375–94
95. Whipple, F. L., Hughes, R. F. 1955. On the velocities and orbits of meteors, fireballs, and meteorites. In *Meteors,* ed. T. R. Kaiser, 145–56. *Spec. Suppl. 2, J. Atmos. Terr. Phys.*
96. Whipple, F. L. 1967. On maintaining the meteoritic complex. *Smithson. Astrophys. Observ. Spec. Pap. 239,* 409–26. NASA SP-150
97. Whipple, F. L. 1967. The meteoritic environment of the moon. *Proc. Roy. Soc.* 296A:304–15
98. Williams, J. G. 1969. *Secular perturbations in the solar system.* Ph.D. dissertation, Univ. California, Los Angeles. 273 pp.
99. Williams, J. G. 1971. Proper elements, families, and belt boundaries. In *Physical Studies of Minor Planets,* ed. T. Gehrels, 177–81. NASA SP-267
100. Williams, J. G. 1973. Meteorites from the asteroid belt? *EOS* 54:233
101. Williams, J. G., Wetherill, G. W. 1973. Physical studies of the minor planets XIV, Long term orbital evolution of 1685 Toro. *Astron. J.* 78:510–15
102. Zähringer, J. 1968. Rare gases in stony meteorites. *Geochim. Cosmochim. Acta* 32:209–37
103. Zimmerman, P. D., Wetherill, G. W. 1973. Asteroidal source of meteorites. *Science* 182:51–53

THE ATMOSPHERE OF MARS

Charles A. Barth
Department of Astro-Geophysics and Laboratory for Atmospheric and Space Physics, University of Colorado, Boulder, Colorado 80302

INTRODUCTION

Our knowledge of the Mars atmosphere has increased rapidly during the past several years because of intensive spacecraft exploration of that planet and because of earth-based observations using improved spectroscopic techniques. The atmosphere of Mars is essentially a pure carbon dioxide atmosphere that contains a small and seasonally varying amount of water vapor. Observations have revealed a number of minor constituents that arise from the interaction of solar radiation with water vapor and carbon dioxide: namely, carbon monoxide, atomic oxygen, molecular oxygen, ozone, and atomic hydrogen. At the surface of Mars the atmospheric pressure is less than one hundredth of the pressure at the surface of the earth and, in fact, corresponds in magnitude to the pressure in the earth's stratosphere. Near the surface the temperature of the atmosphere is low and decreases further with increasing height and reaches the condensation temperature of carbon dioxide under many circumstances. Elevation differences on Mars are large and, consequently, there are large variations in the surface pressure around the planet. Because of the thinness of the atmosphere, the temperature varies greatly from day to night. Winds blow and at times blow very intensely. Dust storms are created by the winds and on rare occasions envelope the entire planet. Dust in the atmosphere absorbs heat and changes the temperature structure of the atmosphere, even when present in small amounts. Frozen carbon dioxide is the major constituent of the polar caps, but the caps also contain water ice, perhaps even in large amounts. The amount of water vapor in the atmosphere appears to be controlled, in part, by seasonal conditions at the polar caps and also by the temperature of the surface at all locations on Mars. Extensive cloud systems appear on Mars and perhaps all those that have been observed are composed of water ice crystals. The amount of ozone in the atmosphere varies seasonally and from day to day. It appears that when ozone is present, water vapor is not; and when water vapor is present, ozone disappears. In the upper atmosphere the temperature is low, but not low enough to prevent the escape of atomic hydrogen through thermal evaporation. The ionosphere of Mars is composed of ions that result from the photochemistry of the carbon dioxide–water vapor atmosphere. Recombination

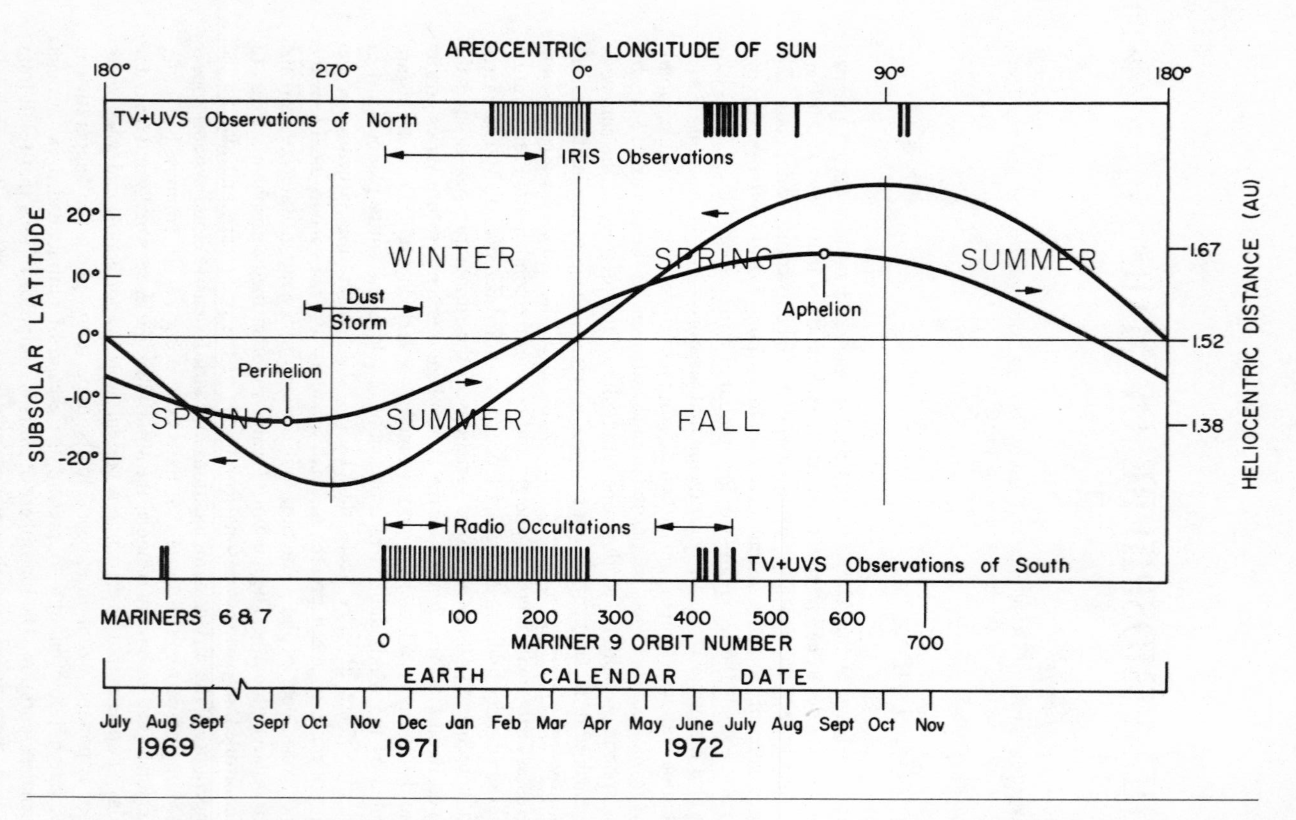

Figure 1 A Mars seasonal calendar. The dates of Mariner observations are shown related to the Mars local season.

of these ions provides nonthermal mechanisms for atoms to escape from the upper atmosphere of Mars. Since the pressure and even the composition of the Mars atmosphere is intimately connected to the temperature of polar caps, long-term changes in the heat input to the polar caps may lead to dramatic changes in the pressure and temperature of the atmosphere and even to the amount of water on the surface of the planet.

SPACECRAFT OBSERVATIONS

Observations of Mars from planetary spacecraft began in July of 1965 with the Mariner 4 "fly-by" mission and continued in July and August of 1969 with two additional fly-by missions, Mariner 6 and 7. In 1971 and 1972, extensive planetary observations were conducted with three spacecraft in orbit about Mars: Mariner 9 operating from November 1971 until October 1972, and Mars 2 and 3 observing between December 1971 and March 1972. In comparing observations from the different spacecraft missions to each other and to earth-based observations, it is appropriate to take into account the season on Mars at the time. In particular, Mariner 9 operated in Mars orbit for almost half of a Mars year and acquired measurements during all four seasons while observing both the northern and southern hemispheres. Figure 1 indicates the seasonal time of the Mariner 9 observations and it may be used to interrelate other observations as well. Since on Mars the seasons are produced by the inclination of the spin axis to the orbital plane—the obliquity—and by the eccentricity of the orbit, both properties are plotted in Figure 1. Mariner 9 began observations in the southern hemisphere where it was early summer, shortly after going into orbit. Later in the mission when observations were begun in the north polar region, it was past mid-winter. All observations stopped shortly after the beginning of the spring-fall seasons with resumption in mid-fall in the north. A few observations were made at the beginning of summer in the north polar region. The equatorial region and the south polar region during early spring were observed from Mariner 6 and 7; Mariner 4 flew by Mars during the late summer-winter season. The Mars 2 and 3 observational period was from mid to late summer-winter. Several Soviet Mars spacecraft are scheduled to arrive at Mars in 1974, near the beginning of the spring-fall season.

COMPOSITION

The atmosphere of Mars is essentially a pure carbon dioxide atmosphere. The evidence for this has accumulated from observations made from spacecraft and from earth-based telescopes. The first direct measurement of the Mars atmosphere, made by the Mariner 4 radio occultation experiment in July 1965, showed that the total atmospheric pressure was small, somewhere between 4.1 and 6.2 mbar (41).

Since this experiment measures the index of refraction of the atmosphere, a knowledge of the atmospheric composition is necessary to determine a unique value for the pressure. However, when the radio occultation results are combined with spectroscopic measurements of the carbon dioxide abundance, useful constraints are placed upon the composition of the Mars atmosphere. High resolution spectroscopy of the 1.038 and 1.050 μ bands in the Mars atmosphere from an earth-based telescope during the 1967 opposition showed that the amount of CO_2 in the Mars atmosphere was 78 ± 11 m-atm,[1] an amount corresponding to a partial pressure between 4.7 and 6.3 mbar (14). With the caution that the two observations—the radio occultation and the high resolution spectrocopy—were made at different locations on the planet and at different times, the results may be combined to show that the Mars atmosphere must contain at least 75% carbon dioxide.

At the time of the 1967 opposition of Mars, high resolution spectra were obtained in the near-infrared using the technique of Fourier spectroscopy (18). This was a significant advance in earth-based planetary spectroscopy. Analysis of these observations showed that all of the features in the Mars spectrum are produced by carbon dioxide and carbon monoxide in the Mars atmosphere (39). A detailed analysis of the high resolution Mars spectra yielded both the total pressure of the Mars atmosphere and the partial pressure of carbon dioxide (93). Within the errors of the analysis, the partial pressure of CO_2 was found to be equal to the total pressure of 5.2 ± 0.3 mbar (abundance 70 ± 4 m-atm). An analysis of the 2.35 μ CO band showed an abundance of carbon monoxide of 13 ± 8 cm-atm (94) showing that the percentage composition of carbon monoxide in the Mars atmosphere is only 0.2% (limits 0.07–0.32%). Some implications of this small ratio will be discussed later.

Water vapor in the Mars atmosphere is measurable with earth-based telescopes using high resolution spectroscopy in the near infrared portion of the spectrum (8200 Å water vapor band) (86). Telescopic observations show that the amount of water is variable, ranging from the detection limit of about 10 μm precipitable water to nearly 60 μm, depending upon the season (4). During the summer season in either hemisphere, maximal amounts of water vapor are found. Following the equinoxes, the amount of water vapor in the Mars atmosphere has been below the detection limit of earth-based telescopes.

The Mariner 6 and 7 infrared spectrometer observations of Mars during the 1969 fly-by detected only three gaseous constituents: carbon dioxide (71 m-atm), carbon monoxide, and water vapor (37). Although these spectrometers had the sensitivity and resolution to detect parts per million of a number of molecules, no additional constituents were found. No ammonia, no nitrogen dioxide, and no nitric oxide were detected with an instrument sensitivity of less than a part in a million. With a sensitivity of a few parts in a million, the following hydrocarbons were also absent: methane, ethane, ethylene, and acetylene. A compilation of upper limits of minor constituents is given in Table 1. The low limits determined by the

[1] The amount of the atmospheric constituent is expressed in meter-atmospheres, the thickness of the column in meters when compressed to standard temperature and pressure on the earth; 1 m-atm = 2.69×10^{21} molecules per square centimeter, 1 cm-atm = 2.69×10^{19} molecules per square centimeter, 1 μm-atm = 2.69×10^{15} molecules per square centimeter.

Table 1 Upper limits for minor constituents

	Molecule	Abundance
Mariner 6 and 7		(μm-atm)
Infrared spectrometer	NO_2	16
(curve of growth) (37)	NH_3	31
	C_3O_2	32
	SO_2	37
	OCS	40
	NO	50
	CH_4	260
	N_2O	1300
Mariner 6 and 7		(mm-atm)
Infrared spectrometer	C_2H_4	0.12
(absorption coefficient) (37)	CS_2	0.14
	C_2H_6	0.25
	C_6H_6	0.40
	C_2H_2	0.64
	HCN	0.96
Earth-based		(μm-atm)
Infrared interferometer (13)	HCl	11
	HCOH	50
	HCOOH	70
OAO		(μm-atm)
Ultraviolet spectrometer (72)	SO_2	10
	H_2S	10
	NH_3	10
	NO_2	10
	N_2O_4	30

Mariner 6 and 7 infrared spectrometer experiment were the result of an elaborate laboratory program (37).

Additional earth-based infrared observations of Mars, made during the 1969 opposition using a Connes-type interferometer (13) were specifically designed to search for hydrochloric acid, which had been discovered on Venus with the same type instrument. A search was also made for formaldehyde and formic acid in the atmosphere of Mars. None of these constituents were found with the detectability limit one part per million or better. The upper limits for these molecules are listed in Table 1.

During the 1971 opposition of Mars, earth-based high resolution spectroscopic observations led to the discovery of molecular oxygen in the atmosphere of Mars (3, 17). Two rotational lines in the A-band of molecular oxygen at 7635 Å were observed at a time when the Doppler shift permitted them to be distinguished

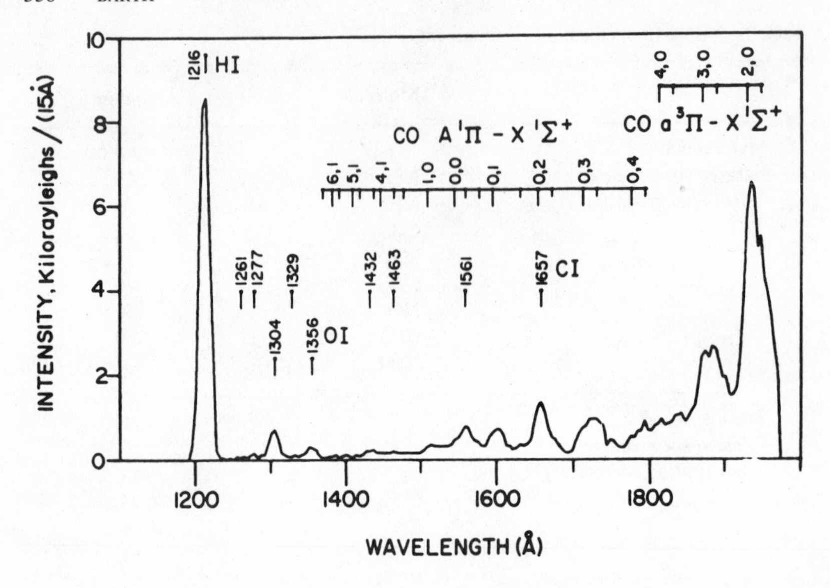

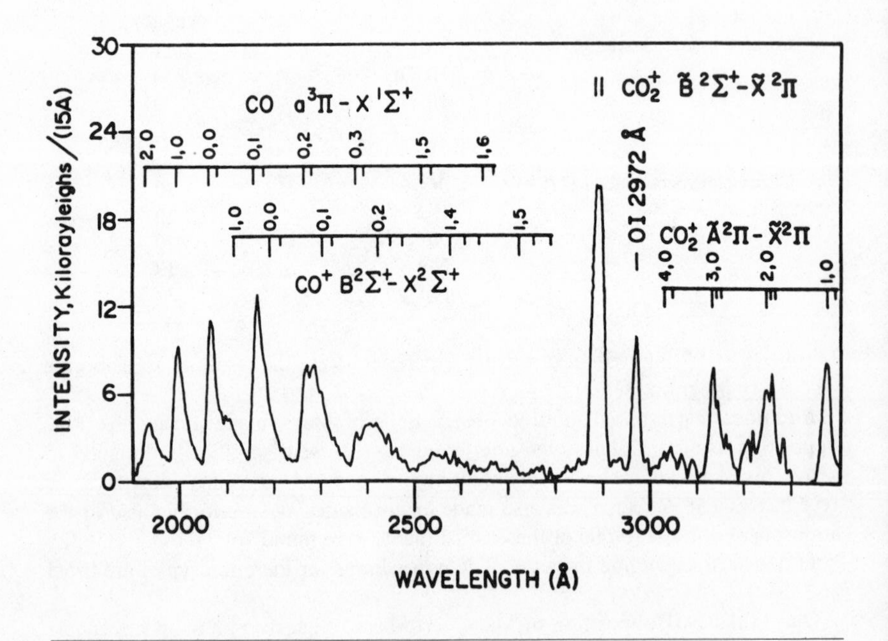

Figure 2 Mars airglow spectrum between 1100 Å and 3400 Å obtained by the ultraviolet spectrometer on Mariner 9 (9).

from the earth's molecular oxygen. The amount of oxygen observed was small, 10.4 ± 1.0 cm-atm, corresponding to a percentage composition of 0.13%.

The Mariner 6 and 7 ultraviolet spectrometer experiment measured the emission spectrum of the upper atmosphere of Mars (6). Ultraviolet observations show Mars has an essentially pure carbon dioxide atmosphere. With two exceptions, all of the observed spectral features can be explained by the direct or indirect action of solar ultraviolet radiation on carbon dioxide (7). The two exceptions are resonance lines of atomic hydrogen at 1216 Å and atomic oxygen at 1304 Å. The atomic hydrogen and atomic oxygen in the Mars atmosphere are produced by the photodissociation of water vapor and carbon dioxide. A radiative transfer treatment of the atomic hydrogen Lyman alpha 1216 Å data shows the amount of atomic hydrogen at the top of the Mars atmosphere (250 km) is 3×10^4 atoms cm^{-3} (2). An elaborate analysis of the atomic oxygen 1304 Å data concludes that the concentration of atomic oxygen in the upper atmosphere at 135 km lies between 0.5 and 1% of the total density at that altitude (90). Molecular nitrogen was searched for in the ultraviolet data and not found (6). The sensitivity of the ultraviolet spectrometer for detecting molecular nitrogen is estimated to be approximately 1% (5). A comparison of the ultraviolet data with calculations based on a number of model atmospheres leads to estimates of the nitrogen to carbon dioxide mixing ratio that lie between 0.5 and 5%, depending on the degree of mixing in the atmosphere (22).

Hundreds of ultraviolet spectra were obtained of the Mars upper atmosphere during the long operational lifetime of the Mariner 9 Mars orbiter. The Mars airglow spectrum in Figure 2 is the sum of a large number of individual observations (9). This spectrum, with higher signal to noise ratio than the Mariner 6 and 7 data, shows all of the spectral features first identified in the 1969 experiment and no others. Atomic hydrogen and atomic oxygen were also measured by the ultraviolet photometer on Mars 3 (23, 47).

Ozone was first detected on Mars by the Mariner 7 ultraviolet spectrometer (8) and has been measured extensively from the Mariner 9 orbiter. Figure 3 shows the

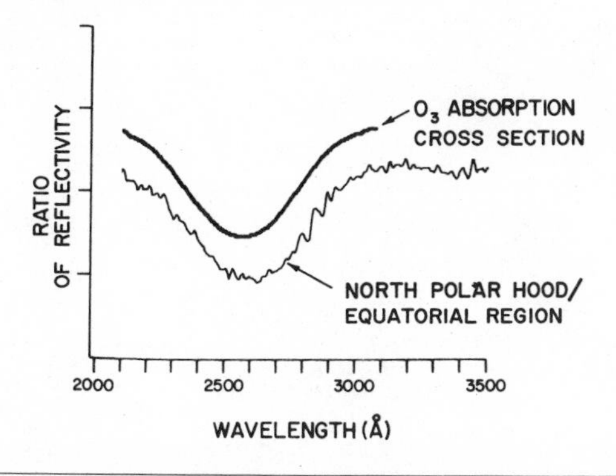

Figure 3 Mars absorption spectrum compared with laboratory ozone curve (48).

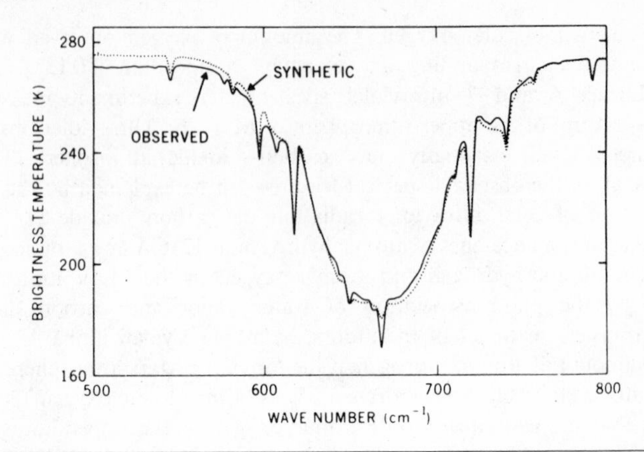

Figure 4 Mars absorption spectrum in the 12.5 to 20 μ region measured by the infrared interferometer spectrometer experiment on Mariner 9 compared to carbon dioxide synthetic spectrum (19).

distinctive spectral absorption feature in the 2000 to 3000 Å wavelength region that is used to detect and measure ozone (48). The amount present in the atmosphere is small and variable. Over most of the planet, the amount is less than the detection limit of 3 μm-atm or less than 0.5 parts in 10 million. In the polar regions, detectable amounts occur varying from 3 μm-atm to 60 μm-atm (10). The seasonal variation and correlation with other properties of the Mars atmosphere will be discussed in a subsequent section.

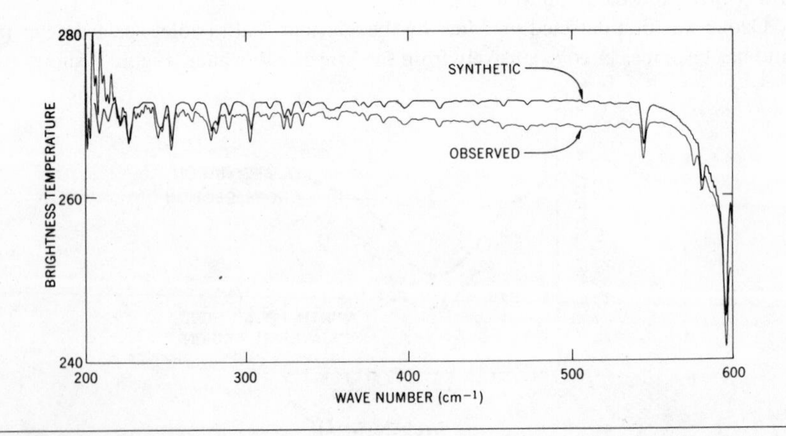

Figure 5 Mars absorption spectrum in the 16.7 to 50 μ region observed by Mariner 9 infrared interferometer spectrometer compared to water vapor synthetic spectrum (19).

The Mariner 9 infrared interferometer spectrometer obtained many tens of thousands of spectra of Mars in the wavelength region 5 to 50 μ (200 to 2000 cm^{-1}). The thermal emission of the surface and atmosphere of Mars was observed over a number of different Mars environmental conditions; for example, the planetwide dust storm, the subsequent clear atmosphere, the receding polar cap, and the changing polar hood. The only spectral features observed by this instrument were those produced by carbon dioxide (including its isotopes), by water vapor, and by the atmospheric dust. No other molecules were observed (33). In the wavelength region observed, the most prominent CO_2 spectral feature is the fundamental 15 μ band. Figure 4 shows this observed band and numerous other CO_2 bands as well, including the isotopes $O^{16}C^{13}O^{16}$, $O^{16}C^{13}O^{18}$, and $O^{16}C^{12}O^{18}$. The isotopic ratios for C^{12}/C^{13} and O^{16}/O^{18} are in general agreement with ratios found on earth (19). Figure 5 shows the spectral absorption produced by 10 μm of precipitable water in the Mars atmosphere. The amount of water is variable and depends on the location and season. This seasonal variation will be discussed in a subsequent section.

STRUCTURE OF THE LOWER ATMOSPHERE

Following the first direct measurement of the structure of the lower atmosphere of Mars by Mariner 4 in 1965 (41), the radio occultation experiments on Mariner 6 and 7 again probed the atmosphere in 1969. Temperature profiles from the four 1969 measurements, which included locations ranging from equatorial to polar latitudes, are shown in Figure 6 (42, 43, 75). The temperature of the atmosphere

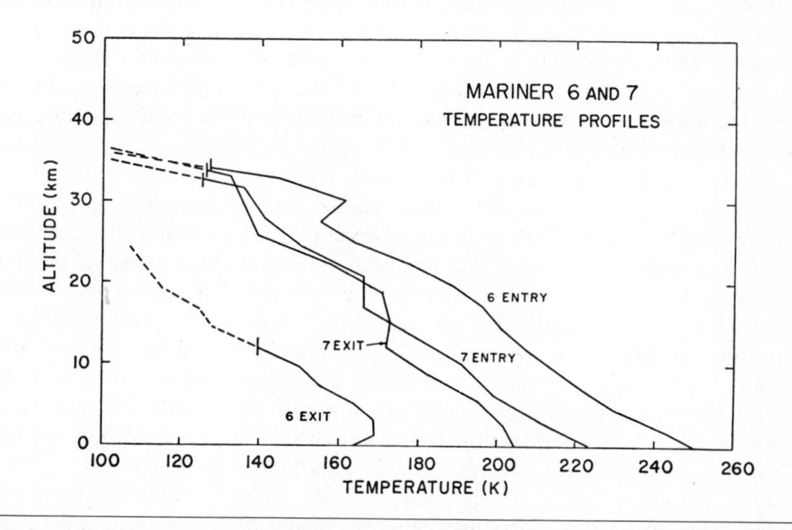

Figure 6 Atmospheric temperature profiles for the four occultation points of Mariner 6 and 7(75). (Figure reproduced by permission of *J. Atmos. Sci.*)

was 250°K at the warmest location (afternoon near the equator) and 170°K at the coldest location (before midnight in the north polar region). With increasing altitude, the temperature dropped with a lapse rate of about 3°K/km for the nonpolar measurements. For the north polar region, the atmospheric temperature at 12 km altitude decreased to a value of 140°K. Above this altitude, the atmosphere was cold enough for carbon dioxide, the major constituent of the atmosphere to condense. Observations by the Mariner 6 and 7 infrared spectrometer of the 4.3 μ "reflection-absorption" band of solid carbon dioxide have been interpreted as direct evidence for the presence of solid carbon dioxide in the atmosphere of Mars (34a).

At the six locations probed by the Mariner 4, 6, and 7 radio occultation experiments, the measured values of the surface pressure varied from 4.2 to 7.3 mbar. The variations are real, due to variations in elevation at the different locations. On earth, pressures of this magnitude are reached in the stratosphere at an altitude of about 33 km.

In 1971 and 1972, measurements of the Mars lower atmosphere were made by the Mariner 9 radio occultation experiment during two periods: first, during mid-summer in the southern hemisphere (mid-winter in the north), while a planetwide dust storm was in progress; second, during spring in the northern hemisphere (fall in the south) while the atmosphere was relatively clear (see Figure 1). Measurements during the first period were in two groups: daytime measurements between latitudes 40°S and 20°N and nighttime (or twilight) measurements near 65°N. During the first nine orbits of Mariner 9, the atmosphere was isothermal in the 40°S to 20°N latitude band in the lowest 10 km (44). Throughout the first 40 days (80 orbits), while occultation measurements were possible, the magnitude of the temperature gradient gradually increased and at the same time, the temperature of the atmosphere directly above the surface increased (45). However, the lapse rate varied from measurement to measurement, and the systematic increase apparently reversed toward the end of this period. The magnitude of the mean lapse rate was always less than 2.5°K/km.

Measurements made at night or in the early morning in the 65°N latitude band showed that the atmospheric temperature reached the condensation temperature of carbon dioxide at an altitude of 10 km or lower (45). The season at this time was mid-winter, when the carbon dioxide polar cap was forming beneath the polar hood.

During the second period that Mariner 9 radio occultation measurements were possible, essentially all latitudes were sampled, including both polar regions. At this time, the atmosphere was clear; the planetwide dust storm had subsided four months earlier. Measurements at equatorial and mid-latitudes showed that the atmosphere was no longer isothermal. The temperature did decrease with increasing altitude, but the lapse rate surprisingly had a mean value of only 2.3°K/km, still less than the mean lapse rate of the few measurements by Mariner 6 and 7 (46).

Over the north polar cap in mid-spring, the radio occultation experiment measured atmospheric temperatures between 178 and 191°K on different days; all temperatures above the sublimation temperature of carbon dioxide (147°K at 6.105 mbar). The lapse rate was 2.2°K/km. Measurements of pressure in the north polar region showed values between 4.4 and 7.4 mbar, due to elevation differences, with a mean pressure of 5.7 mbar. The south polar cap has a mean elevation of 3 to 4 km higher than the north polar cap. The temperature of the atmosphere above the south polar cap (season mid-fall) was consistent with temperatures necessary for carbon dioxide condensation (46).

Approximately 222 individual radio occultation measurements were made by Mariner 9. Because of elevation differences, the pressure at the surface varied for all of these observations. At the lowest elevation, in the 65°N latitude band, the pressure was over 10 mbar (46). Near the top of Middle Spot, the highest elevation measured by the radio occultation experiment, the pressure was 1 mbar. The mean of all of the pressures measured by the radio occultation experiment was 6.6 mbar.

Temperature profiles of the lower atmosphere of Mars were obtained by the Mariner 9 infrared interferometer spectrometer experiment. Analysis of the spectral shape of the 667 cm^{-1} (15 μ) carbon dioxide band leads to the determination of the temperature at a number of discrete pressure levels in the altitude range from the surface to 40 km. Over 17,000 temperature profiles were measured during the first 210 orbits of Mariner 9, a period which lasted from early summer to late summer in the southern hemisphere (early winter to late winter in the north). This period included the abatement of the planetwide dust storm and the clearing of the atmosphere. From this vast resource of data, examples of a number of the important results have been reported (19, 32, 33).

Changes in the temperature structure of the atmosphere at mid-latitudes are illustrated in Figure 7. Early in summer, during the dust storm (Revolution 20), the atmosphere was relatively warm from the surface all the way up to 40 km and the temperature gradient in the lowest 10 km was small. At mid-summer, when the dust storm was beginning to subside, the atmosphere above 10 km began to cool. Below the 10 km level, the atmosphere warmed up and a temperature gradient of −3°K/km was established from the surface up to 20 km. Above that altitude, the

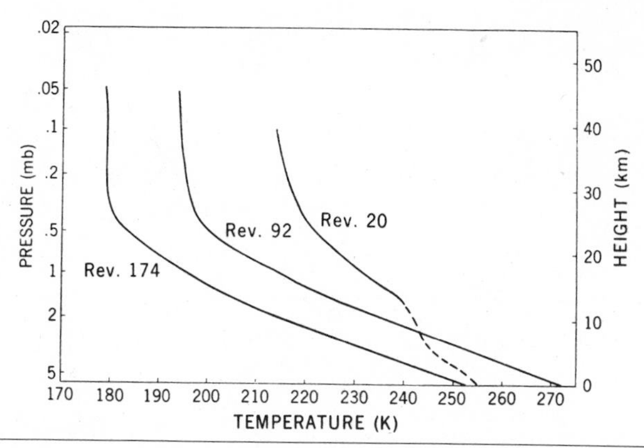

Figure 7 Atmospheric temperature profiles for three orbits during the Mariner 9 mission, showing changes detected for differing atmospheric conditions (33).

atmosphere became isothermal. Later in the summer when the atmosphere was essentially clear, the temperature dropped further and the temperature gradient remained near −3°K/km (33).

Temperatures over the south polar region were measured by the infrared interferometer spectrometer during the summer and over the north polar region during the winter and spring. Temperature profiles over the south polar region for three dates in summer are shown in the lefthand side of Figure 8. Early in summer (Rev 30), the temperature structure was relatively warm and nearly isothermal. In mid-summer (Rev 116), the temperature dropped and a temperature inversion appeared with a maximum near the 2 mbar (10 km) level. At the end of summer (Rev 188), the temperature dropped still further. Above the 0.5 mbar (25 km) level, the temperature was isothermal at about 155°K. In all three of these cases, the atmospheric temperature was greater than the surface temperature. Over the north polar region in mid-winter (Rev 102), the temperature of the atmosphere was higher above 25 km than at lower altitudes (righthand side of Figure 8). Below the 3 mbar (7 km) level, the temperature was approximately 150°K, just above the condensation temperature of carbon dioxide. The winter temperature of the polar cap was 145°K, the condensation temperature of dry ice. In late spring (Rev 528), the temperature of the atmosphere directly above the polar cap rose, leading to a negative

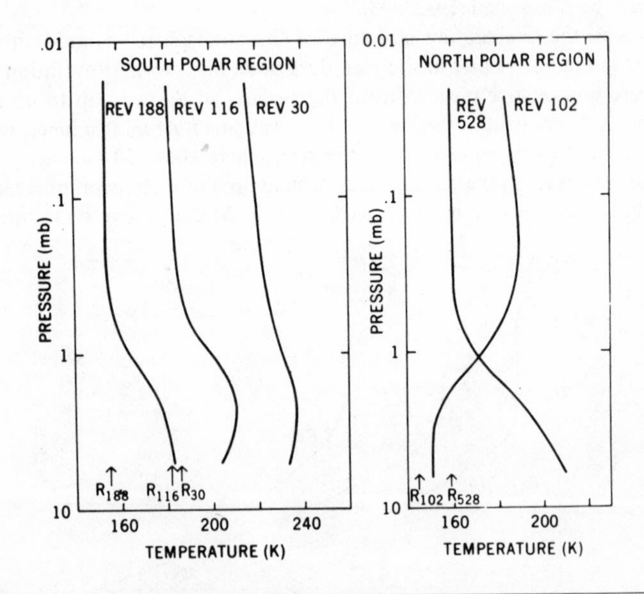

Figure 8 Atmospheric temperature profiles for the south polar cap region in summer and for the north polar cap during winter and spring. These profiles were obtained by the infrared spectrometer on Mariner 9. The arrows near the bottom of the figure indicate the surface temperatures (19).

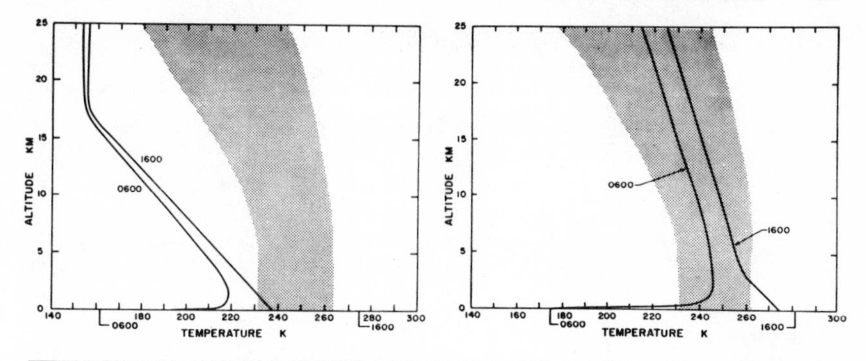

Figure 9 Theoretical and observed temperature profiles for Mars. The stippled areas include the Mariner 6 and 7 observations: the theoretical profiles are for a pure CO_2, atmosphere (left) and an atmosphere containing dust (right) (28). (Figure reproduced by permission of *J. Atmos. Sci.*)

temperature gradient. The atmosphere was isothermal with a temperature of 160°K above the 0.5 mbar (25 km) level.

Theoretical calculations of the temperature structure of the Mars lower atmosphere predict a lapse rate of 5°K/km for a clear carbon dioxide atmosphere under conditions of radiative-convective equilibrium (27). Mariner 6 and 7 and the extensive Mariner 9 measurements of the temperature profile showed the lapse rate to be less than this value when the atmosphere was clear and much less when the atmosphere was dusty. When dust is suspended in the atmosphere, theoretical calculations show that the dust heats the atmosphere while absorbing solar radiation (28). Even a very small amount of dust causes the lapse rate to be less than the value calculated for a clear atmosphere. Large amounts of dust cause the atmosphere to become very stable and nearly isothermal from the surface to an altitude of 25 km. The dusty atmosphere is much hotter than the clear atmosphere. These results are illustrated in Figure 9 (28). Inclusion of large-scale dynamics into the calculation of the temperature profile of the Mars atmosphere also causes the calculated lapse rate to be less than the value calculated considering only radiation and convection (15, 49a, 89).

VARIATIONS IN THE LOWER ATMOSPHERE

Variations occurring in the atmosphere of Mars can be separated into diurnal and seasonal changes. Superimposed on the regular changes are unexpected events that characterize Mars weather. Mars rotates at a rate slightly slower than the earth: the day is 24 hr and 39 min long. While the Mars year is almost twice as long as the earth's, the Mars obliquity of 25° produces seasons similar to those on earth. However, on Mars at the present time the greater eccentricity of the orbit accentuates the seasons in the southern hemisphere and moderates them in the north.

Variations in the temperature of the atmosphere as a function of local time and

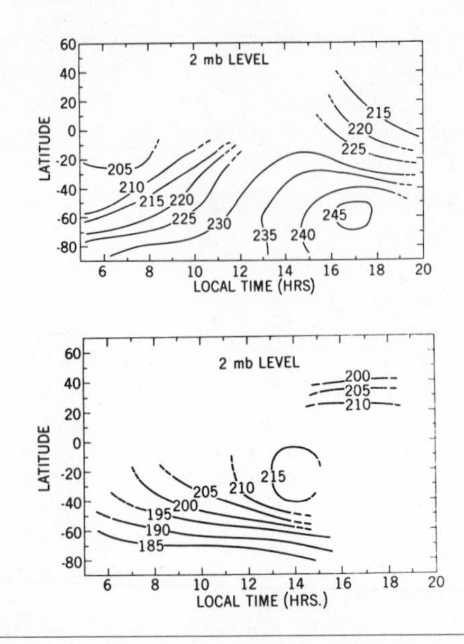

Figure 10 Atmospheric temperature on Mars showing latitude and local time variations. The upper figure is from observations taken during the dust storm, the lower figure is from observations taken when the atmosphere was relatively clear (33).

latitude are illustrated in Figure 10. The diurnal behavior of the atmosphere at the 2 mbar level ($\sim$ 10 km) during the dust storm is shown by the isotherms in the upper part of the figure. The temperature maximum occurs late in the day, about 5 PM local time, and it lies far south of the subsolar latitude of 20°S. The diurnal behavior of the clear atmosphere is illustrated in the lower part of the figure. The isotherms now show the temperature maximum further north and occurring earlier in the day, about 2 PM local time. The subsolar point at this time was 8°S. Also, the clear atmosphere at the 2 mbar level was cooler than the dusty atmosphere during the storm. Diurnal variations in the atmospheric temperature lead to a tidal effect in the pressure distribution. A calculation of the planetwide pressure distribution based on the temperatures measured during the dust storm is shown in Figure 11 (19). The surface pressure is at its lowest value late in the afternoon, at a latitude between the equator and the maximum temperature location (see Figure 10). Regions far from the place of maximum heating have a surface pressure 25% higher than the minimum surface pressure. When similar pressure calculations are performed for the clear atmospheric temperature distribution, the pressure minimum occurs earlier in the day and is closer to the equator in a manner analogous to the change in location of the temperature maxima in Figure 10. Winds that should result from the measured temperature fields and the tidal pressure variation lead to calculated patterns shown in Figure 12 (19). These results, which apply to conditions during the dust storm (summer in the southern hemisphere), show strong northeasterly winds north of the equator and strong northwesterly winds south of the equator in the afternoon. The magnitude of the mean meridional flow crossing the equator depends upon the magnitude of the friction between the atmosphere and the surface,

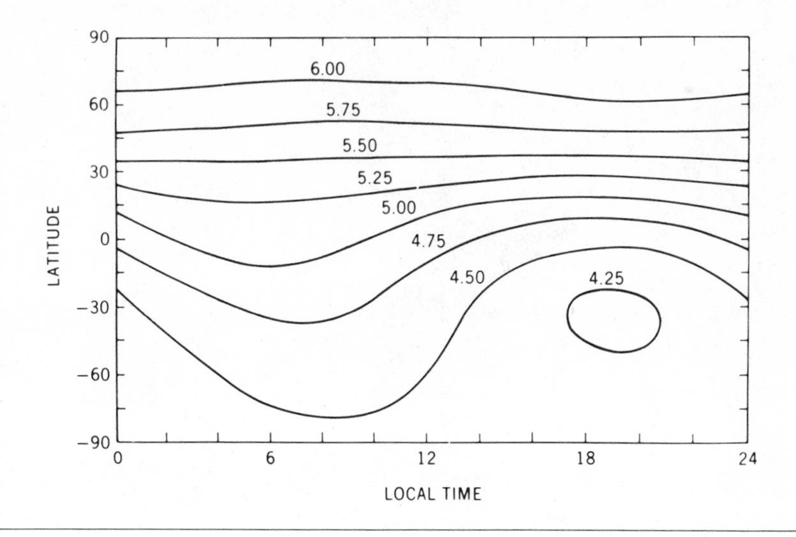

Figure 11 Calculated atmospheric pressure distribution showing tidal effect (19).

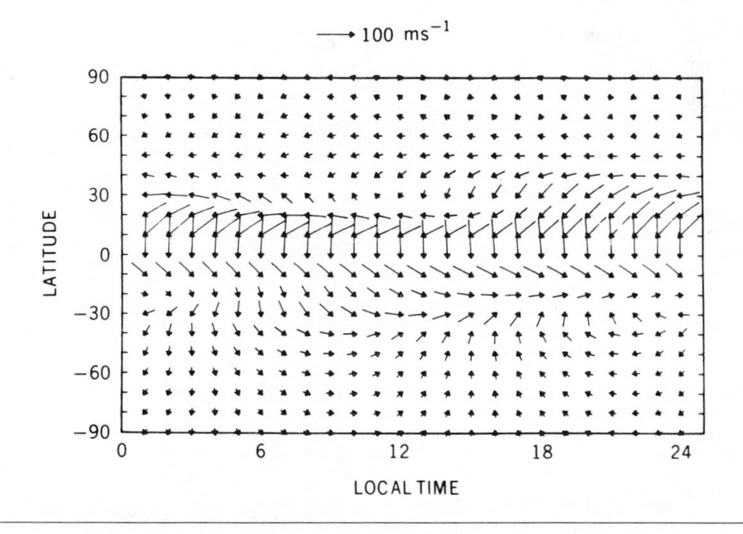

Figure 12 Near surface winds during the great dust storm (19).

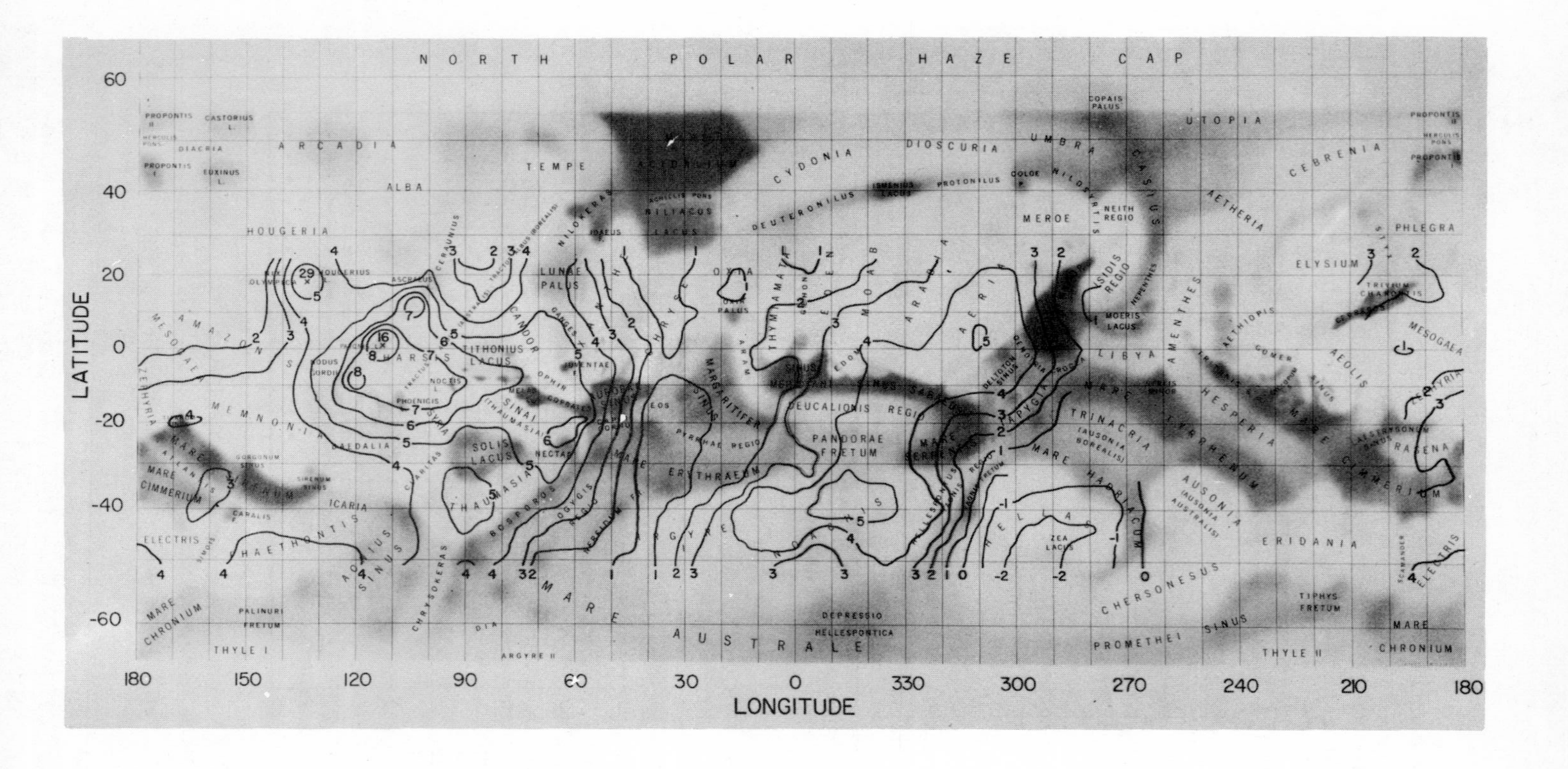

Figure 13 Elevation contours of the surface of Mars measured by the ultraviolet spectrometer experiment (36).

and is thus highly uncertain (53). These Mariner results confirm the general picture of temperature distributions and wind patterns given by general circulation model calculations (49a).

In the latitude band between 20°N and 20°S, observed wind streak patterns, formed on the surface of Mars from blowing dust, point in directions arising from winds following the calculated patterns (80). At other latitudes, the wind streak patterns do not conform to the calculated wind directions. Topography, an important factor that affects wind patterns, has not been taken into account in the above calculations (27a). On Mars, elevation variations are extreme compared to the scale height of the atmosphere.

In addition to the radio occultation experiment, two other Mariner 9 experiments measured elevation variations on Mars: the infrared interferometer spectrometer (19, 33), and the ultraviolet spectrometer (35, 36). In the lowest measured location on the planet, the Hellas basin, the surface pressure is over 8 mbar, while over the highest large-scale feature, the Tharsis ridge, it is only 3 mbar (19). An elevation contour map based on the ultraviolet spectrometer data, Figure 13, shows an elevation change of over 10 km between these two locations (36). This contour chart also shows that the major high ridges and low depressions are oriented in a north-south direction and, hence, would severely modify the atmospheric winds, particularly those flowing in an easterly or westerly direction.

Seasonal variations of water vapor on Mars were first recorded by spectroscopic observations through earth-based telescopes (81). A summary of earth-based measurements averaged over the disc of Mars is shown in Figure 14. Water vapor is apparently present in the atmosphere in late spring or summer in either hemisphere (4). The maximum amount observed was equivalent to 60 μm of

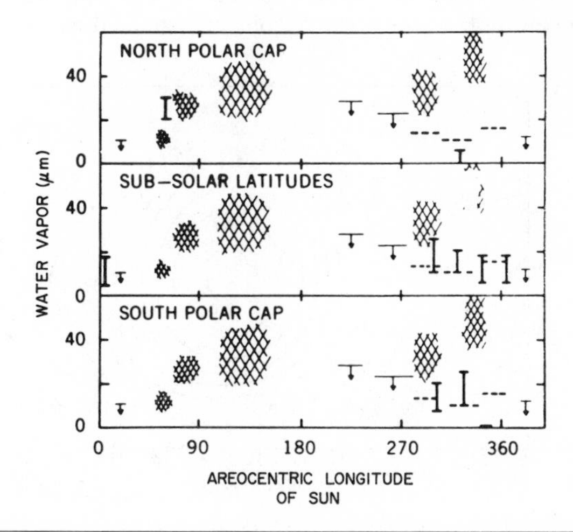

Figure 14 Seasonal variations of water vapor on Mars. Results from earth-based observations are shown with cross hatching, horizontal dash lines, and horizontal lines with arrows (upper limits). Results from Mariner 9 infrared interferometer spectrometer experiment are shown by vertical bars (19).

precipitable water while the minimum was less than the detection limit of 10 μm. Mariner 9 infrared interferometer spectrometer measurements of water vapor are also included in Figure 14 (19). During summer in the southern hemisphere, between 10 and 20 μm of precipitable water was present over the summer polar region and at subsolar latitudes but water vapor was not detectable over the northern winter polar region. During the northern hemisphere spring, water vapor of between 20 and 30 μm appeared in the atmosphere over the north polar cap and was also observed at subsolar latitudes.

Measurements by the infrared photometer on Mars 3 showed smaller amounts of water vapor in the atmosphere (66). On all observations between December 15, 1971 and February 28, 1972, the maximum amount observed was 3 μm precipitable water. In the polar region, the amount was less than 0.5 μm.

If the sole source of water vapor in the Martian atmosphere were the release of water from subliming carbon dioxide caps, the maximum injection of water vapor into the atmosphere would occur in a relatively short period following the spring equinox when the polar cap begins to recede (51). Measurements by both Mariner 9 and earth-based telescopes show that the seasonal variation is more moderate than expected from a single seasonal injection from the polar caps. Surface temperature measurements of Mars, made by the Mariner 9 infrared radiometer experiment, fall to very low values at night (40). Figure 15 shows surface isotherms calculated from data obtained shortly before the equinox when the atmosphere was clear. At all latitudes, the temperature dropped below 190°K, the condensation temperature of 10 μm of precipitable water. It appears that the observed amount of water in the Mars atmosphere at low and mid-latitudes is equal to the saturation vapor pressure of water at the daily minimum temperature. Water vapor in the atmosphere may exchange with low and mid-latitude permafrost on a daily basis (4).

Ozone is present only in the polar regions of Mars and has a strong seasonal

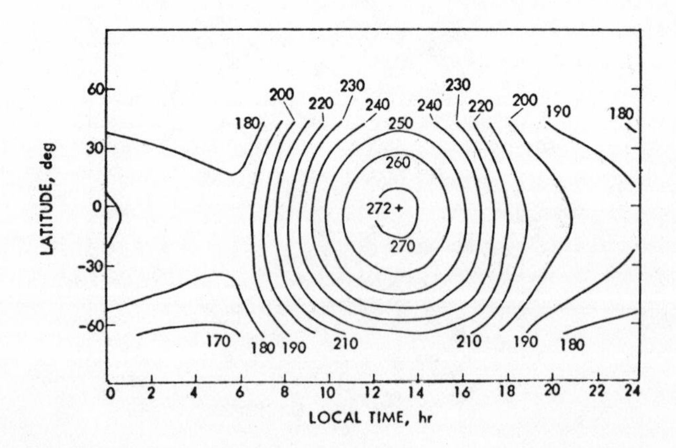

Figure 15 Surface temperatures on Mars calculated for a clear atmosphere based on Mariner 9 infrared radiometer observations (40).

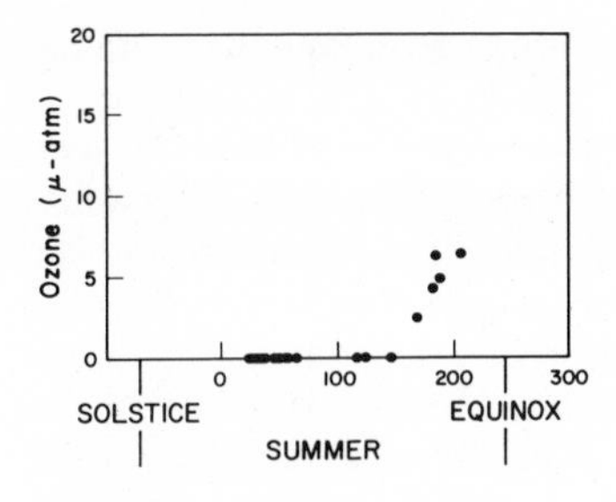

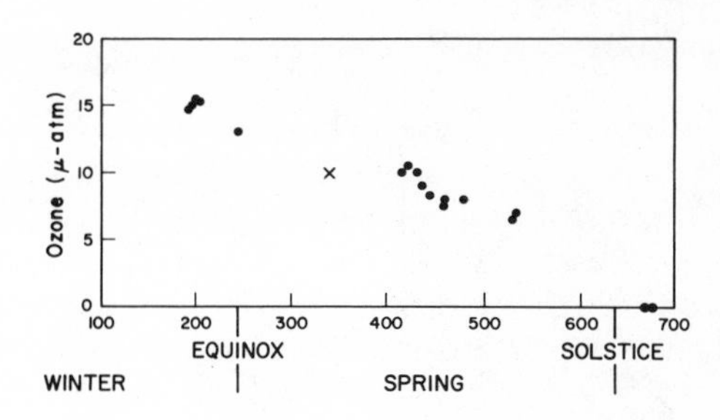

Figure 16 Seasonal ozone over the polar cap of Mars measured by the ultraviolet spectrometer experiments (10). (Copyright 1973 by the American Association for the Advancement of Science.)

variability. Over the polar cap itself, ozone is absent during early and mid-summer. The amount increases during late summer, reaches a maximum in winter, and then decreases in the spring. This behavior is illustrated by the Mariner 9 ultraviolet spectrometer measurements presented in Figure 16 (10). Ozone also appears in the Martian atmosphere in association with the polar hood. In the fall and winter hemisphere, the atmosphere is sufficiently cool so that the entire region poleward of 45° latitude is covered by clouds of water ice crystals. This is the polar hood that has been observed by astronomers for hundreds of years (12). In the latitude range of the hood, ozone actually appears in the atmosphere in late summer, before the polar hood appears. During fall and winter, ozone and the visible hood are both present. Ozone remains in the atmosphere in the spring, after the visible hood has disappeared. The amount of ozone occurring in conjunction with the polar hood is variable, just as the hood itself is variable in its appearance and extent. The conclusion has been drawn that ozone is present when the atmosphere is cold and dry and it disappears, or in fact, never appears, when water vapor is present (11).

During the winter season north of 45° latitude, the Mariner 9 television cameras observed extensive cloud systems (50). This is the location and expected time of appearance of the polar hood, which has been observed from earth-based telescopes for many years (12). In some cases the clouds are diffuse, in other cases there is a distinctive wave-like structure with a spacing of about 5 km. Bands of the wave-like clouds move in the manner of cold fronts on the earth—in one case moving 500 km in a southeasterly direction in one day (50). When these cloud systems flow over topographical features, lee waves are formed in the downwind direction, typically with a spacing of about 30 km. The infrared interferometer spectrometer made atmospheric temperature measurements at the time and place of the occurrence of the cloud systems of the polar hood, and found the temperature is that of water ice crystals (33).

Cloud patterns with wave-like structure were also observed by the Mariner 9 television cameras south of 50° latitude during the fall season (52). From observations both in the southern fall and in the northern winter, the clouds are more diffuse closer to the poles.

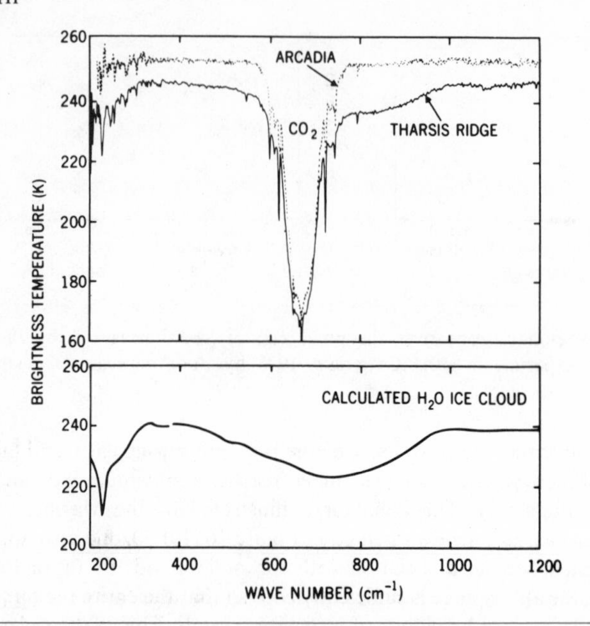

Figure 17 Mars absorption spectrum of water ice clouds over the Tharsis ridge. This spectrum is compared to another Mariner 9 infrared interferometer spectrometer spectrum obtained of a clear atmosphere over the Arcadia region which shows only absorption produced by carbon dioxide and water vapor. A theoretical calculation of the spectrum of a water ice cloud is produced in the lower part of the figure (20). (Copyright 1973 by the American Association for the Advancement of Science.)

Mariner 9 television pictures taken of the Tharsis-Nix Olympica region in late spring and early summer show clouds on the western slopes of the four volcanoes (52). These volcanoes are very high; Nix Olympica rises 25 km above the surrounding terrain (36). Mariner 9 infrared interferometer spectrometer observations of these clouds detected a spectrum characteristic of water ice crystals (20). Figure 17 shows that the spectrum observed over the Tharsis region contains two broad absorption features that are reproduced in a theoretical calculation of the spectrum of a water ice cloud. From earth-based telescopes, clouds in this region of Mars are observed to brighten in the afternoon; they are known as the W-clouds (84). They may be formed by the upslope flow of air from east to west containing perhaps 10 μm of precipitable water. Adiabatic cooling could lead to water ice crystal condensation (74). In addition, convection and even local outgassing might contribute to the cloud formation (52).

During the last few months of 1971, the great planetwide dust storm covered Mars. It began on a high plateau in the southern hemisphere, the Noachis region (see Figure 13), on September 22, 1971, a time when Mars had just passed perihelion and just before summer solstice in the southern hemisphere (see Figure 1) (16). In November, during the first weeks of Mariner 9 observations, the entire planet was covered with dust up to an altitude of 50 km (1, 50), except for the south polar region (55).

Theoretical consideration has been given to various aspects of the great dust storm (30, 31, 53). Since a maximum amount of insolation is apparently required to initiate and drive planetwide dust storms, they occur at the time there is a near coincidence between perihelion and southern summer solstice (31). An elevated region such as the Noachis plateau under maximum summer heating would develop a low pressure region with cyclonic flow (30). It has been hypothesized that dust would be raised to substantial heights over an extended area in the region. As dust accumulates in the atmosphere, solar radiation would be directly absorbed, heating the atmosphere still more (28). As more and more energy is absorbed, the dust storm may rotate and grow in a manner analogous to a terrestrial hurricane (30). In addition, the perihelic-summer solstice is a time of sublimation of carbon dioxide from the south polar cap. A large mass outflow from the polar cap could create winds to further drive the dust storm (53).

UPPER ATMOSPHERE AND IONOSPHERE

In the upper atmosphere, above approximately 100 km, the temperature structure is determined in part by heating processes such as photodissociation and photo-ionization. Two observational techniques utilizing Mariner spacecraft yield direct observational information on the density structure of the upper atmosphere from which temperature profiles can be calculated. On Mariner 4, 6, 7, and 9, the radio occultation experiments measured the electron density directly as a function of altitude (25, 26, 44). Without additional information, however, the interpretation of these data in terms of neutral density and temperature may lead to ambiguous results (25). Ultraviolet spectrometer observations of the Mars airglow from Mariner 6, 7, and 9 yield direct measurements of the rate of photodissociation and photoionization of carbon dioxide as a function of temperature (7). Above the altitude where the maximum rate of photodissociation occurs, the measurement of the intensity variation of the ultraviolet airglow as a function of altitude leads to the determination of the density variation of carbon dioxide. From the scale height (logarithmic variation of density as a function of altitude) the temperature of the atmosphere can be determined. Profiles of the density of carbon dioxide and the temperature as a function of altitude are given in Figure 18 (9). Density distributions for atomic oxygen, carbon monoxide, and molecular oxygen, which are shown in the figure, were calculated assuming diffusive equilibrium above 100 km. This model atmosphere pertains to conditions at the time of the Mariner 6 and 7 observations in 1969. In July and August of 1969, the temperature at the top of the Mars thermosphere, determined from the scale height of the ultraviolet airglow, was 350°K (87), a value far less than anticipated from theoretical calculations (58, 59). In November and December of 1971, the mean thermospheric temperature, determined from Mariner 9 ultraviolet spectrometer measurements of the Cameron bands, was 325°K (88). However, while the mean value of the scale height was 17.8 km, individual values varied from 14.8 km to 24.3 km, corresponding to

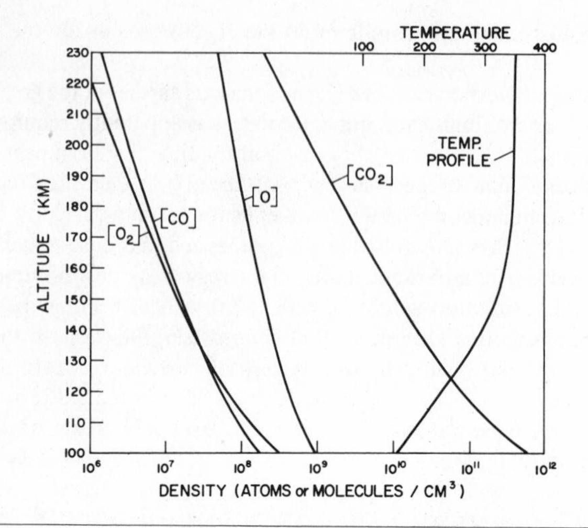

Figure 18 Mars model atmosphere. Density distributions are based on ultraviolet spectrometer, radio occultation and earth-based telescope observations (9).

temperatures of 270 to 445°K. The intensity of the ultraviolet airglow is correlated with an indicator of solar activity, the 10.7 cm solar radio flux; however, the airglow scale height, the measure of temperature, does not appear to correlate with solar activity (88).

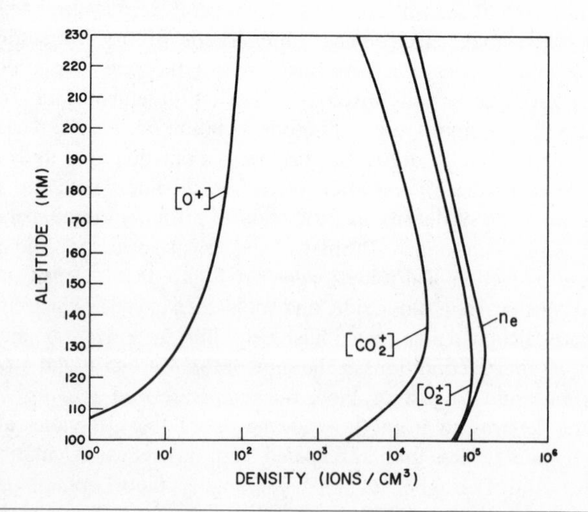

Figure 19 Mars model ionosphere. The model is calculated using radio occultation data together with laboratory measurements of ion chemistry (9).

Electron density measurements by the Mariner radio occultation experiment may be compared to neutral density measurements by the Mariner ultraviolet spectrometer experiment through the development of an ionospheric theory. Since carbon dioxide is the dominant neutral constituent of the atmosphere, it is not unexpected that it is the major source of ionization in the upper atmosphere. However, carbon dioxide ions are not the most abundant ions in the Mars ionosphere. The ultraviolet airglow does not show the characteristic fluorescence spectrum of carbon dioxide ions, a result first obtained by the Mariner 6 and 7 ultraviolet spectrometers (7) and verified by the more extensive observations of Mariner 9 (9). Laboratory studies have shown that atomic oxygen reacts rapidly with carbon dioxide ions to produce ionized molecular oxygen (24). Theoretical calculations show that, even with the small amounts of atomic oxygen in the Mars upper atmosphere, the most abundant ion in the Mars ionosphere is ionized molecular oxygen (87). A model ionosphere calculated for conditions of photochemical equilibrium is presented in Figure 19 (9). Dissociative recombination of molecular oxygen ions is the major mechanism for loss of ionization. Under the assumption of simple photochemical equilibrium, the scale height of the electron density should be approximately twice the scale height of the major ionizable constituent (87).

Mariner radio occultation experiments measured the scale height and electron density distribution in the Mars atmosphere: once in 1965, twice in 1969, 60 times in 1971, and 25 times in 1972. An altitude profile of the electron density obtained early in the Mariner 9 mission is shown in Figure 20 (45). In 1965, Mariner 4

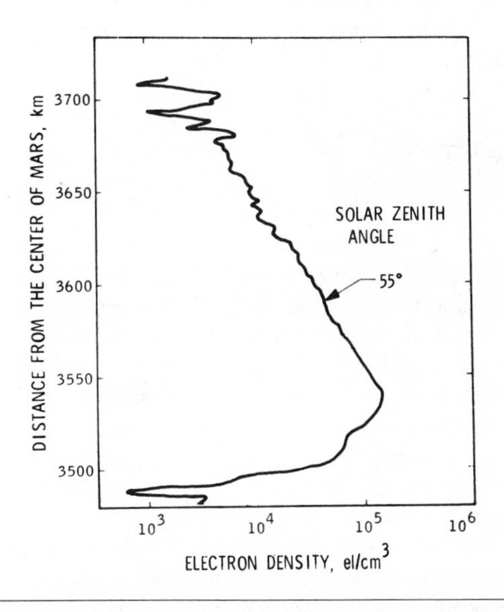

Figure 20 Electron density profile obtained from Mariner 9 radio occultation data (45).

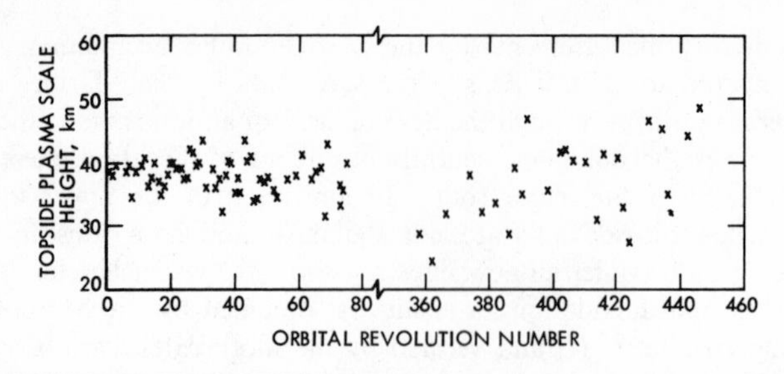

Figure 21 Plasma scale heights from Mariner 9 radio occultation measurements (46).

measured a plasma scale height of 30 km (25); in 1969, Mariner 6 and 7 both measured a plasma scale height of 49 km (26); in 1971 and 1972, Mariner 9 measured a mean scale height of 39 km (45, 46). If the changes in the plasma scale height are solely the result of changes in the temperature of the neutral atmosphere, then assuming simple photochemical equilibrium, the temperatures corresponding to the above scale heights are 270°, 450°, and 350°K, respectively. Individual measurements of the plasma scale height during the Mariner 9 mission are shown in Figure 21 (46). During the first 80 revolutions, the season was summer-winter and a planetwide dust storm existed in the lower atmosphere; between revolutions 360 and 460, the season was spring-fall and the atmosphere was relatively clear (see Figure 1). For these two very different periods, the mean plasma scale height apparently did not change. Daily variations in scale height do not appear to be correlated with indicators of solar activity, such as the 10.7 cm radio flux index (45, 46).

PHOTOCHEMISTRY

During the past several years, there have been a number of papers published on the photochemistry of the Mars atmosphere pertinent to data acquired by planetary spacecraft. The observational findings—the relatively undissociated character of the upper atmosphere and the presence of atomic hydrogen in the Mars exosphere—are related to the photochemistry of both the lower and upper atmosphere. Theoretical studies have dealt with the photochemistry of a dry, carbon dioxide atmosphere (62, 63) and also with the photochemistry of a carbon dioxide atmosphere containing various amounts of water vapor (38, 64, 73).

Photodissociation of carbon dioxide, to produce carbon monoxide and atomic oxygen, takes place from the top of the atmosphere all the way down to the surface (62). In the upper atmosphere, the known recombination reactions are not rapid enough to balance the photoproduction of atomic oxygen to explain the low observed abundances. Eddy diffusion has been proposed as the mechanism for removing atomic oxygen from the upper atmosphere, by transporting it down into the lower atmosphere (83). However, calculations show that the rate of mixing would have to be very rapid; the time constant would be 1 hr for a characteristic length of one scale height (63).

Water vapor is photodissociated in the Mars lower atmosphere, mostly in the lowest 10 km (38). The products of this photodissociation are: atomic hydrogen, H; the hydroxyl radical, OH; and, when a small amount of molecular oxygen is present, the hydroperoxyl radical, HO_2. Laboratory studies have suggested that the recombination of carbon monoxide and atomic oxygen is catalyzed in the presence of these radicals (76). Two recent model calculations incorporate the concepts of transport by eddy diffusion from the upper to the lower atmosphere, where catalytic recombination of atomic oxygen and carbon monoxide takes place (64, 73). The primary sequence of reactions is:

$$H + O_2 + CO_2 \rightarrow HO_2 + CO_2$$
$$HO_2 + O \rightarrow OH + O_2$$
$$OH + CO \rightarrow H + CO_2$$

with the net result of carbon monoxide and atomic oxygen combining to form carbon dioxide

$$CO + O \rightarrow CO_2$$

The results of the two model calculations are shown in Figure 22. Details of the two calculations are not identical; namely, some of the reaction rate coefficients differ, different eddy diffusion coefficients are used, and the mixing ratio of water vapor is different in the two models. However, the major results are similar and the differences illustrate the range of possible results depending on the actual conditions. In the upper part of Figure 22 the density distributions with and without water vapor are shown. In the lowest 10 km of a dry carbon dioxide atmosphere, atomic oxygen is calculated to have a concentration of several parts in one hundred million. (The volume density of the Mars atmosphere at the surface is 2×10^{17} molecules cm^{-3}.) Ozone, formed from atomic and molecular oxygen, is more abundant than atomic oxygen, with a concentration of several parts in ten million. When water vapor is present, the densities of both atomic oxygen and ozone decrease markedly. Figure 22 shows that of the reactive radicals, the density of hydroperoxyl, HO_2, is greater than the density of atomic hydrogen, H, which in turn is greater than the density of the hydroxyl radical, OH. In the model with the larger amount of water vapor, hydrogen peroxide, H_2O_2, is present with a concentration of one part in ten million. Photolysis of H_2O_2 is an important source of OH for the reaction with CO (73). Even at these low concentrations, the cyclic reactions involving H, OH, and HO_2 catalytically recombine CO and O while not being consumed themselves.

These model calculations also illustrate the manner in which the presence of water vapor in the atmosphere inhibits the formation of ozone. In the regions of Mars where the Mariner 9 infrared interferometer spectrometer experiment measured 10–20 μm of precipitable water, the ultraviolet spectrometer determined that the amount of ozone was less than 3 μm-atm (10, 33). In the polar regions, where the infrared experiment measured the atmospheric temperature to be less than

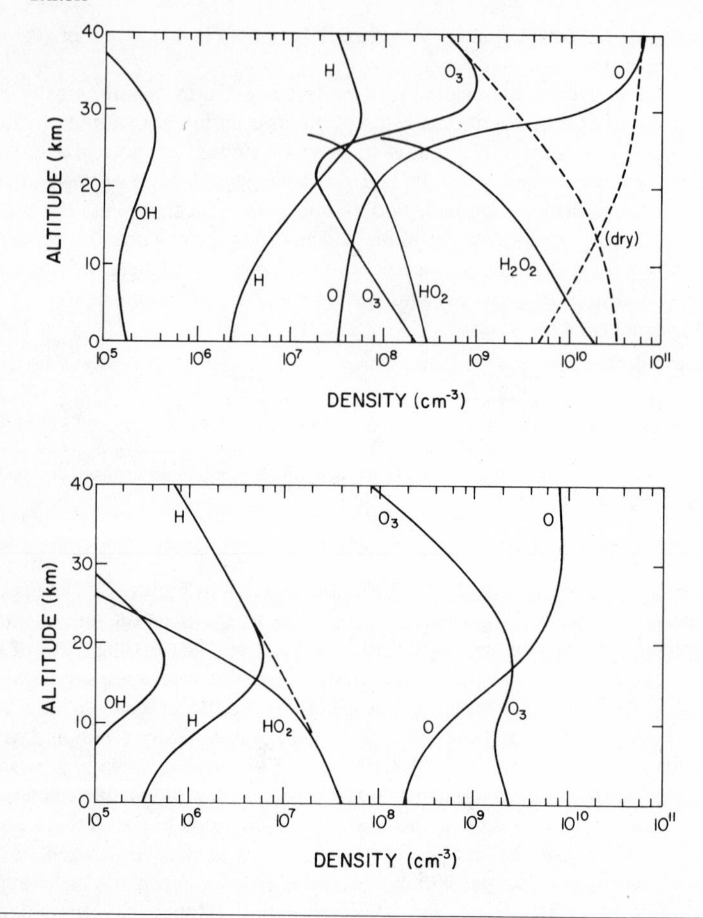

Figure 22 Photochemistry of Mars atmosphere. Density distributions are based on calculations by Parkinson & Hunten (73) (upper figure) and by McElroy & Donahue (64) (lower figure). (Upper Figure reproduced by permission of *J. Atmos. Sci.*; lower figure copyright 1972 by the American Association for the Advancement of Science.)

180°K, the ultraviolet experiment measured ozone in amounts varying between 10 and 60 μm-atm. Since the vapor pressure of water at 180°K is equivalent to 1 μm of precipitable water, the polar atmosphere contains much less than this amount (19). In the upper part of Figure 22, the calculated amount of ozone for the dry atmosphere is approximately 20 μm-atm (73). For atmospheres containing 15 μm of precipitable water, the upper part of the figure shows a calculated amount of ozone equivalent to 0.1 μm-atm (73), while in the lower part of the figure, for the same amount of water vapor, the calculated amount of ozone is approximately 2 μm-atm (64). Both results agree with the ultraviolet observations. It is the photodissociation products of water vapor, hydroperoxyl and hydroxyl radicals, and atomic hydrogen, that control the amount of ozone and atomic oxygen, even though their densities are much less than the density of the ozone. The photochemistry occurring in the lower atmosphere of Mars is similar to that which occurs in the stratosphere of the earth.

ATMOSPHERIC ESCAPE

Atomic hydrogen was discovered in the upper atmosphere of Mars in 1969 by the Mariner 6 and 7 ultraviolet spectrometer experiment (6). From the temperature of the upper atmosphere measured by the same experiment (350°K), it was recognized that evaporative escape of atomic hydrogen must occur in the Mars exosphere and that the time constant for escape was approximately 4 hr (7, 9). A radiative transfer analysis of the Mariner 6 Lyman alpha data determined that the density of atomic hydrogen at the bottom of the exosphere (250 km altitude) is 3×10^4 atoms cm^{-3} and that the escape flux is 2×10^8 atoms cm^{-2} sec^{-1} (2). The amount of water vapor observed in the Mars lower atmosphere, 20 μm of precipitable water, is sufficient to supply this amount of atomic hydrogen for 21,000 years.

In 1971, during the first 30 days of Mariner 9 observations, the temperature of the upper atmosphere was somewhat less than in 1969, but the density of atomic hydrogen appeared to be somewhat greater (9). The escape fluxes remained approximately the same as the 1969 Mariner values.

While the photodissociation of water vapor is the most probable ultimate source of hydrogen, it has been suggested that the atomic hydrogen source in the upper atmosphere is an ionospheric reaction between molecular hydrogen and ionized carbon dioxide (61).

$$H_2 + CO_2^+ \rightarrow H + CO_2H^+$$

Molecular hydrogen is produced in the lower atmosphere by the reaction between the water vapor photodissociation products, hydroperoxyl radicals, and atomic hydrogen.

$$HO_2 + H \rightarrow H_2 + O_2$$

It is transported into the ionosphere where it is converted into atomic hydrogen (38). Since the escape flux of atomic hydrogen must be supplied by an upward flux from the chemical production source, any change in the production should be reflected in a change in the escape flux. Since the production of molecular hydrogen is determined by the oxidation state of the atmosphere, the escape of atomic hydrogen from the exosphere is regulated by the amount of atomic oxygen in the Mars atmosphere, although the time constant for this feedback is very long (60). Interestingly, the escape flux of atomic hydrogen, on the two occasions in 1969 measured by Mariner 6 and 7, and during the thirty days in 1971 measured by Mariner 9, remained essentially constant (9). Also, the atomic oxygen density measured by the ultraviolet spectrometer experiments on Mariner 6, 7, and 9 is essentially the same in 1969 and in 1971; namely, between 0.5 and 1.0% at the ionization peak (90, 91).

As a consequence of reactions in the Mars ionosphere, atomic oxygen is converted into ionized molecular oxygen, through the reaction with ionized carbon dioxide (87).

$$O + CO_2^+ \rightarrow O_2^+ + CO$$

When either molecular oxygen ions or carbon dioxide ions undergo dissociative recombination, oxygen atoms are produced with excess kinetic energy.

$$O_2^+ + e \rightarrow O + O + \text{excess energy}$$
$$CO_2^+ + e \rightarrow CO + O + \text{excess energy}$$

It has been suggested that many of these "hot" oxygen atoms are able to escape from the low gravitational field of Mars (60). Furthermore, the escape flux of atomic oxygen by this mechanism would balance the escape flux of atomic hydrogen, considering that the dissociation of water vapor is the ultimate source of both atoms in the Mars atmosphere (64).

Other ionospheric reactions can lead to the nonthermal escape of carbon atoms and nitrogen atoms from Mars (60). Dissociative recombination of carbon monoxide ions and molecular nitrogen ions can produce carbon atoms and nitrogen atoms, respectively, with energy greater than the gravitational energy of these atoms in the Mars upper atmosphere.

$$CO^+ + e \rightarrow C + O + \text{excess energy}$$
$$N_2^+ + e \rightarrow N + N + \text{excess energy}$$

However, whether or not nonthermal escape by these reactions is significant in depleting the Mars atmosphere of carbon and nitrogen depends upon the precise composition of the Mars ionosphere. If, for example, the ionized molecular nitrogen is converted into ionized nitric oxide, the nitrogen atoms produced in the dissociative recombination of this ion would not have sufficient energy to escape from the gravitational field of Mars.

$$NO^+ + e \rightarrow N + O$$

INTERACTIONS BETWEEN THE ATMOSPHERE AND THE POLAR CAPS

Observational evidence that the major portions of the polar caps of Mars are composed of frozen carbon dioxide was acquired by the infrared instruments on Mariner 7 (34, 70, 71). Measurements by the infrared radiometer experiment of the southern polar cap in the spring showed the temperature was 148°K, the temperature of subliming carbon dioxide at a pressure of 6.4 mbar (71). Spectra of the polar cap obtained by the infrared spectrometer experiment contained absorption features at 3.0 and 3.3 μ that are attributed to frozen carbon dioxide (34). High resolution earth-based observations of the Mars south polar cap have revealed a number of sharp absorption features between 1.2 and 2.4 μ that are identified as solid CO_2 absorption (48a). Observational evidence that carbon dioxide is the principal constituent of the atmosphere and that the polar caps consist of frozen carbon dioxide supports the idea that the current state and evolution of the Mars atmosphere is controlled by the physical state of the polar caps (49). Carbon dioxide sublimes from the cap in the spring-summer hemisphere, and at the same time it is precipitating and forming a polar cap in the fall-winter hemisphere. Model calculations predict that the amount of carbon dioxide precipitated in the polar regions is so great that a seasonal variation in atmospheric pressure will result with maxima occurring just before the solstices and minima near the equinoxes (49). If the minimum temperature on Mars were lower in the past, or future, and all of the carbon dioxide atmosphere condensed, the amount of solid would be equivalent to a layer 10 cm thick if uniformly distributed over the planet. Conversely, at times of higher minimum temperature, the carbon dioxide polar caps would completely sublime away and the atmospheric pressure would increase. It has been estimated that if the remnant north polar cap were 1 km thick, an atmospheric pressure of 1 bar, the sea level pressure on earth, would result from the complete vaporization of the cap (77). Because the triple point pressure of carbon dioxide is 5.1 bars and the triple point temperature 217°K, the thickness of carbon dioxide in the polar cap cannot exceed a few kilometers. At greater depths, the temperature and pressure would be greater than the triple point and the carbon dioxide would liquify and flow away (78).

Mariner 9 obtained television pictures of the north polar cap just after summer solstice (85) and of the south polar cap throughout the summer in that hemisphere (67). One analysis of these observations proposes that there is a large mass of carbon dioxide equal to 2–5 times that of the present atmosphere buried beneath the north polar cap, but not under the south polar cap (68).

While the seasonal waxing and waning of the polar caps plays an important role in controlling the amount of water vapor in the atmosphere, the existence of a permanent polar cap is of great consequence in the evolution of the Mars atmosphere (49). Each fall and winter water vapor condenses out of the atmosphere over the polar regions, forms the clouds of the polar hood, and precipitates onto the polar cap. At colder temperatures, carbon dioxide snow accumulates on the polar cap in an amount far greater than the water snow. In the spring and summer, the reverse sequence takes place. The carbon dioxide sublimes first, followed by the sublimation of water vapor. Water ice may make up a large part of the polar cap remnant. In fact, the interpretation of the Mariner 9 observations of the south polar cap suggests that the summer remnant is made up of water ice (67). If in the past, water was present on the planet in substantial amounts, it may have migrated to the polar cold trap (49).

Certain of the surface erosional channels observed by Mariner 9 (56) are attributed, with varying degrees of certainty, to be evidence of running water in the past history of Mars (54, 57, 65, 82). Bodies of liquid water cannot exist on Mars today because the atmospheric pressure is so close to the triple point pressure of water, 6.105 mbar. If a large amount of liquid water were placed on the surface of Mars today, it would either rapidly freeze or vaporize, or do both simultaneously. If in the past, the atmospheric pressure were higher and there were larger amounts of water, then flowing water could have existed (77).

Evidence of climatic variability on Mars is found in the laminated terrain discovered in the polar regions of Mars by Mariner 9 (67). A series of thin layers some tens of meters thick are arranged in plates that are offset one from the other (21). Long term periodic variations in solar radiation reaching the surface have been suggested as producing periodicity in the formation of the layers and plates (69). The growth and recession of the polar caps each year is very sensitive to the amount of insolation that reaches the polar region which, in turn, is determined by the obliquity of Mars (the angle between the spin axis and the normal to the orbit plane). Calculations show that the Mars obliquity could vary from its present value of 25° to as small as 15° and as large as 35° with a 160,000-yr period superimposed on a 1,200,000-yr period (92). These variations are shown in Figure 23. If there were no other changes, at 35° the annual polar insolation would be 35% greater than its present value, while at 15° it would be only 60% of the current value. Variations of this magnitude would produce substantial climatic changes: there is even the possibility of polar cap—atmosphere system instabilities so that dramatic climatic changes could occur as well (79). Calculations of pressure of the Mars atmosphere for different values of insolation show that atmospheric heat transport may make the atmospheric pressure very sensitive to changes in the amount of solar radiation reaching the polar regions (29). Under present conditions, the atmosphere is so thin that heat conduction by the atmosphere is not very effective. However, if the polar insolation increases, carbon dioxide vaporizes, increasing the atmospheric density which makes the atmosphere more effective in transporting heat from the equator to the pole. Increased heating of the pole would vaporize

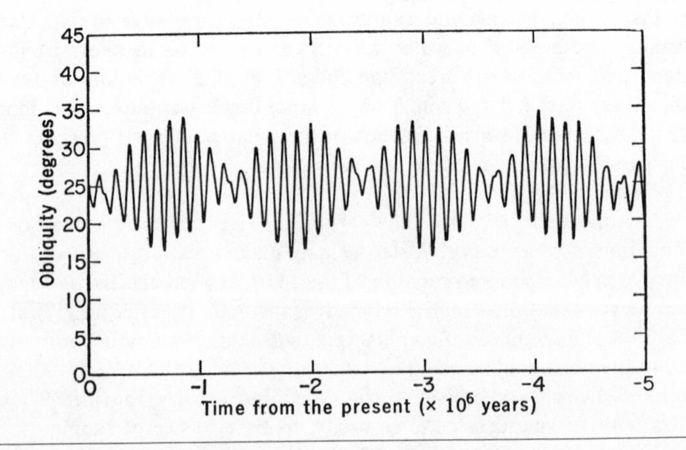

Figure 23 Variations in the obliquity of Mars for the past five million years (92). (Copyright 1973 by the American Association for the Advancement of Science.)

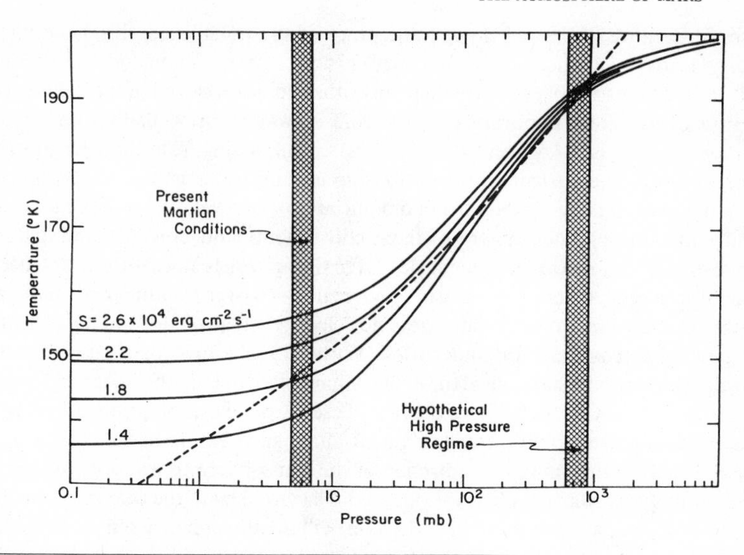

Figure 24 Calculations of Mars polar cap temperature with atmospheric pressure for several values of absorbed solar radiation (79). (Copyright 1973 by the American Association for the Advancement of Science.)

more carbon dioxide and the possibility of a runaway would exist. Figure 24 shows the results of these calculations for a set of conditions that would allow the atmospheric pressure to increase to 1 bar if the polar insolation increased by 14% over its present value (79). If the amount of carbon dioxide in the polar reservoir were not sufficient to produce 1 bar of atmospheric pressure, the pressure increase would stop when all of the carbon dioxide had sublimed. If the temperature of the polar cap continued to increase, then other volatiles, including water could continue to vaporize. A permanent polar cap has been recognized as a cold trap not only for water but for other atmospheric volatiles as well (49).

QUESTIONS STILL TO BE ANSWERED

Our knowledge of the Mars atmosphere has increased vastly during the past several years, but as very often happens, this increased knowledge leads to new questions about the present state of the atmosphere, its history, and origin. Since carbon dioxide is the principal constituent of the atmosphere and since the temperature of the polar cap is sufficiently low to freeze carbon dioxide, a very important question that pertains to both the present state and history of the atmosphere is: "What is the size of the frozen carbon dioxide reservoir at the polar caps?" A related question is: "Is there an annual variation in the amount of carbon dioxide in the atmosphere?" Another important question is: "How much water is stored in the polar caps and how much of it is released and recovered each season?" Because of the very low temperature of the polar caps, this same line of inquiry leads to a question whose answer may reveal the history of chemical evolution of the atmosphere: "Are there any other volatiles stored in the polar caps?" Because of the great importance of the role of water vapor in the atmosphere, another question is: "Are there sources of water outgassing from the interior of the planet today?" The intricate interrelationship among water vapor, ozone, and the undissociated character of the carbon dioxide atmosphere leads to a desire to understand better the photochemistry of these constituents and how their abundances vary seasonally and on longer time scales. Even though the temperature and structure of the upper atmosphere was measured several times over a number of years, and then intensively for several months, it is not clear whether or not the basic heating and cooling mechanisms are understood. Without this knowledge, it is difficult to extrapolate our understanding of the evolutionary state of the atmosphere into either the past or future. Observations of the rate of escape of all escaping constituents over several Mars seasons may provide data to allow such extrapolation to be made. Measurement of the amount of nitrogen in the atmosphere to levels below the 1% upper limit set by Mariner observations, and measurement of the rare gas content of the atmosphere, particularly neon and argon, will greatly enhance our knowledge of the history and origin of the atmosphere. The immense problem of understanding the general circulation on Mars needs to be solved, not only for its own sake, but because the circulation plays a very important role in transporting volatiles from one part of the planet to another. As a final question, we return to the first question asked in this section, but stated in a much more general way: "Did Mars have a substantially different climate in the past, and, if so, how and why did it change?"

Acknowledgments

Much of the information and many ideas in this review were acquired from my Mariner 9 experimenter colleagues, both from the published material cited in the references and from personal conversation. Many of the ideas of what are important unanswered questions were developed through interaction with Professor C. B. Leovy of the University of Washington and Professor P. J. Gierasch of Cornell University. Mrs. S. Schaffner and Mrs. R. A. Goettge aided in the preparation of the manuscript. The National Aeronautics and Space Administration sponsored this research activity.

Literature Cited

1. Ajello, J. M., Hord, C. W., Barth, C. A., Stewart, A. I., Lane, A. L. 1973. Afternoon terminator observations of Mars. *J. Geophys. Res.* 78:4279–90
2. Anderson, D. E. Jr., Hord, C. W. 1971. Mariner 6 and 7 ultraviolet spectrometer experiment: Analysis of hydrogen lyman alpha data. *J. Geophys. Res.* 76:6666–73
3. Barker, E. S. 1972. Detection of molecular oxygen in the Martian atmosphere. *Nature* 238:447–48
4. Barker, E. S., Schorn, R. A., Woszczyk, A., Tull, R. G., Little, S. J. 1970. Mars: Detection of atmospheric water vapor during the southern hemisphere spring and summer season. *Science* 170:1308–10
5. Barth, C. A. 1969. Planetary ultraviolet spectroscopy. *Appl. Opt.* 8:1295–304
6. Barth, C. A. et al 1969. Mariner 6: Ultraviolet spectrum of Mars upper atmosphere. *Science* 165:1004–5

7. Barth, C. A. et al 1971. Mariner 6 and 7 ultraviolet spectrometer experiment: Upper atmosphere data. *J. Geophys. Res.* 76:2213–27
8. Barth, C. A., Hord, C. W. 1971. Mariner ultraviolet spectrometer: Topography and polar cap. *Science* 173: 197–201
9. Barth, C. A., Stewart, A. I., Hord, C. W., Lane, A. L. 1972. Mariner 9 ultraviolet spectrometer experiment: Mars airglow spectroscopy and variations in Lyman alpha. *Icarus* 17:457–68
10. Barth, C. A. et al 1973. Mariner 9 ultraviolet spectrometer experiment: Seasonal variation of ozone on Mars. *Science* 179:795–96
11. Barth, C. A., Dick, M. L. Ozone and the polar hood of Mars. *Icarus*. Submitted for publication
12. Baum, W. A., Martin, L. J. 1973. Behavior of the Martian polar caps since 1905. *Bull. Am. Astron. Soc.* 5:296
13. Beer, R., Norton, R. H., Martonchik, J. V. 1971. Astronomical infrared spectroscopy with a Connes-type interferometer: II—Mars, 2500–3500 cm^{-1}. *Icarus* 15:1–10
14. Belton, M. J. S., Broadfoot, A. L., Hunten, D. M. 1968. Abundance and temperature of CO_2 on Mars during the 1967 opposition. *J. Geophys. Res.* 73: 4795–806
15. Blumsack, S. L., Gierasch, P. J. 1973. The vertical thermal structure of the Martian atmosphere: Modification by motions. *Icarus* 18:126–33
16. Capen, C. F., Martin, L. J. 1972. Mars' great storm of 1971. *Sky Telesc.* 43: 276–79
17. Carleton, N. P., Traub, W. A. 1972. Detection of molecular oxygen on Mars. *Science* 177:988–92
18. Connes, J., Connes, P., Maillard, J. P. 1969. *Atlas des spectres infrarouges de Venus, Mars, Jupiter et Saturne*. Paris: Editions Centre National Recherche Scientifique
19. Conrath, B. et al 1973. Atmospheric and surface properties of Mars obtained by infrared spectroscopy on Mariner 9. *J. Geophys. Res.* 78:4267–78
20. Curran, R. J., Conrath, B. J., Hanel, R. A., Kunde, V. G., Pearl, J. C. 1973. Mars: Mariner 9 spectroscopic evidence for H_2O ice clouds. *Science* 182:381–83
21. Cutts, J. A. 1973. Nature and origin of layered deposits of the Martian polar regions. *J. Geophys. Res.* 78:4231–49
22. Dalgarno, A., McElroy, M. B. 1970. Mars: Is nitrogen present? *Science* 170:167–68
23. Dementyeva, N. N., Kurt, V. G., Smirnov, A. S., Titarchuk, L. G., Chuvahin, S. D. 1972. Preliminary results of measurements of uv emissions scattered in the Martian upper atmosphere. *Icarus* 17:475–83
24. Fehsenfeld, F. C., Dunkin, D. B., Ferguson, E. E. 1970. Rate constants for the reaction of CO_2^+ with O, O_2 and NO; N_2^+ with O and NO; and O_2^+ with NO. *Planet. Space Sci.* 18: 1267–69
25. Fjeldbo, G., Eshleman, V. R. 1968. The atmosphere of Mars analyzed by integral inversion of the Mariner IV occultation data. *Planet. Space Sci.* 16: 1035–59
26. Fjeldbo, G., Kliore, A., Seidel, B. 1970. The Mariner 1969 occultation measurements of the upper atmosphere of Mars. *Radio Sci.* 5:381–86
27. Gierasch, P., Goody, R. 1968. A study of the thermal and dynamical structure of the Martian lower atmosphere. *Planet. Space Sci.* 16:615–46

27a. Gierasch, P. J., Sagan, C. 1971. A preliminary assessment of Martian wind regimes. *Icarus* 14:312–18

28. Gierasch, P. J., Goody, R. M. 1972. The effect of dust on the temperature of the Martian atmosphere. *J. Atmos. Sci.* 29: 400–2
29. Gierasch, P. J., Toon, O. B. 1973. Atmospheric pressure variation and the climate of Mars. *J. Atmos. Sci* 8:1502–8
30. Gierasch, P. J., Goody, R. M. 1973. A model of a Martian great dust storm. *J. Atmos. Sci.* 30:169–79
31. Golitsyn, G. S. 1973. On the Martian dust storm. *Icarus* 18:113–19
32. Hanel, R. A. et al 1972. Infrared spectroscopy experiment on the Mariner 9 mission: Preliminary results. *Science* 175:305–8
33. Hanel, R. et al 1972. Investigation of the Martian environment by infrared spectroscopy on Mariner 9. *Icarus* 17: 423–42
34. Herr, K. C., Pimentel, G. C. 1969. Infrared absorptions near three microns recorded over the polar cap of Mars. *Science* 166:496–99

34a. Herr, K. C., Pimentel, G. C. 1970. Evidence for solid carbon dioxide in the upper atmosphere of Mars. *Science* 167: 47–49

35. Hord, C. W., Barth, C. A., Stewart, A. I., Lane, A. L. 1972. Mariner 9 ultraviolet spectrometer experiment: Photometry and topography of Mars. *Icarus* 17:443–56
36. Hord, C. W., Simmons, K. E.,

McLaughlin, L. K. 1974. Mariner 9 ultraviolet spectrometer experiment: Pressure altitude measurements on Mars. *Icarus*. In press
37. Horn, D., McAfee, J. M., Winer, A. M., Herr, K. C., Pimentel, G. C. 1972. The composition of the Martian atmosphere: Minor constituents. *Icarus* 16:543–56
38. Hunten, D. M., McElroy, M. B. 1970. Production and escape of hydrogen on Mars. *J. Geophys. Res.* 75:5989–6001
39. Kaplan, L. D., Connes, J., Connes, P. 1969. Carbon monoxide in the Martian atmosphere. *Ap. J.* 157:L187–92
40. Kieffer, H. H., Chase, S. C., Miner, E., Münch, G., Neugebauer, G. 1973. Preliminary report on infrared radiometric measurements from the Mariner 9 spacecraft. *J. Geophys. Res.* 78:4291–4312
41. Kliore, A. et al 1965. Occultation experiment: Results of the first direct measurement of Mars' atmosphere and ionosphere. *Science* 149:1243–48
42. Kliore, A., Fjeldbo, G., Seidel, B. L., Rasool, S. I. 1969. Mariners 6 and 7: Radio occultation measurements of the atmosphere of Mars. *Science* 166:1393–97
43. Kliore, A. J., Fjeldbo, G., Seidel, B. L. 1971. Summary of Mariner 6 and 7 radio occultation results on the atmosphere of Mars. *Space Res.* XI:165–75
44. Kliore, A. J., Cain, D. L., Fjeldbo, G., Seidel, B. L., Rasool, S. I. 1972. Mariner 9 S-band Martian occultation experiment: Initial results on the atmosphere and topography of Mars. *Science* 175:313–17
45. Kliore, A. J. et al 1972. The atmosphere of Mars from Mariner 9 radio occultation measurements. *Icarus* 17:484–516
46. Kliore, A. J., Fjeldbo, G., Seidel, B. L., Sykes, M. J., Woiceshyn, P. M. 1973. S-band radio occultation measurements of the atmosphere and topography of Mars with Mariner 9: Extended mission coverage of polar and intermediate latitudes. *J. Geophys. Res.* 78:4331–51
47. Kurt, V. G., Smirnov, A. S., Titarchuk, L. G., Chuvahin, S. D. 1974. Observation of scattered radiation in the OI triplet lines at 1300 Å in the Martian atmosphere. *Icarus*. 21:35–41
48. Lane, A. L., Barth, C. A., Hord, C. W., Stewart, A. I. 1973. Mariner 9 ultraviolet spectrometer experiment: Observations of ozone on Mars. *Icarus* 18:102–8
48a. Larson, H. P., Fink, U. 1972. Identification of carbon dioxide frost on the Martian polar caps. *Science* 153:136–44
49. Leighton, R. B., Murray, B. C. 1966. Behavior of carbon dioxide and other volatiles on Mars. *Science* 153:136–44
49a. Leovy, C. B., Mintz, Y. 1969. Numerical simulation of the atmospheric circulation and climate of Mars. *J. Atmos. Sci.* 26:1167–90
50. Leovy, C. B. et al 1972. The Martian atmosphere: Mariner 9 television experiment progress report. *Icarus* 17:373–93
51. Leovy, C. B. 1973. Exchange of water vapor between the atmosphere and surface of Mars. *Icarus* 18:120–25
52. Leovy, C. B., Briggs, G. A., Smith, B. A. 1973. Mars atmosphere during the Mariner 9 extended mission: Television results. *J. Geophys. Res.* 78:4252–66
53. Leovy, C. B., Zurek, R. W., Pollack, J. B. 1973. Mechanisms for Mars dust storms. *J. Atmos. Sci.* 30:749–62
54. Masursky, H. 1973. An overview of geological results from Mariner 9. *J. Geophys. Res.* 78:4009–30
55. Masursky, H. et al 1972. Mariner 9 television reconnaissance of Mars and its satellites: Preliminary results. *Science* 175:294–304
56. McCauley, J. F. et al 1972. Preliminary Mariner 9 report on the geology of Mars. *Icarus* 17:289–327
57. McCauley, J. F. 1973. Mariner 9 evidence for wind erosion in the equatorial and mid-latitude regions of Mars. *J. Geophys. Res.* 78:4123–37
58. McElroy, M. B. 1967. The upper atmosphere of Mars. *Ap. J.* 150:1125–38
59. McElroy, M. B. 1969. Structure of the Venus and Mars atmospheres. *J. Geophys. Res.* 74:29–41
60. McElroy, M. B. 1972. Mars: An evolving atmosphere. *Science* 175:443–45
61. McElroy, M. B., Hunten, D. M. 1969. Molecular hydrogen in the atmosphere of Mars. *J. Geophys. Res.* 74:5807–9
62. McElroy, M. B., Hunten, D. M. 1970. Photochemistry of CO_2 in the atmosphere of Mars. *J. Geophys. Res.* 75:1188–201
63. McElroy, M. B., McConnell, J. C. 1971. Dissociation of CO_2 in the Martian atmosphere. *J. Atmos. Sci.* 28:879–84
64. McElroy, M. B., Donahue, T. M. 1972. Stability of the Martian atmosphere. *Science* 177:986–88
65. Milton, D. J. 1973. Water and processes of degradation in the Martian landscape. *J. Geophys. Res.* 78:4037–47
66. Moroz, V. I., Ksanfomaliti, L. V. 1972. Preliminary results of astrophysical

observations of Mars from Mars-3. *Icarus* 17:408–22
67. Murray, B. C. et al 1972. Geological framework of the south polar region of Mars. *Icarus* 17:328–45
68. Murray, B. C., Malin, M. C. 1973. Polar volatiles on Mars—Theory versus observation. *Science* 182:437–43
69. Murray, B. C., Ward, W. R., Yeung, S. C. 1973. Periodic insolation variations on Mars. *Science* 180:638–40
70. Neugebauer, G. et al 1969. Mariner 1969: Preliminary results of the infrared radiometer experiment. *Science* 166: 98–99
71. Neugebauer, G., Münch, G., Kieffer, H., Chase, S. C. Jr., Miner, E. 1971. Mariner 1969 infrared radiometer results: Temperature and thermal properties of the Martian surface. *Ap. J.* 76:719–28
72. Owen, T., Sagan, C. 1972. Minor constituents in planetary atmospheres: Ultraviolet spectroscopy from the orbiting astronomical observatory. *Icarus* 16:557–68
73. Parkinson, T. D., Hunten, D. M. 1972. Spectroscopy and aeronomy of O_2 on Mars. *J. Atmos. Sci.* 29:1380–90
74. Peale, S. J. 1973. Water and the Martian W cloud. *Icarus* 18:497–501
75. Rasool, S. I., Hogan, J. S., Stewart, R. W., Russell, L. H. 1970. Temperature distributions in the lower atmosphere of Mars from Mariner 6 and 7 radio occultation data. *J. Atmos. Sci.* 27: 841–43
76. Reeves, R. R. Jr., Harteck, P., Thompson, B. A., Waldron, R. W. 1966. Photochemical equilibrium studies of carbon dioxide and their significance for the Venus atmosphere. *J. Phys. Chem.* 70: 1637–40
77. Sagan, C. 1971. The long winter model of Martian biology: A speculation. *Icarus* 15:511–14
78. Sagan, C. 1973. Liquid carbon dioxide and the Martian polar laminae. *J. Geophys. Res.* 78:4250–51
79. Sagan, C., Toon, O. B., Gierasch, P. J. 1973. Climatic change on Mars. *Science* 181:1045–49
80. Sagan, C. et al 1973. Variable features on Mars, 2: Mariner 9 global results. *J. Geophys. Res.* 78:4163–96
81. Schorn, R. A., Spinrad, H., Moore, R. C., Smith, H. J., Giver, L. P. 1967. High-dispersion spectroscopic observations of Mars II. The water-vapor variations. *Ap. J.* 147:743–52
82. Sharp, R. P. 1973. Mars: Troughed terrain. *J. Geophys. Res.* 78:4063–72
83. Shimizu, M. 1968. The recombination mechanism of CO and O in the upper atmospheres of Venus and Mars. *Icarus* 9:593–97
84. Smith, S. A., Smith, B. A. 1972. Diurnal and seasonal behavior of discrete white clouds on Mars. *Icarus* 16:509–21
85. Soderblom, L. A., Malin, M. C., Cutts, J. A., Murray, B. C. 1973. Mariner 9 observations of the surface of Mars in the north polar region. *J. Geophys. Res.* 78:4197–210
86. Spinrad, H., Münch, G., Kaplan, L. D. 1963. The detection of water vapor on Mars. *Ap. J.* 137:1319–21
87. Stewart, A. I. 1972. Mariner 6 and 7 ultraviolet spectrometer experiment: Implications of CO_2^+, CO, and O airglow. *J. Geophys. Res.* 77:54–68
88. Stewart, A. I., Barth, C. A., Hord, C. W., Lane, A. L. 1972. Mariner 9 ultraviolet spectrometer experiment: Structure of Mars' upper atmosphere. *Icarus* 17:469–74
89. Stone, P. H. 1972. A simplified radiative-dynamical model for the static stability of rotating atmospheres. *J. Atmos. Sci.* 29:405–18
90. Strickland, D. J., Thomas, G. E., Sparks, P. R. 1972. Mariner 6 and 7 ultraviolet spectrometer experiment: Analysis of the O I 1304- and 1356-Å emissions. *J. Geophys. Res.* 77:4052–68
91. Strickland, D. J., Stewart, A. I., Barth, C. A., Hord, C. W., Lane, A. L. 1973. Mars atomic oxygen 1304 Å emission. *J. Geophys. Res.* 78:4547–59
92. Ward, W. R. 1973. Large-scale variations in the obliquity of Mars. *Science* 181: 260–62
93. Young, L. D. G. 1969. Interpretation of high-resolution spectra of Mars I. CO_2 abundance and surface pressure derived from the curve of growth. *Icarus* 11: 386–89
94. Young, L. D. G. 1971. Interpretation of high resolution spectra of Mars-III. Calculations of CO abundance and rotational temperature. *J. Quant. Spectrosc. Radiat. Transfer* 11:385–90

CURRENT VIEWS OF THE DEVELOPMENT OF SLATY CLEAVAGE

×10027

Dennis S. Wood
Department of Geology, University of Illinois, Urbana, Illinois 61801

INTRODUCTION

A knowledge of the nature and origin of slaty cleavage is of paramount importance to the understanding of deformed rocks in general. This results from a number of facts: cleavage is a phenomenon of very widespread occurrence; it clearly has a close geometrical and genetic relationship to processes of rock folding; and the cleaved rocks of the lower metamorphic grades frequently contain indicators which may be used to determine the total strain entailed in the formation of particular geological structures. Slates are among the best materials available for accurate evaluation of natural ductile deformation of rocks. In addition, cleavage has considerable practical significance in terms of its use in the field for elucidation of large-scale geological structures and recognition of the polyphase deformational history of many regions. It is therefore not surprising that investigations of cleavages were undertaken by many of the more prominent earth scientists of the nineteenth century. Notable among these were Greenough, Sedgwick, De la Beche, Phillips, Darwin, Sorby, Harker, and Van Hise.

The axial planar relationship of cleavage to folding was fully appreciated by Sedgwick (50), Darwin (9), and Rogers (47), all of whom believed that slates had undergone recrystallization during deformation. Furthermore, Darwin (9) thought that slaty cleavage was at one end of a spectrum of continuous metamorphic alteration, to which coarser schistose and gneissose textures also belong. The timing relationships of cleavage formation were considered by Sedgwick (50), Phillips (39), and Sharpe (52), each of whom believed that cleavage was posthumous to folding. Tyndall (66), on the other hand, believed both features to be essentially contemporaneous. The intermediate viewpoint of Fisher (14) and Harker (22) was that cleavage and folding are both responses to the same compressional forces, but that cleavage does not appear until well after the initiation of folding.

Initial knowledge concerning the finite amount of shortening undergone by cleaved rocks was provided by studies of deformed fossils from southwestern England, North Wales, and Ireland. Phillips (39) was the first person to notice that there was a close relationship between fossil distortion and cleavage. The significance of this was appreciated by Sharpe (52) who stated the "general rule that the shells are most distorted in those beds which are most slaty," and who also recognized that "rocks affected by slaty cleavage have suffered a compression of their mass in a direction everywhere perpendicular to the plane of cleavage and an expansion of their mass in the direction of cleavage dip." This was the earliest referral of cleavage to a process of flattening and to the existence of one preferred direction of resultant extension. Sharpe (53) also clearly demonstrated the inequidimensional grain shapes that are characteristic of cleaved rocks. Sorby (54, 55) subsequently utilized ellipsoidal reduction bodies from the Cambrian rocks of North Wales to show that the ellipsoid of distortion or deformation ellipsoid had one of its principal planes precisely parallel to the cleavage, which is the plane normal to the direction of maximum finite shortening. In order to verify these relationships, Sorby (55) and Tyndall (67) conducted experiments with clay and wax, respectively, and both succeeded in producing fissility perpendicular to the applied force. Sorby (54) believed that the Cambrian rocks of North Wales, in which he found a typical deformation ellipsoid to have the form 1.6/1/0.27, had undergone shortening of 75% perpendicular to the cleavage.

The processes whereby preferred shape and crystallographic orientations of minerals are achieved and give rise to cleavage could only be seriously considered after Sorby's development of microscopic petrography. The microscope assisted Sorby in developing a hypothesis of mechanical rotation of initially inequidimensional grains such that their long dimensions come to lie in the cleavage. Sorby also believed that preferred orientation achieved by mechanical rotation could be further enhanced by the growth of new micaceous minerals in the plane of cleavage. Furthermore, he hinted that intracrystalline gliding could be a contributing mechanism (55) and he was first to demonstrate situations in which pressure-solution-transfer had been important during cleavage formation (56–58). Nevertheless, Sorby believed that during the formation of cleavage in argillaceous rocks, all other processes were subsidiary to rotation induced by compression. This was questioned by Van Hise (68) who followed Sorby in believing cleavage to have formed perpendicular to the principal compression direction, but differed in thinking that the primary factor in cleavage development was the growth of new minerals in a plane of flattening. Throughout the present century there have been many proponents of both Sorby's rotation hypothesis (e.g. Maxwell, 35) and Van Hise's crystallization hypothesis (e.g. Kamb, 29; Swanson, 62). Some workers have favored a combination of both hypotheses. For example, Ramsay (45) has suggested that mechanical rotation may have the dominant effect at the lowest metamorphic grades, with recrystallization being of greatest importance under higher temperature conditions. Oertel (38) has theoretically demonstrated how a mechanically produced preferred orientation can be enhanced by late-tectonic or post-tectonic crystal growth.

There has been more controversial discussion between those who believe cleavage to be a shearing phenomenon and those who believe it to be a flattening phenomenon which formed precisely perpendicular to the direction of maximum finite shortening. A shearing mechanism was advocated by Phillips (39), Laugel (32), Fisher (14), Becker (4, 5), Sander (48), Turner (66), Hoeppener (28), Wunderlich

(77), and Voll (69); whereas cleavage has been referred to a flattening process by Sharpe (52, 53), Sorby (54, 55), Tyndall (67), Haughton (24), Harker (21, 22), Van Hise (68), Leith (33), Ramsay (45), Dieterich (10), and Wood (75, 76). The substance of this controversy, within the present century, resulted from the incorrect extension by Becker (4) of the concept of the strain ellipsoid (63) or ellipsoid of distortion which was introduced into geology as a purely descriptive tool for describing deformational shape changes by Harker (22). Becker wrongly treated slaty cleavage as a fracture phenomenon and contended that only exceptional irrotational strains could produce flattening at right angles to the line of force, so that if fissility were produced by flattening it would be a mistake to infer that the direction of force was normal to the fissility. Becker considered that the circular sections of the strain ellipsoid were surfaces of no finite distortion and coincident with the surfaces of maximum shearing strain. He was incorrect with respect to both of these relationships. The circular cross sections will only be equivalent in area to the original spherical section provided that the intermediate axis of the ellipsoid remains similar in length to the original sphere diameter. As was pointed out by Griggs (19), the surfaces of no distortion in the triaxial ellipsoid are generally elliptical cones rather than planes of circular cross section. Furthermore, neither the planes of circular cross section nor the surfaces of no-finite distortion coincide with the surfaces of maximum shearing strain. As a result of these errors, Becker concluded that the direction of principal compression or maximum shortening would bisect the obtuse angle between the surfaces of maximum shearing strain. This is in direct contradiction to all experimental evidence, which shows that shear fractures are inclined at angles of less than 45° to the direction of principal compression. Thus even if Becker had been correct in his view that cleavage was a fracture phenomenon, his interpretation of its origin by shear is supported neither by theory nor experiment. Despite this, and notwithstanding all the evidence from natural strain indicators which clearly shows that cleavage is perpendicular to the short dimension of the deformation ellipsoid, the shear hypothesis has found continuing support.

The use of natural strain indicators in the evaluation of strain magnitudes relies upon the comparison of measurable axes of the deformation ellipsoid with some original parameter. Two alternative assumptions have commonly been made: either that the intermediate axis of the ellipsoid has remained unchanged (e.g. Cloos, 8), or that the volume has remained constant (e.g. Wood, 76). In some cases either or both of these assumptions may be valid, but in the general case both are invalid. It is clearly important to determine the volume changes that have occurred during deformation, but these are extremely difficult to evaluate. Apart from exceptional situations it is most unlikely that increase of volume occurs during natural deformational processes. The first consideration of volume change was by Sorby (54), who proposed that volume loss of almost 60% had accompanied the formation of cleavage in the Cambrian rocks of North Wales. Sorby arrived at this figure by assuming that the intermediate axis of the deformation ellipsoid had undergone no dimensional change. In the case considered this would have meant that compression parallel to the short axis of the ellipsoid was incompletely compensated by resultant extension parallel to the long axis of the ellipsoid. Fisher (14), on the other hand, believed that volume changes of the magnitude suggested by Sorby were completely unrealistic and, instead, considered the volume change to be totally negligible. Becker (5), from a consideration of the compressibilities of various rock materials, also thought that it was virtually impossible for significant volume changes to have occurred. Because ductile deformation is not dependent upon any elastic property, the relevance of Becker's argument is lost. Sorby's later viewpoint was that volume reduction of approximately 10% had occurred in the Cambrian slates of Wales (59), but Cloos (8) was of the opinion that there had been no deformational volume change in the central Appalachian fold belt. The somewhat novel view that cleavage is the result of a volume increase has been expressed by Kirillova (31) and explained in terms of expansion induced by the adsorption of water by clay minerals. More recently Ramsay (45) and Ramsay & Wood (46) have drawn attention to the consequences of ignoring volume change and have demonstrated that the deformation of lithified materials may well involve volume losses of up to 20%. Unless appropriate allowance is made for such changes, strains that are determined from natural indicators will be in error.

Considerations of volume change during deformation raise fundamental questions concerning the conditions of the materials at the time of deformation. Most workers have either considered or assumed that the rocks which now possess cleavage were completely lithified materials prior to their deformation. The person to first suggest that this might not be the case was Sedgwick (50, p. 486), who believed that cleavage developed as a last stage in alteration before the rock in question became completely lithified. The opinion that deformation occurred prior to complete consolidation was also expressed by Cloos (8) in connection with the folding at South Mountain. The possibility of cleavage being a prelithification deformation feature was more strongly suggested by Maxwell (35) for the Martinsburg slates in the Delaware Water Gap area of New Jersey and Pennsylvania. Maxwell's view has been followed in other areas by Moench (37), Powell (41), Braddock (6), and Clark (7). The idea that tectonic overpressures may create abnormally high pore water pressures such as to cause dewatering which may assist the formation of cleavage by facilitating mechanical rotation of grains, is at variance with most opinions of cleavage genesis. If such a process were to have been of universal significance, the effects of post-tectonic metamorphism would be much more widespread than has previously been suspected.

Although it is well over a hundred years since Sharpe observed that the most severely distorted fossils occur in the most intensely cleaved rocks, there has been little attempt to relate measurable variations in strain intensity to variations of either the microscopic or megascopic features of slaty rocks. Much information should be forthcoming concerning the relationship between variation in large-scale regional structure and variation in strain. Similarly, the relationship between variation in finite total strain and such microscopic features as the degree of preferred shape and crystallographic orientation need to be fully evaluated. These relationships, together with some of the other interesting aspects of rocks which possess slaty cleavage, will be discussed in the light of recent work.

STRAIN AND SLATY CLEAVAGE

Despite the stimulus which should have been provided by earlier workers such as Sorby, the systematic investigation of natural strain in rocks was long neglected. A major exception is the work of Cloos (8) which stands as one of the great contributions to structural geology. Other important additions to knowledge concerning the intensity and variation of natural strain included the works of Heim (26, 27) and Wettstein (72) using deformed oolites and fossils, and of Strand (61) and Flinn (15) using deformed conglomerates.

A great impetus to strain studies was provided by Flinn (16) in the paper entitled "On Folding during Three Dimensional Progressive Deformation" which dealt with strain from a theoretical standpoint and predicted the type of fold and boudinage structures that should result from all possible types of strain geometries. This work has had a deep influence upon subsequent theoretical and practical investigations of natural strain. Flinn (16) emphasized that the development of structures such as folds was subsidiary to the basic distortion of rocks. He also urged that the conventional tectonic axes (a, b or B, and c) of Sander should be discarded in favor of the deformation ellipsoid. In so doing, Flinn focused attention upon the true internal deformation of rocks, rather than on partial expressions of the deformation, and placed the onus for determining the deformation ellipsoid upon the field geologist.

The deformation ellipsoid is best obtained from measurements of formerly spherical bodies which possess mechanical properties that are similar to their enclosing matrix. Because such materials are relatively uncommon, various methods have recently been developed which enable objects of nonspherical and different initial shapes (8) to be used for strain determination (11, 13, 18, 45).

Finite strain states are most satisfactorily represented by plotting the deformation ellipsoid according to its axial ratio on a rectangular coordinate graph, termed the deformation plot (15, 16). Some terminological confusion has resulted from differing definitions of the three mutually perpendicular axes. Flinn used the system $Z > Y > X$, whereas Ramsay (45) and Wood (76) have used the system $X > Y > Z$. The only advantage of the former system is that it conforms to the practice used in describing the optical indicatrix, and if for any reason it is useful to know the angle between the circular cross sections of the ellipsoid, this is a convenient method. In view of the fact that these cross sections have virtually no mechanical relevance to rock deformation, the system $X > Y > Z$ is more logically preferable. The three axes of the deformation ellipsoid, X, Y, and Z define the principal finite strains, $1+e_1$, $1+e_2$, and $1+e_3$, where e_1, e_2, and e_3 are the principal extensions. Thus the ratio $(1+e_1)/(1+e_2)$ is plotted as the ordinate (X/Y); and the ratio $(1+e_3)/(1+e_2)$ is plotted as the abcissa (Z/Y). The simplest and most convenient deformation plot is that in which the ratios X/Y and Z/Y are plotted on logarithmic scales. This has the advantage that all lines representing equal changes of length of the three dimensions X, Y, and Z are straight. Figure 1 shows a logarithmic deformation plot for which a large number of lines of equal

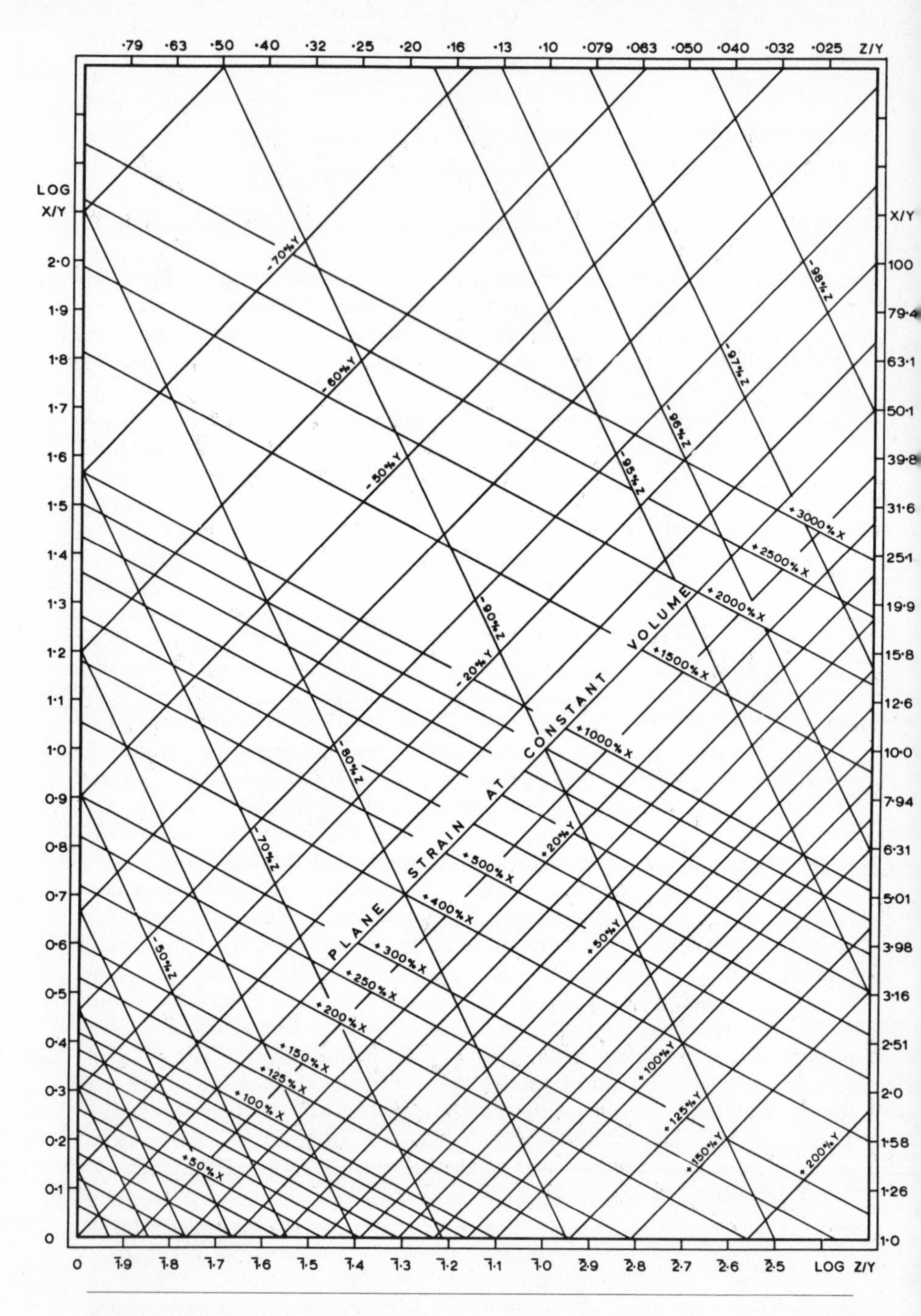

Figure 1 Logarithmic deformation plot ($\log_{10}$) for evaluation of finite strain parallel to the axes of the deformation ellipsoid.

dimensional change have been calculated (76). All deformation ellipsoids which are to be found in nature can be represented on this plot and, provided that volume changes have been insignificant, the strains in the three principal directions are immediately determinable. Oblate spheroids ($X = Y > Z$) of increasing oblateness plot progressively along the abcissa and prolate spheroids ($X > Y = Z$) of increasing prolateness plot along the ordinate. All general triaxial ellipsoids fall into two fields which are separated by the line of unit slope through the origin. Under conditions of deformation at constant volume these two fields are synonymous with deformation of flattening type and constriction type, respectively. Under similar conditions, a rock deforming by a process of plane strain would be represented by ellipsoids which progressively plot along the line of unit slope. Such ellipsoids have the special property that their intermediate axis (Y) would maintain constant length and be equivalent to the geometric mean of the two other axes. Rocks which have deformed by ideal plane strain appear to be relatively rare in nature, as are tectonically deformed rocks with either perfect linear fabrics of pure constrictional origin or perfect planar fabrics of pure flattening origin. Perfect linear fabrics are perhaps more frequently approached in nature than purely planar fabrics; possibly they are more noticeable.

All slates appear to fall in the flattening field of the deformation plot. Figure 2 shows some finite strain states for the Cambrian slate belt of North Wales, a region of essentially upright fold structures. For convenience, the lines of equal dimensional changes are indicated as percentages of the diameter of the equivalent volume sphere (d). In all cases, the principal plane XY of the deformation ellipsoid is parallel to the cleavage. There can be no question that cleavage is precisely perpendicular to the direction of maximum finite shortening.

The mean strain determinations represented in Figure 2 show considerable heterogeneity from one locality to another, although all ellipsoids at any single locality have remarkable shape homogeneity. The average length of the short dimension (Z) of the deformation ellipsoid is 35% of the equivalent sphere diameter, indicating a shortening across the cleavage of 65%. If volume change occurred during deformation, this would be a minimum shortening. The resultant principal extension is such that the long dimension (X) of the deformation ellipsoid has an average value of approximately 220% d. Consequently, the average principal elongation in the plane of cleavage has involved more than doubling of dimension. In this case, the measured extension would be in excess of the real extension in the event that significant volume change had occurred. The intermediate ellipsoid dimension (Y) has undergone apparent average extension by approximately 35%, which would also be reduced if allowance were needed because of volume change.

The variation in dimensional shortening parallel to Z has been between 55% ($Z = 45\%\ d$) and 75% ($Z = 25\%\ d$). Extension parallel to X has been between 60% ($X = 160\%\ d$) and 170% ($X = 270\%\ d$); and extension parallel to the intermediate axis (Y) has been between 20% ($Y = 120\%\ d$) and 45% ($Y = 145\%\ d$). These results from Wales are an interesting comparison with strain data for similar age rocks in the Taconic slate belt of the United States, where the associated fold structures are overturned and tight rather than upright and open as in North Wales. Therefore as might be expected, the deformation ellipsoids lie even farther into the flattening field of the deformation plot, with the average ellipsoid having the form 1.7/1/0.17 for the Taconic slate belt rather than 1.7/1/0.26 for the Welsh Cambrian slate belt. The finite strain data for the Taconic slate belt (Figure 3) shows that the short dimension of the average ellipsoid indicates a shortening of approximately 75% ($Z = 25\%$ *d*) across the cleavage, with concomitant extension of

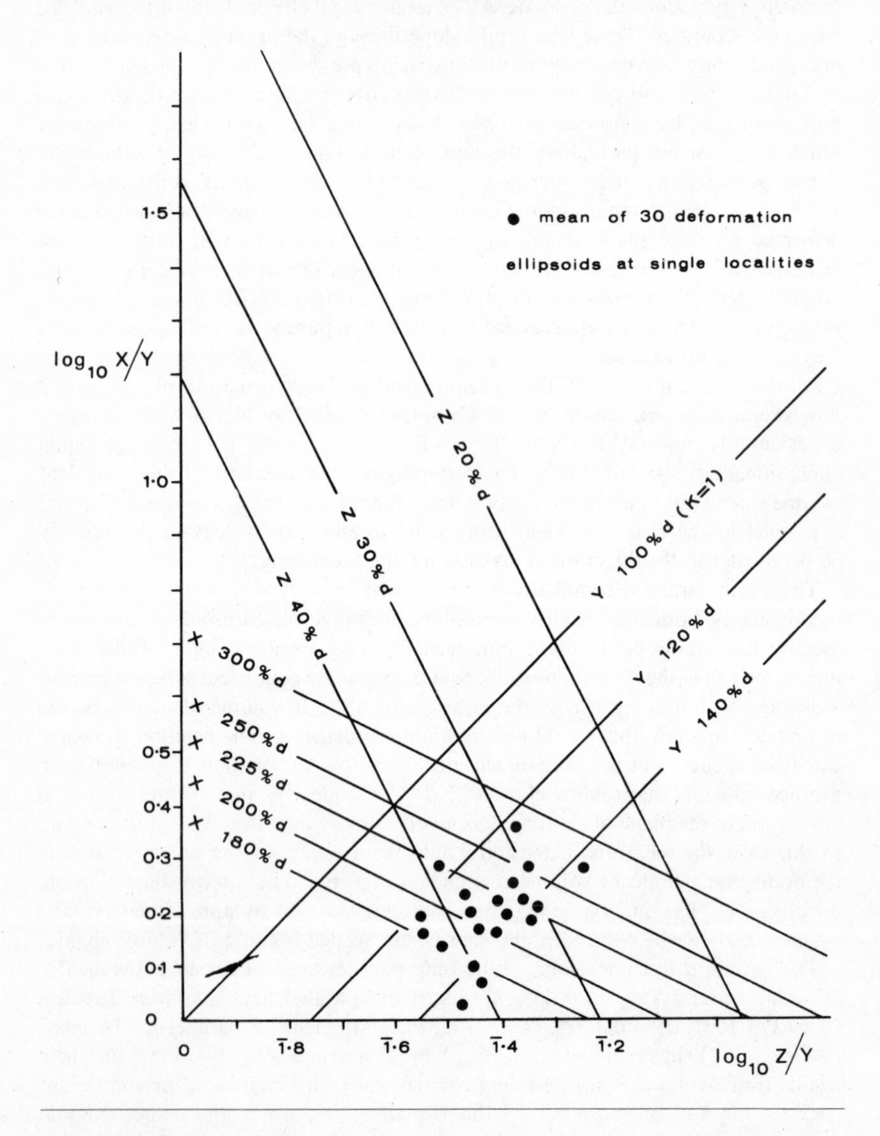

Figure 2 Finite strain data for the Cambrian slate belt of Wales.

the long dimension by 150% ($X = 250\%$ *d*) and of the intermediate axis by 50% ($Y = 150\%$ *d*). Although in this instance the intermediate dimension is reasonably constant from locality to locality, there is 10% variation in the amount of shortening across the cleavage, but almost 70% of strain variation in the direction of greatest finite extension within the cleavage.

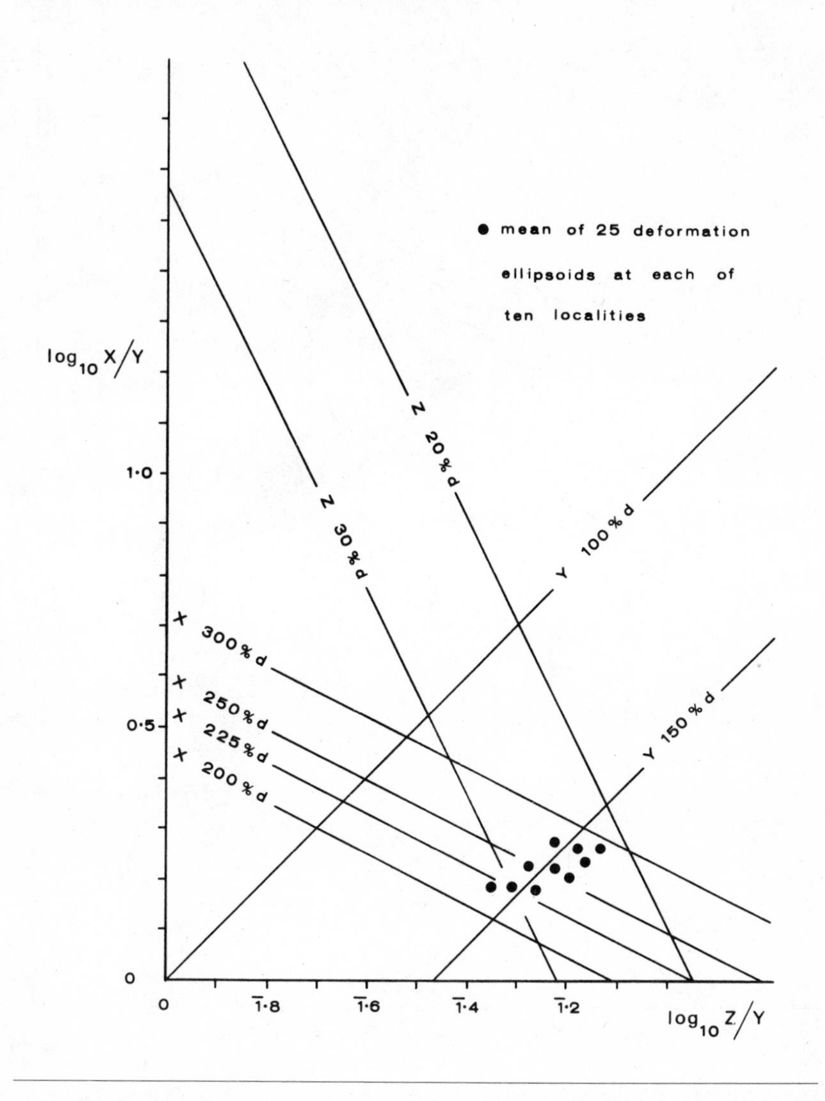

Figure 3 Finite strain data for the Taconic slate belt of the Appalachian region, Vermont, USA.

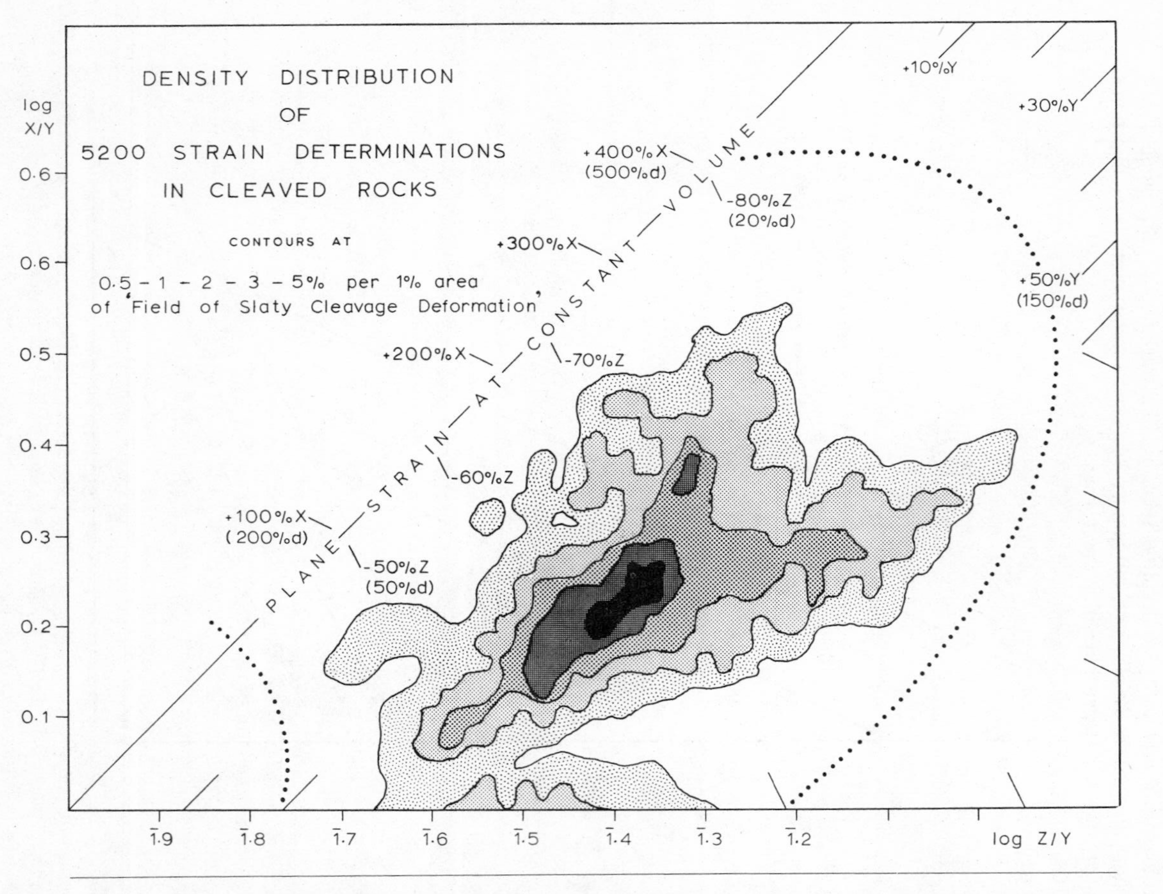

Figure 4 Contoured deformation plot of finite strain in slates of Cambro-Ordovician age from the European Caledonides and their North American equivalents.

When strain obtained from a large number of other slaty rocks is plotted together with that from the slate belts of Wales and the Taconic region, the deformation plot shows an extremely strong concentration of deformation ellipsoids, but represents a large variation of strain magnitudes (Figure 4). The data on this diagram consist of over 5000 strain measurements on slates of Cambrian age from New England, Wales, Scotland, and Norway; and of Ordovician age from New England, Wales, and Northern England. It is reasonable to place conservative limits to a "field of slaty cleavage deformation" (75), and this is outlined in Figure 4. The lower limit to this field depends upon the fact that a certain minimum amount of strain is required before cleavage appears. The upper limits to the field depend upon both geometrical and physical considerations. With continued deformation, cleavage eventually becomes deformed itself, frequently by folding upon axes which bear some close geometrical relationship to the previous strain pattern. Folding on axes parallel to the direction of principal extension in the previously formed cleavage is extremely common, particularly where the cleavage fabric contained an important linear element. In instances where the initial elongation component is less obvious, refolding may occur on axes parallel to the Y axis of the previous deformation ellipsoid; more commonly there is no such simple relationship. Another reason for assigning an upper limit to the field is that there will be a limit to the finite strain which any rock material can sustain and still retain a structure that could reasonably be termed cleavage. There will be no slaty rocks which have suffered shortening much in excess of 90% and resultant elongation in excess of 500%. For such strains to be achieved, ductilities would need to have been greater than those permitted by the temperature and pressure conditions of metamorphism under which slaty cleavage typically develops.

The average deformation ellipsoid for all cleaved rocks has the form 1.76/1/0.24, which implies a shortening perpendicular to the cleavage of approximately 70% ($Z = 30\%\ d$), a principal elongation in the cleavage plane of rather more than 150% ($X = 250\%\ d$) and a linear increase of 35% ($Y = 135\%\ d$) parallel to the intermediate axis of the deformation ellipsoid. All of the best commercial roofing slates have strain values close to these and it is interesting to speculate whether this is a function of sampling or whether there is an optimum strain which gives rise to the premium slate.

VOLUME CHANGES DURING DEFORMATION

It is clearly important that the extent of volume change during deformation should be known. Since the time of the early suggestions by Sorby (54, 59) concerning this matter, it has received little serious attention. This was perhaps a consequence of the fact that Sorby (59) considered reduction in volume of only about 10% to have occurred in the slates which he examined. Such small volume changes would require little correction of the apparent strains measured. This results from the fact that such strains are obtained by comparing the three ellipsoid axes with the diameter of the equivalent volume sphere. If there was volume decrease, the measured ellipsoid must be compared with a greater than equivalent volume sphere. In the event of a 20% volume decrease, the correct comparative sphere diameter is only increased by approximately 6%. Therefore for volume changes of the order of 10%, only small errors are involved in neglecting volume change. If, however, the volume changes approach or exceed 20%, they must be taken into account. The effects of such volume change during deformation have recently been discussed by Ramsay & Wood (46).

Argillaceous slates have densities between 2.7 and 2.85 g/cc. These densities are considerably greater than those of the muds, and appreciably greater than those of the mudstones from which such slates were derived. Thus the magnitude of volume change during deformation depends to a considerable extent upon the condition of the materials at the commencement of deformation. This raises the fundamental question as to whether cleavage is a phenomenon involving the tectonic dewatering of unconsolidated sediments (35), or whether most slates were lithified prior to deformation. From a consideration of pre- and postdeformational densities, it appears possible that volume losses of 10–20% might be involved in the progression from lithified mudstone to slate (46). The relationship between slate densities and predeformational densities for various changes in volume are shown in Figure 5. The average densities of Cambrian slates from Wales and the Taconic region are 2.82 and 2.80, respectively. A decrease in volume of 20% during deformation would require that the initial material have had a density of 2.26. Materials that would fulfill this requirement would include average shales and mudstones with a porosity of about 25%, pure illite, or montmorillonite with a moisture content of about 16%. All of these are unlikely possibilities. No undeformed argillaceous sediments with porosities of 20% have been found from depths greater than 7000 feet in boreholes. Furthermore, Athy (1) showed that Pennsylvanian shales from Oklahoma possessed a density of 2.42 at 1000 feet, rising to 2.66 at 5000 feet. Similarly Hedberg (25) found that undisturbed shales of Tertiary age in Venezuela had a near surface density of about 2.0, rising to 2.35 at 3500 feet and 2.52 at 6100 feet.

It can be shown on structural and stratigraphic grounds that most argillaceous slates, at the time of deformation, were under high confining pressures with an overburden of several thousand feet. It is therefore very unlikely that predeformational densities could be less than 2.5, thereby placing an apparent limit of about 10% upon the amount of volume loss.

Consideration of density alone may be highly misleading. Shackleton (51) pointed out that removal of calcium carbonate and silica during deformation might be very considerable. He cited the fact that Cloos (8), in his study of oolite deformation, found that the more highly deformed oolites were of consistently smaller volume. This could mean that the size difference was a tectonically induced volume change rather than an original sedimentary feature. Volume loss as a result of processes such as continued alignment and closing up of enmeshed clay particles, and dehydration mineral transformations such as the transition from clay minerals to micas would all be taken into account by a simple consideration of densities. This would not be the case, however, for volume loss which had

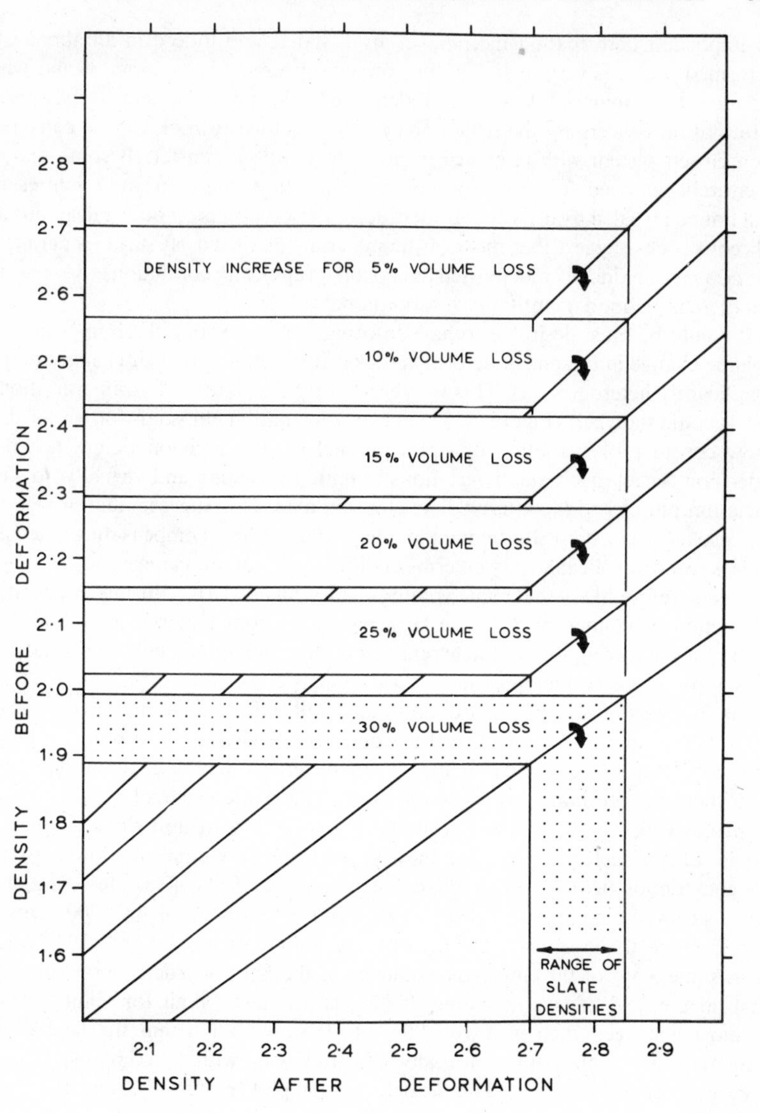

Figure 5 Relationship between slate densities and predeformational densities as a function of volume change.

occurred either as a result of the removal of soluble solids by permeating solutions, or as a result of the removal of solid material at stressed grain contacts by pressure solution transfer. The probability that pressure solution transfer has been an important deformation mechanism in all slates, and indeed in all appreciably deformed rocks, is very high. The fundamental problem is to decide if and where the dissolved material has been redeposited. At present there is no precise information concerning the scale upon which a deformed rock can be considered as a closed system with reference to pressure solution transfer. If some material is entirely removed from the rock body in question, the associated volume loss will not be revealed by density considerations. Observations on both calcareous and siliceous rocks suggest that most of the materials dissolved off surfaces parallel to the cleavage are locally redeposited as crystallographically continuous overgrowths, but there is no good quantitative information.

It would be most desirable to have information concerning the homogeneity of volume change in regions of deformed rocks. It appears that such changes may be significantly heterogeneous. This is suggested by the nature of strain variation in the Taconic slate belt (Figure 3). The mean deformation ellipsoids for ten localities show considerable variation of strain parallel to the direction of greatest finite extension for relatively small variations in finite shortening and virtually no strain variation parallel to the intermediate ellipsoid axis. This suggests that in response to reasonably constant shortening there has been variable compensating extension. This is readily explicable only in terms of differential volume change.

It is therefore likely that some volume change has occurred during deformation. Reduction of volume by 10% can be achieved without material loss. Additional reductions involving the complete removal of material are difficult to evaluate, but it is most unlikely that pressure solution transfer operates over long diffusion paths. It seems reasonable to propose that on all but the most microscopic scales, the total volume loss entailed in the acquisition of cleavage has been no greater than 20%. Ramsay & Wood (46) introduced the terms "field of apparent flattening" and "field of true flattening" to emphasize that, in the event of volume change, ellipsoids which have apparent positions below the line of unit slope through the origin of the deformation plot may in reality fall within the lower part of the constriction field. It is instructive to examine the effects of such lowering of the plane strain line on Figure 4. For a 10% volume loss none of the 5200 ellipsoids would come to lie outside the field of apparent flattening. For a 20% volume loss, some 3.5% of the ellipsoids would be in the field of true constriction rather than in the field of true flattening. If 20% is a realistic limit for volume change in most instances, then well over 90% of all slates fall within the field of true flattening. Also, the strains for any ellipsoid will, within acceptable limits of accuracy, remain as indicated on the deformation plot (Figures 2, 3, and 4).

STRAIN VARIATION AND GEOLOGICAL STRUCTURE

Few studies have attempted to establish the relationship between large-scale geological structures on the one hand and the patterns and variations in finite strain on the other hand. The nature of structural styles as related to both magnitudes of strain and patterns or modes of strain have recently received some preliminary treatment (76). The geometric effects of variation in finite strain

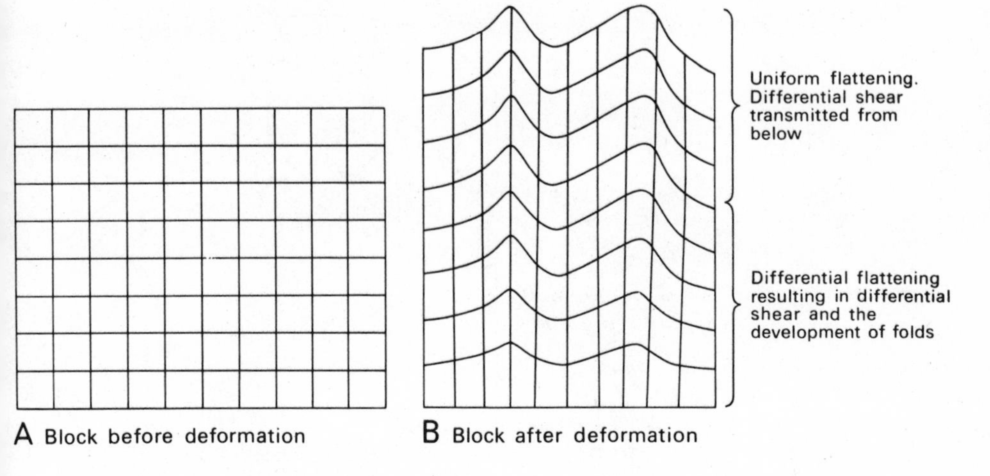

Figure 6 Folding as a result of differential flattening and differential flow (after Ramsay, 43).

related to folding have been more fully discussed by Ramsay (43; 45, p. 435) who demonstrated that differential flattening results in differential flowage which is megascopically expressed as increasingly amplified folding. He further showed that folding can develop in zones which themselves undergo uniform flattening, but where differential flow is transmitted from a neighboring zone (Figure 6). In addition Ramsay (43, 45) extended these theoretical considerations into three dimensions to possibly account for variations in fold plunge (Figure 7).

A practical demonstration of the manner in which inhomogeneous compressive strain along the length of a fold axial surface may lead to the formation of folds with variable plunge is afforded by the Cambrian slate belt of Wales. Along the length of this belt of rocks possessing vertical cleavage, there are four major culmination zones. Finite strain determinations reveal an extremely close relationship between the positions of fold culminations and depressions on the one hand and maximum and minimum values of the flattening component of strain on the other hand (Figure 8). This section, parallel to the cleavage of the belt in question, also shows the variation in vertical thickness of the Cambrian sequence and the variation in shape of the XY principal plane of the deformation ellipsoid. Variations both in fold plunge and in tectonic thickening of the stratigraphic section are clearly dependent upon variation in the amount of flattening across the cleavage. The lowest values of flattening (55%) coincide with the most obvious plunge depression, whereas flattening values in excess of 70% coincide with the major plunge culmination zones. Although all the deformation ellipsoids fall in the flattening field of the deformation plot (Figure 2), those which are more nearly oblate spheroids occur in the plunge depressions, whereas those which demonstrate

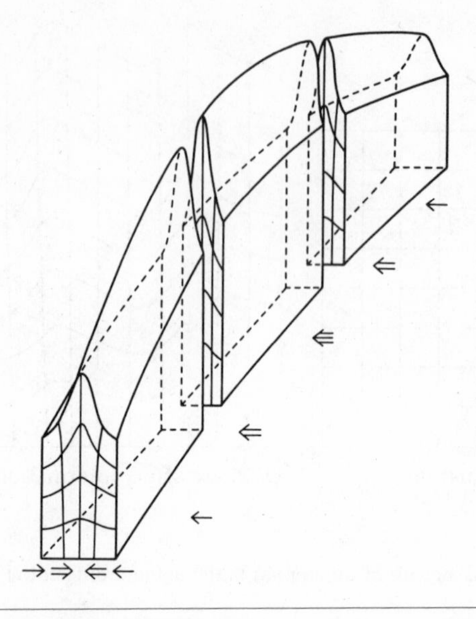

Figure 7 Variation in fold plunge as a result of heterogeneous compressive strain along the length of a fold axial surface (after Ramsay, 43).

the greatest principal extension occur in the culmination zones. Every direction within the plane of cleavage is one of extension, but this extension is more nearly equal in all directions in the case of the lower compressive strains that are characteristic of the plunge depression zones. The response to differential shortening has been such that the present vertical thickness in the depression zones is 50% greater than the initial thickness ($X = 150\%\ d$), whereas in the culmination zones it is 180% greater ($X = 280\%\ d$). The significance of such thickness variations for the stratigrapher is virtually self-evident and means that all isopach maps of strongly deformed regions require most careful evaluation.

STRAIN AND PREFERRED ORIENTATION

Anisotropic crystal growth in metamorphic rocks was clearly described by Becke (3) and Grubenmann (20). Harker (23) and subsequently Ramberg (42) considered that minerals with a growth anisotropy would orientate themselves during deformation so that their direction of most rapid growth would be parallel to the direction of least pressure. This preferred orientation would be thermodynamically controlled and would result in a minimization of internal stress (23). In a nonhydrostatic stress field such as pertains for almost all natural rock deformation, crystal lattices tend to be orientated so that their axes of maximum linear compressibility are parallel to the direction of maximum compressive stress (29, 30), thereby achieving the most thermodynamically stable position. Thus, as has been pointed out by

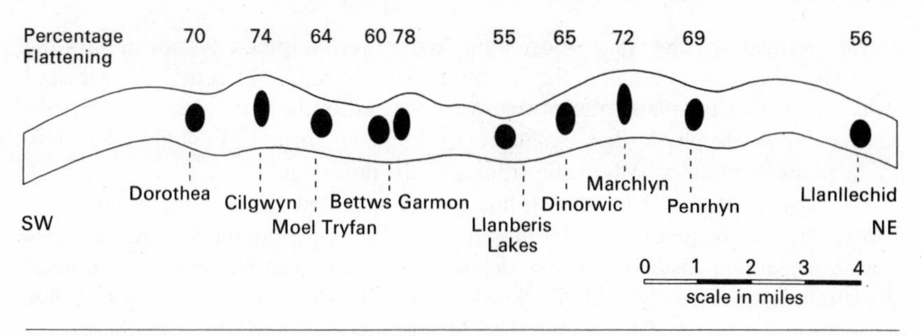

Figure 8 Relationship between variation in fold plunge and tectonic thickening as a function of differential strain along the length of the Cambrian slate belt of Wales.

Flinn (17), for a given mineral and stress condition, the preferred orientation resulting from growth anisotropy is precisely the same as that to be expected if orientation were dependent upon the elastic properties of minerals.

Because cleavage is defined by the preferred shape and crystallographic orientation of phyllosilicates together with the preferred shape orientation of other constituents, it is reasonable to assume that there should be some simple relationship between the degree of preferred orientation and the strain magnitude. In view of the fine grain size of slaty rocks and the fact that at least two crystallographic directions of a mineral must be measured in order to fully define its orientation, optical methods are insufficiently precise to permit the detection of small but significant fluctuations in mineral orientation. For slates, X-ray analysis provides the only satisfactory method for the investigation of preferred orientation as a function of strain. The value of such investigations depends upon whether the relationship between preferred orientation and independently measured total strain can be quantitatively expressed for any given environment.

The investigation of preferred mineral orientation in deformed rocks by X-ray photographic methods dates from the work of Sander & Sachs (49). Improved precision has been obtained in recent years by Wenk (70, 71) and by Starkey (60), the latter having obtained the first X-ray measurement of preferred orientation in fine-grained experimentally deformed quartz aggregates. The first quantitative evaluations of preferred orientation of platy elements in experimentally deformed materials were obtained by Means & Paterson (36). Using a modification of the commercial Norelco pole-figure goniometer, and applying spherical harmonic analytical methods, Baker, Wenk & Christie (2) have subsequently obtained fully quantitative data for quartz in fine-grained experimentally deformed aggregates. Texture or pole-figure goniometry has been applied to the investigation of preferred orientation of phyllosilicates in natural slates by Oertel (38) and by Tullis & Wood (65). In a preliminary single-specimen study of a slate containing natural strain indicators from the English Lake District, Oertel found that the basal planes of chlorite and muscovite had a $9\times$ random preferred orientation parallel to the cleavage and that the rock in question had undergone a tectonic shortening of 50%.

Oertel verified that greatest shortening had occurred precisely normal to the cleavage. He also found that the lineation parallel to the direction of principal extension within the plane of cleavage was reflected in the ultrafabric by a greater scatter of poles to phyllosilicate platelets in the plane normal to the lineation than in the plane containing both the lineation and the pole to cleavage.

The degree of preferred orientation to be expected for a given amount of shortening may be predicted according to the method of March (34). This assumes that the inequidimensional particles defining the cleavage anisotropy, were original constituents with initially random orientation which subsequently underwent passive rotation into a plane of flattening. The March model is basically a mathematical expression of Sorby's first prediction of fabric for a given magnitude of strain (55). Sorby considered the case in which the ratio of the greatest to smallest axes of the deformation ellipsoid was 6: 1. He showed both geometrically and by experiment that grains which were initially inclined at any given angle to the plane perpendicular to the subsequent pressure would rotate such that the tangent of the angle after deformation would be reduced to one sixth of the tangent of the original angle. The validity of the assumption concerning the initially random orientation of components is highly questionable. Sorby (58, pp. 72–74) admitted that the major difficulty with the rotation hypothesis was that he had never been able to find an undeformed rock with random fabric, but he suggested that a parent rock possessing a tendency toward preferred orientation parallel to stratification could, with less than 10% of initial compressional shortening, be transformed into an essentially random starting orientation for the remainder of the strain history. It is doubtful whether this overcomes a problem which will not be satisfactorily solved until there is more precise information concerning both the present and previous mineralogy of slaty rocks.

It is usually assumed that the white mica of slates is muscovite. Many slates are both more sodic and more aluminous than would be expected if this were the case, and it seems likely that considerable amounts of paragonite are commonly present. This means that the starting phyllosilicate also contained sodium as the inter-layer cation. The precise nature of recrystallization of muscovite from illite and of paragonite from brammallite is unknown. Such knowledge would be most valuable to any explanation of the syntectonic achievement of preferred orientation. It is not known how or whether such preferred orientation involves the growth of muscovite, paragonite, and also chlorite by some process of replication upon the lattices of earlier clay minerals which have undergone tectonic rotation and collapse. The enhancement of rotationally achieved preferred orientation by late-tectonic or post-tectonic recrystallization is suggested by Oertel (38), who measured 9 × random preferred orientations for strains in which the March model prediction allowed for only an 8 × random orientation. Oertel has demonstrated that preferential post-tectonic growth of phyllosilicates parallel to their basal planes is more frequently stopped by grain impingement for those grains which deviate most strongly from the plane of preferred orientation. Valuable evidence bearing upon this problem has been provided by Tullis (64), who conducted experiments involving crystallization of micas during deformation and found that the preferred orientations obtained for a given strain value were precisely those predicted by the rotational model, with the greatest concentration of poles to mica basal planes always exactly parallel to the direction of maximum finite compressive strain.

The results of Oertel and Tullis clearly suggested that measurement of preferred orientation might allow not only the directions, but also the magnitudes of principal strains to be determined. Tullis & Wood (65) found that the degree of preferred orientation for three slates, for which the strains had been independently measured, was systematically less than that expected from the March model prediction for those strains. This suggests that other aspects of the total fabric, such as interactions between adjacent grains and the effect of the nonlayer silicates upon the theoretical orientation of phyllosilicates, cannot be neglected. An alternative explanation is that the preferred orientation was inhibited by an initial preferred orientation related to the primary layering.

Thus in the only two published comparisons between predicted strain and measured strain there is contradiction. Oertel's measured preferred orientation was greater than that predicted from the strain measured, whereas the preferred orientations measured by Tullis and Wood were lower than predicted. Accordingly, in the example considered by Oertel, the strains predicted from preferred orientation are higher than the measured strains, whereas the reverse is the case for the slates considered by Tullis and Wood. If significant volume loss had occurred during deformation, the measured compressive strains would need to be increased, and this would be more compatible with the result of Oertel.

A number of cleaved rocks of known strain have recently been examined by X-ray transmission pole-figure goniometry (G. Oertel and D. S. Wood, in preparation) and remarkable agreement is found between the measured deformation ellipsoid and the predicted deformation ellipsoid (Table 1). The largest difference

Table 1 Comparison of strains in slates obtained by measurement of natural strain indicators and strains predicted by the March theory from preferred orientation.[a]

Localities	Measured strain			Strain by March prediction			
	X	Y	Z	X	Y	Z	
Cambrian Slate Belt, Wales	180	131	43	181	138	40	(1)[b]
	200	138	37	174	138	42	(2)
	175	140	40	190	131	40	(3)
	156	147	45	150	150	44	(4)
	180	124	45	176	121	47	(5)
	218	133	35	200	124	40	(6)
	215	132	36	204	126	39	(7)
Ordovician slates, Kentmere	197	126	40	190	124	42	(8)
English Lake District	171	118	49	174	121	48	(9)

[a] Present axial lengths of deformation ellipsoids are given compared to 100 units of original length. The assumption is made that there has been no significant volume change during deformation.

[b] Numbering as on the deformation plot, Figure 9.

found between measurement and prediction of the principal finite compressive strain is 5%; for seven of the nine examples these strains agree to within 3%. The comparative strains in the two other principal directions also show good agreement. In considering strains in the direction of principal finite extension (*X*) the agreement is particularly significant because very small errors in measurement of the short axis of the strain indicators produce much larger errors in the strain computed for the *X*

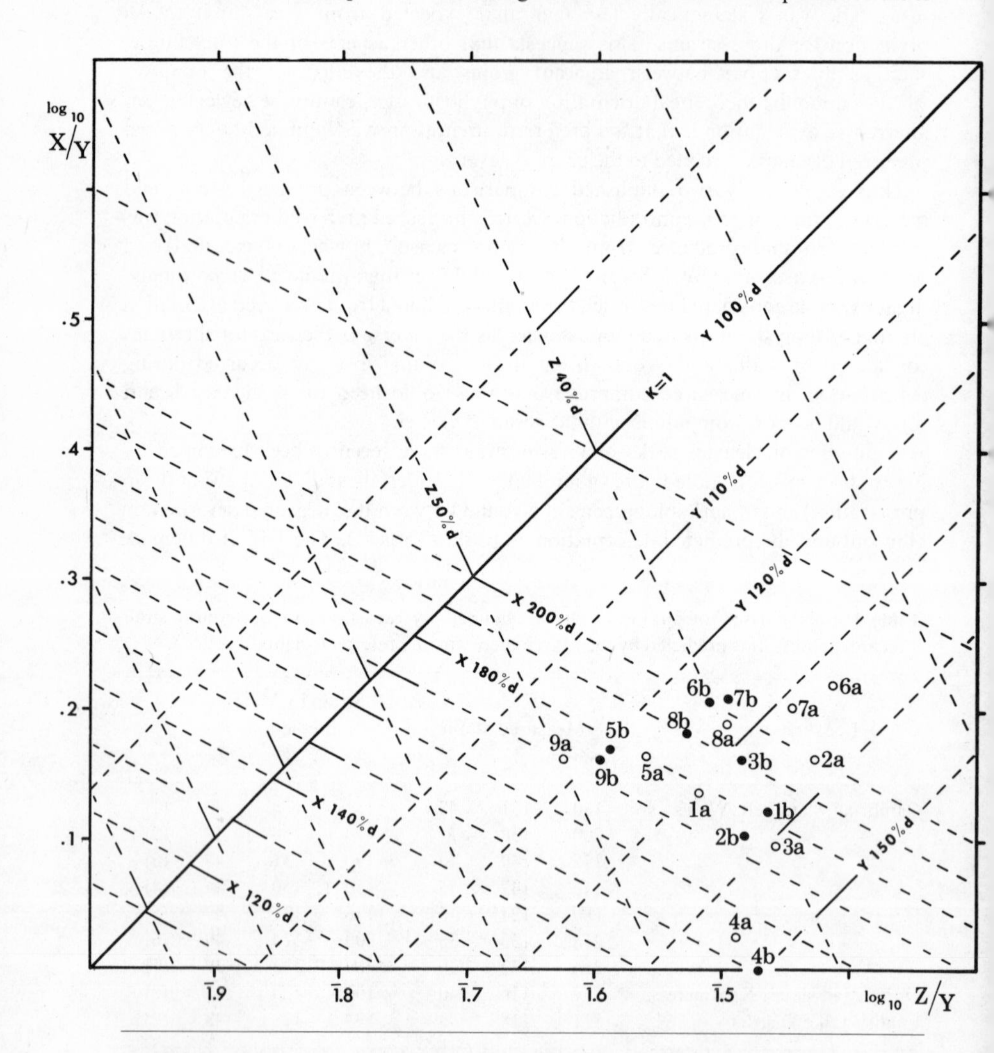

Figure 9 Deformation plot showing finite strains measured from deformed objects (a) and predicted by the March theory from preferred orientation of phyllosilicates (b) for slates from nine localities in North Wales and the English Lake District.

direction by reference to the diameter of the equivalent volume sphere. For the 27 strains compared by Oertel and Wood, 13 strains are virtually identical, 3 show somewhat higher predicted strains, and 11 show lower predicted strains than the measured apparent strains. These results are shown on the deformation plot in Figure 9.

If such agreement can be consistently obtained for a constant lithology in a given environment, it raises the possibility that strains predicted from preferred orientation, and checked against a minimal number of independently determined strains, can be readily and reliably obtained. Since good natural strain indicators are relatively rare in many deformed rocks, texture goniometry now allows detailed assessment of natural strain both on the regional scale and in relation to the evolution of individual geological structures. These methods will permit the production of strain contour maps on various scales.

PATHS OF PROGRESSIVE DEFORMATION IN CLEAVED ROCKS

Almost all measurements of strain in deformed rocks relate only to the finite state of strain. For a full understanding of deformational processes it is necessary to also consider the strain history. This requires knowledge of the incremental strains which have cumulatively resulted in the observed finite strain. Flinn (16) described the manner in which an original sphere deforms progressively through a series of intermediate ellipsoidal shapes until deformation is completed. The locus of the sequential intermediate ellipsoids is the deformation path, which may be considered as resulting from the cumulative effect of superposition of an infinitesimal number of incremental deformation ellipsoids. The simplest deformation paths result when the incremental deformation ellipsoids are of constant shape and the strain is irrotational. Such deformation paths define straight lines on the deformation plot in the three special circumstances where $X = Y > Z$; where $X > Y = Z$; and where $X/Y = Y/Z$ when Y has undergone no length change. For these three situations, Flinn (16, 17) introduced the now widely used terminology $K = 0$, $K = \infty$, and $K = 1$. The parameter K is defined from the deformation plot (Figure 1) as $a-1/b-1$, where $a = X/Y$ and $b = Y/Z$. In the general case all other deformation paths will not define straight lines on the deformation plot (44; 45, p. 329). The precise course of a deformation path is primarily dependent upon rheological behavior as a function of variability of stress values, strain rates, degree of anisotropy, and rotational components of deformation. The resulting complex nature of deformation paths has been theoretically considered by Ramsay (45, pp. 322–32) who pointed out that most natural strain histories have probably involved rotation and variable states of stress. In view of these considerations, and because there is no simple relationship between strain increments and finite strain, or between the finite strain and the deforming stresses, there is no practical justification for assigning intermediate values of K for general ellipsoids which fall within the fields of flattening or constriction. There is more value in assigning deformation fields (45, p. 141) for areas of the deformation plot which contain

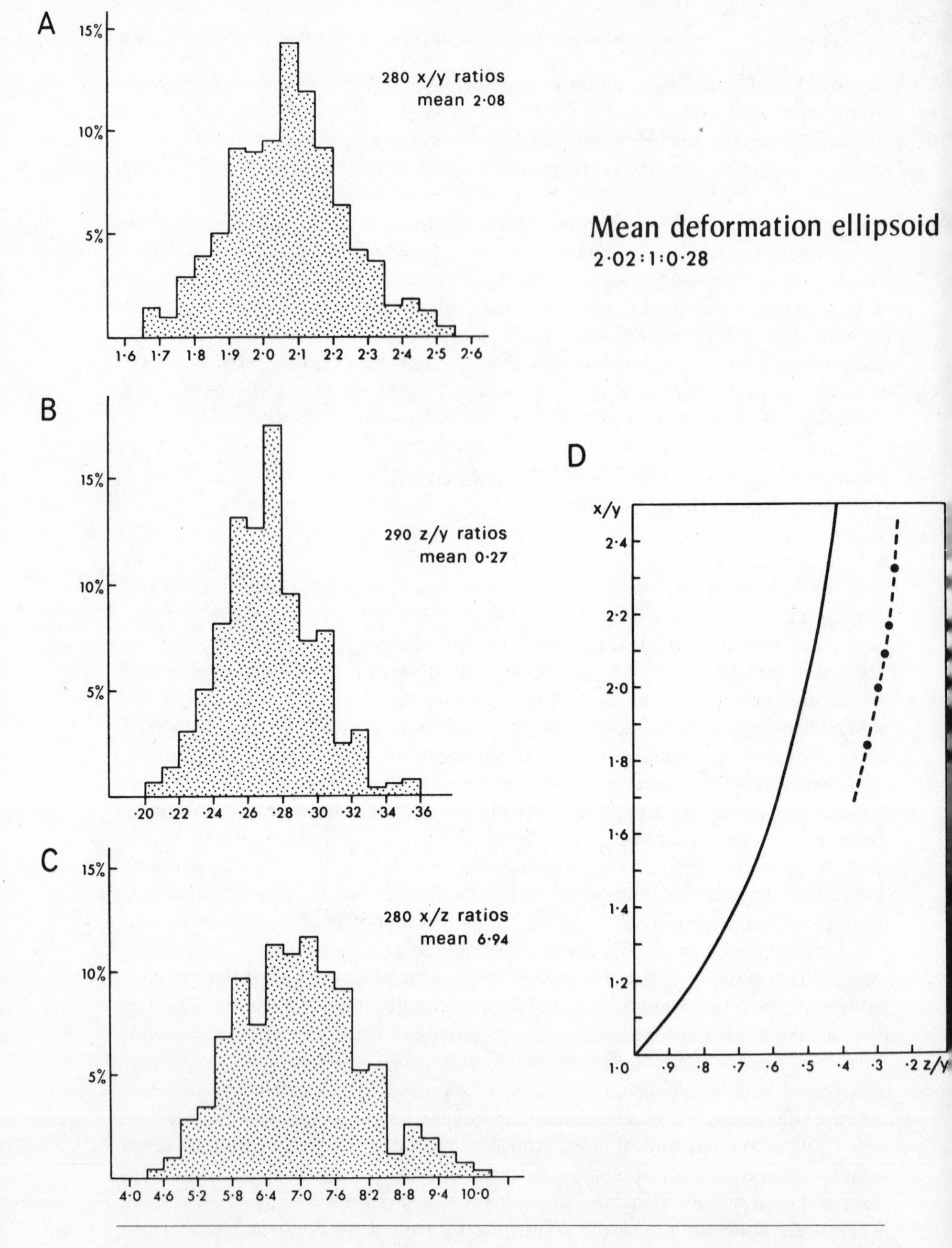

Figure 10 Variation of axial ratios in each plane of the deformation ellipsoid for a single rock specimen (a–c). Nonlogarithmic deformation plot with curve of plane strain at constant volume and deformation path predicted from strain heterogeneity.

various end points of many individual deformation paths relating to a particular structure or region (46).

There is no simple method for the derivation of natural deformation paths from deformation fields. The best approximation for obtaining natural deformation paths entails utilization of the normally heterogeneous nature of finite strain on any specified scale and requires one significant assumption. Figure 10 (a-c) shows three histograms illustrating considerable variation of axial ratios in each plane of the deformation ellipsoid for an Ordovician spotted slate from North Wales (75). The natural strain indicators show a mean deformation ellipsoid having an axial ratio of 2.02/1/0.28. Although the measurements were taken from a uniformly cleaved specimen of only ten cubic inches, it is evident that within this volume the rock has undergone highly nonuniform strain. The measurements indicated on the histograms together comprise the end points of a large number of individual deformation paths. Slight variations in rigidity among the indicators, size variation of the indicators, and variation in frequency of indicators relative to a matrix of slightly different ductility, would account for the observed strain heterogeneity. In view of the small specimen size, it seems reasonable to make the major assumption that despite variations in the state of strain of individual indicators, the general pattern of deformation for the volume in question would have been similar.

If all the components measured underwent strain according to a single overall pattern but achieved differing magnitudes of finite strain, they would lie on a single progressive deformation path. If this were the case, those objects with the highest strain value would have passed through stages represented by the objects with lower strain values. Conversely, if deformation had continued according to the same pattern, objects with the lower strain values would have attained the strain state of the objects with the highest measured strain values. Accepting the assumption, five points on a proposed deformation path may be obtained. By pairing the 20% least deformed ratios of Figure 10a with the 20% least deformed ratios of Figure 10b or 10c, the lowest point on the proposed path is obtained. This point and the remaining four mean 20% sample points are plotted on the nonlogarithmic deformation plot (Figure 10d). Some justification for this procedure is afforded by the fact that the proposed path tends toward the origin of the plot.

Application of the same procedure to the strain data of the Taconic slate belt (Figure 3) gives the result shown in Figure 11. In this example, complete measurements of individual ellipsoids removed the need to pair ratios from different symmetry planes of the ellipsoid. For each of the localities (1–4 on Figure 11), the mean values of the 20% samples of increasing departure from spherical are plotted directly. The resulting paths also tend strongly toward the origin and also tend to include the mean strain values for the remaining localities (5–7 on Figure 11).

This provides a reasonable general method for obtaining some idea of the form of actual deformation paths. It seems to work equally well both on the scale of the hand specimen and on the regional scale. The only justification for this or any other method of obtaining deformation paths will depend upon experimental

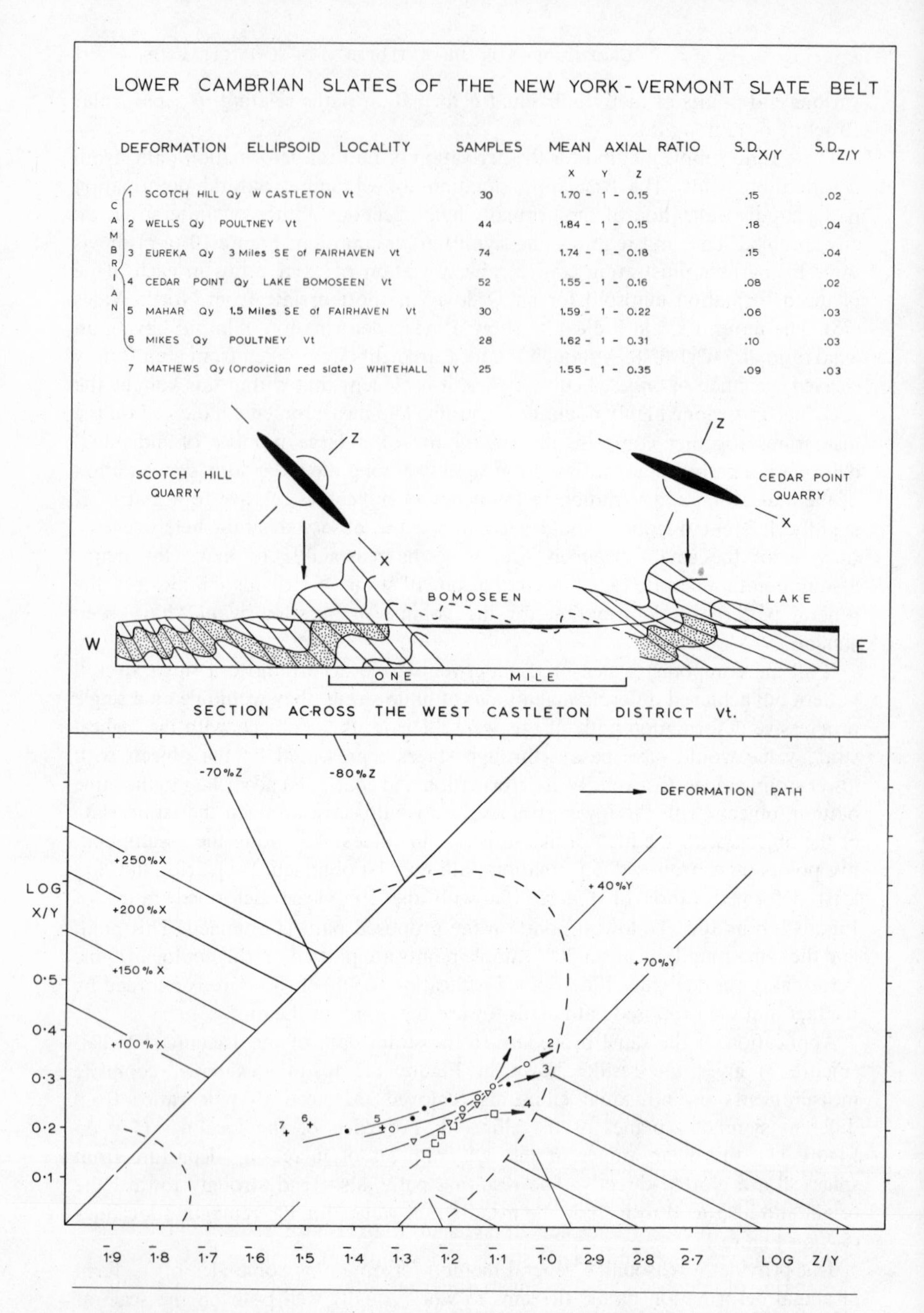

LOWER CAMBRIAN SLATES OF THE NEW YORK - VERMONT SLATE BELT

	DEFORMATION ELLIPSOID LOCALITY	SAMPLES	MEAN AXIAL RATIO X Y Z	S.D.X/Y	S.D.Z/Y
CAMBRIAN	1 SCOTCH HILL Qy W CASTLETON Vt	30	1.70 - 1 - 0.15	.15	.02
	2 WELLS Qy POULTNEY Vt	44	1.84 - 1 - 0.15	.18	.04
	3 EUREKA Qy 3 Miles SE of FAIRHAVEN Vt	74	1.74 - 1 - 0.18	.15	.04
	4 CEDAR POINT Qy LAKE BOMOSEEN Vt	52	1.55 - 1 - 0.16	.08	.02
	5 MAHAR Qy 1.5 Miles SE of FAIRHAVEN Vt	30	1.59 - 1 - 0.22	.06	.03
	6 MIKES Qy POULTNEY Vt	28	1.62 - 1 - 0.31	.10	.03
	7 MATHEWS Qy (Ordovician red slate) WHITEHALL NY	25	1.55 - 1 - 0.35	.09	.03

Figure 11 Deformation paths for the Taconic slate belt predicted from strain heterogeneity of single localities.

verification. Recently F. A. Donath and D. S. Wood (in preparation) have sought to do this by conducting high pressure experiments involving the deformation of oolitic limestone under identical conditions at 10% intervals of finite shortening. The results indicate a significant degree of inhomogeneity of strain among individual ooids for any given average finite strain. The considerable overlap in ooid shape from one deformed sample to another is such that the mean deformation ellipsoid of specimens shortened to 50% could have been precisely predicted from the heterogeneity of strain in samples shortened by 30% and vice versa. This leads to two tentative conclusions. The deformation paths determined by the method proposed above are reasonable approximations and, although natural deformation paths are not straight lines on the deformation plot, they are less complex than might be expected from theoretical considerations.

MECHANISMS OF CLEAVAGE FORMATION

Attempts to account for the origin of slaty cleavage have resulted in three principal controversies. The first of these concerns the respective viewpoints of Sorby who believed that mechanical processes were the dominant factor, and Van Hise who considered that chemical processes were of greater importance. Each admitted that processes of the alternative type had played some part in cleavage formation. The second controversy has been between those who considered cleavage to be essentially a fracture phenomenon produced by a shear mechanism, and others who believed it to be a flow phenomenon produced by a process of flattening. The ill-founded shearing hypothesis which has been previously discussed need not be considered further. A third and more recent controversy has appeared between those who favor tectonic dewatering of unconsolidated materials as an essential early stage in cleavage formation, and others who believe the materials to have been low porosity lithified rocks prior to deformation.

It must now be generally accepted that slaty cleavage is a flow phenomenon that has formed precisely parallel to the direction of maximum finite shortening. This has been fully substantiated by field studies (e.g. 75, 76), by computer-simulated model studies of folding (e.g. 10), and by experimental growth of phyllosilicates during deformation (64). Before considering whether the preferred orientations observed in slates were achieved by new crystal growth in the plane of flattening, or whether it is more likely that pre-existing grains underwent mechanical rotation into the plane of flattening, it is appropriate to consider the tectonic dewatering hypothesis. The unidirectional passage of large volumes of liquid through unconsolidated sediment would provide most efficient assistance for the rotation of inequidimensional particles.

On good evidence from the Orodovician Martinsburg slates, Maxwell (35) adopted the view that, in its early stages of development, cleavage may involve the dewatering of unconsolidated sediment under the influence of tectonically initiated pore water pressures of abnormal magnitude. By this hypothesis Maxwell sought to explain the high degree of preferred orientation of phyllosilicates as being produced by rotation contingent upon the through passage of water. He further suggested that such water would be likely to have effected some redistribution of both silica and carbonate. Selective removal of silica during deformation is strongly suggested by the work of Williams (73). It is difficult to refute Maxwell's explanation for the Martinsburg slates; it is well supported both by the mineralogy of the rocks in question and by the timing relationships between deposition and deformation. In view of the presence of clastic dikes subparallel and parallel to the cleavage of the Martinsburg slates, the occurrence of similar dikes in other cleaved rocks has been taken as proof of a similar origin for cleavage despite the absence of other supporting evidence (e.g. 37, 41). Clastic dikes, which are common in many sedimentary environments certainly demonstrate dewatering, but whether they indicate that this was in any way related to cleavage formation is highly questionable. If a sequence containing such dikes is subsequently affected by deformation leading to the formation of cleavage with high angle to the initial layering, most of the dikes will attain an orientation close to the plane of cleavage after moderate shortening has occurred. For the dikes to be genetically related to the cleavage, they would all need to be precisely parallel to that cleavage. In many instances where dewatering has been invoked to account for slaty cleavage this is clearly not the case. In the Siamo slate of Michigan (41) clastic dikes deviate up to 50° from the cleavage.

There are several reasons for rejecting the dewatering hypothesis as a general explanation for slaty cleavage. Three of these reasons refer to the evidence afforded by fossil distortion, the presence of cleavage in associated igneous rocks, and the development of cleavage in retrograde and retrogressive metamorphic rocks, respectively. In respect to the latter it is quite certain that water has been added rather than subtracted in many instances, for example where greenschist to amphibolite facies retrogressive metamorphism affects older granulite facies rocks.

The presence of distorted fossils in slaty rocks and the fact that they are most severely distorted in the most strongly cleaved rocks, raises one significant objection to the dewatering hypothesis. It is hardly conceivable that the organisms were deformed prior to fossilization and equally unlikely that they were lithified prior to the lithification of the enclosing sediment. Consequently the deformation related to the fossil distortion must have occurred subsequent to lithification of both fossil and matrix. Under such conditions, the expulsion of a certain amount of pore water during the early deformational stages would not be capable of significantly assisting the rotation of inequidimensional mineral components.

Most deformed regions contain pretectonic igneous rocks which possess cleavage or schistosity in common with their sedimentary host rocks. The Archaean greenstone belts consist of folded structures with associated slaty cleavage which passes uniformly through thick basic volcanic sequences as well as arenaceous, rudaceous, and ferruginous sedimentary materials (76). Similarly the extrusive and early intrusive members of the late Precambrian and earliest Paleozoic of the Scottish Caledonides possess the primary Caledonian cleavage structure of that region. The cleavage of the Cambrian slates of North Wales equally affects interbedded and underlying ignimbrites (74).

Because of the ductility contrasts normally involved, it is usually difficult to

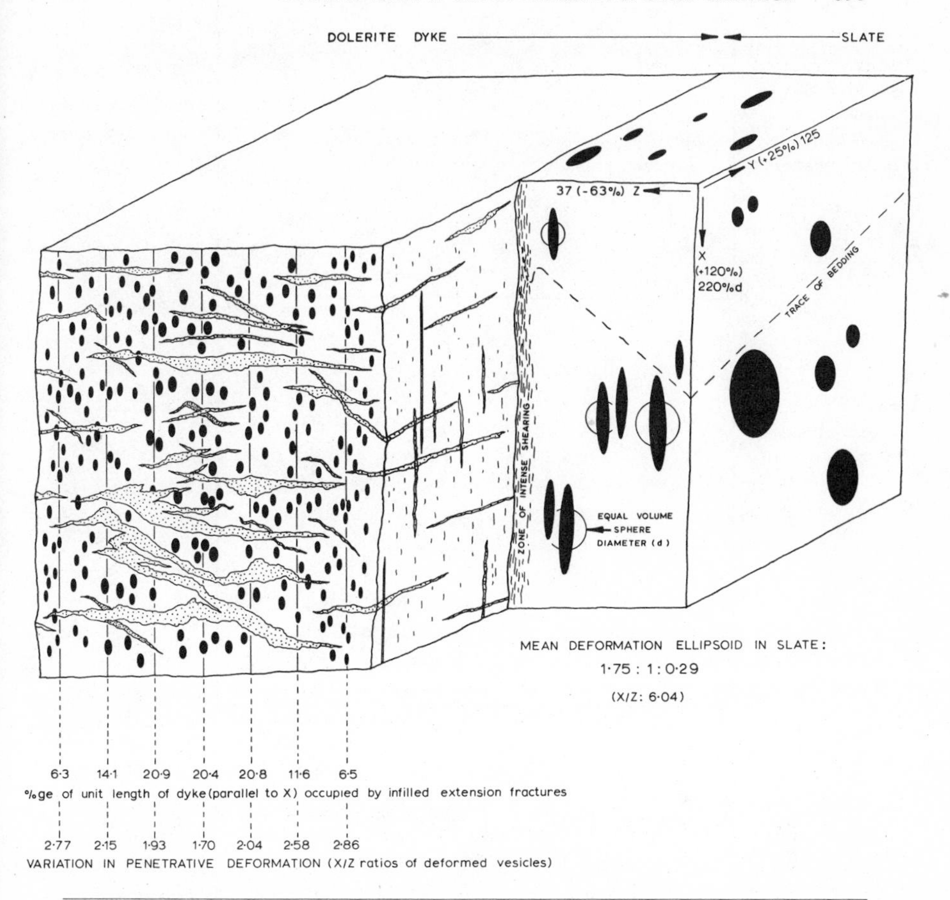

Figure 12 Variation of penetrative strain achieved by flow and discontinuous strain achieved by fracture across a dolerite dike included within slates of known strain (Penrhyn Quarry, Bethesda, North Wales).

obtain a precise relative assessment of the intensity of penetrative deformation in adjacent materials of igneous and sedimentary origin. An attempt to obtain such information is shown in Figures 12 and 13. In this example from Wales, Lower Cambrian spotted slates are cut by Ordovician diabase dikes. The deformation ellipsoid for the enclosing slates has the axial ratio 1.75/1/0.29, which indicates a shortening across the cleavage of 63% ($Z = 37\% d$) and a principal extension in the plane of cleavage of 120% ($X = 220\% d$). The enclosed dike which is about ten feet in width, possesses a cleavage that is everywhere less obvious than in the slates and which is better developed in the finer-grained marginal portions of the dike than in its coarser-grained interior. The dike, however, responded to the deforming stresses by

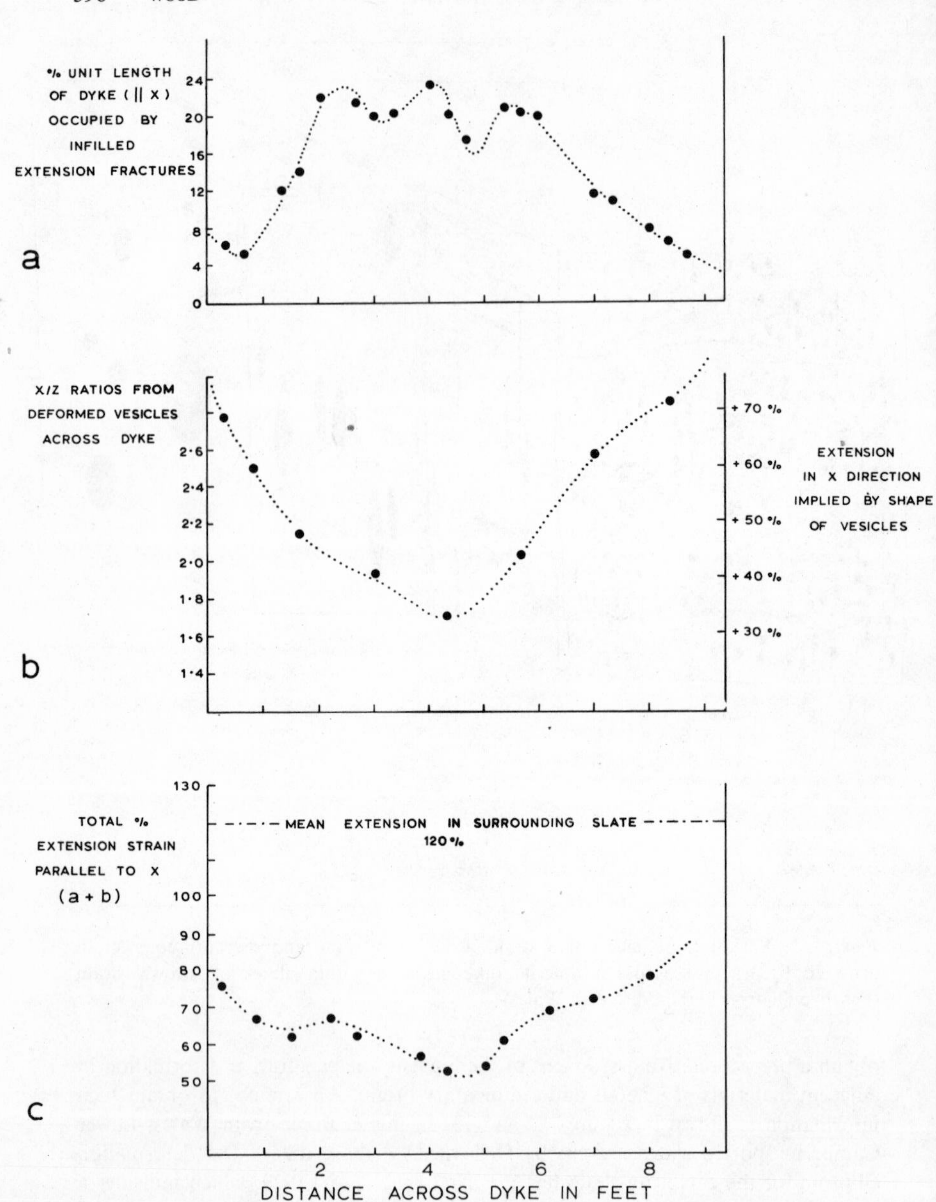

Figure 13 The extensional component of strain in a dolerite dike achieved by a combination of fracture (a) and flow (b), compared to the mean extensional component of strain in enclosing slates (c) (Penrhyn Quarry, Bethesda, North Wales).

adopting two different strain modes. It underwent penetrative nonuniform flow which is reflected by the formation of cleavage, and underwent additional non-penetrative discontinuous deformation which is reflected by the presence of extension fractures. The latter (Figure 12) are of more than one stage of development, with the earlier fractures having undergone folding prior to being cut by later fractures. These extension fractures probably represent periods of increased rate of strain, to which the dike was unable to respond by continuous flow. The variation in the elliptical cross sections of deformed vesicles across the dike is indicated in Figure 12 and the principal extension which this implies is shown in Figure 13b. In assessing the finite extensional strain, the width of the extension fractures must also be taken into account. The percentage of unit distance within the dike, measured parallel to the direction of principal extension, which is occupied by fracture infilling is shown in Figures 12 and 13a. The total principal extensional strain within the dike varies from a minimum of about 50% in the center to over 90% in the marginal zones (Figure 13c). This differential strain within the dike must be compensated elsewhere, either by major extension fractures or by boudinage which is locally common. The important comparison is that between the total strain in the dike margin and the total strain of the host slate (Figure 13c). There is a discrepancy of about 30% strain between these two principal extensional strains. This may be largely accounted for by the difficulty encountered in measuring vesicle distortion in the fine-grained zone of contact strain at the dike margin. It is clear that the strain within the dike increases considerably toward its margin and it is possible that it may increase to the strain level of the slates. It is more likely, however, that there is no continuous increase of strain up to the strain level of the enclosing slates. This is strongly suggested by the fact that the dike margins are always strongly sheared and are often faults. The example demonstrates that there is a considerable degree of compatibility between the strain states of both materials and offers an explanation for faulting as a function of differential ductility and heterogeneous strain. The host rocks to such commonly deformed intrusive rocks must have been considerably lithified to sustain the fracturing which permitted dike emplacement. In view of this, and because the mechanism of cleavage formation must have been generally the same for both materials, the dewatering hypothesis cannot be invoked to account for the cleavage of the Cambrian rocks of North Wales.

Having eliminated the hypothesis of tectonic dewatering as a universal explanation for slaty cleavage, the original mechanical and recrystallization hypotheses remain. Mechanical orientation of inequidimensional grains has undoubtedly contributed to the formation of cleavage in many instances. In a few situations it is difficult to prove that rotation has not been the sole mechanism involved. However, in the great majority of cases there is no doubt that chemical processes have been of incomparably greater importance. The textural relationships, mineralogical composition, and grain size of phyllosilicates in most slates demonstrate that they are the products of syntectonic recrystallization rather than being initial characteristics. The anisotropy possessed by cleaved rocks is only partially the result of preferred orientation of phyllosilicates. The majority of slates contain considerable proportions of highly inequidimensional grains of other minerals, the shape of which is certainly not a detrital feature. The ratio of longest to shortest dimensions of quartz grains in the Cambrian slates of Wales and the Taconic region is commonly 3:1; for the Valdres slates of Norway it is 4:1; and for the Ballachulish slates of Scotland it is about 6:1. Such grains possess syntaxial overgrowths in the direction of principal extension and have undergone dissolution along surfaces situated at high angles to the direction of principal shortening.

The effect of pressure solution as a mechanism for the formation of grain shape anisotropy in cleaved carbonate rocks, which was first described by Sorby (56, 57), has been more recently shown by Plessmann (40) to have been responsible for dimensional shortening of 50% perpendicular to the cleavage. The importance of pressure solution in effecting metamorphic differentiation was emphasized by Voll (69, p. 534), who noted its special significance at quartz-mica grain contacts.

All slates are characterized by the presence of phyllosilicates, carbonaceous material, ore minerals, and other heavy minerals which are concentrated in closely spaced dark seams parallel to the cleavage. These are interpreted as the insoluble residues remaining after pressure solution has occurred (40). These are strictly analogous to, but more uniform and regular than, the stylolite seams of diagenetic origin in sedimentary rocks. It is evident that this process facilitated concentration of primary fine-grained clay minerals, such that syntectonic recrystallization into coarser-grained muscovite and paragonite could occur more efficiently. Consequent upon the appearance of coarser-grained phyllosilicates, further pressure solution resulted in the profusion of plane quartz-mica boundaries that are so important a feature of the fabric of slates. The importance of pressure solution transfer as a deformational mechanism has been emphasized by Durney (12) and explained in terms of nonhydrostatic thermodynamic theory. The special significance of pressure solution for slaty rocks results from the fact that diffusion paths are shorter and rates of solution transfer consequently are higher than would be the case for coarser-grained rocks. The most important current problem is to determine the rock volume over which a particular system may be considered to be chemically closed. The solution to this problem will provide the only satisfactory evaluation of volume changes during deformation. On the scale of individual rock formations it is clear that the system has remained essentially closed, inasmuch as major concentrations of deposited silica and carbonate are not found marginal to such formations.

In the present state of knowledge it must be concluded that the recrystallization hypothesis of Van Hise is valid and that pressure solution transfer provides the mechanism. The fact that measured strain values agree with strain predictions which are based upon preferred orientations and which assume mechanical rotation raises interesting questions. Nevertheless, if the March model prediction of strain provides a sound empirical tool, its practical value is quite undiminished.

CONCLUDING REMARKS

Slaty cleavage is a flowage phenomenon involving flattening. It has formed precisely perpendicular to the direction of maximum finite shortening and has involved some loss in volume. Well-developed slaty cleavage has entailed shortening perpendicular to the cleavage in excess of 50% and commonly between 65 and 75%. In response to compression perpendicular to the cleavage, resultant extension has always occurred within the plane of cleavage. This extension has usually been distinctly greater in one preferred direction in the cleavage.

In some instances, where deformation closely followed deposition, mechanical rotation of inequidimensional particles may have been facilitated by tectonic dewatering which occurred as an initial stage of a continuous deformation from which slaty cleavage resulted. The majority of slates, however, were lithified rocks at the time of acquisition of cleavage. Some degree of rotation of existing inequidimensional components into the plane perpendicular to the direction of maximum finite shortening has undoubtedly occurred in all slates. Rotation has always been subsidiary to the metamorphic growth of new phyllosilicate minerals which are an essential feature of all rocks possessing slaty cleavage. Some redistribution of minerals such as quartz and carbonates by solution processes has occurred in all slates. Pressure solution is the mechanism most likely to have effected the grain shape changes which have occurred during deformation and to have caused the inequidimensional form of minerals other than phyllosilicates in slaty rocks.

Cleavage is therefore the result of a combination of various mechanically and chemically controlled processes which operated over extended periods of time. It is an early stage in a metamorphic continuum of which schistosity and gneissosity may be the end result under appropriate conditions.

Literature Cited

1. Athy, L. F. 1930. Density, porosity and compaction of sedimentary rocks. *Am. Assoc. Petrol. Geol. Bull.* 14: 1–24
2. Baker, D. W., Wenk, H. R., Christie, J. M. 1969. X-ray analysis of preferred orientation in fine-grained quartz aggregates. *J. Geol.* 77: 144–72
3. Becke, F. 1903. Über Mineralbestand und Struktur der Kristallischen Schiefer. *Rep. Int. Geol. Congr. Vienna,* 4: 553–70
4. Becker, G. F. 1896. Schistosity and slaty cleavage. *J. Geol.* 4: 429–48
5. Becker, G. F. 1904. Experiments on schistosity and slaty cleavage. *US Geol. Surv. Bull.* 241: 1–34
6. Braddock, W. A. 1970. The origin of slaty cleavage: evidence from Precambrian rocks in Colorado. *Geol. Soc. Am. Bull.* 81: 589–600
7. Clark, B. R. 1970. Origin of slaty cleavage in the Coeur d'Alene district, Idaho. *Geol. Soc. Am. Bull.* 81: 3061–72
8. Cloos, E. 1947. Oolite deformation in the South Mountain fold, Maryland. *Geol. Soc. Am. Bull.* 58: 843–918
9. Darwin, C. 1846. *Geological Observations in South America.* London: Smith-Elder
10. Dieterich, J. H. 1969. Origin of cleavage in folded rocks. *Am. J. Sci.* 267: 155–65
11. Dunnet, D. 1969. A technique of finite strain analysis using elliptical particles. *Tectonophysics* 7: 117–36
12. Durney, D. W. 1972. Solution-transfer, an important geological deformation mechanism. *Nature* 235: 315–17
13. Elliot, D. 1970. Determination of finite strain and initial shape from deformed elliptical objects. *Geol. Soc. Am. Bull.* 81: 2221–36
14. Fisher, O. 1884. On cleavage and distortion. *Geol. Mag.* (Decade 3) 1: 268–76, 396–405
15. Flinn, D. 1956. On the deformation of the Funzie conglomerate. Fetlar, Shetland. *J. Geol.* 64: 480–505
16. Flinn, D. 1962. On folding during three-dimensional progressive deformation. *Quart. J. Geol. Soc. London* 118: 385–433
17. Flinn, D. 1965. Deformation in meta-

morphism. In *Controls of Metamorphism,* ed. W. S. Pitcher, G. W. Flinn, 46–72. Edinburgh: Oliver & Boyd
18. Gay, N. C. 1968. Pure shear and simple shear deformation of inhomogeneous viscous fluids, 2. The determination of the total finite strain in a rock from deformed objects such as deformed pebbles. *Tectonophysics* 5: 315–39
19. Griggs, D. T. 1935. The strain ellipsoid as a theory of rupture. *Am. J. Sci.* 30: 121–37
20. Grubenmann, U. 1904. *Die Kristallinen Schiefer*. Berlin: Gebrüder Borntraeger
21. Harker, A. 1885. The cause of slaty cleavage. *Geol. Mag.* 2: 15–17
22. Harker, A. 1886. On slaty cleavage and allied rock structures. *Rep. Brit. Assoc. Advan. Sci.* 1885: 813–52
23. Harker, A. 1939. *Metamorphism.* London: Methuen. 2nd ed.
24. Haughton, S. 1856. On slaty cleavage and the distortion of fossils. *Phil. Mag.* 12: 409–21
25. Hedberg, H. D. 1936. Gravitational compaction of clays and shales. *Am. J. Sci.* 31: 241–87
26. Heim, A. 1878. *Untersuchungen über den Mechanismus der Gebirgsbildung.* Basel: Schwarbe
27. Heim, A. 1921. *Geologie der Schweiz.* Leipzig: Tauchnitz
28. Hoeppener, R. 1956. Zum Problem der Bruchbildung, Schieferung and Faltung. *Geol. Rundsch.* 45: 247–83
29. Kamb, W. B. 1959. Theory of preferred crystal orientation developed by crystallisation under stress. *J. Geol.* 67: 153–70
30. Kamb, W. B. 1961. The thermodynamic theory of nonhydrostatically stressed solids. *J. Geophys. Res.* 66: 259–71
31. Kirillova, I. V. 1962. Cleavage as an indicator of the character of mass movement during the folding process. In *Folded Deformations in the Earth's Crust, Their Types and Origins,* ed. V. V. Beloussov, A. A. Sorskii. Moscow: Izdatel'stvo Akad. Nauk S.S.S.R.
32. Laugel, A. 1855. Du clivage de roches. *C. R. Acad. Sci.* 40: 182, 185, 978–80
33. Leith, C. K. 1905. Rock Cleavage. *US Geol. Surv. Bull.* 239: 1–216
34. March, A. 1932. Mathematische Theorie der Regelung nach der Korngestalt bei affiner Deformation. *Z. Kristallogr.* 81: 285–97
35. Maxwell, J. C. 1962. Origin of slaty and fracture cleavage in the Delaware Water Gap area, New Jersey and Pennsylvania. *Geol. Soc. Am. Mem.* Buddington Vol.: 281–311
36. Means, W. D., Paterson, M. S. 1966. Experiments on preferred orientations of platy minerals. *Beitr. Mineral. Petrogr.* 13: 108–33
37. Moench, R. H. 1966. Relation of S_2 schistosity to metamorphosed clastic dikes, Rangeley-Phillips area, Maine. *Geol. Soc. Am. Bull.* 77: 1449–62
38. Oertel, G. 1970. Deformation of a slaty, lapillar tuff in the Lake District, England. *Geol. Soc. Am. Bull.* 81: 1173–88
39. Phillips, J. 1844. On certain movements in the parts of stratified rocks. *Rep. Brit. Assoc. Advan. Sci.* 1843: 60–61
40. Plessman, W. von, 1964. Gesteinslösung, ein Hauptfactor beim Schieferungsprozess. *Geol. Mitt.* 4: 69–82
41. Powell, C. McA. 1969. Intrusive sandstone dykes in the Siamo slate near Negaunee, Michigan. *Geol. Soc. Am. Bull.* 80: 2585–94
42. Ramberg, H. 1952. *The Origin of Metamorphic and Metasomatic Rocks.* Chicago: Univ. Chicago
43. Ramsay, J. G. 1962. The geometry and mechanics of formation of "similar" type folds. *J. Geol.* 70: 309–27
44. Ramsay, J. G. 1964. Progressive deformation in tectonic processes. *Trans. Am. Geophys. Union* 45: 106
45. Ramsay, J. G. 1967. *Folding and Fracturing of Rocks.* New York: McGraw-Hill
46. Ramsay, J. G., Wood, D. S. 1973. The geometric effects of volume change during deformation processes. *Tectonophysics* 16: 263–77
47. Rogers, H. D. 1856. On the laws of structure of the more disturbed zones of the earth's crust. *Trans. Roy. Soc. Edinburgh* 21: 431–72
48. Sander, B. 1930. *Gefügekunde der Gesteine.* Vienna: Springer
49. Sander, B., Sachs, B. 1930. Zur röntgenoptischen Gefügeanalyse von Gesteinen. *Z. Kristallogr.* 75: 529–40
50. Sedgwick, A. 1835. Remarks on the structure of large mineral masses, and especially on the chemical changes produced in the aggregation of stratified rocks during different periods after their deposition. *Trans. Geol. Soc. London Ser. 2,* 3: 461–86
51. Shackleton, R. M. 1962. In discussion of D. Flinn. *Quart. J. Geol. Soc. London,* 118, 430
52. Sharpe, D. 1847. On slaty cleavage. *Quart. J. Geol. Soc. London* 3: 74–105

53. Sharpe, D. 1849. On slaty cleavage. *Quart. J. Geol. Soc. London* 5 : 111–29
54. Sorby, H. C. 1853. On the origin of slaty cleavage. *New Phil. J. Edinburgh* 55 : 137–48
55. Sorby, H. C. 1856. On slaty cleavage, as exhibited in the Devonian limestones of Devonshire. *Phil. Mag.* 11 : 20–37
56. Sorby, H. C. 1863. On the direct correlation of mechanical and chemical forces. *Proc. Roy. Soc.* 12 : 538–50
57. Sorby, H. C. 1879. Structure and origin of limestone. *Quart. J. Geol. Soc. London* 35 : 39–95
58. Sorby, H. C. 1880. On the structure and origin of non-calcareous stratified rocks. *Quart. J. Geol. Soc. London* 36 : 46–92
59. Sorby, H. C. 1908. On the application of quantitative methods to the study of the structure and history of rocks. *Quart. J. Geol. Soc. London* 64 : 171–232
60. Starkey, J. 1964. An X-ray method for determining the orientation of selected crystal planes in polycrystalline aggregates. *Am. J. Sci.* 262 : 735–52
61. Strand, T. 1944. Structural petrology of the Bygdin conglomerate. *Nor. Geol. Tidsskr.* 24 : 14–31
62. Swanson, C. O. 1941. Flow cleavage in folded beds. *Geol. Soc. Am. Bull.* 52 : 1245–63
63. Thomson, W., Tait, P. G. 1879. *Treatise on Natural Philosophy.* Cambridge: Cambridge Univ.
64. Tullis, T. E. 1971. *Experimental development of preferred orientation of mica during recrystallization.* PhD thesis. Univ. California, Los Angeles
65. Tullis, T. E., Wood, D. S. 1972. The relationship between preferred orientation and finite strain for three slates. *Geol. Soc. Am.* (Abstr. with programs) 4(7) : 694
66. Turner, F. J. 1948. Mineralogical and structural evolution of the metamorphic rocks. *Geol. Soc. Am. Mem.* 30 : 1–342
67. Tyndall, J. 1856. Observations on "The theory of the origin of slaty cleavage" by H. C. Sorby. *Phil. Mag.* 12 : 129–35
68. Van Hise, C. R. 1896. Principles of North American Pre-Cambrian geology. *Ann. Rep. US Geol. Surv.* 16: 517–843
69. Voll, G. 1960. New work on petrofabrics. *Liverpool Manchester Geol. J.* 2 : 503–67
70. Wenk, H. R. 1963. Eine Gefüge-Röntgenkamera. *Schweizer. Min. Pet. Mitt.* 43, 707–19
71. Wenk, H. R., 1965. Eine photographische Röntgen - Gefugeanalyse. *Schweiz. Mineral. Petrogr. Mitt.* 45: 517–50
72. Wettstein, A. 1886. Uber die Fischfauna des Tertiären Glarner Schiefers. *Schweiz. Palaontol. Ges. Abh.* 13 : 1–101
73. Williams, P. F. 1972. Development of metamorphic layering in low grade metamorphic rocks at Bermagui, Australia. *Am. J. Sci.* 272 : 1–47
74. Wood, D. S. 1969. The base and correlation of the Cambrian rocks of North Wales. In *The Precambrian and Lower Palaeozoic rocks of Wales,* ed. A. Wood. Cardiff: Univ. Wales
75. Wood, D. S. 1971. *Studies of strain and slaty cleavage in the Caledonides of northwest Europe and the eastern United States.* PhD thesis. Univ. Leeds, England
76. Wood, D. S. 1973. Patterns and magnitudes of natural strain in rocks. *Phil. Trans. Roy. Soc. Ser. A* 274 : 373–82
77. Wunderlich, H. G. 1959. Erzeugung engständiger Scherflächen in plastischem Materiel. *Neues Jahrb. Geol. Palaeontol. Abh.* 1 : 34–44

PHANEROZOIC BATHOLITHS IN WESTERN NORTH AMERICA:

A Summary of Some Recent Work on Variations in Time, Space, Chemistry, and Isotopic Compositions[1]

Ronald W. Kistler
U.S. Geological Survey, Menlo Park, California 94025

INTRODUCTION

Experimental petrologists now favor generation of siliceous plutonic rocks as a consequence of oceanic lithosphere descending along Benioff zones either in an island arc or at a continent-ocean boundary. In the arc environment second-stage melting of oceanic crust forms mixed basalt-andesite-rhyodacite magmas. In the continent-ocean environment melted thickened continental crust overlying the descending lithosphere produces rhyolite or rhyodacite magmas under the influence of water released by dehydration reactions in the descending materials (22).

The large composite Phanerozoic batholiths of western North America provide a testing ground for models of the magmatic phenomena that occur at the junctions between oceans and continents during sea-floor spreading and continental drift. This paper summarizes and reviews some areas of investigation in these batholiths that at the present state of the art have not yet yielded unequivocal conclusions, but at the same time seem to be giving results that are most useful in evaluating models of magmatic processes.

Among these investigations are those of geochronology and isotope geochemistry. The time span of major Phanerozoic Cordilleran siliceous plutonism from the Triassic to the Miocene is fairly well established. Along with the accumulation of geochronologic data, concepts have been developed of both periodic and continuous emplacement of the plutons. Whether emplacement of plutons has been continuous or periodic has to be resolved in order to understand the magmatic process.

In many areas, evaluation of periodic versus continuous emplacement of plutons, however, cannot be done rigorously. This paper will restate the criteria for interpreting available geochronologic data and discuss some examples of interpretation of

[1] Publication authorized by the Director, U.S. Geological Survey.

age data. A summary is also given of the rapidly accumulating data on major element chemistry and initial strontium isotopic compositions of plutons. Within the central Sierra Nevada of western North America there now exist several chemical and isotopic investigations of granitic rocks. A summary of these data demonstrate a correspondence between chemical and isotopic variations. These same data appear to place severe constraints on possible source materials of the plutonic rocks, and these constraints illustrate that a major paradox exists in currently accepted models of magma sources and mechanisms of magma generation for the Sierra Nevada and possibly for most calcic and calc-alkalic plutons in the Cordillera.

AGE OF GRANITIC ROCKS

Any understanding of the chemical and isotopic variations observed in the large Mesozoic and Cenozoic composite batholiths of the North American Cordillera is dependent on the understanding of the time it took to generate and emplace them. Age determinations of igneous rocks by physical methods have been extremely important in the development of concepts of magmatic processes in the last 20 years. However, in spite of the hundreds of dates now available in batholithic rocks, the interpretation of the significance of the dates and the interpreted rates of emplacement of volume increments of batholiths are becoming increasingly active subjects of discussion. It is obvious that conclusions about the chemical evolution of composite batholiths will be different depending on whether the plutons represent an emplacement interval of about half a geologic period (20), or whether they may represent periodic intervals of episodic emplacement of 10 to 15 m.y. (million years) duration (8, 15, 19, 31), or whether they represent continuous emplacement on a continental scale with nonperiodic episodes in different geographic areas (6, 21, 48).

Most of the available age determinations for the Mesozoic and Cenozoic composite batholiths of the North American Cordillera are by the potassium-argon method on mineral separates. Only a few areas have additional dates by the Rb-Sr whole rock technique or by the U-Pb technique on accessory zircons. All of these dating methods at the present time are capable of yielding precise dates with analyses of suitable material. The uncertainty of whether or not a measured date indicates the emplacement age of a pluton generally, therefore, is because datable minerals in plutons are subject to radioactive daughter element loss during subsequent thermal or tectonic events affecting the pluton dated. In some cases this uncertainty is due to geochronologic results reported for impure mineral separates.

Because of the different temperatures required to expel accumulated radiogenic argon from biotite (250 to 450°C) and from hornblende (550 to 650°C) during a period of time of the order of 10^6 years, comparisons of apparent ages of these two minerals from a granitic rock give useful insights into the thermal history of the rock. If the two minerals from a rock yield the same age (concordant), a first interpretation is that the age is either that of the pluton's emplacement and crystallization or it is the age of an immediately adjacent pluton that influenced the isotopic composition of both minerals exactly to the same extent. If the two minerals yield different ages (discordant), the postcrystallization thermal history of the pluton has not been of uniform influence and the generally older hornblende age more closely approximates the crystallization age of the pluton.

A study by Hart (23) was critical to an understanding of K-Ar methods of age dating applied to interpretations of ages of plutonic rocks, although these criteria have not been uniformly applied in geochronologic papers subsequent to 1965. Hart showed that in the contact zone of a small Tertiary stock intruded into Precambrian rocks, concordance of biotite and hornblende K-Ar ages can represent either an initial cooling history, or may represent subsequent degassing of both minerals by a later event. He showed that there was concordance of mineral ages for hornblende and biotite beyond the thermal effect of the younger pluton, but that there was increasing discordance with approach to the younger pluton until the biotite reaches the age of the younger pluton. Concordance of hornblende ages with the biotite ages occurred again only within a few meters of the pluton contact and they were the same as the age of the younger pluton.

Unfortunately, however, even such a straightforward interpretation of concordant biotite and hornblende K-Ar ages is not necessarily unique. Concordance of apparent K-Ar ages may not establish either the absolute age of the pluton dated or of the immediately adjacent pluton. For example, in California, the youngest Mesozoic plutons of the Coast Ranges, Transverse Ranges, and Peninsular Ranges (15) all show concordant K-Ar ages of biotite and hornblende, but Rb-Sr whole rock ages (15, 43) and U-Pb zircon ages (2) are all significantly higher than the concordant K-Ar ages.

Engels (14) has shown that a minor amount of biotite or other high potassium-bearing contaminant in a hornblende separate can cause an apparent concordance of K-Ar ages for biotite and hornblende from a rock even though the two minerals have grossly discordant dates.

All of these factors have to be evaluated when interpreting K-Ar dates and the only real test of the significance of the ages is obtained from dating by one or more other techniques combined with study of the geologic relations of the dated plutons with their enclosing rocks.

Evernden & Kistler (15) reported a large number of K-Ar dates of a variety of minerals, principally biotite and hornblende, from plutons in California and Nevada. In this study, apparent K-Ar ages of plutons ranged from 210 to 77 m.y., and in 46 biotite-hornblende pairs no concordant pair of ages was lacking for any significant time interval after an age of 150 m.y. The K-Ar ages taken alone suggested granite emplacement in California was a continuous process lasting for the period from approximately 210 m.y. to about 80 m.y. ago and did not support the concept of several discrete periods of granite emplacement separated by intervals of no granite emplacement (7, 30). However, evaluation of the K-Ar dates in relation to dates from the same rocks by the Rb-Sr whole rock technique and by the U-Pb technique on zircons, combined with data on observed intrusive relations between granitic rock sequences of different relative age, showed that the volumes of granite emplaced during equal intervals of time was grossly different and that age distributions documented pulses of intrusion representing periodic culminations of activity of a basically continuous process. Five epochs of granitic rock emplacement that took 10 to 15 m.y. to complete and that had beginnings at about 210, 180, 148, 121, and 90 m.y. ago were defined. The diagrammatic granitic rock age profile across central California of Evernden & Kistler (15) is shown in Figure 1 modified with more recent data in the Coast Ranges (25, 43) to reemphasize the varied behavior of mineral ages in complex batholith terranes. The line of section of Figure 1 is shown as line A-A′ on Figure 2. Boundaries between granitic rock series emplaced during the intrusive epochs labeled on Figure 1 are based not on the age dates but on mapped boundaries, while age determinations by K-Ar on biotite and hornblende and Rb-Sr of whole rock specimens from these mapped granitic rock series are projected onto the plane of the cross section. In the Sierra Nevada, mapping shows the rocks intruded during the Cathedral Range intrusive epoch to be younger than those intruded during the Yosemite intrusive epoch and during the Lee Vining intrusive epoch. Biotite and hornblende K-Ar ages are concordant and the same as

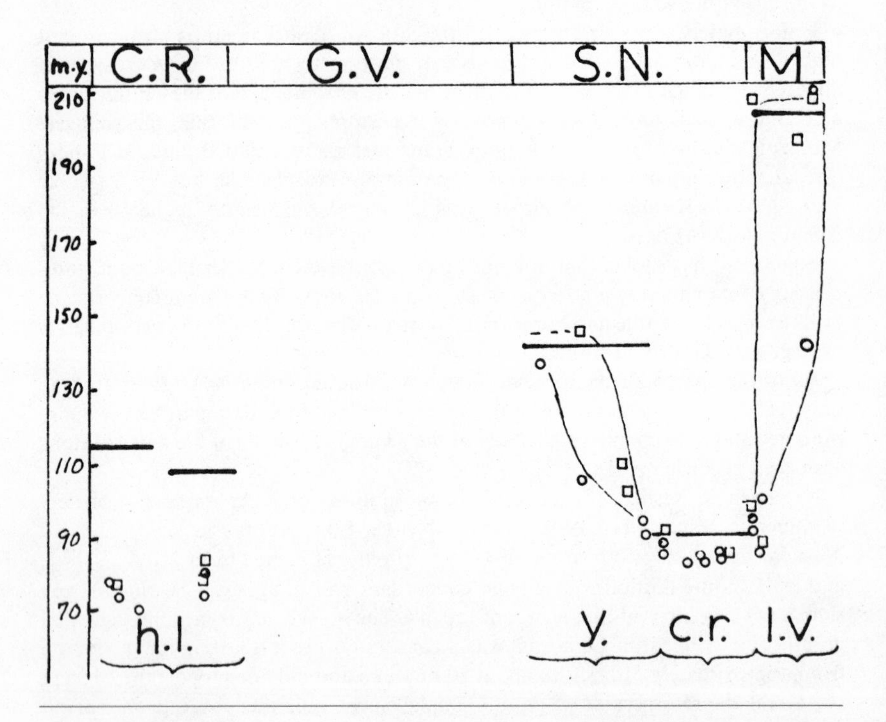

Figure 1 Potassium-argon biotite ages (open circles) and hornblende ages (open squares) and rubidium-strontium whole-rock ages (horizontal lines) for granitic rocks emplaced during the Huntington Lake (h.l.), Yosemite (y.), Cathedral Range (c.r.), and Lee Vining (l.v.) intrusive epochs projected onto a diagrammatic cross section of California. Geographic areas from west to east along the section are Coast Ranges (C.R.), Great Valley (G.V.), Sierra Nevada (S.N.), and Mono Basin (M.). Age in millions of years ago is shown left-hand column of figure. The section is along line A-A′ in Figure 2.

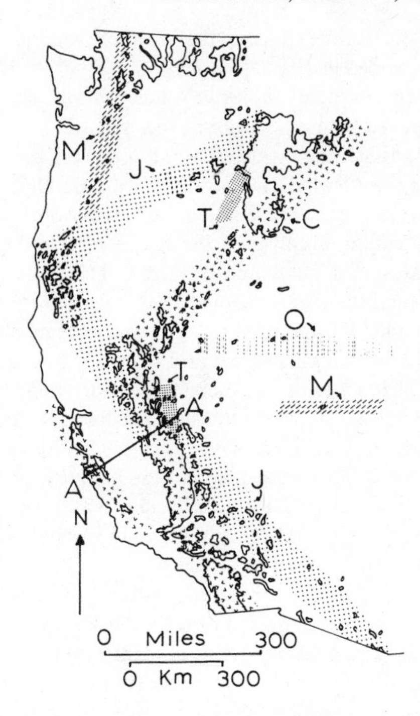

Figure 2 Map showing outlines of Mesozoic and Cenozoic siliceous plutonic rocks in the western United States. Loci of principal plutonic activity of different ages are labeled as follows: T, Triassic; J, Jurassic; C, Cretaceous; O, Oligocene; M, Oligocene-Miocene. The line A-A′ is the trace of the profiles shown in Figures 1 and 4.

the Rb-Sr whole rock ages in the older granitic rocks only at a considerable distance from the younger plutons. This is the pattern of age dates observed by Hart in a contact metamorphic environment. The Coast Range data do not yield a pattern of ages like those found in contact metamorphic zones, and concordant biotite and hornblende K-Ar ages are younger than the Rb-Sr whole rock ages of the plutons.

The beginnings of the intrusive epochs defined by Evernden & Kistler (15) were separated by approximately 30 m.y. intervals. Gabrielse & Reesor (19) using limited K-Ar data of biotites from intrusive rocks and geologically determined periods of magmatism and orogeny also defined an approximate 30 m.y. interval for magma emplacement during the Mesozoic in the Canadian Cordillera. Damon & Mauger (8) using K-Ar dates of biotites from plutons and volcanic rocks in the Great Basin defined two principal periods of magmatism in the Late Cretaceous-early Tertiary and in the middle Tertiary. These two epochs of magmatism also had approximately 30 m.y. intervals between their respective beginnings. Within the limits of the data, the epochs of magmatism in these areas seemed synchronous and the concept of periodic magmatic episodes during the Phanerozoic in the North American Cordillera was developed (e.g. 9, 29, 31, 45).

Subsequent geochronologic studies in other granitic terranes of the Cordillera have cast doubt on the periodic nature of the magmatic episodes if larger areas of the composite batholiths are considered (e.g. 1, 6, 46, 48). Inspection of the analytical data of Armstrong & Suppe (1), however, reveals that many of their dated specimens were not pure mineral separates and many of their reported concordant ages could be misleading because of the age effects of high K_2O impurities in hornblende separates as documented by Engels (14). The data of these workers, therefore, cannot be compared unequivocably with data from the other areas. In two of these areas (46, 48), the relative age sequences of pluton emplacement are poorly or not at all defined in the field and only K-Ar dates, mostly of biotite, are available. The data from these areas, therefore, are insufficient to yield emplacement ages or to define intrusive epochs. On the other hand, Crowder et al (6) give their analytical data, present a map showing intrusive relations between plutons, and show relative ages of plutons determined in the field. For this area in the northern White Mountains in eastern California adjacent to and overlapping the east end of the profile in Figure 1, Crowder et al (6) conclude that evidence for intrusion is lacking in the intervals of about 90 to 130 m.y. and 185 to 200 m.y., that intrusive intervals here are different than those suggested for the Sierra Nevada, and that, for the whole Cordillera, magmatism during the Mesozoic was essentially continuous but episodic from area to area. However, of eight biotite-hornblende pairs of K-Ar ages determined in the study, only two pairs of ages yield concordant results. These two pairs are the only ages that can be considered to represent an intrusive age for any of the plutons, and the ages fall within the time intervals of two of the intrusive epochs of Evernden & Kistler (15). The rest of the ages are not definitive of times of pluton emplacement by themselves. The data of Crowder et al (6) are compatible with, but do not prove the existence of, two of the five epochs of intrusion in the Sierra Nevada. Their data, however, certainly do not define epochs of intrusion different from those defined elsewhere in the Cordillera.

An apparent continuity of Mesozoic and Cenozoic magmatic activity is shown for the Cordillera in papers summarizing age dates from all of the region (21, 33). In these papers large numbers of K-Ar ages, regardless of whether or not they can be documented as intrusion ages, have been plotted on histograms to show that offsets occur in apparent intrusive intervals from area to area or that a K-Ar date can be found for every increment of time from the Triassic to the Recent (21). Even when only concordant biotite-hornblende ages are compiled (33), for the reasons stated above, intrusive histories in regions characterized by multiple intrusions are not revealed by K-Ar ages alone. In view of the complex response of minerals to various geologic phenomena, plots of K-Ar data from different areas on histograms only lead to erroneous conclusions concerning intrusive intervals and reveal nothing toward the solution of the most important question, that of the rates of emplacement of volume increments of granitic magmas.

In different areas of the North American Cordillera, emplacement of plutons of Mesozoic and Cenozoic age occurred in discrete episodes. The episodes have been described as either reflecting a periodic or continuous emplacement of plutons if the whole Cordillera is considered. The periodic emplacement of plutons during the Mesozoic and Cenozoic in the North American Cordillera is by no means an established fact, but neither can it be contradicted with existing age data if these data are interpreted in a complete isotopic and geologic framework. The existence of periodicity or continuity in pluton emplacement has to be resolved for any real understanding of magmatic processes. However, something more than a replot of available data or a few more K-Ar dates without reference to geologic control are required to test the two alternatives. The real test will come as more Rb-Sr whole rock and U-Pb zircon ages become available for those rocks for which intrusive relations are known from field studies.

PLUTONS IN SPACE AND TIME

In spite of the difficulties associated with the interpretation of available age dates, and regardless of the problem of periodic versus continuous emplacement, a spatial pattern of pluton emplacement in time is emerging for the Cordillera. This pattern is shown in Figure 2 along with outlines of outcrops of Mesozoic and Cenozoic intrusive rocks on an outline map of the western United States (modified from Jerome & Cook, 28; plates 9 and 12). The shifting loci of pluton emplacement in time shown on Figure 2 have to be considered in any regional study of chemical variations in batholiths and accounted for in models of Cordilleran magmatic processes.

The loci of Jurassic (180 to 136 m.y. ago) and Cretaceous (121 to 75 m.y. ago) intrusive activity without any attempt to separate rocks emplaced during individual intrusive epochs are from Kistler et al (31). The loci boundaries do not mean that there are no granitic plutons of those ages outside of the boundaries or no granitic plutons of other ages inside the boundaries, but simply outline belts of principal volumes of pluton emplacement of specific ages.

The extent of Triassic pluton emplacement is slowly emerging. Granitic rocks of Triassic age are known in the San Gabriel Mountains (47), central California (15, 38), southeastern California (5), west-central Idaho (18), and southern British Columbia (51). Data are still too scattered to define a locus of plutons of Triassic age like those for the Jurassic and Cretaceous plutons.

Latest Cretaceous and early Tertiary plutons (65 to 45 m.y.) in significant volumes are known in two principal areas. The first is southwestern Arizona (8, 37). The second is northeastern Washington, southeastern British Columbia, northern Idaho, and western Montana. Lipman et al (36) contains a complete list of references of age determinations for this area. The distributions of plutons of this age in these areas do not define narrow loci of emplacement and no attempt is made to outline them on Figure 2.

Middle Tertiary plutons (38 to 20 m.y.) are volumetrically minor relative to those of Mesozoic age. It appears as if narrow loci of Middle Tertiary pluton emplacement of restricted age span shifted southward with decreasing age in the Great Basin (Figure 2). This is consistent with a general southward shift with time of Cenozoic volcanic activity in the same area (36).

Outside of the area of Figure 2, a significant number of age determinations on minerals from granitic rocks have been made in two other areas of the North American Cordillera. In the Canadian Cordillera, most ages of granitic plutons are by K-Ar on biotite. Geologic data suggests that some, if not all, of these ages in the Coast Range batholith of British Columbia have been reduced subsequent to pluton emplacement as a result of extremely complex younger tectonic and thermal events in the region (27); a situation partially comparable to that in western California. If they exist, the data available do not permit delineation of belts of plutons of specific ages in the Canadian Cordillera, but the total time span of pluton emplacement is certainly equivalent to that in the western United States.

In siliceous plutons, K-Ar ages of both mica and hornblende in the Alaska-Aleutian Range batholith (42) indicate three epochs of pluton emplacement. The first and volumetrically most significant group of plutons have dates that range from 179 m.y. to 157 m.y.; the same within experimental error as plutons in California emplaced during the Early and Middle Jurassic. Of the other two groups, one has ages that indicate emplacement of plutons from about 70 to 58 m.y. ago and the other has ages that indicate emplacement of plutons from about 38 to 26 m.y. ago. Supporting age data by other techniques is lacking for the Alaskan plutons, and the interpretation of the dates as pluton emplacement ages must remain tentative. The age data from this area also reveals a shift of pluton emplacement loci with time.

Keeping in mind all the pitfalls and uncertainties in interpretation of the available age determinations, Figure 2 will be used as a reasonable time-space framework for discussion of chemical and isotopic variations in the plutonic rocks in the western United States.

CHEMICAL VARIATIONS

As a result of his studies in the Cordillera of the United States, Lindgren (35) suggested the Mesozoic and Cenozoic plutonic rocks were both compositionally and temporally zoned; older more mafic rocks occurred in the western Cordillera and younger more felsic rocks occurred in the eastern Cordillera. Using a modal mineral classification of plutonic rocks, Moore (39) documented the compositional variation and defined a line (Figure 3) that separated predominantly quartz diorite to the west from predominantly granodiorite and quartz monzonite to the east.

It is obvious from Figure 2 that the suggestion of simple progressive eastward decrease in age of the plutonic rocks was wrong. The exact position of various segments of the quartz diorite boundary have also been modified by more recent studies. Taubeneck (49) suggested that it is not in the correct position in northeastern California and southeastern Oregon but should intersect the Nevada-Oregon boundary at approximately 118°30′W long. Ross (44) has shown that in California quartz diorite is a minor component of the Coast Ranges to the west of the San Andreas fault and their anomalous position relative to the quartz diorite boundary is probably due to offset along the fault.

Regardless of the exact position of the quartz diorite boundary it is a useful concept and the modal classification reflects a general chemical difference that is independent of age in plutons on either side of the line.

Literally hundreds of chemical analyses of plutonic rocks in the region of Figure 2 have been published. Several fine summary articles offer extensive data on chemical variations of regional scale: Klamath Mountains and western Sierra Nevada (24), northwestern Nevada (48), central Sierra Nevada (3), California Coast Ranges (44), Transverse Ranges (44), Southern California batholith (34), and Boulder batholith (50). It has been found useful in these studies to compare data on the basis of the chemical classification of Peacock (40) that utilizes the silica value at which curves on variation diagrams intersect for the total alkalis (Na_2O and K_2O) and CaO; this is the so-called alkali-lime index. Where known, the characteristic alkali-lime index of Mesozoic and Cenozoic plutons are shown on Figure 3. Granitic rocks with an alkali-lime index of less than or equal to 61 (calc-alkalic) or of greater than 61 (calcic) are not separated by the quartz diorite line that is based on modal compositions of granitic rocks. Instead, the chemical separation seems

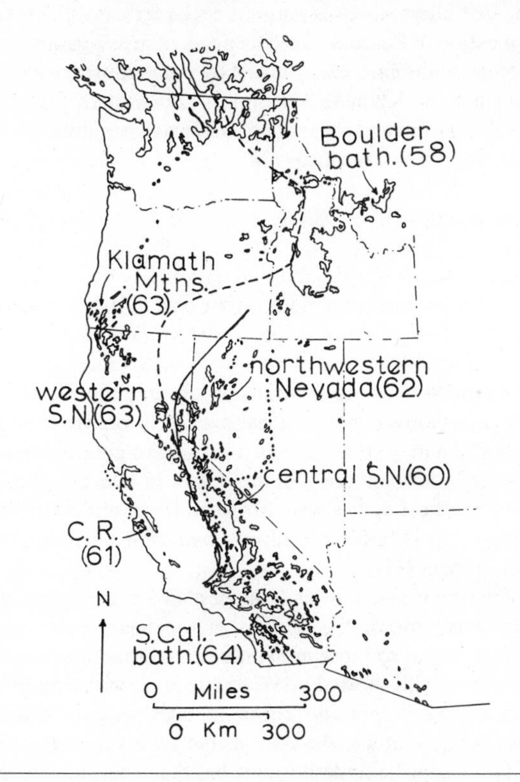

Figure 3 Map showing outlines of Mesozoic and Cenozoic siliceous plutons in the western United States. The alkali-lime index of plutons in various geographic areas is indicated in parentheses. The dashed line is the quartz diorite boundary of Moore (39). The solid line and the dotted line are the initial Sr^{87}/Sr^{86} boundary lines 0.704 and 0.706, respectively. C. R. is the Coast Ranges and S. N. is the Sierra Nevada.

more closely related to initial strontium isotopic compositions of the granitic rocks and is discussed further in the next section of this paper.

In two areas, detailed analysis of major element variation has shown additional systematic geographic variation in elemental concentrations, most notably in potassium. In the central Sierra Nevada, Bateman & Dodge (3) have shown that in granitic rock sequences separated in the field, K_2O increases in a systematic way from west to east across the area of study and that each sequence has a characteristic K_2O index $[K_2O \times 10^3/(SiO_2 - 45)]$. Tilling (50), mainly on the basis of $K_2O - SiO_2$ and $Na_2O - SiO_2$ variation, has separated the plutonic rocks of the Boulder batholith into a main series and a sodic series. These two series seem to be geographically localized and plutonic rocks of adjacent batholiths, including the Idaho batholith, are more akin to and may form a continuation of the sodic series to the west.

Generally the Cordillera siliceous plutonic rocks are calc-alkalic or calcic according to the classification of Peacock. Specific areas of investigation show more subtle variations of composition most easily noted by potassium concentration. However, some large areas like the Klamath Mountains and western Sierra Nevada do not show this type of variation and there is no systematic variation in K_2O concentrations in granitic rocks over large areas (24).

STRONTIUM ISOTOPES

Faure & Powell (17) summarized published data to 1971 on initial Sr^{87}/Sr^{86} values in granitic rocks. They separated plutons on the basis of this ratio into three groups. The first included plutons with low values (0.703 to 0.706) similar to mantle derived basalts. The second included those plutons with high values, greater than or equal to 0.721 with this ratio similar to this ratio in average continental crust. The third group included granitic rocks with values that fall in between the extremes of the first two groups. The data to that time for Cordilleran granitic rocks showed a best value for initial Sr^{87}/Sr^{86} in these plutons was in the intermediate group and about 0.707 ± 0.001. These values were obtained from studies in British Columbia (16), Boulder batholith (12), Sierra Nevada batholith (26), Inyo Mountains, and California Coast Ranges (31).

In reports of Evernden & Kistler (15), Best (4), and Kistler et al (31), initial Sr^{87}/Sr^{86} values for some of the plutons in the Sierra Nevada batholith were reported that were similar to those in mantle derived oceanic basalts.

Kistler & Peterman (32) showed a systematic areal variation of initial Sr^{87}/Sr^{86} from 0.7032 to 0.7082 in granitic rocks in north-central California that was independent of age and that was the same as the areal variation of initial Sr^{87}/Sr^{86} of superjacent late Cenozoic andesites and basalts. This study yielded a threefold characterization of California plutons based on their initial strontium isotopic compositions. One group with alkali abundances equivalent to those of oceanic tholeiitic and alkaline basalts had initial Sr^{87}/Sr^{86} values less than 0.704. Another group had alkali abundances greater than oceanic basalts but had initial strontium isotopic ratios greater than 0.704 but less than 0.706 and similar to these ratios found in oceanic island basalts. A third group with initial Sr^{87}/Sr^{86} ratios greater than 0.706 had this ratio and also alkali abundances greater than those found in oceanic basalts.

The two values of initial Sr^{87}/Sr^{86}, 0.704 and 0.706, not only marked separations of granitic rock chemical data but when contoured coincided with boundaries of paleogeographic significance. Granitic rocks with initial Sr^{87}/Sr^{86} greater than 0.706 intruded only Precambrian or Paleozoic miogeosynclinal rocks, while those with initial Sr^{87}/Sr^{86} less than 0.706 intruded only Paleozoic and Mesozoic eugeosynclinal rocks. Principal exposures of Paleozoic (?) and Mesozoic ultramafic rocks in California are restricted to the area where granitic rocks have initial Sr^{87}/Sr^{86} less than 0.704. Minor exposures of ultramafic rocks in eastern California and west central Nevada occurred in the region where initial Sr^{87}/Sr^{86} of the granitic rocks are greater than 0.704 but less than 0.706 and mostly cropped out close to the line that separates granitic rocks with initial Sr^{87}/Sr^{86} greater than 0.706 from those with this ratio having values less than 0.706. Poole & Desborough (41) have described other inland occurrences of ultramafic rocks, considered to be of Paleozoic age, that lie in a belt that is medial to northern Nevada and that may approximate the eastern extent of plutons with initial Sr^{87}/Sr^{86} less than 0.706 in this area.

Early & Silver (13) and Kistler et al (31) have extended the regional studies of initial Sr^{87}/Sr^{86} ratios of Mesozoic granitic rocks to the California Coast Ranges, Transverse Ranges, western Mojave desert, and Peninsular Ranges.

Although there are large areas with no data, regions where initial Sr^{87}/Sr^{86} are known for Mesozoic and Cenozoic plutonic rocks permit (on Figure 3) for part of the area, separation of plutons into the three isotopic groupings of Kistler & Peterman (32). The quartz diorite line of Moore (39), if modified by the observations of Taubeneck (49) and of Ross (43), is essentially coincident with the line initial $Sr^{87}/Sr^{86} = 0.704$. More specifically, in California the principal granitic rocks characterized by initial Sr^{87}/Sr^{86} less than 0.704 are of quartz diorite and trondjhemite while those characterized by initial Sr^{87}/Sr^{86} greater than 0.704 but less than 0.706 are principally quartz diorite and granodiorite. Quartz monzonite only becomes a major granitic rock phase in the region where initial Sr^{87}/Sr^{86} is greater than 0.706. Using the alkali-lime index of classification of siliceous plutonic rocks, calcic plutonic rock sequences have initial Sr^{87}/Sr^{86} values less than 0.706 while calc-alkalic rock sequences have initial Sr^{87}/Sr^{86} values greater than 0.706.

Available data now show that initial strontium isotopic compositions of Mesozoic Cordilleran plutonic rocks have values that do not average 0.707 ± 0.001 (17), but have a geographically systematic variation from about 0.703 to about 0.709. The isotopic variations observed, like the variations of chemical composition, are not a function of the age of the Mesozoic and Cenozoic granitic rocks investigated but seem to be related only to the geographic position of each sample.

SOURCE MATERIALS FOR THE PLUTONS AND A PARADOX

The variations in space and time of a variety of chemical elements and strontium and lead isotopes are known in the Sierra Nevada at the same latitude as the age profile of Figure 1. Profiles across the Sierra Nevada along the line A-A′ of Figure 1 of averaged radiogenic heat production of both granitic and intruded wall rocks (52), K_2O index of granitic rock sequences (3), and boundaries between the three groups of granitic rocks based on their initial Sr^{87}/Sr^{86} ratios are shown in Figure 4.

Several additional chemical and isotopic parameters have been investigated along the line A-A′ of Figure 1, but the data cannot be easily represented on Figure 4. Rare-earth abundances relative to those in chondrites for apatite specimens (10) from three granitic rocks showed two different distribution patterns. One apatite specimen was from a granitic pluton in the western Sierra Nevada, where initial Sr^{87}/Sr^{86} values of granitic rocks are less than 0.704, and was characterized by an unfractionated rare-earth assemblage. The other two apatite specimens, characterized by fractionated rare-earth assemblages, were from plutons farther to the east where initial Sr^{87}/Sr^{86} values of granitic rocks are greater than 0.704. Potassium, rubidium, and strontium abundances in granitic rocks along the profile could be separated into two general groupings (32). The first of these groupings was in granitic rocks characterized by initial Sr^{87}/Sr^{86} values less than 0.704 and with potassium, rubidium, and strontium abundances the same as those in average tholeiitic and alkaline oceanic basalts. The second grouping was in granitic rocks characterized by initial Sr^{87}/Sr^{86} values greater than 0.704 and with potassium, rubidium, and strontium abundances greater than those in oceanic basalts. Lead isotopic compositions of feldspars from granitic rocks along the profile are of two types (11). The first of these types has lead isotopic composition like that of island volcanic rocks on oceanic ridges and is from granitic rock characterized by initial Sr^{87}/Sr^{86} values of less than 0.704. The second of these types has more radiogenic lead isotopic compositions than the first type and is from granitic rocks characterized by more radiogenic initial Sr^{87}/Sr^{86} values that are greater than 0.704.

The correspondence of chemical and isotopic variations along the profile across the central Sierra Nevada is obvious. Plutons in the western Sierra Nevada with initial Sr^{87}/Sr^{86} values less than 0.704 are depleted in alkali and rare-earth elements

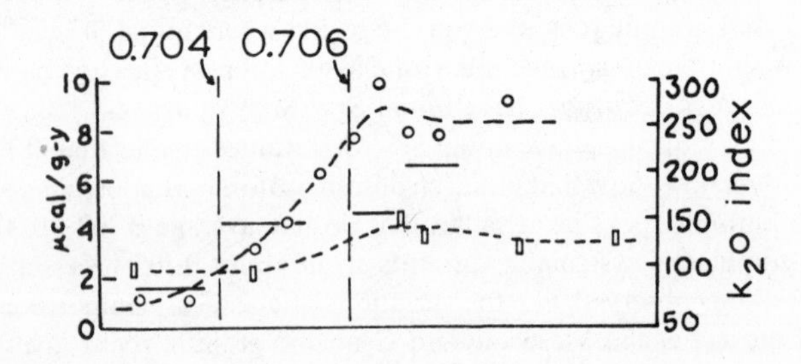

Figure 4 Profiles, along the line A-A′ in Figure 2, from west to east across the central Sierra Nevada of average heat production in μcal/g-y of granitic rocks (open circles) and wall rocks (open rectangles), and of K_2O index (horizontal lines) of granitic rock sequences. The boundaries of initial Sr^{87}/Sr^{86} equal to 0.704 and 0.706 in granitic rocks along the profile are designated as vertical dashed lines.

and, even though they are differentiated rocks, are similar to oceanic tholeiitic and alkaline basalts with respect to concentrations of these elements and with respect to strontium and lead isotopic compositions. Immediately to the east are plutons that have initial Sr^{87}/Sr^{86} values that range to 0.706, and in this respect are similar to some oceanic island volcanic rocks. These same plutons, however, have greater alkali abundances and more fractionated rare-earth assemblages than oceanic volcanic rocks. Finally to the east of this group of plutons are those plutons with both greater initial Sr^{87}/Sr^{86} values and alkali abundances and more fractionated rare-earth assemblages than those found in oceanic basalts.

Each of these chemical and isotopic studies have yielded some insight into possible source material for the granitic rocks investigated. Bateman & Dodge (3) concluded that the increase in K_2O from west to east across the Sierra Nevada batholith was most likely due to lateral changes in composition of source materials of crustal origin for the parent magmas of the granitic rocks. The presence of two different distribution patterns of rare-earth assemblages in apatites from Sierran granitic rocks suggested two chemically distinct parent source materials for the plutons (10). Apatite in western plutonic bodies may have crystallized from magmas derived entirely from mantle material such as oceanic tholeiite characterized by unfractionated rare-earth assemblages. In the other plutonic bodies to the east with apatites having fractionated rare-earth assemblages, the parent magmas were probably derived from crustal material. Doe & Delevaux (11) concluded that likely source materials to yield the lead isotopic compositions in the granitic rocks were also of two types. The first type was a source from oceanic mantle or recycled mantle material for trondjhemite from the Klamath Mountains with lead isotopic composition like that of island volcanic rocks on oceanic ridges. Lead isotopes for the other plutons suggested source materials probably of Precambrian age and of intermediate composition from the lower continental crust and upper continental mantle. The variations in isotopic composition and elemental abundances in the granitic rocks investigated by Kistler & Peterman (32) suggested plutons in the western Sierra Nevada were derived from upper mantle material of oceanic character while plutons in the eastern Sierra Nevada were derived from lower continental crust of Precambrian age.

Some of the isotopic and chemical parameters studied in Cordilleran granitic rocks have been used as evidence in support of their parental magma sources being derived from subducted oceanic crust along Benioff zones. For example, Dickinson (9) suggested the west to east increase in K_2O across the central Sierra Nevada documented by Bateman & Dodge (3) reflected generation of parental magmas of the plutons from oceanic crust at progressively greater depths in a subduction zone. Faure & Powell (17) considering only initial Sr^{87}/Sr^{86} ratios of plutons in Cordilleran batholiths suggested the parental material of the granitic rocks could have originated by partial melting of mainly mantle material with some subducted oceanic crust along subduction zones. As the parental magmas rose, they reacted with crustal rocks and gradually increased their original low strontium isotopic ratios to higher values.

As more and more areas like those considered in this summary are investigated, supporting evidence for calcic and calc-alkalic magma sources along subduction zones becomes less positive. Bateman & Dodge (3) have argued against Dickinson's seismic zone model for an explanation of the observed K_2O variation across the Sierra Nevada because his zone would have to be maintained for more than 100 m.y. and the K_2O concentrations correlate with geographic position and not time of intrusion. Furthermore, there is as much variation in K_2O concentration in granitic rocks along the length of the Sierra Nevada as there is across it (32).

Mixing models, utilizing known compositions of oceanic basalt, Franciscan sediments, and intruded wall and roof rocks, to explain intermediate values of initial strontium isotopic composition of granitic rocks in the Sierra Nevada are eliminated when variations of element abundances are considered along with the isotopic variations (32). Elimination of mixing models to explain isotopic variations require then that these variations are due to compositional inhomogeneities that have existed in the source regions of the granitic rocks since at least the late Precambrian (11, 32). A convincing case cannot be made for increments of magma derived from lithosphere subducted in the Mesozoic in any of these rocks. With the data at hand, it appears as if the pattern of chemical and isotopic variations observed across the Sierra Nevada seems only to reflect a pattern of lateral chemical inhomogeneity that has existed since the late Precambrian in a lithosphere zone that intersects the upper mantle and lower crust. A similar source region for the parent magmas of siliceous plutonic rocks is considered to be the most likely to account for the chemical, lead isotopic, and strontium isotopic variations observed in the Boulder batholith in Montana (50).

Gilluly (21) has estimated the North American continent has overridden a volume of between 100,000 to 250,000 km^3 or more of basaltic and more siliceous rock for every kilometer of its western leading edge in the last 200 m.y. Petrologic theory (22) accounts for some of this huge volume of overridden material by returning it to the upper crust in the great batholiths and volcanic accumulations in the Cordillera. A viable kinematic model supported by many different lines of evidence is available to account for the predominantly calc-alkalic plutonic activity during the Mesozoic and Cenozoic in western North America. However, the large mass of chemical and isotopic data for the plutonic rocks summarized in this report cannot account for any returned material from the overridden oceanic crust. What then has happened to the consumed oceanic lithosphere at the continental margin?

Literature Cited

1. Armstrong, R. L., Suppe, J. 1973. Potassium-argon geochronometry of Mesozoic igneous rocks in Nevada, Utah, and southern California. *Geol. Soc. Am. Bull.* 84:1375–92
2. Banks, P. O., Silver, L. T. 1968. U-Pb isotope analyses of zircons from Cretaceous plutons of the peninsular and transverse ranges, southern California (abstr.). *Geol. Soc. Am. Spec. Pap.* 121:17–18
3. Bateman, P. C., Dodge, F. C. W. 1970. Variations of major chemical constituents across the central Sierra Nevada batholith. *Geol. Soc. Am. Bull.* 81:409–20
4. Best, M. G. 1969. Differentiation of calc-alkaline magmas. *Oreg. Dep. Geol. Miner. Ind. Bull.* 65:65–76
5. Burchfiel, B. C., Davis, G. A. 1971. Clark mountain thrust complex in the Cordillera of southeastern California. *Geological Excursions in Southern California,* ed. W. A. Elders, 1–28. Univ. California, Riverside, Mus. Contr. No. 1. 182 pp.

6. Crowder, D. F., McKee, E. H., Ross, D. C., Krauskopf, K. B. 1973. Granitic rocks of the White Mountains area, California-Nevada: age and regional significance. *Geol. Soc. Am. Bull.* 84:285–96
7. Curtis, G. H., Evernden, J. F., Lipson, J. 1958. Age determinations of some granitic rocks in California by the potassium-argon method. *Calif. Div. Mines Geol. Spec. Rep.* 54:16 pp.
8. Damon, P. E., Mauger, R. L. 1966. Epeirogeny-orogeny viewed from the Basin and Range province. *Soc. Mining Eng. Trans.* 235:99–112
9. Dickinson, W. R. 1970. Relations of andesites, granites, and derivative sandstones to arc-trench tectonics. *Rev. Geophys. Space Phys.* 8:813–60
10. Dodge, F. C. W., Mays, R. E. 1972. Rare-earth fractionation in accessory minerals, central Sierra Nevada batholith. *US Geol. Surv. Prof. Pap.* 800-D:165–68
11. Doe, B. R., Delevaux, M. H. 1973. Variations in lead isotopic compositions in Mesozoic granitic rocks in California: a preliminary investigation. *Geol. Soc. Am. Bull.* 84:3513–26
12. Doe, B. R., Tilling, R. I., Hedge, C. E., Klepper, M. R. 1968. Lead and strontium isotopic studies of the Boulder batholith, southwestern Montana. *Econ. Geol.* 63:884–906
13. Early, T. O., Silver, L. T. 1973. Rb-Sr isotopic systematics in the Peninsular ranges batholith of southern and Baja California. (abstr.). *EOS* (*Trans. Am. Geophys. Union*) 54:494
14. Engels, J. C. 1971. Effects of sample purity on discordant mineral ages found in K-Ar dating. *J. Geol.* 79:609–16
15. Evernden, J. F., Kistler, R. W. 1970. Chronology of emplacement of Mesozoic batholithic complexes in California and western Nevada. *US Geol. Surv. Prof. Pap.* 623:42 pp.
16. Fairbairn, H. W., Hurley, P. M., Pinson, W. H. 1964. Initial Sr^{87}/Sr^{86} and possible sources of granitic rocks in southern British Columbia. *J. Geophys. Res.* 69:4893–99
17. Faure, G., Powell, J. L. 1972. *Strontium Isotope Geology*. New York: Springer. 188 pp.
18. Field, C. W., Bruce, W. R., Henricksen, T. A. 1972. Mesozoic plutonism and mineralization of the Snake River boundary area, Idaho-Oregon. *Geol. Soc. Am. Abstr.* 4, 7:503
19. Gabrielse, H., Reesor, J. E. 1964. Geochronology of plutonic rocks in two areas of the Canadian Cordillera. In *Geochronology in Canada,* ed. F. F. Osborne. *Roy. Soc. Can. Spec. Publ.* 8:96–128
20. Gilluly, J. 1965. Volcanism, tectonism, and plutonism in the western United States. *Geol. Soc. Am. Spec. Pap.* 80:69 pp.
21. Gilluly, J. 1973. Steady plate motion and episodic orogeny and magmatism. *Geol. Soc. Am. Bull.* 84:499–514
22. Green, D. H. 1972. Magmatic activity as the major process in the chemical evolution of the earth's crust and mantle. In *The Upper Mantle,* ed. A. R. Ritsema. *Tectonophysics* 13(1–4):47–71
23. Hart, S. R. 1964. The petrology and isotopic-mineral age relations of a contact zone in the Front Range, Colorado. *J. Geol.* 72:493–525
24. Hotz, P. E. 1971. Plutonic rocks of the Klamath Mountains, California and Oregon. *US Geol. Surv. Prof. Pap.* 684-B:B1–B20
25. Huffman, O. F. 1972. Lateral displacement of upper Miocene rocks and Neogene history of offset along the San Andreas fault in central California. *Geol. Soc. Am. Bull.* 83:2913–46
26. Hurley, P. M., Bateman, P. C., Fairbairn, H. W., Pinson, W. E. Jr. 1965. Investigations of initial Sr^{87}/Sr^{86} ratios in the Sierra Nevada plutonic province. *Geol. Soc. Am. Bull.* 76:165–74
27. Hutchison, W. W. 1970. Metamorphic framework and plutonic styles in the Prince Rupert region of the central coast mountains, British Columbia. *Can. J. Earth Sci.* 7:376–405
28. Jerome, S. E., Cook, D. R. 1967. Relation of some metal mining districts in the western United States to regional tectonic environments and igneous activity. *Nev. Bur. Mines Bull.* 69:35 pp.
29. Johnson, J. G. 1971. Timing and coordination of orogenic, epeirogenic, and eustatic events. *Geol. Soc. Am. Bull.* 82:3263–98
30. Kistler, R. W., Bateman, P. C., Brannock, W. W. 1965. Isotopic ages of minerals from granitic rocks of the central Sierra Nevada and Inyo Mountains, California. *Geol. Soc. Am. Bull.* 76:155–64
31. Kistler, R. W., Evernden, J. F., Shaw, H. R. 1971. The Sierra Nevada plutonic cycle: Part 1. The origin of composite batholiths. *Geol. Soc. Am. Bull.* 82:853–68.
32. Kistler, R. W., Peterman, Z. E. 1973. Variations in Sr, Rb, K, Na, and initial Sr^{87}/Sr^{86} in Mesozoic granitic rocks and

intruded wall rocks in central California. *Geol. Soc. Am. Bull.* 84:3489–3512
33. Lanphere, M. A., Reed, B. L. 1973. Timing of Mesozoic and Cenozoic plutonic events in circum-Pacific North America. *Geol. Soc. Am. Bull.* Vol. 84
34. Larsen, E. S. Jr. 1948. Batholith and associated rocks of the Corona, Elsinore, and San Luis Rey quadrangles, southern California. *Geol. Soc. Am. Mem.* 29:182 pp.
35. Lindgren, W. 1915. The igneous geology of the Cordilleras and its problems. *Problems of American Geology,* 234–86. New Haven, Conn.: Yale Univ. Silliman Found.
36. Lipman, P. W., Prostka, H. J., Christiansen, R. L., 1972. Cenozoic volcanism and plate-tectonic evolution of the western United States. I. Early and middle Cenozoic. *Phil. Trans. Roy. Soc. London Ser. A* 271:217–48
37. Marvin, R. F., Stern, T. W., Creasey, S. C., Mehnert, H. H. 1973. Radiometric ages of igneous rocks from Pima, Santa Cruz, and Cochise counties, southeastern Arizona. *US Geol. Surv. Bull.* 1379:27 pp.
38. McKee, E. H., Nash, D. B. 1967. Potassium-argon ages of granitic rocks in the Inyo batholith, east-central California. *Geol. Soc. Am. Bull.* 78:669–80
39. Moore, J. G. 1959. The quartz diorite boundary line in the western United States. *J. Geol.* 67:198–210
40. Peacock, M. A. 1931. Classification of igneous rock series. *J. Geol.* 39:54–67
41. Poole, F. G., Desborough, G. A. 1973. Alpine-type serpentinites in Nevada and their tectonic significance. *Geol. Soc. Am. Abstr.* 5(1):90–91
42. Reed, B. L., Lanphere, M. A. 1972. Generalized geologic map of the Alaska-Aleutian Range batholith showing potassium-argon ages of plutonic rocks. *US Geol. Surv. Misc. Field Studies Map* MF-372
43. Ross, D. C. 1972. Geologic map of the pre-Cenozoic basement rocks, Gabilan Range, Monterey and San Benito counties, California. *US Geol. Surv. Misc. Field Studies Map* MF-357
44. Ross, D. C. 1972. Petrographic and chemical reconnaissance of some granitic rocks and gneissic rocks near the San Andreas fault from Bodega Head to Cajon Pass, California. *US Geol. Surv. Prof. Pap.* 698:93 pp.
45. Shaw, H. R., Kistler, R. W., Evernden, J. F. 1971. The Sierra Nevada plutonic cycle: Part 2. Tidal energy and a hypothesis for orogenic-epeirogenic periodicities. *Geol. Soc. Am. Bull.* 82:869–96
46. Silberman, M. L., McKee, E. H. 1971. K-Ar ages of granitic plutons in north-central Nevada. *Isochron West* 1:15–32
47. Silver, L. T. 1971. Problems of crystalline rocks of the Transverse Ranges. *Geol. Soc. Am. Abstr.* 3:193–94
48. Smith, J. G., McKee, E. H., Tatlock, D. B., Marvin, R. F. 1971. Mesozoic granitic rocks in northwestern Nevada: a link between the Sierra Nevada and Idaho batholith. *Geol. Soc. Am. Bull.* 82:2933–44
49. Taubeneck, W. H. 1971. Idaho batholith and its southern extension. *Geol. Soc. Am. Bull.* 82:1899–1928
50. Tilling, R. I. 1973. The Boulder batholith, Montana: a product of two contemporaneous but chemically distinct magma series. *Geol. Soc. Am. Bull.* Vol. 84
51. White, W. H., Erickson, G. P., Northcote, K. F., Dirom, G. E., Harakal, J. E. 1967. Isotopic dating of the Guichon batholith, B. C. *Can. J. Earth Sci.* 4:677–90
52. Wollenberg, H. A., Smith, A. R. 1970. Radiogenic heat production in prebatholithic rocks of the central Sierra Nevada. *J. Geophys. Res.* 75:431–38

SATELLITES AND MAGNETOSPHERES OF THE OUTER PLANETS

D. A. Mendis and W. I. Axford
Department of Applied Physics and Information Science, University of California, San Diego, La Jolla, Caifornia 92037

1. INTRODUCTION

In this paper we review the known properties of the satellites of the outer planets, and attempt to provide explanations for some of the more striking features in terms of interaction with the planetary magnetospheres. It is pointed out that all the larger satellites, with the exception of Iapetus, are likely to lie within the magnetospheres of their respective planets at all times and that this may directly or indirectly influence the appearance of the satellites. In particular we pay attention to the possibility that the satellites and other solid bodies situated within the planetary magnetospheres (e.g. interplanetary dust grains) become charged to rather large electrical potentials. We suggest that this may give rise to a number of interesting phenomena, including the Io-associated decametric emissions, and the posteclipse brightening of Io.

In Sections 2 and 3 the observed properties of the satellites of the outer planets are summarized. It is noted that in almost every case the satellites show brightness asymmetries such that the leading face of the satellite in its orbit is brighter than the trailing face or vice versa. In the case of Iapetus, the outermost satellite of Saturn, the effect is most pronounced. In other cases the effect is much weaker and more difficult to observe, but its occurrence appears to be fairly general. The only case in which it is definitely absent is that of Titan which is also the only satellite of the outer planets which is definitely known to have an atmosphere. In the case of the Galilean satellites of Jupiter there is a possibly related effect in that the satellite albedos decrease with increasing distance from the planet. These effects are rather curious, and we believe that they deserve an explanation, although they may be quite superficial and provide no information about the nature of the interiors or origins of the satellites.

The properties of possible magnetospheres of the outer planets are discussed in Section 4. It is argued that there is no reason to expect that extensive magnetospheres with associated radiation belts do not exist, although it is very difficult to make any definite predictions concerning the strength of the magnetic fields involved and the fluxes of energetic particles trapped within the magnetospheres. The distribution of low energy plasma in the magnetospheres is discussed at some length, and it is concluded that even in the case of Jupiter the plasma density probably does not exceed ~ 50 protons cm^{-3}, and that the corresponding figures for Saturn and Uranus are at least a factor 10 smaller.

The nature of satellite atmospheres is discussed briefly in Section 5. It is concluded that except in the case of Titan where an atmosphere is known to exist, most of the satellites may essentially have no atmosphere at all as a result of magnetospheric erosion of any primeval atmosphere that may have existed in the past and the low vapor pressure of water ice which is likely to form the crusts of all the satellites. In the absence of any atmospheres, the magnetospheric plasma and energetic particles will interact directly with the satellites and this will lead to charging of the surface to electrical potentials which may be very large. It is impossible to predict the magnitude and sign of the potentials produced in this way without knowing the precise details of the energetic particle spectra and the density of the low energy magnetospheric plasma. However, if the integral fluxes of energetic electrons above, say, ~ 1 keV substantially exceed the corresponding proton integral fluxes, then the surface potential should be large and negative. It is pointed out in Section 7 that in these circumstances the satellite Io should emit a substantial flux of moderately energetic electrons moving parallel to the local magnetic field direction which could conceivably be the cause of the Io-associated Jovian decametric emissions. In contrast to previous suggestions concerning these emissions, this mechanism does not require that the satellite should have a high effective electrical conductivity, which is unlikely to be the case if it has even a very thin water ice crust. It is also pointed out that high surface potentials may play a role in determining the microstructure of the surface of satellites such as Io, thus providing an explanation for certain peculiar optical properties that have been observed. Furthermore, small grains can be levitated as a result of quite modest surface potentials and we suggest that this may be the cause of the posteclipse brightening of Io without having to invoke the presence of an atmosphere.

Finally in Section 8 we discuss various mechanisms for producing the regular brightness variations of the satellites of the outer planets. It is concluded that the most likely mechanism is a variant of that proposed by Cooke & Franklin (15) for the case of Iapetus, which involves the interaction of micrometeoroids with the satellite surface. However rather than treating the micrometeoroids merely as an eroding mechanism, we suggest that the smallest particles may cause a lowering of the albedo of any surface they cover, while the larger particles tend to restore the high albedo expected of an icy surface by continually overturning the upper layers. In order to explain the variations of brightness in the cases of satellites other than Iapetus, it is necessary once more to invoke charging by energetic particles as discussed in Section 6. We argue that the smallest micrometeoroids will have a sufficiently large charge to mass ratio that they are unable to penetrate very deeply into the magnetospheres of the planets, and hence the inner satellites are shielded from the effects of a directed stream of such particles. With a suitable, and we hope not unreasonable, combination of these processes it appears possible to explain the brightness variations of all the satellites in a straightforward way. Other processes are considered, but none of them appears to be satisfactory.

2. PROPERTIES OF THE SATELLITES OF THE OUTER PLANETS

The basic physical parameters of the regular satellites of Jupiter and Saturn as well as those of Triton are given in Table 1. It is seen that the mean radii and masses (and consequently the bulk densities) of the Galilean satellites of Jupiter are known with reasonable accuracy. In contrast the mean radii and masses of the Saturnian satellites (with the exception of Titan), as well as those of Triton, are known only very roughly. Although Triton seems to have an abnormally large density, the uncertainty in its radius makes it possible for it to have a density only one third as large as that quoted.

Lewis (64) has discussed the major reactions occurring during the cooling of a solar mix containing the most abundant ice-forming and rock-forming elements, assuming that chemical equilibrium is maintained between the condensate and the uncondensed gases. This assumption is probably valid during an early phase of slow accretion into planetesimals. By calculating the bulk density of the condensate

Table 1 Astronomical and physical data on outer planet satellites

Satellite	Distance from planet (planet radius)	Mean radius (km)	Mass (10^{24} g)	Average density (g cm^{-3})
Amalthea (J5)	2.6 R_J	85 ± 40	?	?
Io (J1)	6.0 R_J	1829 ± 3	71.4 ± 5.6	2.8 ± 0.2
Europa (J2)	9.6 R_J	1550 ± 75	47.1 ± 1.0	3.0 ± 0.4
Ganymede (J3)	16 R_J	2640 ± 20	155.0 ± 1.9	2.0 ± 0.1
Callisto (J4)	27 R_J	2500 ± 75	96.7 ± 7.4	1.5 ± 0.4
Janus (S10)	2.7 R_S	?	?	?
Mimas (S1)	3.2 R_S	235 ± 150	0.038 ± 0.001	~ 0.7
Enceladus (S2)	4.0 R_S	275 ± 150	0.085 ± 0.028	~ 1.0
Tethys (S3)	5.0 R_S	600 ± 100	0.648 ± 0.017	~ 0.7
Dione (S4)	6.3 R_S	~ 600	1.05 ± 0.03	~ 1.2
Rhea (S5)	8.8 R_S	725 ± 50	1.80 ± 2.21	~ 1.1
Titan (S6)	20 R_S	2425 ± 150	137 ± 1	2.3 ± 0.4
Hyperon (S7)	25 R_S	~ 200	?	?
Iapetus (S8)	60 R_S	850 ± 100	2.24 ± 0.74	~ 0.9 (Max: ~ 1.7)
Triton (N1-R)	14.3 R_N	1885 ± 650	135 ± 24	~ 4.8 (Min: ~ 1.6)

after each major condensation he has shown that when all the water has condensed the bulk density should be ~1.5 with lower values possible only with an unacceptably large concentration of CH_4. On the basis of these calculations it seems likely that all the regular Jovian and Saturnian satellites listed in Table 1 and possibly even Triton may have bulk densities in the range ~1.5 to ~3. Satellites having densities of ~1.5 should have a large component of water ice, whereas those having densities of ~3 possibly have, at most, only a thin crustal layer of water ice.

Lewis (62, 63) has also given compelling arguments supporting the view that the larger satellites of Jupiter and Saturn (i.e. those with radii greater than about 900 km) have been extensively melted and differentiated as a result of heat liberated by radioactive decay. Tidal heating, which amounts to 10–20 cal g^{-1} for the Galilean satellites, Titan and Triton (but only 1–2 cal g^{-1} for the smaller Saturnian satellites), could have provided a significant supplementary heat source for the larger satellites in the past. Accretional heating may also have been important, but is difficult to estimate. The likely structure of such low density satellites comprises a core composed of mud (i.e. nonvolatiles such as hydrous silicates and iron oxides, etc), a liquid mantle of aqueous ammonia solution, and a thin crust of nearly pure water ice (see Figure 1). The frequency independent, high brightness temperature of Callisto at microwave frequencies has been shown to be entirely consistent with such a model provided the thickness of the icy crust is greater than about 100 km (Kusmin & Losovsky, 61). Small amounts of ammonia hydrate may be present in the thin crust and in the case of the Saturnian satellites, at least, there may be some methane (probably in the form of the clathrate hydrate). Other

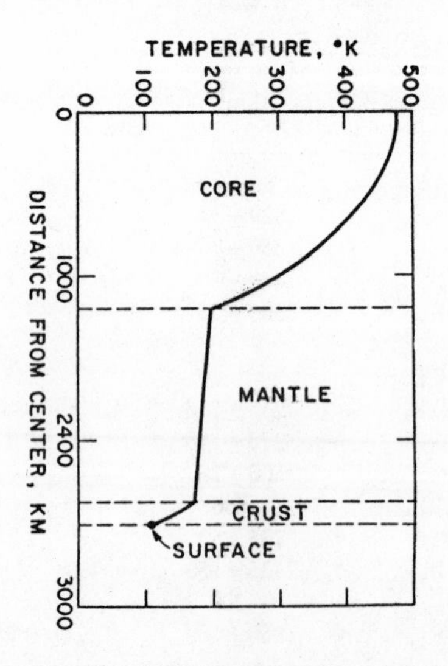

Figure 1 Approximate temperature profile in a Callisto-sized icy satellite (Lewis, 64).

carbonaceous material which would float on water should tend to be concentrated in the outermost surface layers, together with small quantities of nonvolatile dust of external or internal origin. Large meteoroid impacts will continually turn over and mix the surface material and possibly give rise to venting of volatile material (e.g. methane, argon, etc) which may have been retained in the mantle and the core.

The only satellite of the outer planets which is definitely known to possess a substantial atmosphere is Titan. Following the detection of two methane bands on Titan around 6190 and 7260 Å (Kuiper, 59), it was estimated to have an abundance of about 200 m-atm (Kuiper, 60). From the observed saturation of the CH_4 band at 1.1 μ, Trafton (112) has concluded that either the methane abundance in Titan's atmosphere is nearly an order of magnitude larger than the earlier estimate or else that methane is a minor constituent of the atmosphere. Trafton (113) has also reported a weak absorption feature near the 3-OS(1) quadrupole line of H_2. The assumption that this absorption is indeed due to H_2 implies an abundance of 5 km-atm for this gas. The abnormally low brightness temperature of Titan around 20 μ, as compared with lower infrared wavelengths has been interpreted as being due to high atmospheric opacity in this region (Morrison et al, 80; see also Sagan, 95; Sagan & Mullen, 96), resulting from pressure-induced absorption by H_2, since this is the only common gas known to absorb in the 18–25 μ region. The necessary abundance of H_2, however, is extremely large, being of the order of 100 km-atm. It has, however, been pointed out that this greenhouse effect is compatible with the lower abundance of H_2 (5 km-atm) provided the pressure broadening of the H_2 rotational line, responsible for the absorption, is caused by a heavier molecule like N_2 (Hunten, 47) or CH_4 (Lewis & Prinn, 65). This of course implies a rather large abundance of these latter species, being about 50 km-atm in the case of N_2 and 33 km-atm in the case of CH_4. Recently, Gillet et al (32) have suggested that a strong temperature inversion seen at about 10 μ may indicate the presence of another spectroscopically active component, possibly NH_3 which has an absorption band near 10.5 μ. Polarimetric observations to date (Veverka, 115, 117; Zellner, 130) seem to indicate the existence of an opaque cloud deck underlying an optically thin atmosphere.

All attempts to detect an Ionian atmosphere spectroscopically have failed so far. During the recent occultation of β-Scorpii by Io, an attempt was made to detect the effect of a tenuous atmosphere on the occultation curve. The lack of any detectable effect places an upper limit of about 2×10^{-7} atm on the atmospheric pressure (Bartholdi & Owen, 5).

The thermal inertias of the uppermost surface layers of the Galilean satellites have been calculated from recent eclipse observations at 10 μ and 20 μ (Hansen, 40; Morrison & Cruikshank, 79; see also Morrison et al, 81). The very low values obtained imply surfaces of high porosity and rule out atmospheres with surface pressures greater than $\sim$1 mbar. However photoelectric observations from two stations of a recent stellar occultation by Ganymede seem to indicate the presence of an atmosphere with a surface pressure of 10^{-3}–10^{-1} mbar (O'Leary, 86).

Excellent infrared spectra of moderate resolution with good signal to noise ratio have recently been obtained for all of the Galilean satellites (Fink et al, 27). The

Table 2 Satellite photometric data

Satellite	Mag_v	U-B	B-V	Δ Mag	r[a]	p_v[b]	Comments
Amalthea (J5)	13.0	?	?	See comment	?	?	Probably fainter near western elongation (single observation).
Io (J1)	4.80	1.30 (Var)	1.17 (Var)	0.21	0.83	0.73	Reddest of the Jovian satellites. Variation in color larger than variation in brightness: $\Delta(U\text{-}B) = 0.5$, reddest at western elongation. Possible posteclipse brightening. Lack of absorption features in infrared. Visible surface features but no polar caps. Opposition effects.
Europa (J2)	5.17	0.52	0.87	0.34	0.73	0.72	Very small variation in color with ϕ, definite H_2O frost absorptions in infrared. Visible surface features and polar caps. Opposition effects.
Ganymede (J3)	4.54	0.50	0.83	0.16	0.86	0.40	When $\alpha < 1.5°$ no observable brightness variation. Possible H_2O absorption feature in 3–4 μ. Visible surface features and polar caps. Opposition effects.
Callisto (J4)	5.50	0.55	0.86	0.16	1.16	0.21	When $\alpha < 1.5°$ no observable brightness variation. Possible H_2O absorption feature in 3–4 μ. Visible surface features and polar caps. Opposition effects.
Enceladus (S2)	11.77	?	0.62	~0.4*	~1.45	?	* Single observation.
Tethys (S3)	10.27	0.34	0.73	~0.3	~0.77	0.61	Very small variations in color with ϕ (except for Tethys). Reflection spectra in 0.4–1.1 μ consistent with those of H_2O and NH_3 frosts but not with those of common ferrosilicates. (applies to S3, S4, S5)
Dione (S4)	10.44	0.30	0.71	~0.4	~0.70	~0.5	
Rhea (S5)	9.76	0.35	0.76	~0.3	~0.77	0.57	
Titan (S6)	8.35	0.75	1.29	0	1	0.22	Very red object. Measurable atmosphere. CH_4 present, possibly H_2 also.
Iapetus (S8)	11.03	0.28	0.71	2.12	7.0	0.04 L 0.28 T	Very small variation in color despite large variation in brightness. Darker side slightly "redder." Reflection spectrum similar to those of S3, S4, and S5.
Triton (N1-R)	13.55	0.40	0.77	~0.25*	~0.8	?	Limited number of observations (uncertain).

[a] $r = B_T/B_L$ = brightness of the trailing face/brightness of the leading face.
[b] p_v = visual geometric albedo.

analysis of these has set an upper limit of 6×10^{-5} mbar for the surface partial pressure of CH_4 and NH_3 on all four satellites. Although this result sets a very stringent upper limit for the NH_3 and CH_4 abundances in the atmospheres of the Galilean satellites, the presence of some other constituent (possibly a heavy inert gas such as radiogenic argon) cannot be excluded (see Lewis, 62).

Theoretical considerations tend to suggest that it is possible that both Io and Ganymede could have retained substantial atmospheres of CH_4 and N_2 (Binder & Cruikshank, 9). However, the vapor pressures of these constituents at the satellite temperatures may be grossly overestimated. Any CH_4 present is almost certainly locked in a clathrate and consequently its vapor pressure would be very similar to that of H_2O (i.e. much smaller than that of CH_4 ice at the same temperature). Nitrogen is also likely to be locked up in less volatile compounds such as NH_3. Radiogenic argon produced by K^{40} decay in the interior could be an important atmospheric constituent, however, it too may be contained in the crust in clathrate form. We suggest that it is in fact very difficult for a satellite to retain a thin atmosphere for long in the face of photo-ionization and other ionizing agencies, with subsequent erosion of ions by the magnetosphere. In the case of Titan the atmosphere appears to be so thick it is able to protect itself to some extent against such losses by the creation of a magnetic barrier in the upper atmosphere as seems to occur in the cases of Venus and Mars.

The photometric data obtained to date on the satellites under discussion are summarized in Table 2. They appear to be in reasonable accord with Lewis' conclusions with regard to the likely surface compositions. The reflection spectra of the Saturnian satellites Iapetus, Rhea, Dione, and Tethys are available in the spectral range 0.3–1.1 μ (McCord et al, 68). They all show very flat spectra in the range 0.4–0.8 μ, and, with the exception of Iapetus, a slight decrease in reflectivity towards the red (the effect being most marked in the case of Dione where the decrease appears almost as a broad absorption band). These spectra are consistent with those of water and ammonia frosts but not with those of common ferro-silicates.

Water frost absorption features have now been definitely detected in the infrared reflection spectra of the Galilean satellites Europa and Ganymede, and possibly also of Callisto (Fink et al, 27; see also Pilcher et al, 91). The excellent infrared spectra obtained by Fink et al (see Figure 2) leave little doubt of the existence of water frost on Ganymede and Europa. They show broad characteristic water frost absorption features in the ranges 1.5–2 μ, 2–2.5 μ, and 3–4 μ. In fact, in the case of Ganymede the almost total absorption between 3 and 4 μ indicates that water ice in some form may be distributed over the entire surface, for otherwise the uncovered region is likely to reflect light in this band.

Surprisingly the reflection spectrum of Callisto shows very little evidence of surface ice, despite its low density. There is, however, some absorption in the 3–4 μ range and less certainly at the other absorption wavelengths. This could be consistent with most of the surface being covered with some type of dust or other material of low reflectivity, but with a small amount of water ice showing through. The lack of water frost absorption features in the infrared spectrum of Io does not

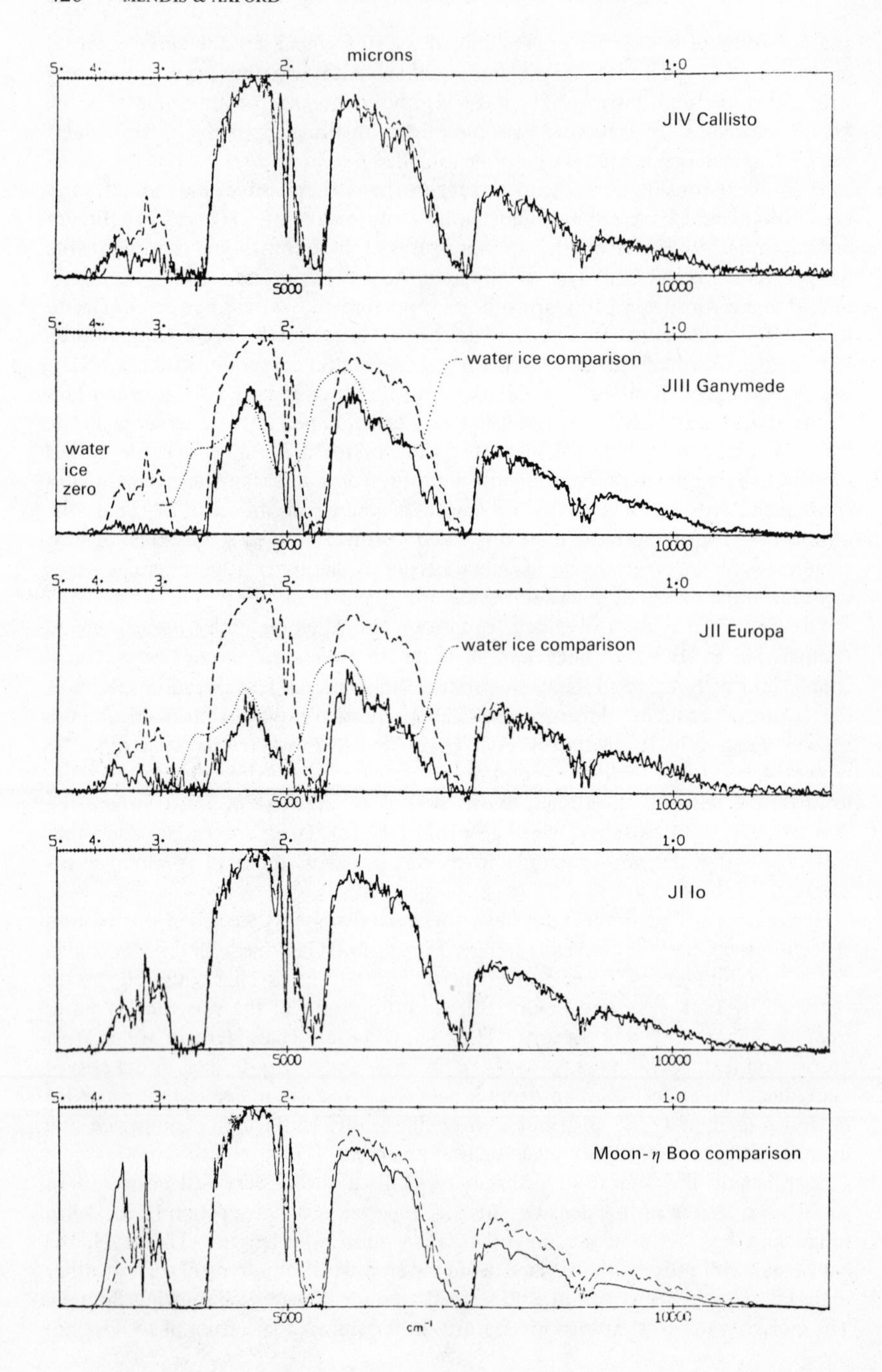
microns
5· 4· 3· 2· 1·0
JIV Callisto
5000
10000
water ice comparison
JIII Ganymede
water
ice
zero
JII Europa
JI Io
Moon-η Boo comparison
cm⁻¹

entirely rule out its presence on the surface of Io, for laboratory studies of the infrared spectral reflections of water frosts have shown that the strengths of the absorption features depend critically on their texture. If the frost is extremely fine grained, the absorption features may be completely obliterated (Kieffer, 54). However, if this is the reason for the lack of water frost absorption features in the infrared spectrum of Io it is necessary to understand its cause.

The surface material of all the Galilean satellites appears to be much more absorbing in the blue and near violet than ordinary water frost, the effect being most marked for Io and least for Ganymede (Johnson, 50; Johnson & McCord, 52). In contrast to the other Galilean satellites, Io also exhibits a spectral feature between 0.5 μ and 0.6 μ in the form of a broad depression. It also shows a dip around 0.8 μ, a feature it shares with Callisto (Johnson & McCord, 51).

Polarization measurements of the Galilean satellites have been interpreted to indicate that the surfaces of Io, Ganymede, and Europa are covered mostly by bright multiple-scattering material, possibly frost, whereas that of Callisto is covered by particulate material sufficiently dark to inhibit multiple scattering (Veverka, 116). It has also been pointed out (Veverka, 116) that the strong opposition effects seen in the phase curves of all four Galilean satellites (Stebbins, 108; Stebbins & Jacobson, 109) indicate surfaces that are "microscopically rough and probably particulate." Observations of the thermal response of satellites to changing insolation during eclipse provide a means of determining the thermophysical properties of their uppermost surface layers. Such observations of all four Galilean satellites at 20 μ (Morrison & Cruikshank, 79; see also Morrison et al, 81) and of Io, Europa, and Ganymede at 10 μ (Hansen, 40), seem to indicate a two-layer surface with a thin (i.e. a few millimeters) upper layer of low thermal conductivity, consistent with rock powder or frost, and an immediate subsurface of high thermal conductivity, consistent with solid rock or dense ice (see Figures 3a and 3b). Morrison & Cruikshank (79) have further pointed out that the upper several tens of centimeters of the satellites are likely to have been extensively fragmented or gardened by repeated meteoritic impacts. Consequently, unless the subsurface rocks can heal or resolidify after fragmentation their thermal conductivity will not show the discontinuity a few millimeters below the surface as indicated by the above observations, and this seems to argue against rock and in favor of ice. Laboratory studies indicate that rock melting upon impact is not great. Whereas in the case of ice, not only will there be a process of slow fusion and crystallization of the fragments, but the melted ice will immediately refreeze thus healing some of the fragmentation, and any vapor which is produced would quickly condense, presumably forming a frosty surface layer.

←

Figure 2 Spectra of the four Galilean satellites with a resolution of 25 cm^{-1} and average telluric air mass of 1.9. A moon comparison (air mass 1.94) is shown for the narrow features. A solar-type star (η Boo, air mass 1.13) is shown as a dashed line for the general albedo characteristics. A laboratory water-frost reflection spectrum is superposed on the spectra of Jupiter II and Jupiter III as a dotted line (Fink et al, 27).

It seems significant that the visual geometrical albedos of the Galilean satellites decrease systematically with distance from the planet (see Table 2). In the case of the Saturnian satellites the albedos are not known with sufficient precision at the present time to draw any definite conclusions of this sort. However, a new technique which combines infrared radiometry with visual photometry has yielded the radii and albedos of Iapetus and Rhea (Murphy et al, 83) and Dione (Morrison, 78) directly. These estimates are much more reliable than the earlier ones, where the large uncertainties in radii led to even larger uncertainties in the albedos. This was particularly true for Dione which until very recently seemed to have abnormally large values both for its albedo ($p_v \approx 1.12$) and density ($\rho \approx 3.6$), as a result of a large underestimate of its radius.

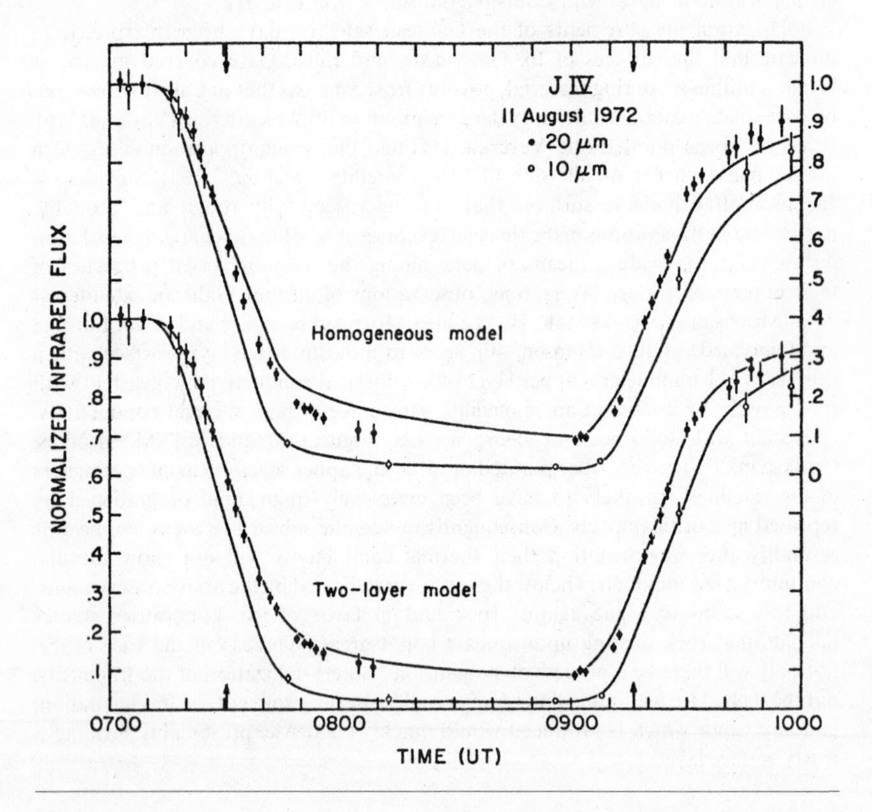

Figure 3a Observations of the eclipse of Callisto of 11 August 1972. Filled circles are 20 μ points; open circles are at 10 μ. The arrows mark the photometric half-intensity times. The upper part of the figure illustrates the best fitting homogeneous model, characterized by a thermal inertia $(Kpc)^{1/2} = 1.2 \times 10^4$ erg cm^{-2} sec$^{-1/2}$ K^{-1}. The lower part illustrates the best-fitting two-layer model, characterized by a thermal inertia of 1.0×10^4 and a surface density of 0.12 g cm^{-2} (Morrison & Cruikshank, 79).

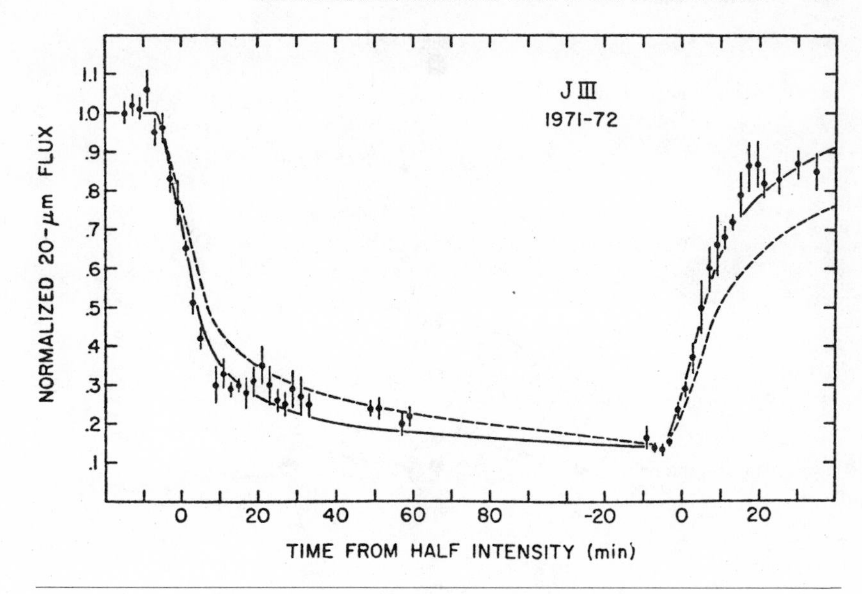

Figure 3b Average of the 20 μ observations of three eclipses of Ganymede. The data have been aligned to correspond to photometric half-intensity at time = 0. The dashed curve illustrates the best-fitting homogeneous model, with thermal inertia 2.0×10^4. The solid curve illustrates the best-fitting two-layer model, with upper layer characterized by thermal inertia 1.4×10^4 and surface density 0.15 g cm^{-2} (Morrison & Cruikshank, 79).

Finally, it should be noted that visual observations of the Galilean satellites show distinct surface features with dark areas interspersed with light areas (see Antoniadi, 2; Dollfus, 19; Sagan, 94). Europa, Ganymede, and Callisto also seem to show the existence of polar caps, while Io does not. In fact, in the case of Io, the polar regions seem to be distinctly darker than the central regions.

3. SATELLITE BRIGHTNESS VARIATIONS

The Galilean satellites of Jupiter and all but one of the larger satellites of Saturn show regular changes of brightness which are associated with their orbital phase (e.g. Harris, 42; Newburn & Gulkis, 85). This suggests that the satellites are in a state of synchronous rotation, which in the case of the Galilean satellites is confirmed by observations of surface markings (e.g. Lyot, 67). The most curious feature of these variations is that the times of maximum and minimum brightness occur when the satellites are at either eastern or western elongation, as indicated in Figure 4 for the cases of Rhea and Iapetus.

The three inner Galilean satellites (Io, Europa, and Ganymede) and three of the four large inner Saturnian satellites (Tethys, Dione, and Rhea) are 10–30%

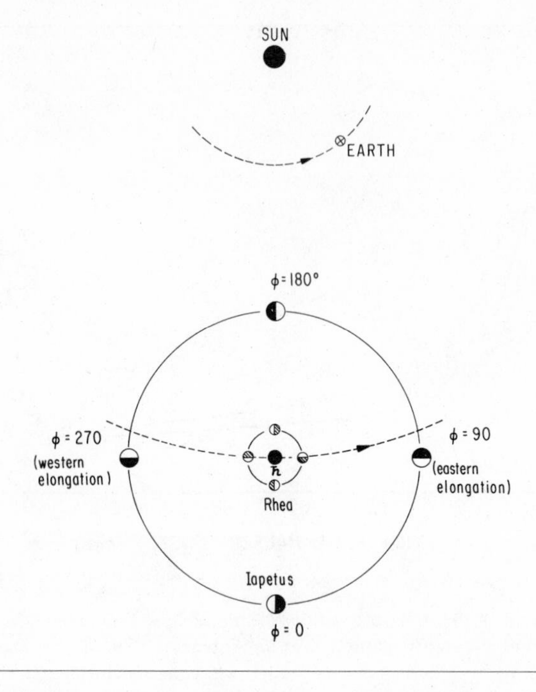

Figure 4 The motion of Iapetus and Rhea around Saturn, as observed from the earth.

brighter at eastern elongation than at western elongation (see Figures 5a, 5b, 5c, and 6a). Amalthea, the small innermost satellite of Jupiter, may also show this effect but the observation (there is only one to date) is made difficult by the close proximity of the very faint satellite to the bright planet (Van Biesbroeck & Kuiper, 114). These observations, taken together with the assumption of synchronous rotation, imply that the leading faces of these satellites reflect more light than do the trailing faces.

In contrast, Callisto, the outermost Galilean satellite, and Iapetus, the outermost regular satellite of Saturn, show maximum brightness at western elongation (see Figures 5d and 6b). While Callisto's variation is only about 0.16 mag, Iapetus shows a remarkably large variation of 2.12 mag, which means that its trailing face reflects about 7 times as much light as its leading face. Indeed, when Cassini first discovered Iapetus in 1671, he was able to observe the satellite only when it was near western elongation. Franz & Millis (29) have recently reported that Enceladus, one of the small inner satellites of Saturn, also shows brightness variations with orbital phase. Due to its proximity to the rings it could only be observed near maximum elongation. It was found to be $\sim$0.4 mag brighter at western elongation than at eastern elongation (Figure 6c), thereby indicating (as is the case of Callisto and Iapetus) that its trailing face is brighter.

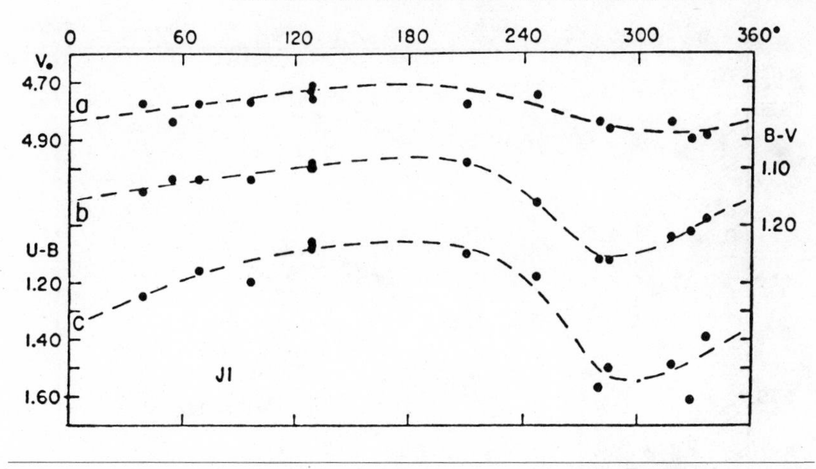

Figure 5a Variation of Io (Jupiter I) with rotational phase angle: (*a*) In yellow light, V_0; (*b*) in the *B-V* color; (*c*) in the *U-B* color; McDonald observations (Harris, 42).

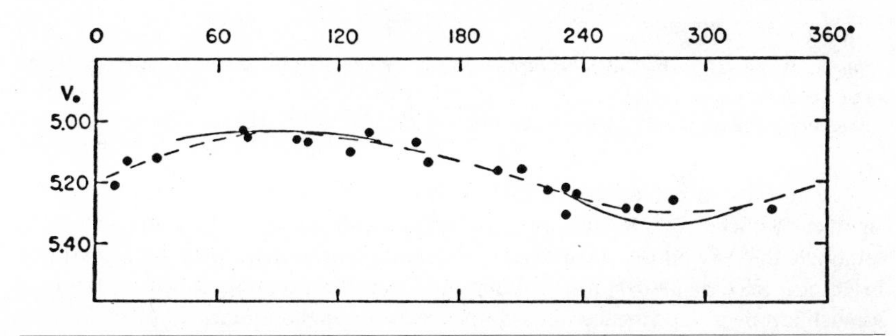

Figure 5b Variation of Europa (Jupiter II) in yellow light, V_0, with rotational phase angle. Full-drawn curves: Stebbins & Jacobsen (109); dashes and heavy dots: McDonald observations (Harris, 42).

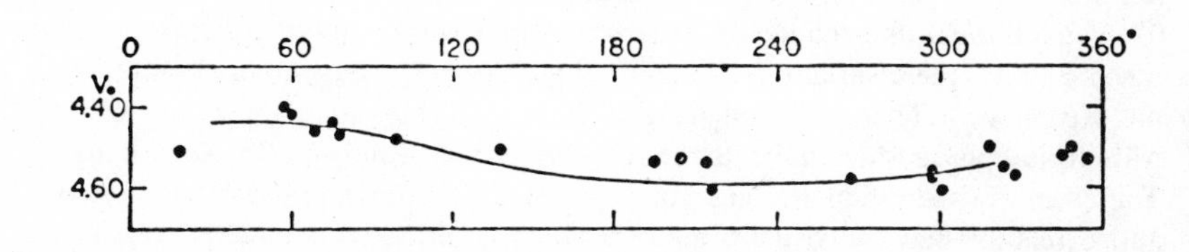

Figure 5c Variation of Ganymede (Jupiter III) in yellow light, V_0, with rotational phase angle. Curve: Stebbins & Jacobsen (109); dots: McDonald observations (Harris, 42).

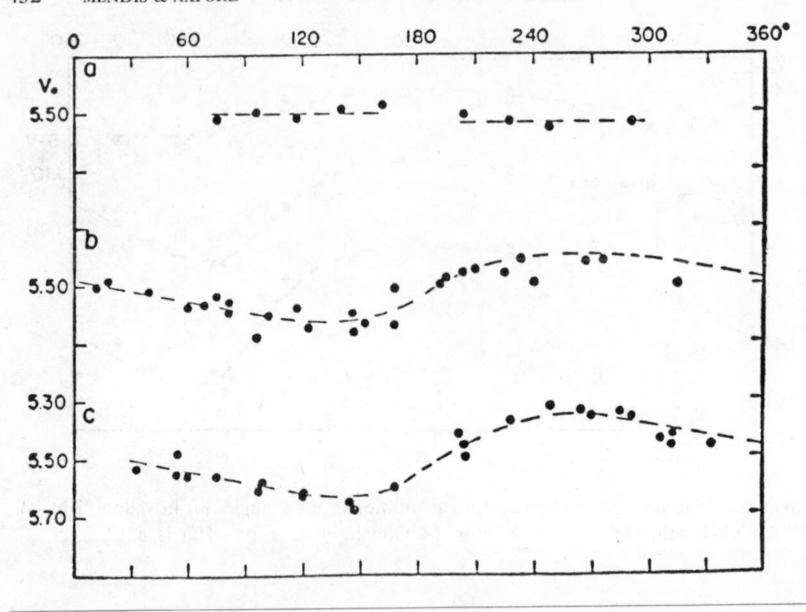

Figure 5d Variation of Callisto (Jupiter IV) in yellow light, V_0, with rotational phase angle. (*a*) Solar phase angle less than 1.5°; (*b*) solar phase angle between 2.7° and 8°; (*c*) solar phase angle greater than 8°. Data by Stebbins & Jacobsen (109) (Harris, 42).

Recent *UBV* observations of Iapetus (Millis, 77) appear to have revealed a significant difference (~0.3 mag) in the depths of two successive brightness minima, although there is hardly a noticeable difference in the heights of the successive brightness maxima (~0.04 mag). While this effect has been interpreted in terms of a much stronger dependence on solar phase angle of the brightness of the darker face as opposed to that of the brighter face (Millis, 77), the limited number of observations (there are only two near the first minimum) require that such an explanation be taken with some reservation at the present time.

Together with the brightness variations, all four Galilean satellites also show color variations with orbital phase in the sense that makes the darker sides redder (Harris, 42; Johnson, 50). The effect is largest for Io where $\Delta(U\text{-}B)$ amounts to about 0.5 mag, although the change in visual brightness is only about 0.2 mag. For the other satellites these variations are quite small. The larger Saturnian satellites (with the exception of Titan and Tethys) also show small but detectable color changes with orbital phase which indicate that the darker face is also redder (see Figure 7). The effect is greatest in the case of Iapetus. This satellite also shows a strong anticorrelation between visual brightness and the infrared flux at 20 μ (Murphy et al, 83).

In the case of the small Jovian and Saturnian satellites similar effects may be present but they are unobservable from the earth as a result of their faintness. Titan, the largest satellite of Saturn, does not show any brightness variation with orbital phase (see Figure 6a), but it is significant that it is the only satellite of Jupiter and Saturn which is definitely known to have an atmosphere (Kuiper, 59; Trafton, 112, 113; Hunten, 47; Lewis & Prinn, 65).

Triton, the retrograde satellite of Neptune, has been reported to be significantly brighter at western elongation than at eastern elongation, which due to its retrograde rotation implies that, like the inner Jovian and Saturnian satellites, its leading face is brighter than its trailing face (Harris, 42). However, due to the limited number

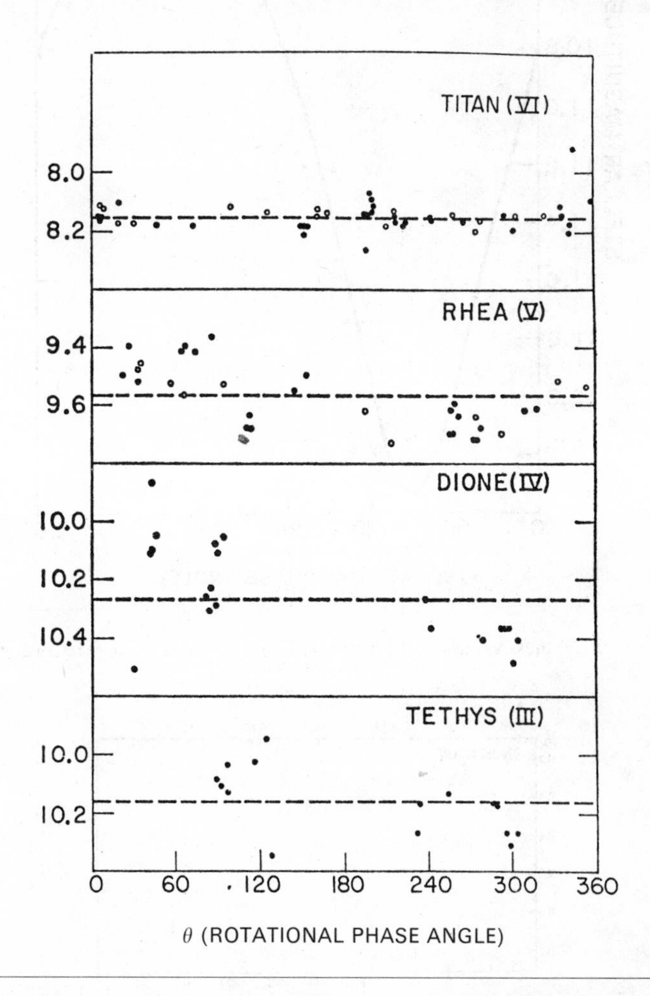

Figure 6a Brightness variations of Tethys, Dione, Rhea, and Titan with orbital phase angle (McCord et al, 68).

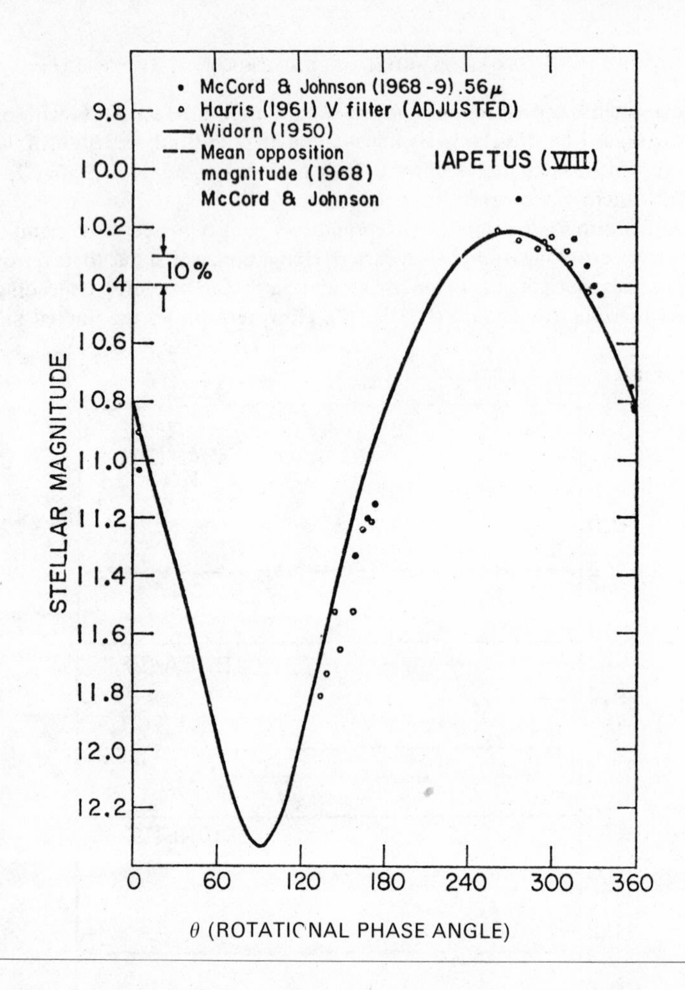

Figure 6b Brightness variation of Iapetus with orbital phase angle (McCord et al, 68).

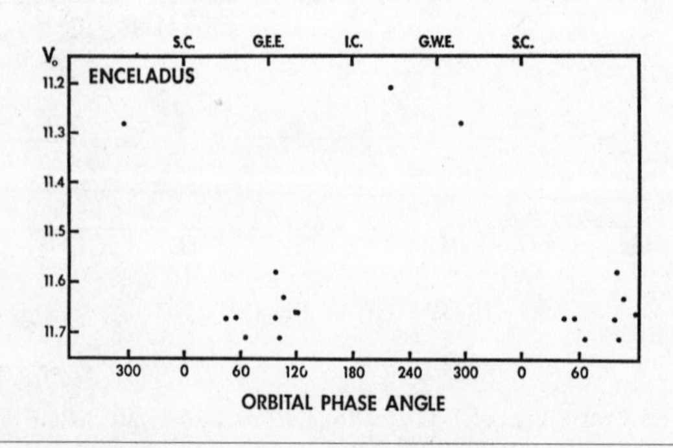

Figure 6c Brightness variation of Enceladus with orbital phase angle (Franz & Millis, 29).

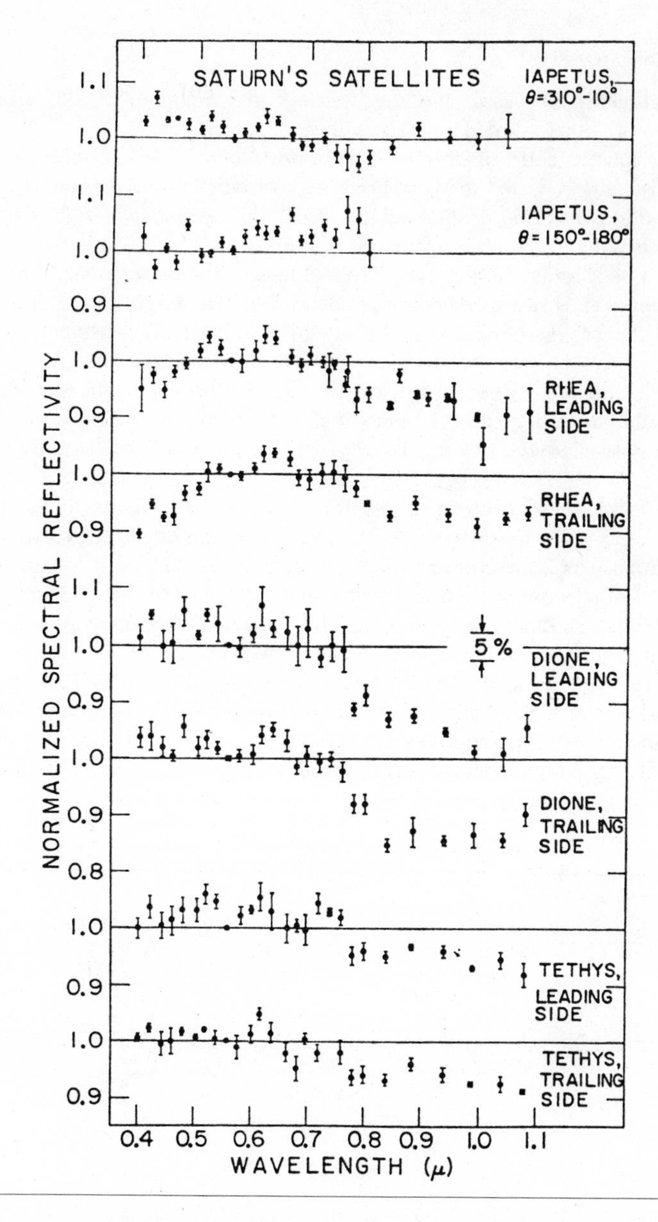

Figure 7 Reflectivities as a function of wavelength of four of Saturn's satellites, scaled to unity at 0.56 μ. Trailing- and leading-side spectra are shown separately. Error bars represent the standard deviation of the average of several measurements. Points with no error bars are from a single measurement. Satellite orbital phase angle is defined in the prograde sense with 0° at superior geocentric conjunction, i.e. on the earth-planet line behind the planet. The leading side is that facing the observer when the satellite is moving in its orbit from behind Saturn to in front of Saturn as viewed from the earth (McCord et al, 68).

of observations in this case, and the difficulty in making them, this apparent variation in brightness should be taken with some reservation.

The outermost satellites of Uranus—Titania and Oberon—and possibly the inner satellite Ariel, are reported to show brightness variations (Steavenson, 106, 107). These variations have been observed not only when the line of sight from the earth is essentially in the plane of the satellite orbits, but also when it is virtually normal to it, suggesting that the axes of spin of these satellites are highly inclined to their rotation axes. However, there is no evidence that these brightness variations are related to the orbital positions of the satellites and a more extended series of observations would be required to determine if there is any periodicity.

In addition to these periodic brightness changes, there are other phenomena which should be considered in the present context. Binder & Cruikshank (9) have reported a posteclipse brightening of Io by ~ 0.1 mag at visible wavelengths, which decays in ~ 15 min. This has been confirmed by O'Leary & Veverka (87), Johnson (50), and Cruikshank & Murphy (16) (see Figures 8a and 8b). Johnson (50), in fact, claims the effect to be much larger (~ 0.5 mag). However, due to the scattering by Jupiter's atmosphere, these observations are difficult to make and indeed careful attempts by other observers have drawn negative results (Fallon & Murphy, 25; Franz & Millis, 28). Binder & Cruikshank (9) interpreted their observation in terms of an Ionian atmosphere which condenses at least partially, during eclipse. Lewis (62) has argued that the condensible gas that could be responsible is NH_3 or possibly an inert gas, but not CH_4. He has calculated that the amount of gas required would correspond to an atmospheric pressure of about 2×10^{-7} atm. Fallon & Murphy (25) have attempted to reconcile both the positive and the negative observations of the posteclipse brightening phenomenon by suggesting that it is an effect associated

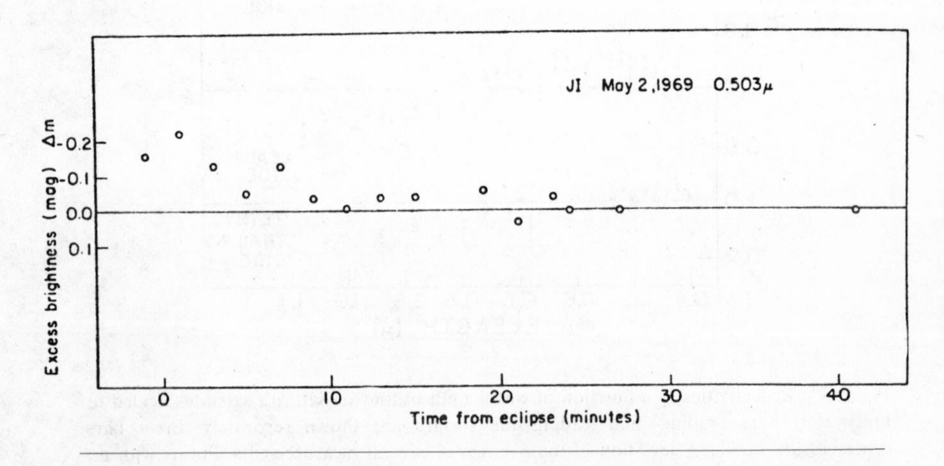

Figure 8a The brightness of Io during eclipse reappearance (Veverka: Harvard 16-in.). In all figures, the zero time is that of the nominal eclipse reappearance as given in the American Ephemeris and Nautical Almanac (O'Leary & Veverka, 87).

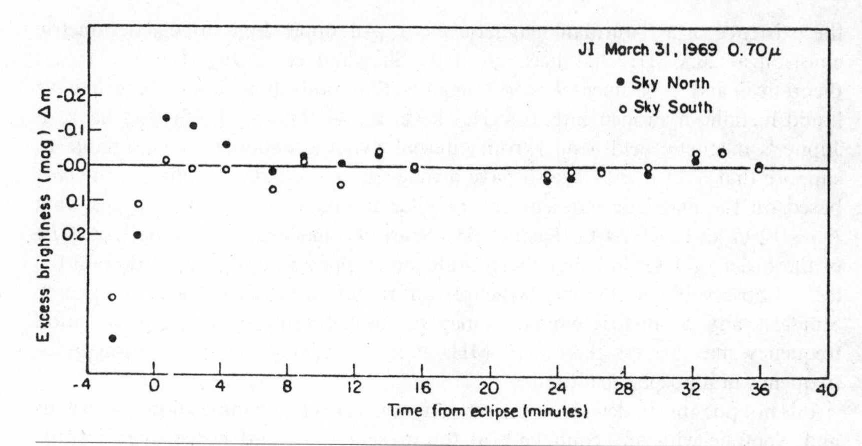

Figure 8b The brightness of Io during eclipse reappearance (O'Leary: Kitt Peak 16-in.), with respect to sky North and South of Io (O'Leary & Veverka, 87).

with a temporary atmosphere caused by sporadic outgassing. It should be noted, however, that there are other possible explanations of this phenomenon which do not require such a substantial atmosphere (see Section 6).

In this review we suggest that it may be possible to account for these brightness changes of the Jovian and Saturnian satellites and also that of Triton (if it is indeed real) by making reasonable assumptions concerning the magnetospheres of the outer planets and the thermal plasma and energetic particles contained within them. We conclude that the very high electric potentials which are likely to be attained by solid bodies within the magnetospheres may play an important role in producing these effects, as well as certain other observed phenomena.

4. THE MAGNETOSPHERES OF THE OUTER PLANETS

There is no doubt that Jupiter has an extensive magnetosphere and that the intensity of energetic trapped particles is very large, especially in the inner regions (e.g. Beck, 7; Newburn & Gulkis, 85; Warwick, 120, 122). Radio-frequency observations in the decimetric and decametric ranges suggest that the magnetosphere and the energetic trapped particles extend to at least the orbit of Io. It has been deduced from polarization measurements of the decimetric emissions that the magnetic field is predominantly dipolar with an equatorial magnetic field strength $B_{0J} \approx 10$–50 G, and with the magnetic moment and angular momentum vectors almost parallel (Berge, 8; Komesaroff et al, 58). Earlier arguments based on observations of the decametric emissions suggest that $B_{0J} \approx 10$ G and that the dipole has the direction described, but is slightly tilted and perhaps substantially offset (Warwick, 119–121).

Apart from some unconfirmed reports of observations of sporadic nonthermal radio emissions in the decametric range (Braude, 11), there is no direct evidence for the existence of a Saturnian magnetosphere. (An upper limit to the decametric emission at 26.3 MHz has been given by Shawhan et al, 99.) However, recent theoretical and experimental work suggests that both Jupiter and Saturn have liquid metallic hydrogen interiors (Hawke et al, 44; Hubbard, 45), and hence if Jupiter's magnetic field results from internal dynamo action it is reasonable to suppose that Saturn should also have a magnetic field. Simple scaling arguments based on the magnetic moment and angular moment of the earth suggest that $B_{0J} \approx 12$ G and $B_{0S} \approx 4$ G (Kennel, 53). Scarf (97) has argued that B_{0S} could be of the order of 1 G and that there could be trapped radiation belts despite the lack of observable nonthermal (synchrotron) radiation associated with the planet. Similarly any decametric emissions may be unobservable since the upper cutoff frequency may be less than ~ 10 MHz if it corresponds to the electron gyrofrequency at ionospheric altitudes.

It is not possible to determine the internal structures and compositions of Uranus and Neptune with any confidence at the present time, and hence there is little basis for deciding on theoretical grounds whether or not these planets have magnetic fields. The scaling arguments suggest that the equatorial magnetic field strengths are $B_{0U} \approx 1.25$ G and $B_{0N} \approx 0.9$ G. On the assumption that the magnetic dipole and rotation axis are roughly aligned, one might expect that the magnetosphere of Uranus has some rather unusual features, since its axis of rotation lies close to the ecliptic plane with one pole pointing more-or-less towards the sun at the present time (Siscoe, 101); however, this should not affect our discussion significantly.

A rough value for the minimum distance to the magnetopause (R_0) can be obtained in the usual way by equating the solar wind ram pressure to the magnetic pressure, allowing for a factor 2 in the magnetic field strength to take into account

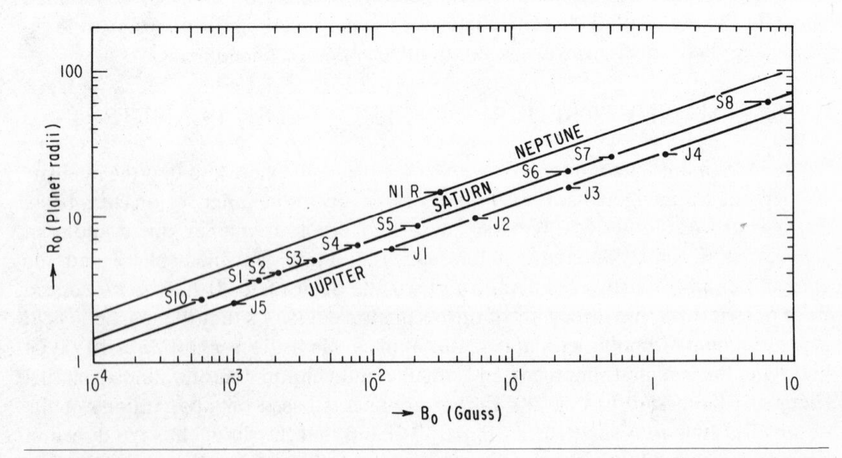

Figure 9 Minimum distance of the magnetopause (in planetary radii) versus equatorial surface magnetic field for Jupiter, Saturn, and Neptune.

the confinement of the field. The values of R_0 corresponding to a range of values of B_0 are shown in Figure 9 for the case of Jupiter, Saturn, and Neptune. It is evident on the basis of the above estimates of the equatorial magnetic field strengths that the regular satellites of Jupiter, and also Triton, and all the Saturnian satellites with the exception of Iapetus and Phoebe, might be expected to be situated within the magnetospheres of their parent planets at all times. [This is also the case for the five satellites of Uranus, although it is not shown in the figure.] It is important to note that all satellites which show bright leading faces lie well within their planetary magnetospheres; this remains true even if the assumed equatorial magnetic field strengths are overestimated by a large factor. In contrast, the satellites which most clearly show bright trailing faces are expected either to be situated outside the magnetosphere most of the time (as in the case of Iapetus), or to lie in the outer part of the magnetosphere at all times (as in the case of Callisto). Enceladus, however, appears to be anomalous in this respect.

Despite much effort on the part of theorists (e.g. Beck, 7), we are quite ignorant of the nature of the Jovian trapped radiation. It is usually assumed that the magnetosphere is filled with energetic particles of solar wind origin which are transported inwards from the magnetopause by processes involving convection and radial diffusion. It is presumed that the particles are eventually lost as a result of absorption by satellites, precipitation into the atmosphere of the planet, and (in the case of electrons) by synchrotron radiation (e.g. Beard, 6; Brice & McDonough, 14; Luthey, 66; Mead & Hess, 73; Neubauer, 84; Thorne & Coroniti, 111). In the absence of losses, the characteristic energy (E^*) of particles on field lines crossing the equatorial plane at a distance L planetary radii would be

$$E^* = E_0(R_0/R_J\,L)^{4(\gamma-1)} \qquad 1.$$

where E_0 is the mean energy of solar wind particles in the magnetosheath (or perhaps in the plasmasheet), γ is the adiabatic index, and it is assumed that the pitch angle distribution remains nearly isotropic. Similarly, the flux would vary approximately as $L^{-16/3}$ if $\gamma = 5/3$. In fact losses should modify these simple results significantly, but in the absence of any knowledge of the rate of radial transport and of the loss processes it is difficult to make any quantitative estimate of the characteristics of the particle distributions. It has been shown, however, that by using a form for the radial diffusion coefficient appropriate to the case where diffusion is caused by random electric fields induced at ionospheric levels (Brice, 12), the distribution of electrons required to produce the observed decimetric radiation can be accounted for (Brice & Ioannidis, 13).

It should be noted that protons drift eastwards and electrons drift westwards in the Jovian magnetosphere, but as a consequence of the rapid rotation of the planet the net drift is eastwards for all particles other than very high energy electrons. Thus energetic particles tend to be absorbed on the trailing faces of the satellites, and if the relative motion between any particular group of particles and the satellites is sufficiently slow they may be totally lost in this way. It might seem that electrons are particularly prone to such absorption losses since not only do they have higher bounce frequencies than protons with the same energy, but their longitudinal drift carries them past the satellites more slowly. As we will see, however, there is some compensation for this in that the satellites without atmospheres tend to be negatively charged, so that protons and other nuclei are efficiently absorbed, whereas electrons with insufficient energy to overcome the potential barrier are reflected. Those electrons with sufficiently high energy to reach the satellite surface are in part replaced by freshly emitted electrons with energies determined by the total potential drop involved. The overall result of satellite absorption of energetic particles is to remove protons and electrons at equal rates provided effects due to absorption of low energy plasma are negligible. The energy spectra of the remaining populations are distorted during this process, and it is not completely obvious that the mean energy of the electrons should decrease more rapidly than that of the protons.

Kennel (53) has pointed out that the limiting flux for stable trapping, assuming that electromagnetic waves induced by loss-cone instabilities cause precipitation of particles, is roughly the same (viz. $3\text{–}10 \times 10^{10}/L^4$ cm^{-2} sec^{-1}) for all the outer planets and the earth assuming the scaling of the magnetic field is valid. However this flux refers only to particles with energies exceeding a resonant energy (E_{res}) which is roughly $B_0^2/8\pi N_{eq} L^6$, where N_{eq} is the total electron density in the equatorial plane, which is of course also unknown. Thus there may be very intense fluxes of stably trapped particles with energies less than the resonant value, which can be large if N_{eq} is small.

There is no reason to believe that the magnetospheres of Saturn, Uranus, and Neptune do not contain energetic particle fluxes comparable to those existing in the magnetospheres of the earth and even Jupiter, except of course in the vicinity of Saturn's rings where rapid absorption would occur. We will proceed on the assumption that this is the case, and that the relationship in equation 1 is approximately correct, at least in the outer parts of the magnetospheres. It is necessary to await the results which will be obtained from the Pioneer 10 fly-by of Jupiter in December 1973, and of fly-bys of Saturn planned for the early 1980s, before we can expect to have any definite information concerning these points.

In addition to energetic particles, any magnetosphere should contain thermal plasma which originates in the ionosphere and accumulates in regions which are not strongly convective or unstable (e.g. Brice & Ioannidis, 13; Ioannidis & Brice, 48). Except in regions where externally or internally induced convective motions are dominant, the plasma must to a first approximation corotate with the planet. Hence, since the magnetospheres of the giant planets lie mostly outside the synchronous orbit, centrifugal effects are very important and consequently in diffusive equilibrium the thermal plasma should tend to be concentrated towards the equatorial plane (Ellis, 24; Gledhill, 33; Melrose, 74). It is not easy to calculate the distribution of thermal plasma in these circumstances, because it depends on a rather complicated manner on the structure of the ionosphere which is itself not well understood, especially at low and high altitudes (Hunten, 46; Owen & Westphal, 88). The electrical conductivity of the ionosphere determines the rate at which magnetospheric interchange motions can occur, and the high altitude thermal structure (including the photoelectrons) determines the flux of plasma into the magnetosphere. Furthermore the density of thermal plasma near the equatorial plane plays a very important role in driving rotational and interchange instabilities, and since this is in turn determined by the sources and sinks of plasma, as well as by the heating and cooling mechanisms, it is very difficult to construct a self-consistent model.

The first model for the distribution of thermal plasma in the Jovian magnetosphere was that of Gledhill (33), who assumed that the plasma is isothermal with a temperature equal to that of the ionosphere at high altitudes, and that convection (and therefore the interchange instability) does not occur. By invoking the rotational instability as a means of limiting the plasma density at the equatorial plane to a value given by

$$N_{eq} = B_{0J}^2/(4\pi m_p \Omega^2 R_J^2 L^8) \qquad 2.$$

one can easily determine (see Appendix A) the density distribution in the magnetosphere for any assumed plasma temperature as shown in Figure 10a. Apart from some

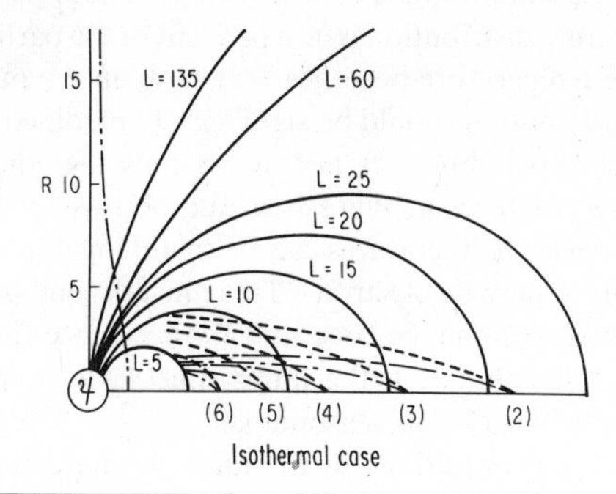

Figure 10a Distribution of thermal plasma density in the Jovian magnetosphere, with the plasma density limited by rotational instability (isothermal case). The numbers within parenthesis indicate $\log_{10} n$.

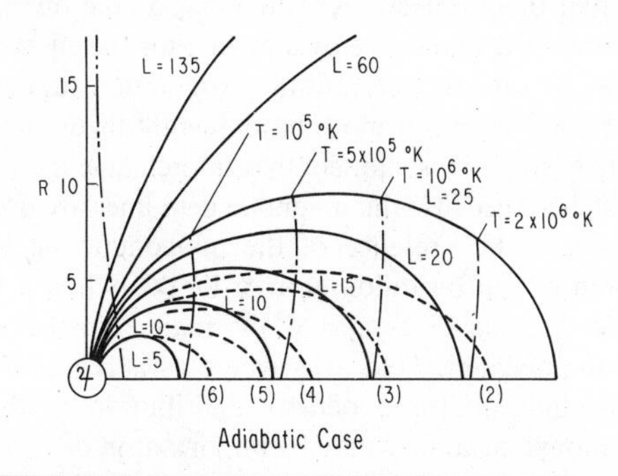

Figure 10b Distribution of the thermal plasma density and the plasma temperature in the Jovian magnetosphere with the plasma density limited by rotational instability (adiabatic case). The numbers within parenthesis refer to $\log_{10} n$.

questionable assumptions that are built into this model, there is an obvious shortcoming of the calculation in that the plasma distribution obtained does not join properly on to the ionosphere. This criticism has been made in another way by Ioannidis & Brice (48) who point out that electron-ion recombination would prevent the density from attaining the values given by equation 2, at least in the inner parts of the magnetosphere ($L \leqq 10$), and that in the inner magnetosphere the requirements of diffusive equilibrium are such that N_{eq} must be quite small.

As an alternative to the Gledhill model, it is worth considering the case in which the plasma behaves adiabatically (details of the calculation are given in Appendix A). Since protons escaping from the ionosphere should have essentially zero energy at the surface upon which the components of the gravitational and centrifugal forces parallel to the magnetic field are equal, it is appropriate to assume that the temperature is zero on this surface. Temperature and density distributions obtained on the basis of this assumption together with equation 2, are shown in Figure 10b. Note that the temperature distribution is independent of the particular form assumed for N_{eq}, and that the temperature becomes very large in the outer magnetosphere. Clearly the density distribution should be significantly modified in the inner regions for the reasons given above, but it is not as obvious that the connection to the ionosphere is incorrect. However, there is a question as to the extent to which such large magnetospheric temperatures can be maintained in the presence of heat conduction to the ionosphere by electrons. This must depend on the relative values of the ion-electron energy exchange time and the residence time of the plasma in the face of rotational instability; if the latter time scale is the smaller then the assumption of adiabatic behavior is reasonable.

There are observational objections to the high plasma densities implied by the above models in which N_{eq} is given by equation 2. These involve the lack of any detectable free-free and free-bound (e.g. H_α) emission from the plasma (Gulkis, 38), and the absence of Faraday rotation of the decimetric emission (McCulloch, 69) and also of polarized bursts of decametric radiation (Warwick & Dulk, 123). It is doubtful, however, that these objections hold as far as the outer Jovian magnetosphere ($L \geqq 10$) is concerned, and hence by allowing for the effects of recombination any disagreement between the models and the observations to date could probably be removed. Nevertheless there is a more fundamental theoretical objection which appears to invalidate both models, namely that interchange motions are possible in the Jovian magnetosphere because the magnetic field lines are not strictly frozen to the planet as a result of the presence of the nonconducting lower atmosphere. Consequently, as pointed out by Piddington & Drake (90) and Ioannidis & Brice (48), it is possible for plasma to escape to the magnetopause without stretching and breaking open the magnetic field everywhere as is required by the rotational instability, and thus the equatorial density distribution need not be given by equation 2. The rotational instability may be important in disrupting the outermost regions of the magnetosphere, but since other processes such as reconnection to the interplanetary magnetic field are likely to play a dominant role in controlling the behavior of the plasma, it does not seem profitable to speculate further at this stage. Presumably however, the plasma is ultimately lost into the tail of the magnetosphere where there are no containing influences associated with the solar wind.

It is very difficult to construct a satisfactory model to describe the processes by which plasma is fed into and escapes from the Jovian magnetosphere. Ioannidis & Brice (48) have made a rough calculation which appears to contain most of the important physical effects, and their result for N_{eq} is shown in Figure 11a. We have analyzed the problem on essentially the same lines (see Appendix B for details) and we find the plasma distribution in the outer parts of the equatorial plane is given by

$$n = \frac{B_0}{C_0^{1/2} h m_p} \frac{x^{1/2}}{L^3} \qquad 3.$$

where

$$C_0 = \frac{\beta_0 B^2}{T^{3/4} h m_p f_0} \qquad 4.$$

and $x = C_0(M/\Phi)^2$ satisfies the differential equation

$$\frac{dx}{dL} = \frac{C_0 \Sigma_0 f_0}{\Omega^2} \cdot \frac{(2L-1)[L^3-(L-1)^{1/2}x]}{(L-1)^2(L^3-L_s^3)L^{7/2}} \qquad 5.$$

(all the symbols are defined in Appendix B). The variation of n with L, calculated from equations 3, 4, and 5 subject to the boundary condition assumed by Ioannidis & Brice (48), viz. $n = 0.25$ cm^{-3} (which corresponds to the solar wind density) at $L = 50$, and also assuming $kT = 10$ eV and $h = R_J = 7 \times 10^9$ cm, is shown by the

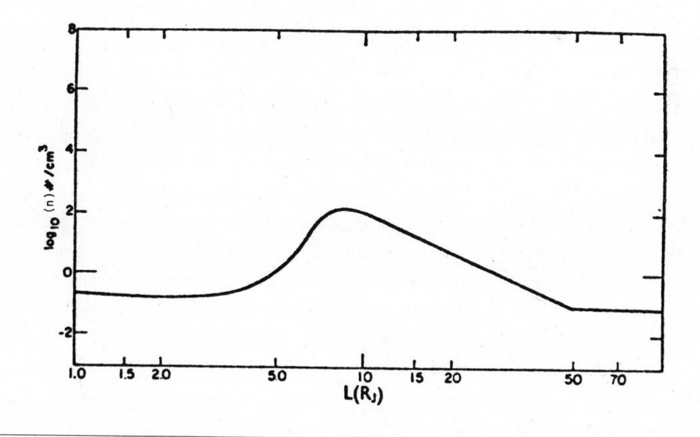

Figure 11a The "present best estimate" of the equatorial plasma density profile in the Jovian magnetosphere, including all anticipated losses of plasma. Near Jupiter, the dominant loss is diffusion into the ionosphere but beyond about $7R_J$, interchange instabilities dominate, and the density decreases as L^{-4}, until the solar wind density is reached (Ioannidis & Brice, 48).

dotted part of the curve in Figure 11b. The variation of n with L in the inner magnetosphere ($2 \lesssim L \lesssim 10$) is computed on the basis of diffusion equilibrium with the ionosphere and assuming the minimum value at the synchronous orbit ($L \approx 2$) to be 10^{-2} cm^{-3}. The maximum density of $n \approx 30$ cm^{-3} around $L \approx 10$ is in fair agreement with the best estimate of Ioannidis & Brice (48). However, due to the uncertainties in the values of several parameters we have used in the calculation, and the arbitrariness of the boundary conditions, the results of this calculation cannot be regarded as being definitive in any sense.

It has been pointed out by Coroniti (personal communication) that even though these interchange motions are energetically possible they do not necessarily occur. A more careful examination of the stability of the magnetosphere (allowing for the damping effects of energetic particles) is required to provide an answer to this question. It should be noted that all interchange motions, whether driven from low altitudes, or as a consequence of the interaction with the solar wind, will affect the distribution of low energy magnetospheric plasma even if the plasma is itself stable as far as corotation effects are concerned.

The characteristic time scale τ for the interchange process, is given by

$$\tau = \frac{B_0 \Sigma_0 C_0^{1/2}}{2\Omega^2} \cdot \frac{(2L-1)L^{1/2}}{(L-1)^{3/2}(L^3 - L_s^3)x^{1/2}} \qquad 6.$$

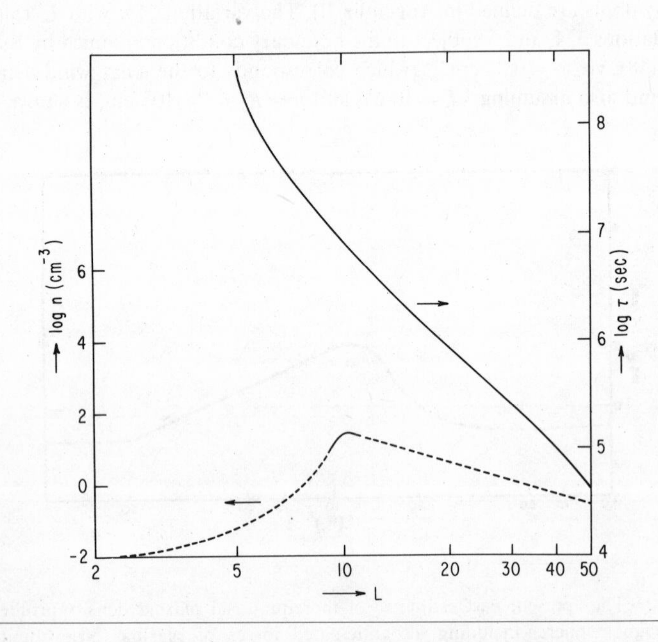

Figure 11b The equatorial plasma density profile in the Jovian magnetosphere and the time-scale for the interchange instability subject to the boundary conditions $n = 0.25$ cm^{-3} at $L = 50$ (Appendix B).

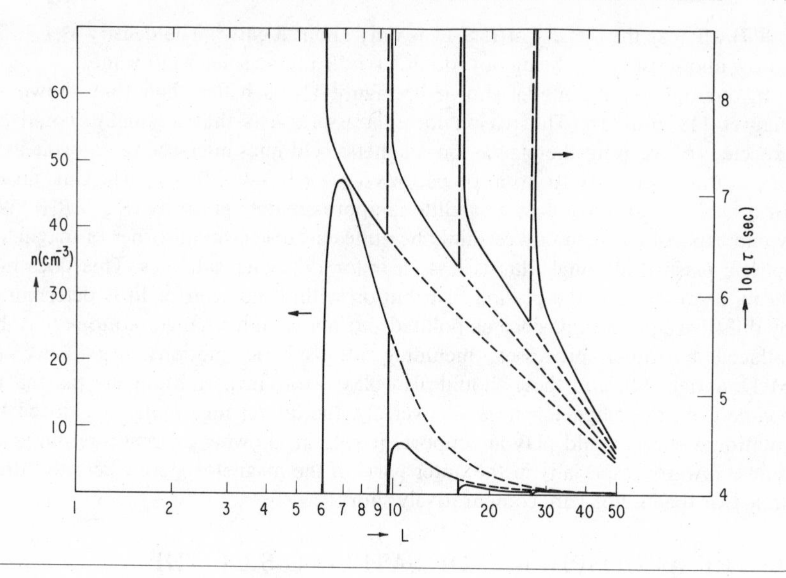

Figure 11c The equatorial plasma density profile in the outer Jovian magnetosphere and the time-scale for the interchange instability, taking into account absorption by the satellites.

and the variation of τ with L is shown by the solid line in Figure 11b. While τ is only about 12 hr when $L = 50$, it increases to about 4 months at $L = 10$, and is almost 10 yr when $L = 5$, under the conditions assumed. If these time scales exceed those required for inwards diffusion of energetic particles, then driven interchange motions must determine the distribution of the low energy plasma.

Since $\Sigma_0 f_0$ should vary approximately as the inverse third power of the heliocentric distance (see Appendix B), and, if recombination is neglected, $n_{max} \sim \Sigma_0^{1/2} f_0^{1/2} B_0/\Omega h$, we can find the maximum plasma densities in the magnetospheres of Saturn and Neptune by assuming the scaled values of the magnetic field strength and that h is proportional to the planetary radius. These maxima are found to be roughly 3 cm^{-3} for Saturn and 0.8 cm^{-3} for Neptune. An immediate consequence of these relatively low plasma densities is that the resonant energy $B_0^2/8\pi n L^6$ may be large, and indeed it is by no means certain that it is anywhere less than the characteristic energy E^* for any of the outer planets as argued by Brice (12), Thorne & Coroniti (111), and Scarf (97).

A serious limitation in the above calculation (as well as that of Ioannidis and Brice) of the plasma distribution in outer Jovian magnetosphere is the neglect of the plasma absorption by the satellites. We have attempted to take this effect into account by redoing the previous calculation subject to a different set of boundary conditions, viz. $n = 0$ at the orbits of Io, Europa, Ganymede, and Callisto. The profiles of n and τ in this case are shown in Figure 11c. It is interesting to note that the maximum density ($n_{max} \approx 47$ cm^{-3}) is now reached very close to Io (at

$L = 7$) whereas the density at $L \approx 11$ is only about 8 cm^{-3}. The density at $L = 50$ is now about 0.4 cm^{-3}, being not too different from the solar wind value.

In our opinion the model shown in Figure 11c is better than that shown in Figures 11a and 11b. The reason for this assertion is that a satellite could be expected to sweep up the plasma on magnetic field lines intersecting its orbit in a time of the order of 10–10^2 rotation periods (i.e. 4×10^5–4×10^6 sec). The time taken for a field line to move past a satellite is approximately given by $(R_{sat}/LR_J)\tau$. We would expect that Io should certainly be quite effective as an absorber of magnetospheric plasma, although this is less clear for the outer satellites. This does not mean, however, that the plasma distribution within the orbit of Io is determined by diffusive equilibrium, since as pointed out above, interchange motions may be induced by other processes, including atmospheric motions (e.g. Brice & McDonough, 14), and these should also play a role in permitting the plasma to escape from the magnetosphere. Conversely, the interchange motions induced by rotational effects could play an important role in allowing energetic particles to diffuse inwards, especially in the outer parts of the magnetosphere where the time scales for the motions are comparatively short.

5. THE ATMOSPHERES OF SATELLITES OF THE OUTER PLANETS

The effects of the plasma and magnetic field on the atmospheres of satellites contained within the magnetospheres of the outer planets must in many respects be similar to the effects of the solar wind and interplanetary magnetic field on the atmospheres of Mars, Venus, and the moon (e.g. Banks et al, 3; Michel, 75, 76). On the basis of our estimates for the plasma densities and magnetic field strengths in the magnetospheres of the outer planets, it is concluded that the relative velocity between the corotating magnetospheric plasma and its satellites is sub-Alfvénic in every case except Callisto and Iapetus (see Table 3). Consequently it should be expected that in general there is no bow shock produced as a result of the interaction between satellites and magnetospheric plasma as there is in the case of the interaction between Mars and Venus and the solar wind. However, the question of whether or not there is a shockwave is not of primary importance, since the essential nature of the interaction is determined by the electrical conductivity of the satellite and its atmosphere (if any). At one extreme if the satellite is a poor electrical conductor and has no atmosphere (or at most a thin exosphere), the interaction should be similar to that occurring between the moon and the solar wind. In this case the plasma impinges directly on the surface of the satellite and is absorbed. Otherwise very little disturbance is produced in the flow, and a void should exist in the wake of the satellite since the thermal speed of the ions is small compared with the relative velocity between the satellite and plasma. At the other extreme, if the atmosphere is fairly dense as it is in the case of Titan, a strong interaction will occur with the magnetic field lines wrapped around, and to some extent imbedded in the satellite ionosphere. There may be a plasma tail formed of ions of atmospheric origin in the wake of the satellite similar to that observed in the case of Venus.

Table 3 Satellite magnetospheric parameters

Satellite	Distance from planet (planet radii)	Orbital velocity (km/sec)	Corotational velocity (km/sec)	Δv (km/sec)	$\Delta v/v_{\text{Alfvén}}$	$\lvert\varphi\rvert(1\ \mu)$ (volts)
Synch (J)	2.3 R_J	29.1	29.1	0	0	2.3
Amalthea	2.6 R_J	26.4	32.9	6.5	3.3×10^{-7}	3.0
Io	6.0 R_J	17.3	76.4	59.1	1.3×10^{-4}	15.8
Europa	9.6 R_J	13.8	122	108	2.2×10^{-2}	40.5
Ganymede	16 R_J	10.9	201	190	1.2×10^{-1}	113
Callisto	27 R_J	8.2	348	340	1.0	321
Synch (S)	1.8 R_S	19.2	19.2	0	0	5.2
Janus	2.7 R_S	15.3	29.0	13.7	8.0×10^{-7}	11.7
Mimas	3.2 R_S	14.3	33.3	19.0	6.1×10^{-6}	16.6
Enceladus	4.0 R_S	12.6	42.0	29.4	5.8×10^{-5}	25.6
Tethys	5.0 R_S	11.4	52.5	41.1	1.8×10^{-4}	40.0
Dione	6.3 R_S	10.0	66.5	56.5	7.1×10^{-4}	63.8
Rhea	8.8 R_S	8.5	92.5	84.0	8.4×10^{-3}	124
Titan	20 R_S	5.6	209	203	1.6×10^{-1}	640
Hyperon	25 R_S	5.0	262	257	3.0×10^{-1}	1000
Iapetus	60 R_S	3.3	627	624	7.2	5760
Synch (N)	3.5 R_N	9.8	9.8	0	0	187
Triton	14.3 R_N	(−4.4)	40.2	44.6	10^{-2}	3100

It is to be expected that the interaction with the magnetosphere tends to remove ionized gas from the atmospheres of the satellites at a rate which is comparable to the total ionization rate in the case of weak (i.e. moon-like) interaction, and perhaps considerably less in the case of strong (i.e. Venus-like) interaction. The observed upper limits on the atmospheric pressures of the outer satellites are well above the very low value (equivalent to $\sim 10^{15}$ mol cm^{-2} column density, or $\sim 3\times10^{-9}$ mbar surface pressure) which would permit a weak interaction, and hence it is not possible to make any definite predictions concerning the nature of the interaction. However, if the atmospheric pressure is controlled by the vapor pressure of water, as would be the case if the surface is covered with either pure water ice or clathrate hydrates, then the surface pressures for the atmospheres of all satellites other than Titan would be such that the interaction is weak (see Figure 12b). If a substantial amount of nitrogen or methane is present (other than in a clathrate hydrate) the atmospheric pressures may be many orders of magnitude larger than the pressure of water vapor at the appropriate temperatures, and in these cases the interaction must be strong.

Primeval atmospheres of the Galilean satellites, if comprised of molecular species with molecular masses exceeding ~ 15, should still be present if the usual evaporative processes are the main cause of loss (see Figure 12a). However, this conclusion is misleading, since interaction with the magnetospheric plasma provides a much more effective loss mechanism. As a consequence of the motion of magnetic field lines past the satellite, plasma in the topside ionosphere is continually eroded away at a rate which is determined by the rates of photo-ionization by sunlight, charge exchange with low energy magnetospheric ions, and ionization by energetic magnetospheric particles. This process could lead to an escape rate as large as $\sim 10^8$ mol cm^{-2} sec^{-1} in the absence of compensating factors such as partial shielding of the ionosphere by trapped magnetic field and the return of magnetospheric plasma to the ionosphere. Thus it would in principle be possible for an initially quite substantial atmosphere (i.e. $\lesssim 1$ mbar surface pressure) to be completely eroded away in a period comparable with the age of the solar system, regardless of the molecular mass of the most abundant constituents. This would not be enough to remove atmospheres formed of methane and/or molecular nitrogen from the Galilean satellites if these gases were freely available from their respective ices (see Figure 12b). However, atmospheres comprised of ammonia and/or water would certainly be lost even if the efficiency of the process is as low as 10^{-6}.

In the case of the weak interaction, atmospheric ions, once they are formed, immediately begin to move on trochoidal trajectories with amplitudes depending on the gyroradius of the ion when it is moving at the corotational velocity (e.g.

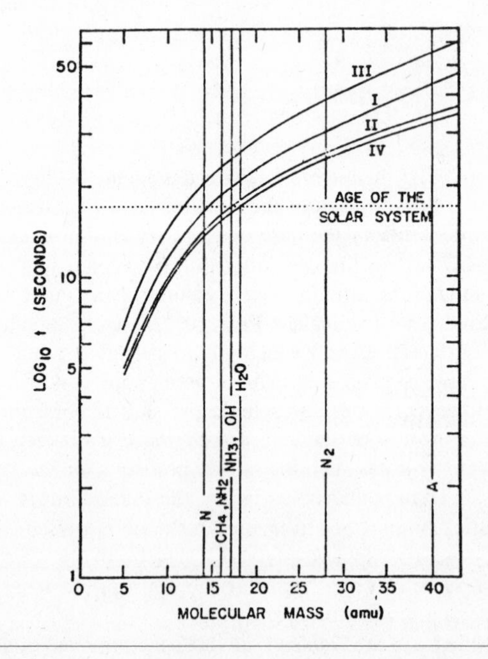

Figure 12a Escape times for gases of different molecular weights for each of the Galilean satellites. The intersection of the satellite curves with lines of different molecular weights gives the time for reduction of the abundance of that gas by a factor $1/e$ (Binder & Cruikshank, 9).

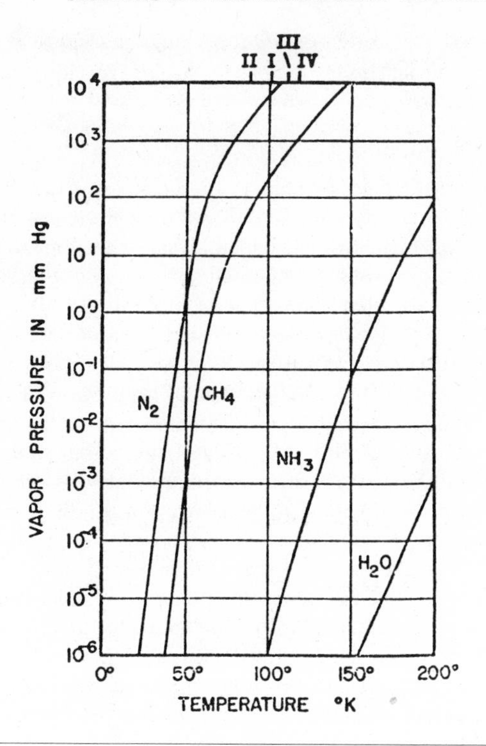

Figure 12b Vapor pressures of various gases under consideration. The Roman numbers at the top indicate the position of the Galilean satellites on this diagram (Binder & Cruikshank, 9).

Michel, 76; Banks et al, 3). Ions whose trajectories intersect the satellite are mostly absorbed and neutralized on the surface. However, a substantial fraction may escape and become part of the magnetospheric plasma. The same process occurs in the case of the strong interaction, but it is limited by the fact that the inertia of newly introduced ions may be so great that the motion of the magnetic field line is impeded. Consequently only the topside of the ionosphere can be affected significantly, and the loss rate may be substantially smaller than the total ionization rate. A simple analysis of this process has been given by Johnson & Axford (49).

It should be noted that the argument put forward by Webster et al (124) leading to their conclusion that the ionosphere of a satellite such as Io cannot be stripped away in this manner, is incorrect. They have compared the $\mathbf{j} \times \mathbf{B}$ force required to lift ionospheric plasma against gravity, to the $\mathbf{j} \times \mathbf{B}$ force required to move the plasma through the background neutrals, and found that the former is very much larger in the topside ionosphere where the process should be expected to occur. However, this is irrelevant since the only important question is whether or not a $\mathbf{j} \times \mathbf{B}$ force can exist which is sufficient to lift the plasma against gravity. As a rough guide we should require that

$$B^2/4\pi R_{sat} \geqq mnAg_{sat} \quad 7.$$

where B is the magnetic field strength at the position of the satellite, R_{sat} is the radius of the satellite, g_{sat} is the gravitational acceleration at the surface of the satellite, n is the number density of ionospheric ions, A their mass in amu, and m the mass of a proton. In the case of Io we find that this requires $nA \lesssim 3 \times 10^{11}$ amu cm^{-3}, which far exceeds the expected value $nA \lesssim 10^5$–10^6 amu cm^{-3}.

On the basis of these arguments, it seems reasonable to conclude that if the Galilean satellites have water ice crusts, they do have substantial atmospheres unless there is some venting of highly volatile material from below. The observed upper limits on the surface pressures of ammonia and methane are quite low ($<6 \times 10^{-8}$ atm according to Fink et al, 27), and since this amount of material could easily be removed during 4×10^9 yr, it seems improbable that primeval atmospheres of these gases could remain. Thus if atmospheres with surface pressures in the range 10^{-5}–10^{-9} mbar exist they are likely to be associated with venting, or the presence of nonhydrous ices on the surface. [The minimum atmospheric pressure may be controlled by sputtering due to the impact of magnetospheric plasma and energetic particles on the surface, rather than by the vapor pressure of the solid.] On the other hand, apart from the unconfirmed measurement of O'Leary (86) and the arguments concerning the posteclipse brightening of Io (Binder & Cruikshank, 9), there are no compelling reasons for believing that even the large Galilean satellites have atmospheres with surface pressures exceeding 10^{-9} mbar. Thus, on the basis of present observations, one may, if one wishes, assume that the atmospheres of all the satellites of the outer planets other than Titan are exospheric in nature, and that their magnetospheric interactions are weak. It should be noted that an observation of the nature of the interaction with the magnetosphere is one of the more sensitive ways for determining the presence of a very thin atmosphere (e.g. Siscoe & Mukherjee, 102, 103).

It is important to note that there is one significant difference between the solar wind/planet interactions, and the satellite/magnetosphere interactions. In the former case the plasma impinging on the atmosphere is of external origin, and plasma lost from the atmosphere can never return. In contrast, in the latter case, the plasma lost from the atmosphere remains temporarily trapped in the magnetosphere, and hence the plasma impinging on the atmosphere should be partly of external origin (i.e. planetary or solar wind), and partly originating from the satellite itself. This reduces the efficiency of the atmospheric erosion process we have described, but since magnetospheric plasma can escape by interchange and other instabilities as described in Section 4, and also by recombination, the net effects of the satellite/magnetosphere interaction is to cause erosion of the satellite atmosphere.

Although plasma interactions as described above may be the main cause of loss of atmospheric atoms and molecules, evaporative escape of neutrals is of some interest, especially in the case of Titan where there is likely to be a copious supply of neutral hydrogen atoms. It has been pointed out by McDonough & Brice (70, 71) that at the exospheric temperatures expected for Titan only a small percentage of the neutral hydrogen atoms evaporating from the atmosphere will have enough energy to escape from the Saturnian gravitational field. As a consequence, these atoms will remain in orbit about Saturn, forming what is in effect an extended exosphere of Titan which is of toroidal form. Note that this applies only to hydrogen atoms that are evaporated by the usual Jeans process, and not to those that are given relatively large escape energies as a result of dissociative recombination or photodissociation (e.g. McElroy, 72). The latter effects may be important as far as the loss of constituents other than hydrogen is concerned since thermal escape of heavy atoms and molecules must be very slow at the exospheric temperatures expected on Titan.

It is in fact quite difficult to determine the mean density of hydrogen in the toroidal exosphere of Titan since losses due to photo-ionization and charge exchange with magnetospheric ions are likely to be important. Since Titan lies well beyond the synchronous orbit of Saturn, charge exchange must produce hydrogen atoms with energies far exceeding the escape energy for Saturn (i.e. their initial velocity is essentially the corotational velocity), and hence such atoms are lost from the system. The ions which remain can be recaptured by Titan, but since the interaction with the magnetospheric plasma and magnetic field is strong the capture efficiency may be quite low. McDonough & Brice (70, 71) have estimated that the mean density of the ring may be $1\text{–}10^3$ atoms cm^{-3} depending on the temperature of the Titanian exosphere. The upper limit corresponds to a situation in which the mean free path for charge exchange between magnetospheric protons and hydrogen atoms is less than the circumference of Titan's orbit. In this situation the magnetospheric protons would continually be undergoing charge exchange and their thermal motion would quickly become comparable to the corotational velocity at the orbit of Titan.

6. ELECTRICAL POTENTIALS OF SOLID BODIES IN PLANETARY MAGNETOSPHERES

Very large negative electrical potentials (i.e. -1 to -10 kV) have been measured occasionally on earth satellites situated within the magnetosphere in solar eclipse and in regions where the ambient plasma density is low (DeForest, 17; Fredricks & Scarf, 30). The lunar surface potential on the sunlit side is positive (Freeman et al, 31) and may be relatively large when the moon is inside the high latitude magnetotail ($\gtrsim 200$ V according to Reasoner & Burke, 92). When the moon is immersed in the earth's plasmasheet the surface potential on the dark side is likely to be of the order of -1 kV or larger (Knott, 55). It has been pointed out by Scarf (97) that high surface potentials may be the rule within the Jovian magnetosphere, since even in full sunlight the solar EUV flux is diminished by a factor ~ 27 in comparison with that at the earth. This could constitute a serious hazard for spacecraft such as Pioneers 10 and 11, and Mariner-Jupiter-Saturn 1977.

In general the potential of a body is determined by the condition that there should be no net current to its surface; thus if the potential ϕ is negative

$$\mathbf{J}_{-}(E > e\phi) - \mathbf{J}_{+}(E > 0) = \Phi_p + \Phi_s + \Phi_f \qquad 8.$$

Here e is the charge on an electron; $\mathbf{J}_{-}(E > e\phi)$ and $\mathbf{J}_{+}(E > 0)$ are the integral electron and proton fluxes of particles with energies at infinity greater than $e\phi$ and 0 respectively; Φ_p is the photoelectron flux from the surface; Φ_s the flux of electrons due to secondary emission; and Φ_f the average electron flux due to field emission.

As an illustration of the effectiveness of charging of solid surfaces in an environment of the sort expected in the magnetospheres of the outer planets, let us consider the case of a plane surface facing a Maxwellian plasma with proton and electron temperatures T_{p1} and T_{e1}, respectively, and number density N_1 (cf. Whipple, 125). In this case, if the terms in equation 8 representing electron emission from the surface are negligible, it is easily shown that the surface potential is given by

$$e\phi = \tfrac{1}{2}kT_{e1}\log_e(m_p T_{e1}/m_e T_{p1}) = \tfrac{1}{2}kT_{e1}[7.65 + \log_e(T_{e1}/T_{p1})] \qquad 9.$$

Thus the surface potential is in general such that $e\phi$ is of the order of the characteristic electron energy unless $T_{p1} \gg T_{e1}$. If we add a second cooler plasma with proton and electron temperatures T_{p2} and T_{e2}, respectively, and number density N_2, then assuming $e\phi \gg kT_{e2}$, we find that the surface potential becomes

$$e\phi = \tfrac{1}{2}kT_{e1}\left\{\log_e(m_p T_{e1}/m_e T_{p1}) - 2\log_e\left(1 + \frac{N_2 v_{p2}}{N_1 v_{e1}}\right)\right\} \qquad 10.$$

where v_{e1} and v_{p2} are the electron and proton thermal speeds in the hot and cool plasmas, respectively. Evidently the cool plasma cannot have a large effect on the surface potential unless $N_2 v_{p2} \gtrsim N_1 v_{e1}$; that is, the proton flux in the cool plasma must be comparable with, or exceed, the electron flux in the hot plasma. For example, if the electron flux in the hot plasma is 10^9 cm^{-2} sec^{-1}, and $T_{p2} \simeq 10^4$ °K, then the cool plasma has very little effect on the surface potential unless $N_2 \gtrsim 10^3$ cm^{-3}.

It seems, therefore, that with the values of N_{eq} expected in the magnetospheres of the outer planets, the thermal plasma is likely to have very little effect on the surface potentials of solid bodies unless the energetic particle flux is quite low. Thus, as a first approximation, we suggest that the potentials are such that $e\phi$ is of the order of E^* or E_{res}, whichever is the lesser. Note, however, that some modulation of the potential can occur even if the temperature of the thermal plasma is relatively low, since v_{p2} in equation 10 may be replaced by the relative speed between the plasma and the surface, and this may range from ~6 km sec^{-1} to ~600 km sec^{-1} depending on position in the magnetosphere (see Table 3). Furthermore the effects of electron emission from the surface are probably not negligible, and could suppress the potential substantially below the value suggested. Secondary emission is expected to be most important in this respect as far as the outer planets are concerned, but in the Jovian magnetosphere in particular photoelectron emission may strongly influence the potentials of surfaces which are fully illuminated by the sun.

In order for any of the terms on the right hand side of equation 8 to be important, they must not be negligible in comparison with the integral proton flux $\mathbf{J}_{+}(E > 0)$. Under quiet solar conditions, the photoelectron fluxes to be expected on fully illuminated surfaces at the heliocentric distances of Jupiter, Saturn, and Neptune, are of the order of, or less than, 10^9 cm^{-2} sec^{-1}, 2.5×10^8 cm^{-2} sec^{-1}, and 3×10^6 cm^{-2} sec^{-1}, respectively (Grard, 36; Grard et al, 37; Shawhan et al, 100; Walbridge, 118; Wyatt, 128). The average photoelectron fluxes from dust grains, which can be only partly illuminated, are about one quarter of these values, and of course the photoelectron fluxes from unilluminated surfaces of the satellites are essentially zero. It is possible that photoelectron emission is important in keeping the surface potentials of the sunlit sides of the Jovian satellites in particular to relatively modest values ($|\phi| \lesssim 1$ kV?). It should be expected that the outer satellites are most strongly affected in this way, and Io the least affected. However, without detailed knowledge of the energetic particle populations in the magnetospheres of the outer planets, no definite predictions can be made.

Secondary emission is likely to be at least of comparable importance to photoelectron emission, but it is impossible to calculate whether or not it has a dominant effect on the surface potentials without knowing the materials concerned and the energy spectra of the incident particles (Grard et al, 37; Knott, 55, 56; Knudson & Harris, 57; Whipple, 125). The secondary emission yield for electron impact is given approximately by the empirical relationship suggested by Sternglass (110):

$$\delta(E') = 7.4\,\delta_{max}(E'/E_{max})\exp\{-2(E'/E_{max})^{1/2}\} \qquad 11.$$

and the emitted flux is

$$\Phi_s = \int_0^\infty \delta(E')\mathrm{j}(E')\,dE' \qquad 12.$$

where E' is the energy of the particles striking the surface, and $\mathrm{j}(E')$ their differential energy spectrum. For many materials δ_{max} ranges between 0.75 and 5.0, and E_{max} between 300 eV and 500 eV (see Grard et al, 37): thus Φ_s depends quite sensitively upon the form of the energy spectrum of the incident electrons and especially on their flux in the energy range near E_{max}. If the characteristic energy of the incident electrons exceeds E_2, the upper energy at which $\delta = 1$, then one should expect that $e\phi$ is of the order of the characteristic energy if other emission processes are relatively inefficient. Otherwise, the potential may be reduced substantially, and in circumstances where the net yield exceeds unity it will become slightly positive (Knott, 55).

Energetic ions can produce secondary electrons very efficiently for incident energies exceeding several kilo electron volts. For example, in the case of proton impact, $\delta \gtrsim 1$ if $E' \gtrsim 2 \times 10^3$ eV for aluminum, and if $E' \gtrsim 2.5 \times 10^4$ eV for tungsten, with $\delta_{max} \approx 4.5$ and 1.5, respectively (Whipple, 125). This suggests that in the absence of substantial photoelectron and field emission fluxes, electron emission due to ion impact could make an important contribution to the current balance when the surface potentials are ~ -10 kV.

The electron flux due to field emission from metallic surfaces is given by the Fowler-Nordheim formula (e.g. see Gomer, 35). In particular, the total flux F from a hemispherical area of radius a centimeters and potential ϕ volts is given by,

$$F = \frac{(\mu/w)^{1/2}}{\mu + w}\phi^2 \exp\left(60.9 - 6.8 \times 10^7 w^{3/2}\frac{a}{\phi}\right)\text{sec}^{-1} \qquad 13.$$

where μ is the Fermi level and w is the work function. Typically $w \approx 4$ for metals and thus field emission becomes important when the electric field $E(=\phi/a) \gtrsim 10^7$ V cm^{-1} (see Figure 13). Even for insulators (e.g. ice) it seems unlikely that field strengths greatly exceeding this value could be supported, since the potential drop across an atom would then be comparable to the ionization potential. In general, field emission should be taken into account if the surface has spiky projections with sufficiently small radii of curvature at their tips to allow the local field strength to exceed the above value. The potential of micrometeoroids can be strongly affected by field emission, and even with large incident energetic electron fluxes one should expect that the potential is limited to values less than $\sim 10^3 s$ V (negative), where s is the particle radius in microns (e.g. Spitzer & Savedoff, 105). It should also be noted that the electric field stresses acting on small particles of irregular shape may be sufficient to distort or disrupt them completely, even with relatively modest surface potentials.

In the above discussion we have assumed that the integral flux of energetic electrons exceeds that of energetic protons, and thus that the potential of a solid body is always negative. In these circumstances the background plasma in the Jovian magnetosphere would be expected to play an insignificant role in determining the surface potential. However, it has been pointed out by Beck (personal

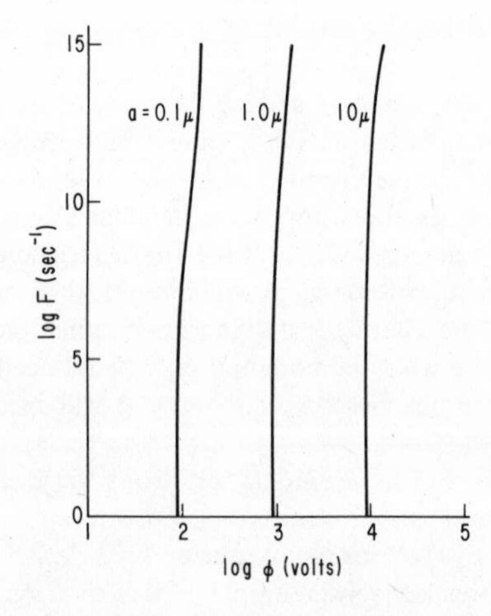

Figure 13 Field emission flux from a hemispherical "pin head" as a function of the potential calculated from equation 13.

communication) that if one accepts the necessarily simple models for the radiation belts that have been developed to date, the surface potentials may be positive because it is assumed that the particles are monoenergetic, and in most cases that the electron losses are rather severe. In our opinion, however, one should not take this result too seriously, since it implies that the integral flux of electrons in the Jovian magnetosphere ($E > 1$ keV, say) is very much less than that found in the outer parts of the earth's magnetosphere. It is in fact impossible to make any realistic predictions on this point, and we must wait instead for the results from the first Jupiter fly-bys before any definitive answers can be given.

7. EFFECTS OF SURFACE CHARGING OF IO

We do not wish to discuss all the possible consequences of very high surface potentials on satellites in this review. However it is obvious that if the surface of Io is in part charged to a potential of the order of -10^3 to -10^5 V, secondary electrons emitted from the surface and accelerated by the electric field in the satellite sheath, could conceivably produce the Io-associated decametric emissions. In effect, as a consequence of surface charging, some fraction (perhaps 10%) of the energy dissipated by energetic magnetospheric particles impacting the surface of Io is converted into a stream of energetic electrons. The average pitch angles of the electrons after they have been accelerated should not exceed $|\phi(V)|^{-1/2}$ radians since the initial energies of the particles are typically only a few eV. It is presumed that on interacting with the upper atmosphere and ionosphere of Jupiter the electrons produce a directed beam of decametric radio noise as discussed by numerous authors (e.g. 21, 24, 34, 69, 90, 104, 120, 121, 126, 131).

In terms of the power requirements, there is no obvious difficulty with this suggestion since the Io-associated decametric emissions require only about 10^{15} erg sec^{-1} (e.g. Warwick, 121). Whereas a flux of 10^8–10^9 electrons emitted from the surface of the satellite with energies of the order of 10 keV carries $\sim 4 \times 10^{17}$–4×10^{18} erg sec^{-1}. The variability of the emissions should be linked directly to variations in the energetic electron fluxes trapped in the Jovian magnetosphere. Since the latter are likely to be related to variations in the solar wind and interplanetary magnetic field near Jupiter, it is therefore not surprising that the Io-associated emissions show a short-term correlation with solar activity (Barrow, 4). This should also be the case for the decametric emissions not associated with Io, if they are the result of ordinary auroral type processes, including substorms (e.g. Duncan, 23). The apparent lack of decametric emissions associated with the other Galilean satellites (Dulk, 22) can be accounted for since the surface potentials of these satellites should be significantly less than at Io, since the energetic trapped particle fluxes and hence the secondary fluxes are also less, and possibly because any beamed radiation produced by the secondary electrons might not intercept the earth. In the case of Amalthea these arguments are probably invalid, however the small size of this satellite (see Table 1) suggests that the radiated power might be less than that associated with Io by a factor ~ 400, and is perhaps not detectable.

The potential at any point on the surface of Io must be strongly dependent on orbital phase, since photoelectron emission occurs only on the sunlit face of the satellite, while as a consequence of magnetospheric rotation, the absorption of thermal protons and energetic electrons takes place mainly on the trailing face. Thus at western elongation, where the trailing face of the satellite is also sunlit, the magnitude of the average surface potential on this face should be a minimum. In contrast, at eastern elongation where the trailing face is in darkness, the magnitude of its average surface potential should be a maximum. Similarly the magnitude of the average potential on the leading face of the satellite should be a minimum at eastern elongation and a maximum at western elongation. It is very unlikely that the net effect of these changes as far as the decametric radiation is concerned, is that the latter should show no variation with orbital phase. However, it is not possible to predict what variation of the decametric radiation should occur without knowing the details of all the processes involved. It is evident for example that if the thermal plasma plays only a minor role in determining the surface potential, the most energetic electrons should be produced at eastern elongation. An asymmetry of this type appears to be required to account for the differences between the Io-associated components of the early and main sources of decametric radiation (e.g. Dulk, 21; Warwick, 121).

One of the most important implications of this mechanism is that no hydromagnetic effects are involved, and in particular there is no requirement for electric currents. Thus the electrical conductivity of Io plays no role in producing the radio emissions, and it does not matter if (as is quite probably the case) the satellite is covered with a thick (1–10^2 km) layer of water ice or has a crust composed of any other poorly conducting material. If the conductivity of the crust is very small [e.g. $\sim 10^{-25}$ mho cm^{-1} for ice at 100°K, and perhaps as little as $\sim 10^{-31}$ mho cm^{-1} for some olivines (Dermott, 18)], then almost all other suggestions for producing the Io-associated emissions are invalid. On the other hand if some current is induced in the interior of the satellite and closes externally, as usually assumed, our mechanism is not affected, but may be dominant or supplementary, depending on the currents involved.

Gurnett (39) and Shawhan et al (100) have suggested that photoelectrons emitted from the surface are accelerated in a sheath surrounding Io. These authors assume that the sheath is in effect an electrostatic double layer (e.g. Block, 10), modified by the emission of photoelectrons, and produced by the electromotive force associated with the relative motion of the satellite and the corotating magnetospheric plasma. It is not clear why the double layer has to occur next to the surface of the satellite (or indeed, anywhere), but apart from this the model appears to require (as do most others) that the satellite has a rather large electrical conductivity ($\gtrsim 5 \times 10^{-11}$ mho cm^{-1}). If the satellite has a crust comprised of water ice or any other poorly conducting material the double layer cannot exist since any currents that might be induced in the mantle must close inside the crust. Note that it is not relevant that the conductivity of the mantle could be large (e.g. ~ 0.1 mho cm^{-1} for aqueous ammonia solution) as argued by Lewis (62), since this means only that the magnetic field in the mantle does not respond to short-term fluctuations in the external field.

Schatten & Ness (98) have also considered the problem of producing decametric emissions in the case where Io is effectively nonconducting, and have reached conclusions which are similar to ours. However, it is not clear from their very brief discussion whether or not they envisage the surface of the satellite to be charged to potentials as high as suggested here, and that the electrons concerned are emitted from the surface as secondaries or as a result of field emission. Furthermore they make a distinction between the cases of super- and sub-Alfvénic relative motion between the satellite and the magnetospheric plasma, which in our view is of no great significance since the surface potential is presumably controlled by relatively energetic radiation belt particles rather than by the low energy plasma. The relative motion is important mainly because it prevents the energetic particle population in the flux tube intersecting Io from being exhausted, and to a lesser degree because some modulation of the surface potential must result from the impact of low energy plasma on the surface. However, we certainly agree with the general philosophy expressed by Schatten and Ness, and in particular with their emphasis on a non-hydromagnetic interaction which does not disturb the large-scale structure of the Jovian magnetosphere.

Several other suggestions have been advanced as to how a nonconducting satellite could induce radio emissions. Mozer & Bogott (82) have pointed out that the result of exchanging energetic radiation belt electrons absorbed by the satellite for photoelectrons emitted from its surface may be to reduce the characteristic energy of electrons on the flux tube intersecting Io, and thus by some means to destabilize the plasma and produce particle acceleration and precipitation. Of course the electron distribution function (but not necessarily the density) may be distorted by these processes of absorption and emission from Io, but it is by no means clear that any instabilities that might result can produce the required effects. Other suggestions along these lines have been made by Smith et al (104) and Wu et al (126), however, these have been criticized by Wu (127) who has, in turn, suggested that the radial density gradient of energetic protons produced by inwards diffusion and absorption by Io leads to an instability which excites waves with frequencies slightly above the local electron gyrofrequency. Finally, Webster et al (124) have argued that even if Io is effectively nonconducting, the conducting path required in theories such as those of Piddington & Drake (90) and Goldreich & Lynden-Bell (34), could be provided by a satellite ionosphere. However, as pointed out in Section 5, it seems quite probable that Io has essentially no atmosphere, in which case an ionosphere could not be present.

The likelihood that Io has a high surface potential, especially where it is not sunlit, may provide the basis for explanations for some of the other peculiar properties of the satellite. In particular, the posteclipse brightening may be evidence for a change in the scattering and reflecting properties of the surface resulting from the very high potentials that might occur during eclipse. The average electric field at the surface of the satellite (neglecting small-scale features) is approximately ϕ/D, where D is the effective Debye length or sheath thickness of the plasma. One cannot estimate D without knowing the properties of the plasma in detail, however, a rough value can be obtained if it is assumed that the electron energy density is

$\sim 10^{-2}B^2/8\pi$, and hence $B \simeq (10^{-2}B^2/32\pi^2 n^2 e^2)^{1/2} \simeq 10^5$–$10^6$ cm if $n \simeq 2$–10 cm^{-3} and $B_{0J} = 10$ G. Thus the average surface electric field can easily be $\sim 10^{-2}$–1 V cm^{-1} if $|\phi| \simeq 10^4$–10^5 V.

One might explain the posteclipse brightening of Io if the surface is fine-grained and loose on a microscopic scale (which is consistent with the thermal inertia measurements and the lack of features in the infrared reflection spectrum). The structure of the top millimeter of the surface in these circumstances would be similar to the fairy castle structure of the lunar surface described by Hapke (41). It would not be unreasonable for the structure to be changed in the presence of strong electrical stresses, and one would expect the fairy castle effect to become more pronounced. Alternatively, it should be noted that small charged grains can be raised against gravity by such electrical fields. The condition for levitation is $\phi \simeq 0.75s\sqrt{(\rho D)}$ V, where ρ is the density of the grain material, s is the radius of the grain in microns, and it is assumed that the grain and surface potentials are the same. Taking $\rho = 1$ g cm^{-3} and $D = 10^6$ cm, we find that micron-sized particles are levitated if $\phi \gtrsim 800$ V. Since the surface potential could be quite large at all times, the smallest particles may be permanently suspended, however, they will not necessarily escape from the satellite altogether because the electric field drops to a relatively small value at altitudes of the order of the Debye length, and furthermore, the potentials achieved by such grains are limited by field emission and similar effects. At times when the surface potential is very large it is possible that the surface becomes shrouded by ice fog which temporarily covers regions of lower than average albedo, and thus causes the apparent brightness of the satellite to increase. The time scale for decay of the phenomenon would in this case be related to the time taken for the suspended grains to fall back on the surface (i.e. $\sim 10^2$ sec to fall a distance of 10 km). The time taken for the potential to be reduced by photoelectron emission, which appears to be the only other significant time scale, is very much shorter.

8. THE EFFECT OF MICROMETEOROIDS ON SATELLITE ALBEDOS

Efforts to explain the brightness variations of the larger satellites of Jupiter and Saturn have so far been concentrated on Iapetus owing to the remarkably large amplitude of the effect in this case. Dollfus (20) has suggested that the variation and brightness of the Iapetus is caused by geometrical effects associated with an elongated shape. This suggestion has been criticized by Hartman (43) who has pointed out that the internal gravitational stresses in a body with the dimensions required (viz. a spheroid with principal axis 3000 km and 340 km) would appear to exceed the crushing strength of meteoroidal material. Furthermore, for this explanation to be valid, the satellite would have to rotate with a period equal to half its orbital period, whereas one would expect that tidal forces on such an elongated object would be large and that it should be in a state of synchronous rotation (in which case brightness variations with the observed phase and period would not occur). It is now accepted that the brightness variation is caused by an

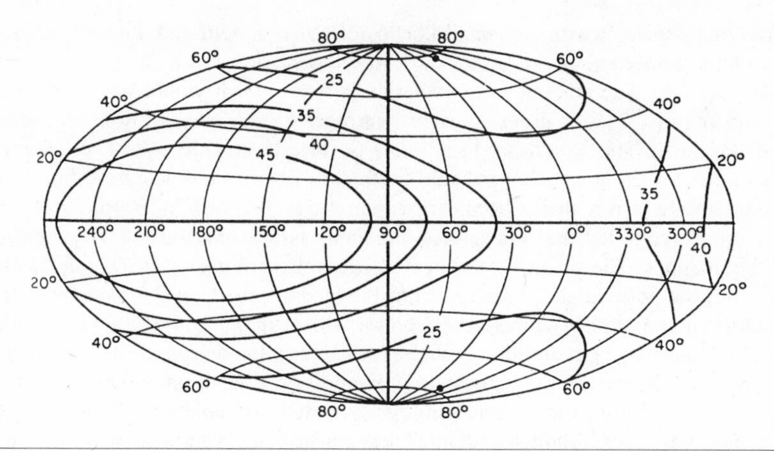

Figure 14 Lines of constant erosion rate on the complete sphere of Iapetus. Numbers are the factors by which the arrival rate of meteoroidal mass in a system at rest relative to the Sun must be multiplied to give the rate of material erosion. The center of the leading hemisphere of Iapetus in orbit about Saturn is at 180°, the trailing at 0° (Cook & Franklin, 15).

albedo difference between the leading and trailing faces; the observed anticorrelation between the infrared and visual brightness of the satellite (Murphy et al, 83) as well as polarimetric properties of the two faces (Zellner, 129) support this view.

Cook & Franklin (15) have suggested that the albedo difference is the result of differential meteoroid erosion, and this, with a slight modification, is in our opinion the most reasonable explanation of the effect as far as Iapetus was concerned. Until recently it was believed that the mean density of Iapetus was ~ 3 g cm^{-3}, and hence that the satellite has a composition similar to that of the moon. Cook and Franklin proposed that the rocky surface of the satellite was originally covered with a thin layer of snow or ice which has become preferentially eroded on the leading face as a consequence of impacts with interplanetary meteoroids, as shown in Figure 14. In view of recent observations which suggest that the mean density of Iapetus is probably ~ 1 g cm^{-3} and almost certainly less than 1.7 g cm^{-3}, this suggestion appears untenable since the satellite must be composed almost entirely of ice and hence it does not seem reasonable that an albedo change can be produced merely by eroding a very thin layer of the surface.

We suggest that a more likely alternative is that micrometeoroids have accumulated in the top few centimeters of the surface, and with the evaporation of a small amount of the icy material, the leading face of the satellite has become largely covered with a thin layer of dark, nonvolatile material of external origin. In these circumstances the albedo, but not necessarily the color, of the surface is changed. Micrometeoroid impacts may cause some evaporation, but it should also be noted that if the particles are initially mixed in a layer several centimeters deep they will eventually be exposed and collected into a very thin layer since temperature variations associated with even small albedo differences should lead to a net transport of volatile surface material from the dark to the light side and to the polar regions of the satellite. The amount of nonvolatile material which would be collected in 4×10^9 yr is 10^{-3}–10^{-4} g cm^{-2}, using micrometeoroid concentrations measured near the Earth. This would produce a dark surface layer 1–10 μ thick of micron-sized particles. In fact the layer could be thicker if the micrometeoroid flux near Jupiter and Saturn is, or has been, larger than it is at the earth at present.

It should be noted that the surface darkening is not uniform if it is produced in this manner. The micrometeoroid flux per unit surface area decreases as the cosine of the polar angle, measured from the leading point on the equator of the satellite, and the surface temperature decrease with latitude. Thus the albedos of the polar regions may be reduced only slightly, whereas a band around the equator on the leading face may be very dark indeed. In addition irregular nonuniformities are likely to result from cratering and venting associated with large meteoroid impacts.

Similar processes, although with much less efficacy, may be operative on Callisto, the outermost Galilean satellite, which shows a similar but much smaller effect.

The brightness variations with orbital phase of the three inner Galilean satellites of Jupiter and the three large inner satellites of Saturn (S3, S4, and S5) too are much smaller than that observed in the case of Iapetus. Furthermore, the phase of the variations is reversed so that the leading faces of the satellites are brighter than their trailing faces. It is less easy to dismiss the suggestion that in some cases at least these variations are caused by geometrical effects than it is for the case of Iapetus, however, the large sizes of the satellites make this explanation rather unlikely. The significant differences observed in the infrared reflectivity curves of the two faces of Ganymede (Pilcher et al, 91) argue in favor of an albedo difference, and the Galilean satellites in any case appear quite spherical.

It seems unlikely that differential meteoroid erosion, or the modified version of the process described above, could lead to the brightness variations of the three inner Galilean satellites or of the Saturnian satellites S3, S4, and S5, as has been pointed out by Cook and Franklin themselves. Evidently we must find a different explanation for the brightness variations in these cases and at the same time show that the process which works so well for Iapetus is quite inefficient for these satellites. Since the higher orbital velocities of the three inner Saturnian satellites (S3, S4, and S5) should lead to a greater rate of meteoroidal impact than at Iapetus, it appears that a very effective compensating factor must be at work and that the brightness variations are likely to be caused by a quite different mechanism.

We suggest that charging of micrometeoroids by interaction with energetic particles as they enter the magnetospheres of Jupiter and Saturn reduces the effectiveness of the Cook and Franklin mechanism as far as the inner Saturnian and Galilean satellites are concerned. The effect of the Lorentz force acting on charged micrometeoroids is to alter their trajectories from straight lines to curves, with the curvature depending on the charge-to-mass ratio and the magnetic field strength. If the charge-to-mass ratio of the micrometeoroids is sufficiently large they will be unable to penetrate very far into the magnetosphere and satellites which are well inside are thus protected from bombardment. Particles with intermediate charge-to-mass ratio which are able to penetrate far enough into the magnetospheres to impact one of the satellites can have their directions of approach changed so much that they are incident more or less isotropically and hence do not systematically change the albedo of either the leading or trailing faces. Finally, micrometeoroids with a very small charge-to-mass ratio will indeed impact preferentially on the leading faces of the satellites, but since the masses of the particles concerned may be quite large their flux is correspondingly low if the mass spectrum has a sufficiently steep slope, and hence they need not produce a large effect on the satellite albedos.

It is difficult to estimate how effective this mechanism for shielding satellites from micrometeoroid impact might be since it depends on the electrostatic potential attained by the particles which in turn must vary with position in the magnetosphere. Roughly one might argue that the directions of arrival of micrometeoroids are large randomized if the Larmor radius of the particles is comparable to the satellite orbital radius. The Larmor radius, r_L (km) of a micrometeoroid of radius s microns, density ρ (g cm^{-3}), speed v (km sec^{-1}), and at a potential ϕ (V) is given by

$$r_L = 3.8 \times 10^4 (s^2 \rho v r^3)/(B_0 R^3 |\phi|) \text{ km} \qquad 14.$$

where r (km) is the distance from the planet and R (km) is its radius. The most extreme case is that of Callisto, where if we take $\rho = 1$ g cm^{-3}, $v = 10$ km sec^{-1}, and $B_{0J} = 12$ G, we require that $|\phi| > 321$ V if $r_L < r$ for 1 μ particles. At Saturn, the worst case is Rhea where, if $B_{0S} = 4$ G, we require that $|\phi| > 124$ V if $r_L \lesssim r$ for 1 μ particles (see Table 3). Potentials considerably in excess of these values may well be the rule, and hence we suggest that it is not unreasonable to expect that as a consequence the Cook and Franklin mechanism is rendered ineffective. In fact the problem of determining the characteristics of the orbits of charged micrometeoroids in rapidly rotating magnetospheres deserves a more detailed treatment than given above: an analysis of the restricted problem where the particles are assumed to move in the equatorial plane is given in Appendix C.

It should be noted that it is not essential that the magnetopause of Saturn lie within the orbit of Iapetus for the modified version of the Cook and Franklin mechanism to be effective in producing its brightness variations. It is sufficient if the micrometeoroids do not become highly charged before encountering Iapetus: that is, $|\phi| > 2300$ V if we require $r_L < r$ for 1 μ particles with $B_{0S} = 10$ G. Accordingly for Iapetus to be shielded from a directed flux of micrometeoroids it would be necessary that the surface magnetic field strength of Saturn is rather large, and that there should be intense fluxes of electrons with energies exceeding a few kilovolts in the outer parts of the magnetosphere. The possibility that the inner Jovian and Saturnian satellites may be incompletely shielded from 1–10 μ micrometeoroids is worth considering. Such particles would impact the satellites more or less isotropically, and thus lead to a general lowering of their albedos. In these circumstances, since the shielding efficiency should increase with decreasing radial distance, there should be a distinct trend such that the visual albedo decreases with increasing distance from the planet. The Galilean satellites indeed show such a trend.

In the case of the inner Saturnián satellites, the precise values of the visual albedos are too uncertain to permit us to draw any definite conclusions. Furthermore, the range of radial distances involved is not very large, and any trend that exists may be difficult to discern.

The recently reported brightness variation with orbital phase of the inner Saturnian satellites, Enceladus (Franz & Millis, 29), shows that its behavior is exactly opposite to that of all the other inner Saturnian and Jovian satellites in that its leading face is darker than its trailing face. Consequently an altogether different explanation is required in this case. It seems possible that this anomalous behavior of Enceladus may be a direct result of its proximity to the Saturnian ring system. If grains sufficiently large ($r \gtrsim 100\ \mu$) not to be strongly affected by electromagnetic forces are perturbed by mutual collisions into elliptical orbits having their aphelia in the vicinity of Enceladus, they will preferentially impact on its leading face as they are overtaken by the faster moving satellite. Since Enceladus is at distance of 4 R_S from the planet whereas the outer extremity of the A-ring is at a distance of only 2.28 R_S, this requires the energy of the grain to increase by over 25%, which is considerable. However, if collisions lead to fragmentation, with only a small fraction of the total mass involved thrown into such energetic orbits the energy requirement is not so formidable. Also there is some observational evidence for a significant amount of material outside the A-ring and perhaps extending all the way to Enceladus (Feibelman, 26). These grains should similarly effect Mimas (S1) which is even closer to the ring system than Enceladus, and it would be very interesting to see if this satellite too shows the same behavior as Enceladus.

It has been suggested in the past that changes in the color and/or albedo of satellite surfaces could be the result of the production of free radicals by energetic particle bombardment (Papazian, 89; Rice, 93; Sagan, 94; Veverka, 115). Accordingly we have considered ways in which energetic particle bombardment could produce the brightness variations of the satellites of the outer planets. The essential requirement that there should be an asymmetry between the leading and trailing faces can be met in various ways: 1. the distribution of energetic particles trapped in the magnetospheres of the outer planets might have steep radial gradients at the positions of the satellites which give rise to a net azimuthal current in a direction which depends on the sense of the magnetic field and the charge of the particles concerned. 2. Low energy plasma corotating with the magnetospheres generally impacts the trailing faces of the satellites, whereas the leading faces are affected only by energetic particles and solar XUV radiation. 3. The surface electrostatic potentials differ on the leading and trailing faces of the satellites as a consequence of differences in the low energy plasma flux impacting the surface. However, we believe that there are difficulties with all of these suggestions, and hence they must be discarded as being the prime cause of the brightness variations, although they may play a secondary role.

If the energetic particles trapped in the magnetospheres of the outer planets originate in the solar wind and are convected or diffused inwards then they should exhibit positive radial density gradients on magnetic shells defined by field lines which interesect the various satellites as the magnetospheres rotate. The losses must be most pronounced at the positions of the innermost satellites where the particle energies are greatest and the diffusion rates are low. A radial density gradient leads to a net azimuthal streaming with a velocity **u** given by

$$\mathbf{u} = \frac{1}{3}\left[\frac{mcv^2}{ZeB^2}\right]\mathbf{B}\times\nabla\log n \qquad 15.$$

where Ze is the charge, m the mass, v the speed, and n the density of the particles concerned. In the case of the Jovian magnetosphere where the magnetic field appears to be directed southwards, the direction of streaming is westwards for positively charged particles and eastwards for negatively charged particles if the radial density gradient is positive. That is, protons and heavier nuclei should preferentially impact the leading faces of the satellites, and electrons the trailing faces. Thus, if electrons and protons are able to discolor the surfaces with differing efficiencies, this could conceivably explain the brightness variations. However, it is difficult to understand how such a mechanism could produce the general decrease of the mean albedos of the Galilean satellites with increasing distance from Jupiter. Similarly the fact that Callisto has brightness variations with opposite phase to those of the other Galilean satellites appears to be very difficult to explain on this basis alone.

The second suggestion concerning the impact of corotating low energy plasma on the trailing faces of satellites appears to have more appeal than the first. The relative velocity between the plasma and the larger satellites ranges from 41 km sec^{-1} in the case of Tethys to 340 km sec^{-1} in the case of Callisto. With reasonable values for the plasma densities we see that the corresponding proton fluxes to be expected are of the order of 4×10^7–10^9 cm^{-2} sec^{-1}. Since the relative kinetic energy of the protons ranges from as little as 10 eV to almost 1 keV, one might at first think that the resulting effect on the surfaces of the satellites should be most pronounced for the outer satellites and hardly noticeable at all for the inner ones. However, if as we have argued in Section 6, the surfaces of the satellites are charged to very high negative potentials, then these kinetic energies do not have much significance. In fact the inner satellites are favored in this respect because the energetic particle fluxes are likely to be much larger in the inner parts of the magnetosphere and hence the magnitudes of the surface potentials should be much greater than for the outer satellites. The chief difficulty with this suggestion is that once again it is difficult to understand the change in phase of the brightness variations between the three inner Galilean satellites and Callisto. Furthermore, one would expect the brightness variations of Triton to be similar to those of the inner Galilean satellites despite the fact that it is in a retrograde orbit, and this does appear to be the case. Finally, the proton fluxes incident on the surfaces of the satellites are probably of the same order as the energetic particle fluxes and perhaps even of the solar XUV flux, and hence it is difficult to understand why there should be such a pronounced effect.

The third suggestion is somewhat related to the second in that the corotating thermal plasma is used to provide a leading-trailing asymmetry of surface conditions on the satellites. In this case, however, we draw attention to the fact that the surface potentials must be asymmetric, since the fluxes of charged particles incident on the leading and trailing faces are different. If there is any process by which the surface characteristics can be altered on a microscopic scale by electric fields, then we may be able to account for asymmetries in the reflecting properties of the satellite surfaces. It should be recalled that similar suggestions have been made with regard to the lunar surface, notably as a means of accounting for erosion. Once again, however, there are difficulties in accounting for the phase of the brightness variations observed on Callisto and Triton.

In view of the various objections to these suggestions, we feel that our modified version of the Cook-Franklin mechanism is probably the most reasonable suggestion for explaining the brightness variations of the satellites, despite the fact that the case of Enceladus remains somewhat puzzling. Nevertheless the possibility that magnetospheric plasma and energetic charged particles play some role in affecting the color, albedo, and other optical properties of the surface of the satellites must be taken seriously if, as we have argued, most of the satellites have negligible atmospheres. In the case of Iapetus, effects of this sort (e.g. the small color differences between the leading and trailing faces) could be produced by the solar wind as well as by energetic plasma contained in the tail of the Saturnian magnetosphere.

ACKNOWLEDGMENTS

We wish to thank Mr. W-H. Ip for substantial help with the programming. This work was supported partly by the Planetary Program Office, Office of Space Sciences, National Aeronautics and Space Administration, under Contract NGR-05-009-110 and partly by NASA under Contract NGR-05-009-081.

APPENDIX A: PLASMA DISTRIBUTION IN THE JOVIAN MAGNETOSPHERE

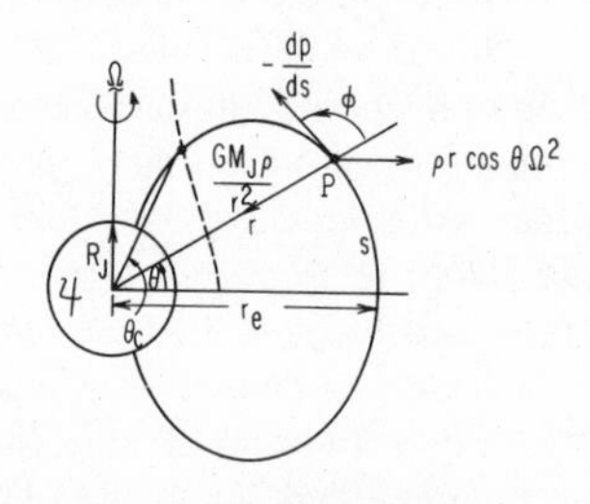

Figure A

Resolving along the field line for the equilibrium of a gas element in the rotating frame of Jupiter:

$$\frac{1}{\rho}\frac{dp}{ds} = -\frac{GM_J}{r^2}\cos\varphi + r\Omega^2\cos\theta\cos(\theta+\varphi) \qquad \text{A1.}$$

Assuming that gas behaves adiabatically (i.e. $p = k\rho^{\gamma}$) and noting that for a dipole field $r = r_e\cos^2\theta$ and $\cot\varphi = -2\tan\theta$, we see that equation A1 can be rewritten after some manipulation as:

$$k\gamma\rho^{\gamma-1}\frac{d\rho}{d\theta} = \frac{2GM_J}{r_e}\frac{\sin\theta}{\cos^3\theta} - 3r_e^2\Omega^2\sin\theta\cos^5\theta \qquad \text{A2.}$$

which on integration yields

$$\frac{\gamma}{\gamma-1}\frac{kT_0}{\bar{m}}\left[\left(\frac{\rho}{\rho_0}\right)^{\gamma-1}-1\right]=-\frac{GM_J}{r_e}\left(\frac{1}{\cos^2\theta}-1\right)+\frac{r_e^2\Omega^2}{2}(\cos^6\theta-1) \qquad \text{A3.}$$

where the suffix 0 refers to the value in the equatorial plane.

Taking $\Omega = 1.9\times10^{-4}$, $\gamma = 5/3$, and $\bar{m} = 2\times10^{-24}$ (to allow for He) and putting $r_e/R_J = L$, we obtain from equation A3:

$$\frac{n}{n_0}=\left[1+\frac{10^5}{T_0}\frac{1}{L}\{\tan^2\theta+0.05L^3(\cos^6-1)\}\right]^{3/2} \qquad \text{A4.}$$

and

$$\frac{T}{T_0}=\left(\frac{n}{n_0}\right)^{2/3} \qquad \text{A5.}$$

The locus of the points at which the field aligned components of the gravitational and centrifugal forces balance is given by $-(GM_J\rho/r^2)\cos\varphi+\rho r\Omega^2\cos\theta\cos(\theta+\varphi)=0$, whence

$$\frac{r}{R_J}=\left(\frac{2GM_J}{3\Omega^2R_J^3}\right)^{1/3}\sec^{2/3}\theta\approx1.9\sec^{2/3}\theta \qquad \text{A5.}$$

The value of T_0 for different values of L can be obtained by assuming that $n=0$ on the locus A5. Then solving A5 with $r=r_e\cos^2\theta$ gives $\cos^2\theta_c=(1.9/L)^{3/4}$, which on substitution in equation A4 gives:

$$T_0=\frac{10^5}{L}(1-0.83L^{3/4}+0.05L^3) \qquad \text{A6.}$$

If we assume that the value of n_0 is set by the condition of rotational instability, i.e.

$$\tfrac{1}{2}n_0\bar{m}r_e^2\Omega^2=\frac{B_{0J}^2}{8\pi}\left(\frac{R_J}{r_e}\right)^6$$

we get

$$n_0=[B_{0J}^2/(4\pi\bar{m}\Omega^2R_J^2L^8)]\approx(4\times10^{12})/L^8 \qquad \text{A7.}$$

(assuming $B_{0J}=12\ \Gamma$).

The isodensity and isothermal lines obtained from equations A4, A5, A6, and A7 are shown in Figure 10b.

Alternately, if the gas behaves isothermally, it is easily shown, using equation A1 that:

$$\frac{n}{n_0}=\exp\left[\frac{\bar{m}}{kT}\left\{\frac{GM_J}{r_e}\left(\frac{1}{\cos^2\theta}-1\right)+\frac{r_e^2\Omega^2}{2}(\cos^6\theta-1)\right\}\right]$$
$$=\exp\left[\frac{3.6\times10^{-11}}{(kT)L}\{\tan^2\theta+0.05L^3(\cos^6\theta-1)\}\right] \qquad \text{A8.}$$

The isodensity lines obtained from equation A8 for $kT=1$ eV and 10 eV are shown in Figure 10a.

APPENDIX B: LIMITS SET ON THE PLASMA DENSITY BY INTERCHANGE INSTABILITIES

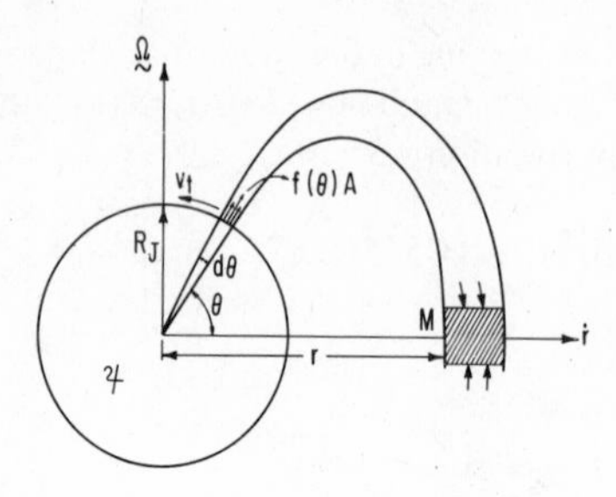

Figure B

Consider a flux tube containing a mass M concentrated near the equatorial plane interchanging with an adjacent empty flux tube. The rate of energy gain is given by

$$\frac{dW}{dt} = M\left(\Omega^2 r - \frac{GM_J}{r^2}\right)\frac{dr}{dt} \qquad \text{B1.}$$

The movement of the flux tube is mapped down onto the ionosphere where an electric field $\mathbf{E} = \mathbf{v} \times \mathbf{B}$ appears. Consequently the rate of ionospheric (Joule) dissipation associated with the interchange movement is given by

$$\frac{dD}{dt} = A\int^h \mathbf{j}\cdot\mathbf{E}\,dh = AE^2\Sigma = AR_J^2\left(\frac{d\theta}{dt}\right)^2 B_R^2\,\Sigma. \qquad \text{B2.}$$

where A is the cross-sectional area of the flux tube at the ionosphere, Σ $(=\int^h \sigma\,dh)$ is the height integrated Pedersen conductivity of the ionosphere, and B_R is the radial component of the magnetic field near the surface.

Equating B1 and B2 we have

$$M\left(\Omega^2 r - \frac{GM_J}{r^2}\right)\frac{dt}{dr} = AR_J^2 B_R^2\,\Sigma\left(\frac{d\theta}{dr}\right)^2 \qquad \text{B3.}$$

Also

$$\frac{dM}{dt} = 2f(L)A - \beta n^2 m_p A'h \qquad \text{B4.}$$

Here $f(L)$ is the mass flux into the tube from the ionosphere, β is the recombination coefficient, A' is the equatorial area of the tube, and h is the scale height of the magnetospheric plasma.

Using $B_R A = BA' = \Phi =$ constant, whereby we get

$$AA' = \frac{\Phi^2}{B_r B} = \frac{\Phi^2 L^{7/2}}{2B_0^2(L-1)}$$

(B_0 being the equatorial surface field), and noting that $M = A'hnm_p$, we obtain from equation B4

$$\frac{dM}{dt} = A\left[2f(L) - \frac{\beta}{hm_p}\frac{2B_0^2(L-1)^{1/2}}{\Phi^2 L^{7/2}}M^2\right] \qquad \text{B5.}$$

Using equations B3 and B5, putting $r/R_J = L$ and noting that $B_R A = \Phi$, we obtain after some manipulation

$$\frac{d}{dL}\left(\frac{M}{\Phi}\right)^2 = \frac{4\Sigma(d\theta/dL)^2[f(L)-(\beta B_0^2/hm_p)[(L-1)^{1/2}/L^{7/2}](M/\Phi)^2]L^2}{\Omega^2[L^3-(GM_J/R_J^3\Omega^2)]} \qquad \text{B6.}$$

since

$$L = \frac{1}{\cos^2\theta}, \qquad \left(\frac{dL}{d\theta}\right)^2 = 4(L-1)L^2 \qquad \text{B7.}$$

Also

$$\frac{GM_J}{R_J^3\Omega^2} = L_S^3 \qquad \text{B8.}$$

where L_S denotes the value of L at the synchronous orbit. Hence, from equation B6, B7, and B8 we obtain

$$\frac{d}{dL}\left(\frac{M}{\Phi}\right)^2 = \frac{\Sigma(L)[L^{7/2}f(L)-(\beta B_0^2/hm_p)(L-1)^{1/2}(M/\Phi)^2]}{\Omega^2(L-L_S^3)(L-1)L^{7/2}} \qquad \text{B9.}$$

$\Sigma_\oplus$ (noon) is available in tabular form (see Akasofu & Chapman, 1, p. 247) and to a reasonable approximation may be written in the form $\Sigma_{\oplus,n} = 2\times10^{-9}[2+(1/L-1)]$ emu for $L \gtrsim 2$. The longitude averaged solar ionizing flux at Jupiter is given by,

$$\bar{F}_{♃} = \frac{1}{\pi R^2}F_\oplus \text{ (noon)} \qquad \text{B10.}$$

where R is the heliocentric distance of Jupiter in AU

Since $\Sigma \propto n_e \propto F^{1/2}$,

$$\bar{\Sigma}_{♃} = \frac{1}{R\sqrt{\pi}}\Sigma_{\oplus,n} = \Sigma_0\cdot\frac{2L-1}{L-1} \qquad \text{B11.}$$

where $\Sigma_0 \approx 2\times10^{-10}$.

Further,

$$f(L) = f_0\cos\theta = \frac{f_0}{L^{1/2}} \qquad \text{B12.}$$

where $f_0 = 7.5\times10^{-17}$ g cm^{-2} sec^{-1} (see Ioannidis & Brice, 48).

Substituting from equation B11 and B12 in equation B9 and taking $x = C_0(M^2/\Phi^2)$, where

$$C_0 = \frac{\beta B_0^2}{hm_p f_0} = \frac{\beta_0 B_0^2}{T_0^{3/4}hm_p f_0}$$

we obtain

$$\frac{dx}{dL} = \frac{C_0 \Sigma_0 f_0}{\Omega^2} \cdot \frac{(2L-1)[L^3 - (L-1)^{1/2} x]}{(L-1)^2 (L^3 - L_S^3) L^{7/2}} \qquad \text{B13.}$$

Also, using $x = C_0(M^2/\Phi^2)$ we obtain

$$n = \frac{B_0}{C_0^{1/2} h m_p} \cdot \frac{x^{1/2}}{L^3} \qquad \text{B14.}$$

The plasma density in the outer magnetosphere ($L \gtrsim 10$) is likely to be determined by such interchange motions. Taking $B_0 = 12\,\Gamma$, $kT_0 \approx 10$ eV, and $h \approx R_J \approx 7 \times 10^9$ cm, and using as a boundary condition, $n = 0.25$ cm^{-3} (which corresponds to the solar wind density) at $L = 50$ (see Ioannidis & Brice, 48), we can numerically integrate equation B13 from $L = 50$ to smaller values of L. The variation of n with L can then be determined using equation B14, and this is shown by the dotted part of the curve in Figure 11b. In the inner magnetosphere ($2 \lesssim L \lesssim 10$) the density will be determined by diffusion from the ionosphere, along the field lines. Using the equations of Appendix A and taking $kT_0 \approx 10$ eV and the minimum density at the synchronous orbit $\approx 10^{-2}$ cm^{-3}, we obtain the variation of n with L in the inner part of the magnetosphere shown by the broken curve. We join these two curves smoothly (by the solid line) around $L \approx 10$ to obtain what we consider to be the most likely plasma distribution in the Jovian magnetosphere in $2 \lesssim L \lesssim 50$.

The speed of the interchange process in the outer magnetosphere is obtained from the relation: $dL/dt = (dL/dM)(dM/dt)$. Substituting for dM/dt from equation B5 in the above relation, we obtain after some manipulation

$$\frac{dL}{dt} = \frac{2 f_0 C_0^{1/2}}{B_0} \frac{[L^3 - (L-1)^{1/2} x \, x^{1/2}]}{(L-1)^{1/2} L^3} \frac{dL}{dx} \qquad \text{B15.}$$

Substituting for dL/dx from equation B13 we finally have

$$\frac{dL}{dt} = \frac{2\Omega^2}{B_0 \Sigma_0 C_0^{1/2}} \left[\frac{(L-1)^{3/2} (L^3 - L_S^3) L^{1/2} x^{1/2}}{(2L-1)} \right] \qquad \text{B16.}$$

The characteristic time scale τ for the interchange process is given by $\tau = L(dt/dL)$ and the variation of τ with L is shown by the solid curve in Figure 11b.

The effects of absorption of the outer magnetospheric plasma by the satellites (which have been neglected in the above treatment) may be roughly accounted for by redoing the above integration stepwise subject to the boundary condition $n = 0$ at the orbits of Io, Europa, Ganymede, and Callisto. The numerical results obtained are shown in Figure 11c.

APPENDIX C: MOTION OF CHARGED MICROMETEOROIDS IN THE EQUATORIAL PLANE OF THE PLANET

In a rest frame centered on the planet the motion of a charged particle of mass m carrying a charge Q is described by the equation:

$$m\ddot{\mathbf{r}} = Q[-(\mathbf{\Omega}\times\mathbf{r})+\mathbf{v}]\times\mathbf{B}-\frac{GMm}{r^3}\mathbf{r} \tag{C1.}$$

(the term $-(\mathbf{\Omega}\times\mathbf{r})\times\mathbf{B}$ being the electric field induced by the rotation of the planet with angular velocity $\mathbf{\Omega}$).

Assuming that the axis of the dipole magnetic field of the planet coincides with its axis of rotation the radial and transverse components of equation C1 can be written as:

$$\ddot{r}-r\dot{\theta}^2 = \frac{A(\dot{\theta}-\Omega)}{r^2}-\frac{GM}{r^2} \tag{C2.}$$

and

$$\frac{1}{r}\frac{d}{dt}(r^2\dot{\theta}) = -A\frac{\dot{r}}{r^3} \tag{C3.}$$

where

$$A = \frac{R^3B_0Q}{m} = 700\frac{R^3B_0}{m}\,se\phi \tag{C4.}$$

Consider a neutral particle injected into the magnetosphere which penetrates to a radial distance r_0 and attains a radial velocity v_{0r} before it instantly gets charged to a constant potential ϕ. The integration of C3 then gives

$$r^2\dot{\theta} = A\left(\frac{1}{r}-\frac{1}{r_0}\right) \tag{C5.}$$

Substituting for $\dot{\theta}$ from equation C5 in equation C2 and integrating we obtain

$$r^2-v_{0r}^2 = \frac{2(A\Omega+GM)}{r_0}\frac{1-Y}{Y^4}(Y^3+aY-a) \tag{C6.}$$

where

$$a = \frac{A^2}{2r_0^3(A\Omega+GM)} \quad \text{and} \quad Y = \frac{r}{r_0} \tag{C7.}$$

Considering the case of Saturn, we see that for particles charged to potentials of over a decavolt $|A\Omega| \gg GM$. Consequently neglecting the gravitational term we have

$$\dot{r}^2-v_{0r}^2 = (2r_0\Omega)^2a\frac{1-Y}{Y^4}(Y^3+aY-2) \tag{C8.}$$

where

$$a = A/2r_0^3\Omega \tag{C9.}$$

also

$$v_\theta = r\dot{\theta} = (2a\Omega r_0)\frac{1-Y}{Y^2} \tag{C10.}$$

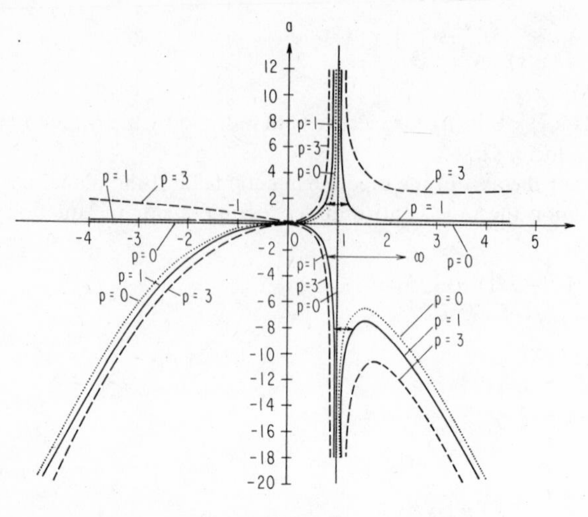

Figure C

If the density of the meteoroid ρ is taken ≈ 1 from equation C4 and equation C9 we have

$$a = \frac{\phi B_0}{120 s^2}\left(\frac{R}{r_0}\right)^3 \qquad \text{C11.}$$

The nature of the trajectories can be deduced from equation C8. Since $\dot{r}^2 \geqq 0$, we obtain

$$F(a, Y) = a^2 + \frac{Y^3}{Y-1}a - \frac{pY^2}{2(Y-1)} \leqq 0 \qquad \text{C12.}$$

where $p = v_{0r}/v_0\Omega$.

The solution of $F(a, Y) = 0$ for different values of p is shown in Figure C; the regions in the a, Y plane in which particles can be trapped are indicated by the arrows. In the limiting case $p = 0$ (shown by dotted lines) the curves in the first and third quadrants degenerate into the pair of asymptotes $Y = 1$, $a = 0$. It is seen that when $a > 0$ the particle is trapped and oscillates in a region around $Y = 1$ as it moves around. For a given value of a, particles with larger p values will penetrate deeper into the magnetosphere. Alternatively particles with given p will penetrate deeper for smaller values of a. If a is negative and numerically sufficiently large once again trapping would occur. However, if a is negative and numerically small the particle will penetrate deeper into the magnetosphere before bouncing off to infinity. This behavior for different values of a (with $p = 1$) is indicated in the figure by the arrows.

The foregoing analysis is qualitatively valid for Jupiter as well. For Jupiter the magnetic and rotation axes are nearly parallel so that B_0 is negative. Consequently negatively charged particles will have positive values for a and so will be trapped.

In the case of Saturn the magnetic and rotation axes may be parallel as in the case of Jupiter or they may be antiparallel as in the case of the earth (in which case B_0 is positive). Consequently, negatively charged particles may or may not be trapped. In either case it is seen that the penetration into the magnetosphere will not be deep for numerically large values of a. For numerically smaller values of a the penetration may be deep enough to impact one of the inner satellites. However, the directions of approach will be changed so much that they will be incident more or less isotropically, and hence will not effect the leading of trailing face preferentially.

Literature Cited

1. Akasofu, S. I., Chapman, S. 1972. *Solar-Terrestrial Physics.* Oxford: Clarendon
2. Antoniadi, E. M., Lyot, B., Carmichel, H., Gentil, M. 1959. *Larousse Encyclopedia of Astronomy,* 216–17. New York: Prometheus
3. Banks, P., Johnson, H., Axford, W. I. 1970. The atmosphere of Mercury. *Comments Astrophys. Space Phys.* 2:214
4. Barrow, C. H. 1972. Decameter-wave radiation from Jupiter and solar activity. *Planet. Space Sci.* 20:2051
5. Bartholdi, P., Owen, F. 1972. The occultation of Beta Scorpii by Jupiter and Io.II.Io. *Astron. J.* 77:60
6. Beard, D. B. 1972. Discussion of the reliability of electron densities and energies interpreted from data and limits on the proton energy and density. *Proc. Jupiter Radiation Belt Workshop, Tech. Memo. 33-543, JPL, Calif.,* p. 251
7. Beck, A. J., Ed. 1972. *Proc. Jupiter Radiation Belt Workshop, Tech. Memo. 33-543, JPL, Calif.*
8. Berge, G. L. 1965. Circular polarization of Jupiter's decimeter radiation. *Ap. J.* 142:1688
9. Binder, A. B., Cruikshank, D. P. 1964. Evidence for an atmosphere on Io. *Icarus* 3:299
10. Block, L. 1972. Potential double layers in the ionosphere. *Cosmic Electrodynamics* 3:349
11. Braude, S. Ya. 1972. Private communication to T. D. Carr
12. Brice, N. M. 1972. Energetic protons in Jupiter's radiation belts. *Proc. Jupiter Radiation Belt Workshop, Tech. Memo. 33-543, JPL, Calif.,* p. 283
13. Brice, N. M., Ioannidis, G. A. 1970. The magnetospheres of Jupiter and Earth. *Icarus* 13:173
14. Brice, N. M., McDonough, J. R. 1972. Jupiter's radiation belts. *Icarus* 18:206
15. Cook, A. F., Franklin, F. A. 1970. An explanation of the light curve of Iapetus. *Icarus* 13:282
16. Cruikshank, D. P., Murphy, R. E. 1973. The post-eclipse brightening of Io. *Icarus* 20:7
17. DeForest, S. E. 1972. Spacecraft charging at synchronous orbit. *J. Geophys. Res.* 77:651
18. Dermott, S. F. 1970. Modulation of Jupiter's decametric radio emission by Io. *MNRAS* 149:35
19. Dollfus, A. 1962. *Handbuch der Physik,* ed. S. Flügg, Band LIV, p. 180. Berlin: Springer
20. Dollfus, A. 1970. *Surfaces and Interiors of Planets,* 46. New York: Academic
21. Dulk, G. A. 1965. Influence of the satellite Io on the decametric radio emission from Jupiter. *Astron. J.* 70: 137
22. Dulk, G. A. 1967. Lack of effects of satellites Europa, Ganymede, Callisto, and Amalthea on the decametric radio emission of Jupiter. *Ap. J.* 148:239
23. Duncan, R. A. 1970. A theory of Jovian decametric emission. *Planet. Space Sci.* 18:217
24. Ellis, G. R. A. 1965. The decametric radio emissions of Jupiter. *Radio Sci.* 69D:1513
25. Fallon, F. W., Murphy, R. E. 1971. Absence of post-eclipse brightening of Io and Europa in 1970. *Icarus* 15:492
26. Feibelman, W. A. 1967. Concerning the "D" ring of Saturn. *Nature* 214:793
27. Fink, U., Dekkers, N. H., Larson, H. P. 1973. Infrared spectra of the Galilean satellites of Jupiter. *Ap. J.* 179:L155
28. Franz, O. G., Millis, R. L. 1971. A search for an anomalous brightening of Io after eclipse. *Icarus* 14:13
29. Franz, O. G., Millis, R. L. 1973. UBV photometry of Enceladus, Tethys, and Dione. *Div. Planet. Sci., Meeting AAS, Tucson, March 1973*
30. Fredricks, R. W., Scarf, F. L. 1972.

Photon and particle interaction with surfaces in space. *Pap. presented ESLAB Symp., Noordwijk, Holland, September 1972*
31. Freeman, J. W. Jr., Fenner, M. A., Hills, H. K. 1973. The electric potential of the moon in the solar wind. *J. Geophys. Res.* 78:4560
32. Gillet, F. C., Forrest, W. J., Merrill, K. M. 1973. 8-13 Micron spectra of NGC 7027, BD+30° 3639, and NGC 6572. *Astrophys. J. Lett.* 183:87
33. Gledhill, J. A. 1967. Magnetosphere of Jupiter. *Nature* 214:155
34. Goldreich, P., Lynden-Bell, D. 1969. Io, a Jovian unipolar inductor. *Ap. J.* 156:59
35. Gomer, R. 1961. *Field Emission and Field Ionization.* Cambridge, Mass.: Harvard Univ. Press
36. Grard, R. J. L. 1973. Properties of the satellite photoelectron sheath derived from photoemission laboratory measurements. *J. Geophys. Res.* 78:2885
37. Grard, R. J. L., Knott, K., Pederson, A. 1973. The influence of photoelectron and secondary electron emission on electric field measurements in the magnetosphere and solar wind. *Internal Working Pap. No. 733, Space Sci. Dep., Eur. Space Res. Technol. Centre, Noordwijk*
38. Gulkis, S. 1972. A note on the Gledhill model of the magnetosphere of Jupiter. *Proc. Jupiter Radiation Belt Workshop, Tech. Memo. 33-543, JPL, Calif.*, p. 251
39. Gurnett, D. A. 1972. Sheath effects and related charged-particle acceleration by Jupiter's Satellite Io. *Ap. J.* 175:525
40. Hansen, O. L. 1973. Ten-micron eclipse observations of Io, Europa and Ganymede. *Icarus* 18:237
41. Hapke, B. 1967. Surveyor I and Lunar IX pictures and the lunar soil. *Icarus* 6:254
42. Harris, D. L. 1961. *The Solar System III Planets and Satellites,* ed. G. P. Kuiper, B. M. Middlehurst, 272–342. Chicago: Univ. Chicago Press
43. Hartman, W. K. 1971. Physical studies of minor planets, ed. T. Gehrels, 162. *NASA SP-267*
44. Hawke, R. S., Duerre, D. E., Huebel, J. G., Keller, R. N., Klapper, H. 1972. Isentropic compression of fused quartz and liquid hydrogen to several mbar. *Phys. Earth Planet. Interiors* 6:44
45. Hubbard, W. B. 1972. A strategy for investigation of the outer solar system (The Science Advisory Group). *Space Sci. Rev.* 14:347
46. Hunten, D. M. 1969. The upper atmosphere of Jupiter. *J. Atmos. Sci.* 26:826
47. Hunten, D. M. 1972. The atmosphere of Titan. *Comments Astrophys. Space Phys.* 4:149
48. Ioannidis, G. A., Brice, N. M. 1971. Plasma densities in the Jovian magnetosphere: plasma slingshot or Maxwell demon? *Icarus* 14:360
49. Johnson, H., Axford, W. I. In preparation
50. Johnson, T. V. 1971. Galilean satellites: Narrowband photometry 0.30–1.10 microns. *Icarus* 14:94
51. Johnson, T. V., McCord, T. B. 1970. Galilean satellites: the spectral reflectivity 0.30–1.10 micron. *Icarus* 13:37
52. Johnson, T. V., McCord, T. B. 1971. Spectral geometric albedo of the Galilean satellites, 0.3 to 2.5 microns. *Ap. J.* 169:589
53. Kennel, C. 1973. Magnetospheres of the planets. *Space Sci. Rev.* 14:511
54. Kieffer, H. 1970. Spectral reflectance of CO_2-H_2O frosts. *J. Geophys. Res.* 75:501
55. Knott, K. 1972. The equilibrium potential of a magnetospheric satellite in an eclipse situation. *Planet. Space Sci.* 20:1137
56. Knott, K. 1973. Electrostatic charging of the lunar surface and possible consequences. *J. Geophys. Res.* 78:3172
57. Knudsen, W. C., Harris, K. K. Secondary electron emission. *J. Geophys. Res.* In press
58. Komesaroff, M. M., Morris, D., Roberts, J. A. 1970. Circular polarization of Jupiter's decimetric emission and the Jovian magnetic field strength. *Astrophys. Lett.* 7:31
59. Kuiper, G. P. 1944. Titan: A satellite with an atmosphere. *Ap. J.* 100:378
60. Kuiper, G. P. 1952. *The Atmospheres of the Earth and Planets,* 306. Chicago: Univ. Chicago Press.
61. Kusmin, A. D., Losovsky, B. Ya. 1973. On the radio emission of Callisto. *Icarus* 18:222
62. Lewis, J. S. 1971. Satellites of the outer planets: Their physical and chemical nature. *Icarus* 15:174
63. Lewis, J. S. 1971. Satellites of the outer planets: thermal models. *Science* 172:1127
64. Lewis, J. S. 1972. Chemistry of the outer solar system. *Space Sci. Rev.* 14:401
65. Lewis, J. S., Prinn, R. G. 1973. Titan revisited. *Comments Astrophys. Space Phys.* (V) 1:1
66. Luthey, J. L. 1972. Equatorial electron

energy and number densities in the Jovian magnetosphere. *Proc. Jupiter Radiation Belt Workshop, Tech. Memo. 33-543, JPL, Calif.*, p. 47
67. Lyot, B. 1953. Aspect des planètes au Pic du Midi dans une lunette de 0^m, 600 d'ouveture, 3. *L'Astronomie* 67:3
68. McCord, T. B., Johnson, T. V., Elias, J. H. 1971. Saturn and its satellites and narrow band spectrophotometry (0.3–1.1 μ). *Ap. J.* 165:413
69. McCulloch, P. M. 1971. Theory of Io's effect on Jupiter's decametric emissions. *Planet. Space Sci.* 19:1297
70. McDonough, T. R., Brice, N. M. 1973. New kind of ring around Saturn? *Nature* 242:513
71. McDonough, T. R., Brice, N. M. 1973. Jupiter's radiation belts. *Icarus* 18:206
72. McElroy, M. B. 1972. Mars an evolving atmosphere. *Science* 175:443
73. Mead, G. D., Hess, S. L. 1973. Jupiter's radiation belts and the sweeping effect of its satellites. *J. Geophys. Res.* 78:2793
74. Melrose, D. B. 1967. Rotational effects on the distribution of thermal plasma in the magnetosphere of Jupiter. *Planet. Space Sci.* 15:381
75. Michel, F. C. 1971. Solar wind interaction with planetary atmospheres. *Rev. Geophys. Space Phys.* 9:427
76. Michel, F. C. 1971. Solar-wind-induced mass loss from magnetic field-free planets. *Planet. Space Sci.* 19:1580
77. Millis, R. L. 1973. UBV Photometry of Iapetus. *Icarus* 18:247
78. Morrison, D. 1973. Albedos and densities of the inner satellites of Saturn. *Div. Planet. Sci., Meeting AAS, Tucson, March, 1973*
79. Morrison, D., Cruikshank, D. P. 1973. Thermal properties of Galilean satellites. *Icarus* 18:224
80. Morrison, D., Cruikshank, D. P., Murphy, R. E. 1972. Temperatures of Titan and the Galilean satellites at 20 microns. *Ap. J.* 173:L143
81. Morrison, D. et al 1971. Thermal inertia of Ganymede from 20-micron eclipse radiometry. *Ap. J.* 167:L107
82. Mozer, F. S., Bogott, F. H. 1972. Photoelectron emission from Io as the cause of enhancements of the Jovian decametric radiation. *Ap. J.* 177:L93
83. Murphy, R. E., Cruikshank, D. P., Morrison, D. 1972. Radii, albedos, and 20-micron brightness temperatures of Iapetus and Rhea. *Ap. J.* 177:L93.
84. Neubauer, F. 1972. A resonant instability of model proton radiation belts in the Jovian atmosphere. *Proc. Jupiter Radiation Belt Workshop, Tech. Memo. 33-543, JPL, Calif.*, p. 405
85. Newburn, R. L. Jr., Gulkis, S. 1973. A survey of the outer planets Jupiter, Saturn, Uranus, Neptune, Pluto and their satellites. *Space Sci. Rev.* 14:179
86. O'Leary, B. 1973. Io's triaxial figure. *Div. Planet. Sc., Meeting AAS, Tucson, March, 1973*
87. O'Leary, B., Veverka, J. 1971. On the anomalous brightening of Io after eclipse. *Icarus* 14:265
88. Owen, T., Westphal, J. 1972. The clouds of Jupiter: observational characteristics. *Icarus* 16:392
89. Papazian, H. A. 1958. Free radical formation in solids by ion bombardment. *J. Chem. Phys.* 29:448
90. Piddington, J. H., Drake, J. F. 1968. Electrodynamic effects of Jupiter's satellite Io. *Nature* 217:935
91. Pilcher, C. B., Ridgway, S. T., McCord, T. B. 1972. Galilean satellites: identification of water frost. *Science* 178:1087
92. Reasoner, D. L., Burke, W. J. 1972. Characteristics of the lunar photoelectron-layer in the geomagnetic tail. *J. Geophys. Res.* 77:6671
93. Rice, F. O. 1956. Colors on Jupiter. *J. Chem. Phys.* 24:1259
94. Sagan, C. 1971. The solar system beyond Mars: an exobiological survey. *Space Sci. Rev.* 11:827
95. Sagan, C. 1973. The greenhouse of Titan. *Icarus* 18:649
96. Sagan, C., Mullen, G. 1973. An elementary greenhouse argument for H_2 in the atmosphere of Titan. *Bull. Am. Astron. Soc.* 4:368 (Abstr.)
97. Scarf, F. L. 1973. Some comments on the magnetosphere and plasma environment of Saturn. *Cosmic Electrodynamics* 3:437
98. Schatten, K. H., Ness, N. F. 1971. The magnetic-field geometry of Jupiter and its relation to Io-modulated Jovian decametric radio emission. *Ap. J.* 165:621
99. Shawhan, S. D., Clark, T. A., Cronyn, W. M., Basart, J. P. An upper limit to the 11.4 m flux of Saturn using VLBI. *Nature Phys. Sci.* In press
100. Shawhan, S. D., Hubbard, G. J., Turnett, D. A. *Photon and Particle Interactions with Surfaces in Space*, ed. R. Grard. Dordrecht, Holland: Reidel. In press
101. Siscoe, G. L. 1971. Two magnetic tail models for 'Uranus'. *Planet. Space Sci.* 19:483
102. Siscoe, G. C., Mukherjee, N. R. 1972. Upper limits on the lunar atmosphere

determined from solar-wind measurements. *J. Geophys. Res.* 77:6042
103. Siscoe, G. C., Mukherjee, N. R. 1973. Solar wind-Mercury atmosphere interaction: determination of the planet's atmospheric density. *J. Geophys. Res.* 78:3961
104. Smith, R. A., Wu, C. S., Zmuidzinas, J. S. 1972. The geometry and dynamic spectra of Io-modulated Jovian decametric radio emissions. *Ap. J.* 177: L131
105. Spitzer, L., Savedoff, M. P. 1950. The temperature of interstellar matter, III. *Ap. J.* 111:593
106. Steavenson, W. H. 1948. Observations of the satellites of Uranus. *MNRAS* 108:183
107. Steavenson, W. H. 1950. The satellites of Uranus. *Brit. Astron. Assoc. J.* 74:54
108. Stebbins, J. 1927. The light variations of the satellites of Jupiter and their application to measures of the solar constant. *Lick Observ. Bull.* No. 385
109. Stebbins, J., Jacobsen, T. S. 1928. Further photometric measures of Jupiter's satellites and Uranus with tests of the solar constant. *Lick Observ. Bull.* No. 401
110. Sternglass, E. J. 1950. Secondary electron emissions and atomic shell structure. *Phys. Rev.* 80:925
111. Thorne, R., Coroniti, F. V. 1972. A self-consistent model for Jupiter's radiation belts. *Proc. Jupiter Radiation Belt Workshop, Tech. Memo. 33-543, JPL, Calif.*, p. 363
112. Trafton, L. M. 1972. The bulk composition of Titan's atmosphere. *Ap. J.* 175:295
113. Trafton, L. M. 1972. On the possible detection of H_2 in Titan's atmosphere. *Ap. J.* 175:285
114. Van Biesbrock, G., Kuiper, G. P. 1946. The 5th satellite of Jupiter. *Ap. J.* 52: 114
115. Veverka, J. 1970. *Photometric and polarimetric studies of minor planets and satellites.* PhD Thesis. Harvard Univ., Cambridge, Mass.
116. Veverka, J. 1971. Polarization measurements of the Galilean satellites of Jupiter. *Icarus* 14:355
117. Veverka, J. 1973. Titan: polarimetric evidence for an optically thick atmosphere? *Icarus* 18:657
118. Walbridge, E. 1973. Lunar photoelectron layer. *J. Geophys. Res.* 78: 3668
119. Warwick, J. W. 1963. The position and sign of Jupiter's magnetic movement. *Ap. J.* 137:1317
120. Warwick, J. W. 1964. Radio emission from Jupiter. *Ann. Rev. Astron. Astrophys.* 2:1
121. Warwick, J. W. 1967. Radiophysics of Jupiter. *Space Sci. Rev.* 6:841
122. Warwick, J. W. 1970. *NASA Report CR-1685*
123. Warwick, J. W., Dulk, G. A. 1964. Faraday rotation on decametric radio emissions from Jupiter. *Science* 145: 380
124. Webster, D. L., Alksne, A. Y., Whitten, R. C. 1972. Does Io's ionosphere influence Jupiter's radio bursts? *Ap. J.* 174:685
125. Whipple, E. C. Jr. 1965. *The equilibrium electric potential of a body in the upper atmosphere and in interplanetary space.* PhD Thesis. NASA-Goddard Space Flight Center, X-615-65-296. Greenbelt, Md.
126. Wu, C. S., Smith, R. A., Zmuidzinas, J. S. 1972. Theory of decametric radio emissions from Jupiter. *Pap. Presented COSPAR Meeting, Symp. A, Madrid, Spain, 1972*
127. Wu, C. S. 1973. Modulation of the Jovian decametric radio emission by Io. Unpublished preprint
128. Wyatt, S. P. 1969. The electrostatic charge of interplanetary grains. *Planet. Space Sci.* 17:155
129. Zellner, B. 1972. On the nature of Iapetus. *Astrophys. J. Lett.* 174:L107
130. Zellner, B. 1973. The polarization of Titan. *Icarus* 18:661
131. Zheleznyakov, V. V. 1965. The origin of Jovian radio emission. *Soviet Astron.–AJ* 9:617

SOME RELATED ARTICLES APPEARING IN OTHER ANNUAL REVIEWS

From the *Annual Review of Astronomy and Astrophysics,* Volume 11 (1973)

Dynamical Astronomy of the Solar System, R. L. Duncombe, P. K. Seidelmann, and W. J. Klepczynski

From the *Annual Review of Materials Science,* Volume 3 (1973)

Theoretical Approaches to the Determination of Phase Diagrams, Larry Kaufman and Harvey Nesor

The Fracture Crack as an Imperfection in a Nearly Perfect Solid, Robb M. Thomson

Some Aspects of Structural Disorder in Solids, Simon C. Moss

From the *Annual Review of Nuclear Science,* Volume 23 (1973)

Photon-Excited Energy-Dispersive X-Ray Fluorescence Analysis for Trace Elements, F. S. Goulding and J. M. Jaklevic

From the *Annual Review of Physical Chemistry,* Volume 24 (1973)

Chemistry of the Planets, John S. Lewis

REPRINTS

The conspicuous number aligned in the margin with the title of each review in this volume is a key for use in the ordering of reprints.

Available reprints are priced at the uniform rate of \$1 each postpaid. Payment must accompany orders less than \$10. A discount of 20% will be given on orders of 20 or more. For orders of 200 or more, any Annual Reviews article will be specially printed.

The sale of reprints of articles published in the Reviews has been expanded in the belief that reprints as individual copies, as sets covering stated topics, and in quantity for classroom use will have a special appeal to students and teachers.

CUMULATIVE INDEXES

CONTRIBUTING AUTHORS VOLUMES 1-2

A

Axford, W. I., 2:419

B

Banerjee, S. K., 1:269
Barnes, I., 1:157
Barton, P. B. Jr., 1:183
Barth, C. A., 2:333
Bathurst, R. G. C., 2:257
Burke, D. B., 2:213
Burnham, C. W., 1:313

C

Clark, G. R. II, 2:77
Coroniti, F. V., 1:107
Crompton, A. W., 1:131
Cruz-Cumplido, M. I., 2: 239

D

Dieterich, J. H., 2:275

F

Fripiat, J. J., 2:239

H

Hargraves, R. B., 1:269
Hem, J. D., 1:157
Hubbard, W. B., 1:85

J

Jeffreys, H., 1:1
Jenkins, F. A. Jr., 1:131

K

Kanamori, H., 1:213
Karig, D. E., 2:51
Kistler, R. W., 2:403

M

Mendis, D. A., 2:419
Millero, F. J., 2:101
Mogi, K., 1:63

P

Pálmason, G., 2:25

R

Riedel, W. R., 1:241
Rubey, W. W., 2:1

S

Saemundsson, K., 2:25
Skinner, B. J., 1:183

T

Thompson, A. B., 2:179
Thompson, G. A., 2:213
Thorne, R. M., 1:107
Toksöz, M. N., 2:151

V

Van Houten, F. B., 1:39
Vonnegut, B., 1:297

W

Walcott, R. I., 1:15
Wetherill, G. W., 2: 303
Wood, D. S., 2:369

Z

Zen, E., 2:179

CHAPTER TITLES VOLUMES 1-2

GEOCHEMISTRY, MINERALOGY, AND PETROLOGY

Title	Author(s)	Vol:Pages
Chemistry of Subsurface Waters	I. Barnes, J. D. Hem	1:157-82
Genesis of Mineral Deposits	B. J. Skinner, P. B. Barton Jr.	1:183-212
Order-Disorder Relationships in Some Rock-Forming Silicate Minerals	C. W. Burnham	1:313-38
Low Grade Regional Metamorphism: Mineral Equilibrium Relations	E. Zen, A. B. Thompson	2:179-212
Clays as Catalysts for Natural Processes	J. J. Fripiat, M. I. Cruz-Cumplido	2:239-56
Phanerozoic Batholiths in Western North America: A Summary of Some Recent Work on Variations in Time, Space, Chemistry, and Isotopic Compositions	R. W. Kistler	2:403-18

GEOPHYSICS AND PLANETARY SCIENCE

Title	Author(s)	Vol:Pages
Structure of the Earth from Glacio-Isostatic Rebound	R. I. Walcott	1:15-38
Interior of Jupiter and Saturn	W. B. Hubbard	1:85-106
Magnetospheric Electrons	F. V. Coroniti, R. M. Thorne	1:107-30
Geophysical Data and the Interior of the Moon	M. N. Toksöz	2:151-77
Solar System Sources of Meteorites and Large Meteoroids	G. W. Wetherill	2:303-31
The Atmosphere of Mars	C. A. Barth	2:333-67
Satellites and Magnetospheres of the Outer Planets	D. A. Mendis, W. I. Axford	2:419-74

MISCELLANEOUS

Title	Author(s)	Vol:Pages
Rock Fracture	K. Mogi	1:63-84
Theory and Nature of Magnetism in Rocks	R. B. Hargraves, S. K. Banerjee	1:269-96
Earthquake Mechanisms and Modeling	J. H. Dieterich	2:275-301
Current Views of the Development of Slaty Cleavage	D. S. Wood	2:369-401

OCEANOGRAPHY, METEOROLOGY, AND CLIMATOLOGY

Title	Author(s)	Vol:Pages
Electrical Balance in the Lower Atmosphere	B. Vonnegut	1:297-312
The Physical Chemistry of Seawater	F. J. Millero	2:101-50

PALEONTOLOGY, STRATIGRAPHY, AND SEDIMENTOLOGY

Title	Author(s)	Vol:Pages
Origin of Red Beds: A Review—1961-1972	F. B. Van Houten	1:39-62
Mammals from Reptiles: A Review of Mammalian Origins	A. W. Crompton, F. A. Jenkins Jr.	1:131-56
Cenozoic Planktonic Micropaleontology and Biostratigraphy	W. R. Riedel	1:241-68
Growth Lines in Invertebrate Skeletons	G. R. Clark II	2:77-99
Marine Diagenesis of Shallow Water Calcium Carbonate Sediments	R. G. C. Bathurst	2:257-74

PREFATORY CHAPTERS

Title	Author(s)	Vol:Pages
Developments in Geophysics	H. Jeffreys	1:1-14
Fifty Years of the Earth Sciences—A Renaissance	W. W. Rubey	2:1-24

TECTONICS AND REGIONAL GEOLOGY

Title	Author(s)	Vol:Pages
Mode of Strain Release Associated with Major Earthquakes in Japan	H. Kanamori	1:213-40
Iceland in Relation to the Mid-Atlantic Ridge	G. Pálmason, K. Saemundsson	2:25-50
Evolution of Arc Systems in the Western Pacific	D. E. Karig	2:51-75
Regional Geophysics of the Basin and Range Province	G. A. Thompson, D. B. Burke	2:213-38